T0192072

Switched Inductor Power IC Design

Gabriel Alfonso Rincón-Mora

Switched Inductor Power IC Design

Gabriel Alfonso Rincón-Mora
School of Electrical and Computer Engineering
Georgia Institute of Technology
Atlanta, GA, USA

ISBN 978-3-030-95901-2 ISBN 978-3-030-95899-2 (eBook)
https://doi.org/10.1007/978-3-030-95899-2

This Springer imprint is published by the registered company Springer Nature Switzerland AG
The registered company address is: Gewerbestrasse 11, 6330 Cham, Switzerland

To Mami and Papi

Minyusita

Summary

The aim of this textbook is to show and illustrate with insight, analysis, examples, and simulations how to design switched-inductor dc–dc power supplies. The book adopts a ground-up approach, from devices to systems. Chapters 1 and 2 are on diodes and transistors, Chaps. 3 and 4 on power transfer and delivery, Chaps. 5 and 6 on frequency response and feedback control, and Chaps. 7 and 8 on system composition and architecture. The emphasis throughout is on design.

Chapter 1 reviews how PN, Zener, and Schottky diodes and bipolar-junction transistors (BJTs) block and conduct current. It starts with how solids and semiconductors behave and how adding impurity dopant atoms alters their behavior. With these concepts in hand, the material then details the operating modalities, characteristics, and response of PN and metal–semiconductor junction diodes and BJTs, including electrostatic behavior, band diagrams, current–voltage translations, capacitances, recovery times, breakdown mechanisms, structural variations, and more.

Chapter 2 reviews how junction and metal–oxide–semiconductor (MOS) field-effect transistors (FETs) block and conduct current. It describes how MOSFETs accumulate, deplete, and invert their channels and how they saturate their currents in cut off, sub-threshold, and inversion. It also discusses body effect, weak inversion, how gate–channel oxide capacitance distributes across operating regions, and short-channel effects, like drain-induced barrier lowering (DIBL), surface scattering, hot-electron injection, oxide-surface ejections, velocity saturation, and impact ionization and avalanche. Discussions extend to varactors, MOS diodes, lightly doped drains (LDD), diffused-channel MOSFETs (DMOS), junction isolation, substrate MOSFETs, welled MOSFETs, and electronic and systemic noise coupling and injection.

Chapter 3 explains how inductors and transformers work and how switching power supplies use them to transfer power. It discusses the applications that demand these switched inductors and the steps and precautions taken when implementing them with complementary MOS (CMOS) integrated circuits (ICs). It also describes how ideal, asynchronous, and synchronous buck–boost, buck, boost, and flyback dc–dc converters operate and how their voltages, currents, duty cycles, and conduction modes relate.

Chapter 4 details how switched-inductor power supplies consume power that is otherwise intended for the output. It discusses the significance of these power losses in voltage regulators, battery chargers, and energy harvesters and the mechanics that govern them. The material explains and quantifies how resistances, diodes, transistors, and gate drivers burn Ohmic, dead-time, current–voltage overlap, and gate-charge power in continuous and discontinuous conduction. Concepts discussed include power-conversion efficiency, fractional losses, maximum-power point, the power theorem, reverse recovery, soft switching, and so on. This chapter also shows how to use these concepts to design power switches and gate drivers and how losses ultimately alter, dominate, and peak conversion efficiency.

Chapter 5 describes how switched-inductor power supplies react and respond across frequency to dynamic fluctuations. It explains the guiding principles that govern two-port models to ultimately show how switched inductors reduce to simple current- and voltage-sourced networks. The chapter also shows how couple, shunt, and bypass capacitors and inductors respond individually and collectively with and without current- and voltage-limiting resistances. With this insight, deriving the frequency response of loaded bucks, boosts, buck–boosts, and flybacks in continuous and discontinuous conduction is more insightful and easier to comprehend and apply. Along the way, this chapter introduces and explains capacitor and inductor poles, in- and out-of-phase left- and right-half-plane zeros, reversal poles and zeros, transitional LC frequency, LC quality and gain, peaking and damping effects, and other relevant concepts that help describe switched LC networks.

Chapter 6 shows how to control and stabilize switched-inductor power supplies. It explains how inverting feedback loops mix, sample, and translate signals across the loop, how they respond across frequency, and how pre-amplifiers, parallel paths, and embedded loops alter their response. This chapter also discusses how power-supply systems use operational amplifiers (op amps) and operational transconductance amplifiers (OTAs) to stabilize feedback systems. With this understanding and insight in hand, the chapter explains how analog and digital, voltage- and current-mode, and voltage and current controllers manage and stabilize switched inductors in continuous and discontinuous conduction. Along the way, it introduces and reviews phase and gain margins, gain–bandwidth product, unity-gain projections, stabilization strategies (Types I, II, and III: dominant pole, pole–zero pair, and pole–zero–zero triplet), non-inverting and inverting feedback and mixed op-amp translations, inherent stability, digital gain and bandwidth, limit cycling, and other relevant concepts that help describe, quantify, and assess feedback controllers.

Chapter 7 explains how feedback loops control switched-inductor power supplies. It describes how pulse-width-modulated (PWM), hysteretic, and constant-time peak/valley loops switch the inductor, offset the current or voltage they control, and respond to fast input or output variations. It also illustrates how summing comparators can contract control loops and remove the loading effect that current-mode voltage loops normally exhibit. The chapter ends with compact, fast, and low-cost resistive, filtered, and voltage-mode voltage-looped (voltage-squared) bucks. Along the way, the material introduces and reviews comparators, hysteretic comparators, summing comparators, pulse-width modulators, set–reset (SR) flip flops, pulse generators, sub-harmonic oscillations, and slope compensation.

Switching power supplies are microelectronic systems with analog, analog–digital, and digital functions that set, manage, and control their outputs. Chapter 8 shows how to implement and design the building blocks needed for this functionality. Some of the blocks covered are current sensors and feedback translations, hysteretic and summing comparators, sawtooth and one-shot generators/oscillators, gate drivers and dead-time logic, zero-current detectors, ring suppressors, switched diodes, and shutdown and starter functions. The material also reviews the circuits used to realize some of these blocks, like low- and high-side supply-sensing comparators, push–pull logic, class-A inverters, SR flip flops, and others.

The appendix is a short reference guide on SPICE simulations. It touches on convergence, models, and structure. It also lists devices, sources, and commands commonly used when simulating switched-inductor power supplies. It ends with behavioral models for common digital blocks and comparators used in this textbook and SPICE code for stabilizers not included in Chap. 6.

Contents

Diodes and BJTs

1

Abbreviations

BJT	Bipolar-junction transistor
FET	Field-effect transistor
MOS	Metal–oxide–semiconductor
A_J	Junction area
α_T	Base-transport factor
α_0	Baseline transport factor
β_0	Baseline base–collector current gain
C_{DEP}	Depletion capacitance
C_{DIF}	Diffusion capacitance
C_J	Junction capacitance
C_{J0}	Zero-bias junction capacitance
D_N	Electron diffusion coefficient
D_H	Hole diffusion coefficient
e^-	Electron
E_B	Energy barrier
E_{BG}	Band-gap energy
E_C	Conduction-edge energy
E_E	Electron energy
E_F	Fermi energy level
E_V	Valence-edge energy
g_m	Small-signal transconductance
γ_E	Emitter injection efficiency
h^+	Hole (missing electron)
i_B	Base current
i_C	Collector current
i_D	Diode current
i_E	Emitter current

G. A. Rincón-Mora, *Switched Inductor Power IC Design*,
https://doi.org/10.1007/978-3-030-95899-2_1

i_F	Forward diode current
i_R	Reverse diode current
i_{RC}	Recombination current
I_S	Reverse saturation current
K_B	Boltzmann's constant
L_N	Electron's average diffusion length
L_P	Hole's average diffusion length
μ_N	Electron mobility
μ_P	Hole mobility
n_E	Electron density
n_H	Hole density
n_I	Intrinsic carrier concentration/ideality factor
N_A	Acceptor doping concentration
N_B	Base doping concentration
N_C	Collector doping concentration
N_D	Donor doping concentration
N_E	Emitter doping concentration
q_E	Electronic charge
q_{FR}	Forward-recovery charge
q_{RR}	Reverse-recovery charge
t_{FR}	Forward-recovery time
t_R	Recovery time
t_{RR}	Reverse-recovery time
T_J	Junction temperature
τ_F	Forward transit time
τ_H	Hole's average carrier lifetime
τ_N	Electron's average carrier lifetime
v_B	Base voltage
v_{BC}	Base–collector voltage
v_{BE}	Base–emitter voltage
v_C	Collector voltage
v_{CE}	Collector–emitter voltage
$v_{CE(MIN)}$	Minimum collector–emitter voltage
v_D	Diode voltage
v_E	Emitter voltage
v_R	Reverse diode voltage
V_{BD}	Breakdown voltage
V_{BI}	Built-in (potential) voltage
V_t	Thermal voltage
w_B	Effective base width
w_{B0}	Zero-bias base width
W_B	Metallurgical base width

Power supplies use switches to draw, steer, and deliver charge from input sources into rechargeable batteries and microelectronic loads. Semiconductor companies use *diodes, bipolar-junction transistors* (BJTs), and *complementary metal–oxide–semiconductor* (CMOS) *field-effect transistors* (FETs) for this purpose. Of these, FETs are oftentimes preferable because they drop lower voltages than diodes and require less current to switch than BJTs, so they consume less power. Still, diodes do not require a synchronizing signal like FETs and BJTs and BJTs cost less money. Plus, MOSFETs incorporate diodes and BJTs that can at times activate and steer some or all of the current. So understanding how diodes and BJTs conduct current is essential.

1.1 Solids

1.1.1 Energy-Band Diagram

Electrons in populated orbits of a material are bound to their home sites around the nucleus in Fig. 1.1. Electrons in the outermost orbit are responsible for bonding with other atoms. These are the *valence electrons* that populate the *valence band* and form covalent bonds. Electrons that break free are available for conduction. These free *charge carriers* are in the *conduction band.*

Although available for conduction, *electron affinity* keeps these electrons in their orbits. Combined potential and kinetic energy is lower for tightly bound electrons. So *electron energy* E_E in the conduction band in Fig. 1.2 is greater than in the valence band. *Conduction-edge energy* E_C is the minimum energy that liberated electrons carry. *Valence-edge energy* E_V is the maximum energy that bound valence electrons hold.

Electrons orbit around the nucleus at discrete energy levels. So no electrons reside between the valence and conduction bands. The energy span between these two bands is the *band gap*. This *band-gap energy* E_{BG} is the energy needed to liberate and promote valence electrons into the conduction band. E_{BG} is therefore the difference between E_C and E_V.

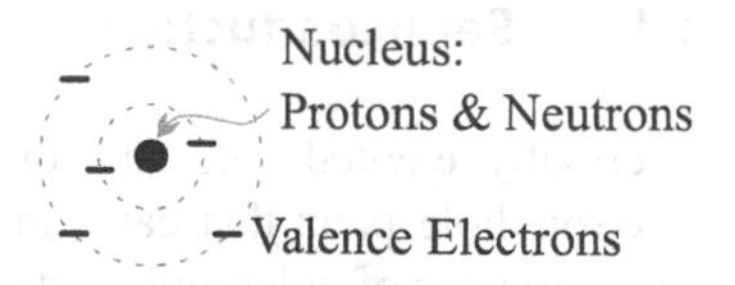

Fig. 1.1 Atom

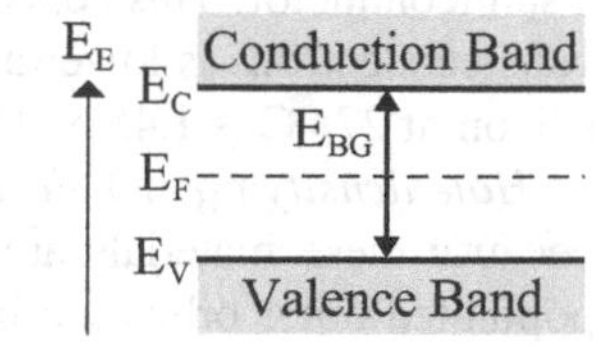

Fig. 1.2 Energy-band diagram

1.1.2 Conduction

Electrons that rise into the conduction band leave voids in the valence band. Once liberated, these free electrons drift easily. Neighbor valence electrons also shift easily into valence *holes*. And as these valence electrons shift positions, holes drift in the opposite direction. So holes in the valence band carry charge like electrons in the conduction band.

Since liberated electrons create holes, the probability of finding electrons in the conduction band of a homogeneous material is equal to the probability of finding holes in the valence band. Because free electrons reside in the conduction band and holes in the valence band, the most probable energy level for a charge carrier (when neglecting that the band gap excludes electrons and holes) is halfway between the bands. The probability that this charge carrier is an electron or a hole is 50%.

This 50% probability is what the *Fermi energy level* E_F indicates and why E_F in homogeneous material is halfway between E_C and E_V. The probability of finding charge carriers above and below this level falls exponentially. E_F is effectively an indicator of charge-carrier density. And like water in a lake, charges do not flow when the concentration across a material is uniform. So E_F is uniform across a material when current is zero and sloped when current is not zero.

1.1.3 Classification

Conductors like metal and aqueous solutions of salts conduct charge easily because valence electrons are so weakly bound to their home sites that they are practically free and available for conduction. This is another way of saying that the valence band overlaps the conduction band. Valence electrons in *insulators* like rubber and plastic, on the other hand, require so much band-gap energy to liberate that they hardly conduct. The band gap in *semiconductors* like *silicon* Si and *germanium* Ge is moderate, so they conduct moderately well. In silicon, the band-gap energy is 1.1 eV or 1.8×10^{-19} J at 27 °C.

1.1.4 Semiconductors

Thermally excited electrons that break free from their valence positions avail electron–hole pairs that can carry charge. These *thermionic emissions* produce an equal number of holes and electrons. This is the *intrinsic carrier concentration* n_i of a semiconductor. This concentration is higher when the band-gap energy that binds valence electrons is lower and the temperature that energizes them is higher. n_i for silicon at 27 °C is 1.45×10^{10} cm^{-3}.

Hole density n_H and *electron density* n_E in *intrinsic semiconductors* equal this n_i because these materials are homogeneous and pure. *Dopant atoms* with partially populated outer orbits are impurities that can alter these concentrations. So *doped semiconductors* are semiconductors with atomic impurities that produce uneven carrier concentrations.

A. N Type

Electrons in the outer orbit of *donor atoms* are so loosely bound in doped semiconductors that they are free in the conduction band. These dopant atoms effectively "donate" negatively charged electrons e^-'s. This is why engineers say material doped with donor atoms is negative or *N type*. In spite of this appellation, the material is electrically neutral because the intrinsic and dopant atoms that comprise it are neutral.

With more electrons in the conduction band, the probability that thermionic holes in the valence band recombine is higher. Electron and hole carrier concentrations are lower as a result. n_H, for example, reduces to the number of thermionic holes n_h that do not recombine. n_h is therefore lower than n_i by the amount that *donor doping concentration* N_D dictates:

$$n_H = n_h = \frac{{n_i}^2}{N_D} < n_i, \tag{1.1}$$

which is what the *mass-action law* states.

The situation for n_E is different because dopants add N_D electrons to the conduction band. Plus, N_D is normally orders of magnitude higher than n_i. So N_D is so much higher than the number of thermionic electrons that do not recombine n_e that n_E is nearly N_D:

$$n_E = N_D + n_e \approx N_D. \tag{1.2}$$

N_D therefore determines the extent to which the material is N type. n_h is so much lower than the resulting n_E that holes in N-type material are *minority carriers* and electrons are *majority carriers*.

Example 1: Find n_H at 27 °C in silicon when doped with $10^{14}/\text{cm}^3$ donor atoms.

Solution:

$$n_H = n_h = \frac{{n_i}^2}{N_D} = \frac{(1.45 \times 10^{10})^2}{10^{14}} = 2.10 \times 10^6\ \text{cm}^{-3}$$

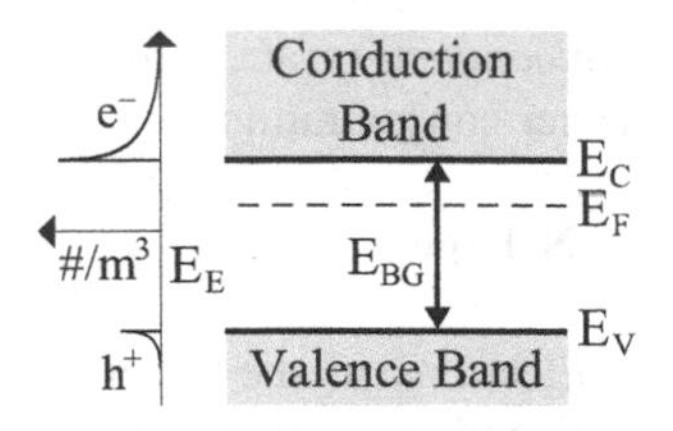

Fig. 1.3 Band diagram of N-type semiconductors

When doped this way, the probability of finding free *electrons* e^-'s is higher than that of finding *holes* h^+'s. So E_F in N-type material is closer to the conduction band in Fig. 1.3 than to the valence band. Since the probability of finding charge carriers drops exponentially away from E_F and the band gap is free of carriers, n_E peaks at the edge of the conduction band and decreases exponentially above E_C. Although to a lower extent, n_H similarly peaks at the edge of the valence band (where electrons are more likely to break free) and decreases exponentially below E_V.

B. **P Type**

Acceptor atoms produce the opposite effect. Electrons in the outermost orbit of acceptor atoms are so tightly bound in doped semiconductors that they are in the valence band. This outer orbit is incomplete, however, with electron vacancies or holes h^+'s. Since these impurities are more likely to "accept" than donate electrons, engineers say material doped with acceptor atoms is positive or *P type*. The material is nevertheless electrically neutral because the intrinsic and dopant atoms that comprise it are neutral.

With more holes in the valence band, the probability that thermionic electrons in the conduction band recombine is higher, so n_E and n_H are lower. n_E therefore reduces to the number of thermionic electrons n_e that do not recombine. n_e is lower than n_i by the amount that *acceptor doping concentration* N_A dictates:

$$n_E = n_e = \frac{{n_i}^2}{N_A} < n_i. \tag{1.3}$$

Since dopants add N_A holes to the valence band and N_A is normally orders of magnitude greater than n_i, N_A holes overwhelm the thermionic holes that do not recombine n_h. As a result, n_H is nearly N_A:

$$n_H = N_A + n_h \approx N_A. \tag{1.4}$$

N_A determines the extent to which the material is P type this way. n_e is so much lower than this n_H that free electrons are minority carriers in P-type material and holes are majority carriers.

When doped this way, the probability of finding holes is higher than that of electrons. So E_F in P-type material is closer to the valence band in Fig. 1.4 than to the conduction band. Since the probability of finding charge carriers drops exponentially

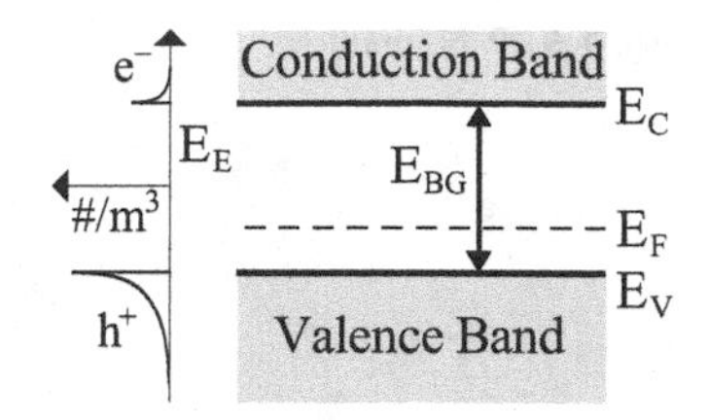

Fig. 1.4 Band diagram of P-type semiconductors

away from E_F and the band gap is free of carriers, n_H peaks at the edge of the valence band and decreases exponentially below E_V. Similarly, but to a lower extent, n_E peaks at the edge of the conduction band and decreases exponentially above E_C.

C. Mobility

Carrier mobility is the ease with which carriers flow when exposed to an electric field. It increases with temperature because thermal energy excites electrons into more mobile states. This higher kinetic energy eases their movement and, with it, conduction.

Bound valence electrons shift into holes in the valence band with less ease than free electrons drift in the conduction band. Valence and nucleic bonds are to blame for this. The effective nucleic mass of holes is therefore higher than that of free electrons. As a result, *hole mobility* μ_P is usually two to three times lower than *electron mobility* μ_N.

D. Notation

Superscripted plus and minus signs normally indicate relative concentration levels. So N_D^+, N_D, and N_D^-, respectively, produce heavily, moderately, and lightly doped N-type material that engineers denote with N^+, N, and N^-. N_D^+ is also usually orders of magnitude greater than N_D and N_D is similarly greater than N_D^-.

The same applies to P-type semiconductors. N_A^+, N_A, and N_A^- produce heavily, moderately, and lightly doped material P^+, P, and P^-. Unless otherwise specified, N_D^+ in N^+ is usually on the same order of magnitude as N_A^+ in P^+ and likewise for N_D in N and N_A in P and N_D^- in N^- and N_A^- in P^-. When doping concentration is so high that Ohmic resistance is comparable to metal, the semiconductor is *degenerate*. So degenerate semiconductors are good Ohmic contacts.

1.2 PN Junction Diodes

A *PN junction diode* is a piece of semiconductor doped so acceptor impurity atoms outnumber donor impurity atoms on one side and donor impurity atoms outnumber acceptor impurity atoms on the other side. The material transitions from one type to the other at the *metallurgical junction* X_J shown in Fig. 1.5. This is where doping concentrations effectively cancel. The doping difference $N_A - N_D$ transitions across

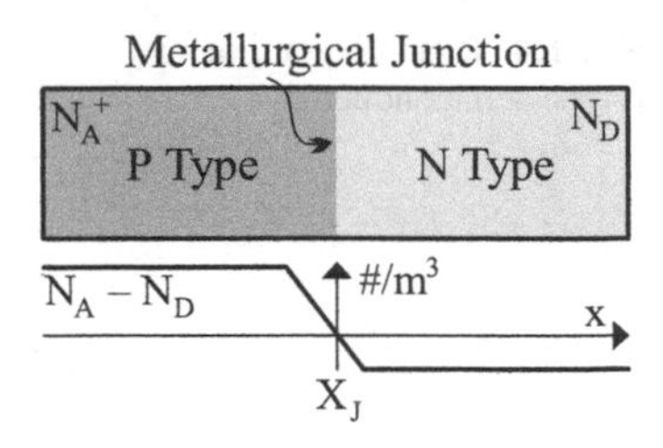

Fig. 1.5 P^+N junction

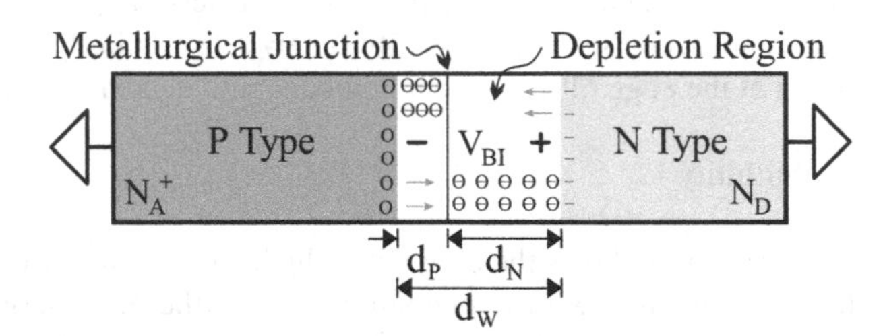

Fig. 1.6 Zero-bias P^+N junction

this zero point abruptly in *step junctions* and more gradually in *linearly graded junctions*.

1.2.1 Zero Bias

A. Electrostatics

Diffusion *Diffusion* is the force in nature that impels motion from dense regions to sparse spaces. In a PN junction, holes in the P side outnumber holes in the N side by orders of magnitude. Electron density in the N side is also much greater than electron density in the P side. Majority carriers therefore diffuse across the junction: holes to the N side and electrons to the P side.

Depletion Diffusing electrons eventually populate holes in the P side when the system reaches *thermal equilibrium*. Diffusing holes similarly capture free electrons in the N side. This *recombination* process depletes the region near the junction of charge carriers. This carrier-free space in Fig. 1.6 is the *depletion region*.

Ionization Parent atoms lose and receive charge as carriers diffuse in and out of their orbits. Departing holes and incoming electrons charge the P side negatively, and departing electrons and incoming holes charge the N side positively. So the P side begins to repel incoming electrons and the N side to repel incoming holes.

Charge carriers nevertheless continue to diffuse until the electric field is strong enough to repel further action, which happens when carrier density (and E_F) is uniform across the junction. The field that results across this *space-charge region* when the system reaches thermal equilibrium establishes a *built-in (potential) voltage* V_{BI}. Held majority carriers must therefore overcome the *energy barrier* E_B that this V_{BI} sets to diffuse across the junction:

$$E_B = q_E V_{BI}, \tag{1.5}$$

where q_E is the *electronic charge*, which refers to the 1.60×10^{-19} Coulombs of charge that each electron carries.

B. Energy-Band Diagram

The band gap is constant throughout the material because both P- and N-type regions are part of the same semiconductor. The probability of finding charge carriers is also uniform because net current flow is zero. Since hole and electron concentrations are high in P and N regions, respectively, E_F is closer to E_V in P material and closer to E_C in N material. So when piecing the band diagram together, E_{BG} and E_F are uniform across the device. E_F in Fig. 1.7 is closer to E_V in the P side than to E_C in the N side because N_A^+ is much higher than N_D. E_C in the P side is higher than in the N side because N-side electrons need additional energy to overcome the energy barrier $q_E V_{BI}$ that impedes further diffusion.

C. Carrier Concentrations

Hole and electron densities n_H and n_E far away from the junction and at the edge of the depletion region are uniform because they do not lose carriers to diffusion. This means that to the left of d_W in Fig. 1.8, where the material is P type, n_H is N_A and n_E is n_e's n_i^2/N_A. n_E is similarly N_D and n_H is n_h's n_i^2/N_D to the right of d_W, where the material is N type.

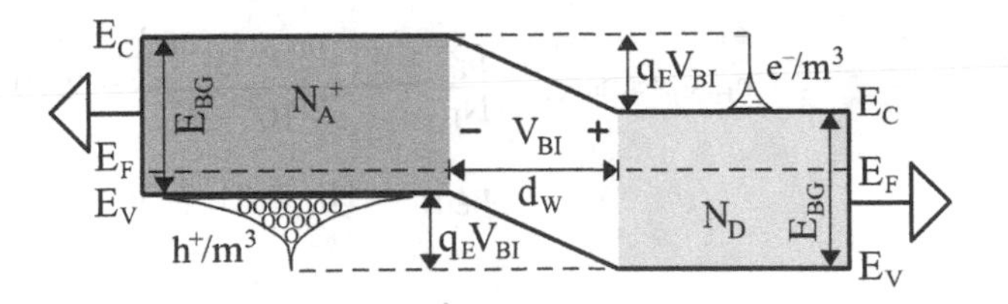

Fig. 1.7 Band diagram of zero-bias P⁺N junction

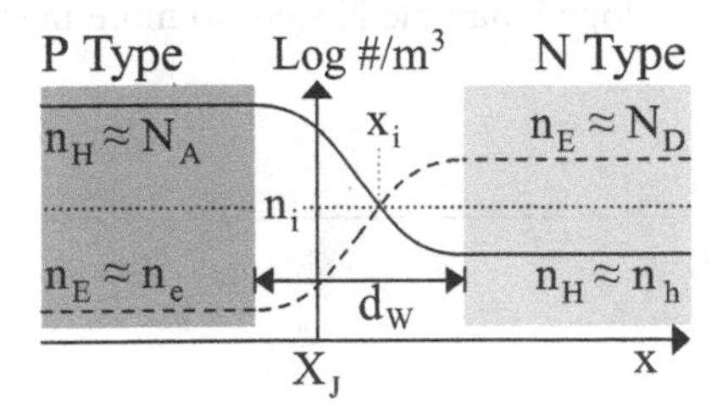

Fig. 1.8 Carrier densities across PN junction

Since fewer dopant carriers reduce the propensity for thermionic carriers to recombine, P-side holes that diffuse away not only reduce n_H but also increase n_E into the depletion region and N-side electrons that diffuse across do the opposite. The material is intrinsic where carrier densities match (at x_i) because dopant electrons and holes neutralize. Here, thermionic emissions avail intrinsic concentrations of holes and electrons, which means n_H and n_E equal n_i.

When asymmetrically doped, the highly doped region diffuses more carriers across the junction than the lightly doped side. In the case of Fig. 1.8, for example, N_A^+ is much greater than N_D. So n_H near the junction X_J is greater than n_E. This is why n_H and n_E crisscross further in the N side at x_i (and not at X_J).

Interestingly, minority carrier concentration in the lightly doped side is greater than in the highly doped counterpart. This is because fewer dopants reduce the number of thermionic carriers that recombine, so more thermionic carriers survive. n_E's n_e in the P side is lower than n_H's n_h in the N side because of this: because N_A^+ is greater than N_D.

Example 2: Find n_H and n_E outside the depletion region at 27 °C for a PN junction doped with $10^{17}/\text{cm}^3$ acceptor atoms and $10^{14}/\text{cm}^3$ donor atoms.

Solution:

$$n_{H(P)} \approx N_A = 10^{17}\ \text{cm}^{-3}$$

$$n_{E(P)} \approx n_{e(P)} = \frac{n_i^2}{N_A} = \frac{(1.45 \times 10^{10})^2}{10^{17}} = 2.10 \times 10^3\ \text{cm}^{-3}$$

$$n_{H(N)} \approx n_{h(N)} = \frac{n_i^2}{N_D} = \frac{(1.45 \times 10^{10})^2}{10^{14}} = 2.10 \times 10^6\ \text{cm}^{-3}$$

$$n_{E(N)} \approx N_D = 10^{14}\ \text{cm}^{-3}$$

Note: $n_{E(P)}$'s $n_{e(P)}$ is lower than $n_{H(N)}$'s $n_{h(N)}$ because the P side is more heavily doped than the N side, so more thermionic electrons recombine.

D. **Depletion Width**

More densely populated regions require less space to neutralize incoming carriers. So depletion distances from the junction are shorter when doping concentrations are higher. Total *depletion width* d_W is shorter when N_A and N_D are higher for this reason:

$$d_W = d_P + d_N \propto \sqrt{\frac{1}{N_A} + \frac{1}{N_D}}. \tag{1.6}$$

Opposing doping concentrations across PN junctions are usually vastly different. In such cases, the highly doped region diffuses more carriers than the lightly doped side. With more incoming carriers and less neutralizing agents, depletion distance in the lightly doped side is far greater than in the highly doped region. So d_N across the P^+N junction that N_A^+ and N_D establish, for example, is usually so much longer than d_P that d_W is nearly d_N. In other words, d_W is largely the depletion distance into the lightly doped region.

Example 3: Draw the band diagram of a zero-bias PN^+ junction and approximate the relative location of the metallurgical junction X_J.

Solution: E_F in Fig. 1.9 is closer to E_C in the N region than to E_V in the P region because N_D^+ avails more free electrons than N_A avails holes. d_W extends more into the P region (from X_J) because the P region is lightly doped and the N region is heavily doped. So the P region requires more space to neutralize all the electrons that diffuse from the N side.

E. **Built-In Barrier Voltage**

The energy barrier across the junction indicates that dopant electrons in the N side need E_B more energy than their equilibrium thermal energy to diffuse. The Fermi

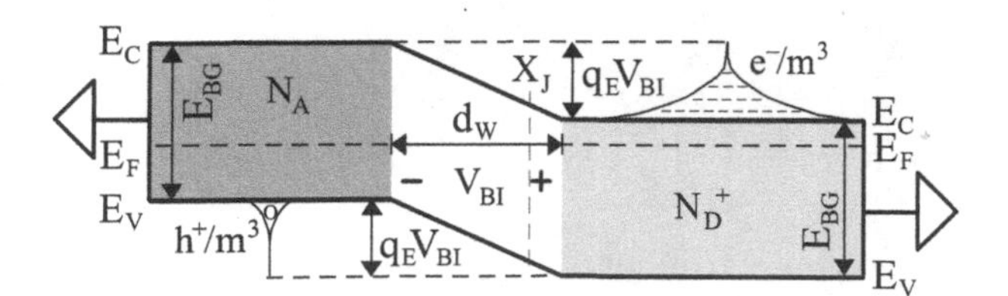

Fig. 1.9 Band diagram of zero-bias PN^+ junction

energy level indicates exponentially fewer electrons in the conduction band carry higher energy. So when combined, $n_{e(P)}$ is lower than N_D by the exponential amount that E_B and V_{BI} overwhelm the thermal energy E_t and *thermal voltage* V_t that *junction temperature* T_J establishes:

$$\begin{aligned} n_{E(P)} \approx n_{e(P)} = \frac{n_i^2}{N_A} = N_D \exp\left(\frac{-E_B}{E_t}\right) = N_D \exp\left(\frac{-q_E V_{BI}}{K_B T_J}\right) \\ = N_D \exp\left(\frac{-V_{BI}}{V_t}\right), \end{aligned} \quad (1.7)$$

where K_B is *Boltzmann's constant* 1.38×10^{-23} J/K, T_J is in Kelvin K, E_t is $K_B T_J$, and V_t is $K_B T_J/q_E$. V_{BI} is therefore a V_t and a logarithmic translation of how much N_A and N_D dwarf n_i^2:

$$V_{BI} = V_t \ln\left(\frac{N_A N_D}{n_i^2}\right). \quad (1.8)$$

Example 4: Find V_{BI} at 27 °C for a PN junction doped with $10^{17}/cm^3$ acceptor atoms and $10^{14}/cm^3$ donor atoms.

Solution:

$$\begin{aligned} V_{BI} &= V_t \ln\left(\frac{N_A N_D}{n_i^2}\right) = \left(\frac{K_B T_J}{q_E}\right) \ln\left(\frac{N_A N_D}{n_i^2}\right) \\ &= \left[\frac{(1.38 \times 10^{-23})(300)}{(1.60 \times 10^{-19})}\right] \ln\left[\frac{(10^{17})(10^{14})}{(1.45 \times 10^{10})^2}\right] \\ &= (25.9\text{m}) \ln\left(0.70 \times 10^{11}\right) = 650 \text{ mV} \end{aligned}$$

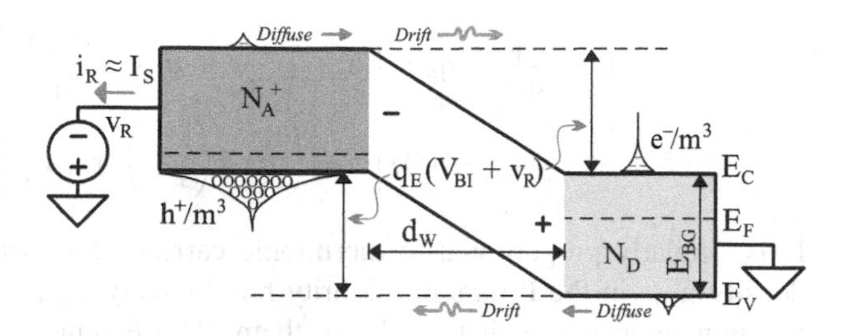

Fig. 1.10 Band diagram of reverse-bias PN junction

1.2.2 Reverse Bias

A. Electrostatics

Dopant Carriers Applying a negative voltage across the PN junction reinforces the built-in electric field. This effectively raises the energy barrier that majority carriers must overcome to diffuse, so dopant carriers do not diffuse. This *reverse voltage* v_R in Fig. 1.10 increases the barrier to $q_E(V_{BI} + v_R)$.

Thermionic Carriers Since the higher barrier deactivates dopant carriers, thermionic emissions avail more minority carriers than dopants in the other side of the junction avail majority carriers. In other words, P-side electrons outnumber N-side electrons and N-side holes outnumber P-side holes. Thermionic carriers therefore diffuse, and with the reverse field that V_{BI} and v_R set, drift across the depletion region. This flow of thermionic carriers establishes a *reverse current* i_R.

B. I–V Translation

i_R peaks and saturates quickly with increasing v_R because thermal energy in semiconductors liberates few electrons. This is why *reverse saturation current* I_S is normally low and why a v_R of only three V_t's raises i_R to 95% of the I_S that is possible:

$$i_R = I_S\left[1 - \exp\left(\frac{-v_R}{n_I V_t}\right)\right]. \qquad (1.9)$$

n_I is the *ideality factor* used to compensate for second-order effects. Structural imperfections in the depletion region, for example, can trap mobile electrons. Since this reduces the effect of v_R, n_I is usually higher (up to two) than in the ideal case, for which n_I is one.

Several components dictate I_S's charge rate dq_S/dt_S. Electronic charge q_E is the most basic of these. Charge-carrier concentrations are next. I_S is proportional to cross-sectional *junction area* A_J, for example, because larger areas supply more carriers:

$$I_S \equiv \frac{dq_S}{dt_S} = q_E A_J \left(n_{e(P)} \sqrt{\frac{D_N}{\tau_N}} + n_{h(N)} \sqrt{\frac{D_P}{\tau_P}} \right) \\ = q_E A_J \left[n_{e(P)} \left(\frac{n_i^2}{N_A} \right) \left(\frac{D_N}{L_N} \right) + \left(\frac{n_i^2}{N_D} \right) \left(\frac{D_P}{L_P} \right) \right]. \tag{1.10}$$

I_S is similarly proportional to thermionic carrier concentration: minority electron density $n_{e(P)}$ in the P region, minority hole density $n_{h(N)}$ in the N region, and the junction temperature that produces them. These minority carrier concentrations ultimately hinge on doping concentrations N_A and N_D.

The number of carriers that cross also depends on *diffusivity*, which is the ability of charge carriers to diffuse. Electron and hole *diffusion coefficients* D_N and D_P quantify this effect. Charge rate also depends on the time that traversing carriers require to recombine. I_S is therefore higher when N- and P-type *average carrier lifetimes* τ_N and τ_P are shorter. Another way to describe the temporal effect of τ_N and τ_P is spatially with *average diffusion lengths* L_N and L_P because L_N and L_P are, respectively, square-root translations of carrier diffusivity and lifetime: $D_N\tau_N$ and $D_N\tau_N$.

C. **Depletion Width**

Note that v_R applies a negative voltage to the P region and a positive voltage to the N region. So v_R attracts dopant holes and electrons away from the junction. v_R therefore widens the depletion region and the depletion distances that quantify this separation: d_P, d_N, and d_W.

1.2.3 Breakdown

A. **Impact Ionization**

When the reverse voltage is very high, v_R accelerates thermionic electrons to such an extent and with such kinetic force that they collide and liberate bound electrons. So one energized electron in Fig. 1.11 drifts, collides, and frees one electron–hole pair that avails another electron and a hole. v_R energizes the two liberated electrons to the same degree, so they generate two other electrons and two other holes. This multiplicative action continues as long as v_R is above the *breakdown voltage* V_{BD} that induces it.

This process of colliding and releasing built-up energy to liberate electrons is *impact ionization*. Since reverse current builds and grows in avalanche fashion, this phenomenon is known as *avalanche breakdown*. The breakdown voltage for this mechanism is higher when doping concentrations are lower because, with the wider depletion regions that result, field intensity is lower. Typical avalanche V_{BD}'s are greater than 5 V.

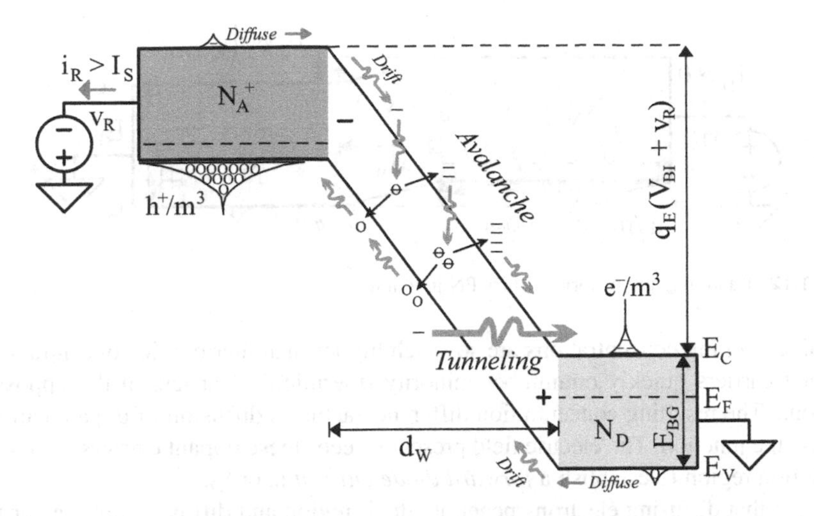

Fig. 1.11 Band diagram of PN junction in reverse breakdown

B. **Tunneling**

When v_R is high and depletion width is very narrow, the field is so intense that valence electrons in the P region in Fig. 1.11 break away and tunnel through the depletion region. This is *Zener tunneling*. Reverse current climbs above I_S this way as long as v_R is higher than the V_{BD} that induces i_R.

For the depletion width to be so narrow, doping concentrations must be very high. This is why *Zener breakdown* normally happens across highly doped junctions. Typical Zener V_{BD}'s are less than 7 V.

C. **Convention**

Irrespective of which mechanism dominates, engineers normally call diodes optimized to operate in breakdown *Zener diodes*. In practice, many of these diodes "break" around 6–7 V. At this level, i_R is the result of both avalanche and tunneling effects. Note, by the way, that neither breakdown mechanism is destructive.

1.2.4 Forward Bias

A. **Electrostatics**

Applying a positive *diode voltage* v_D across the PN junction opposes the built-in electric field. This effectively reduces the energy barrier that dopant carriers must overcome to diffuse to the $q_E(V_{BI} - v_D)$ that Fig. 1.12 shows. As v_D reduces E_B, an exponentially increasing number of dopant carriers become available.

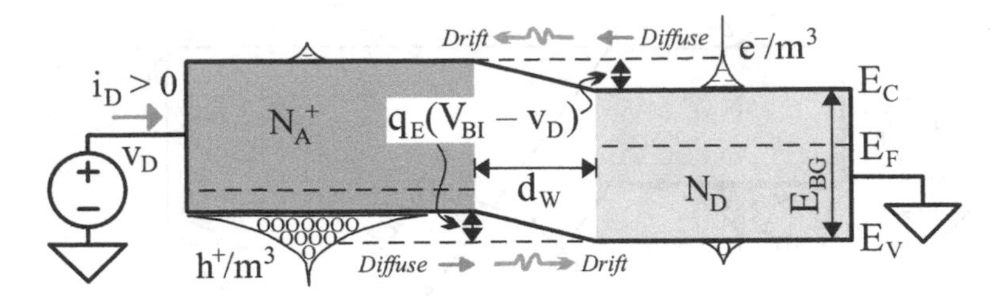

Fig. 1.12 Band diagram of forward-bias PN junction

Since doping concentrations are so much higher than thermionic concentrations, dopant carriers quickly outnumber minority (thermionic) carriers in the opposing regions. The resulting concentration difference actuates diffusion of dopant carriers across the junction. The electric field present sweeps these dopant carriers across the depletion region to establish a *forward diode current* i_F or i_D.

Note that diffusing electrons penetrate the P region and diffusing holes enter the N region to establish i_D. So electrons become minority carriers in the P region, and holes become minority carriers in the N region. In other words, i_D is the result of minority carrier conduction.

B. I–V Translation

i_D is zero with zero bias. i_D climbs when incoming carriers outnumber thermionic carriers. Since I_S is thermionic current and raising v_D avails an exponential number of diffusing carriers, i_D is a scalar translation of I_S that is zero when v_D is zero and increases exponentially with v_D:

$$i_D = I_S \left[\exp\left(\frac{v_D}{n_I V_t}\right) - 1 \right] \propto A_J. \tag{1.11}$$

Doping concentration is so high that three V_t's can establish an i_D that is 20 times greater than the thermionic current that limits i_R to I_S.

When v_D overcomes the built-in potential, the barrier fades and the depletion region shrinks. Since little impedes diffusion, the diode practically becomes a short. So i_D skyrockets past the "knee" that V_{BI} sets.

Ideality The ideality factor when i_D is low is similar to i_R's (greater than one) because imperfections trap a substantial fraction of diffusing electrons. All traps eventually fill, however, so higher current reduces the fraction of electrons lost to traps. Since diffusing electrons increase exponentially with v_D, n_I falls as v_D climbs and approaches one when v_D is roughly $0.5V_{BI}$ to V_{BI}, when diffusion overwhelms second-order effects.

Near and above V_{BI}, incoming carriers can outnumber dopant carriers, so fewer carriers recombine and n_I is again greater than one. This *high-level injection* occurs first in the region with the lowest doping concentration. With these higher current levels, the bulk regions and their contacts drop an Ohmic voltage that compresses (shrinks) v_D. When this happens, i_D reduces to a linear translation of the external voltage applied.

C. Depletion Width

Note that v_D applies a positive voltage to the P region and a negative voltage to the N region. So v_D repels dopant holes and electrons into the junction. v_D therefore narrows the depletion region and the depletion distances that quantify their separation: d_P, d_N, and d_W.

1.2.5 Model

A. Symbol

The PN junction conducts substantial current when forward-biased and hardly any current when reverse-biased. So the symbol that represents it in Fig. 1.13 is an arrow that points in the direction of forward current i_D. To highlight that current does not reverse (by much), a blocking line crosses the tip of the arrow. i_D enters the *anode* terminal of the diode and exits from the *cathode* terminal.

Diodes optimized to operate in breakdown receive the same basic symbol because the overall behavior is the same. But since breakdown conducts substantial reverse current, the blocking line in Fig. 1.14 flares out. These "wings" essentially indicate that the blocking mechanism is conditional.

B. I–V Translation

Notice that the equation for i_D matches i_R when i_D and v_D are negative. So the expression describes both forward and reverse conditions, but not breakdown. So since i_D is high and negative in breakdown, i_D in Fig. 1.15 falls abruptly near $-V_{BD}$, saturates to $-I_S$ in reverse bias, increases exponentially with v_D in forward bias, and skyrockets near V_{BI}.

Fig. 1.13 PN diode symbol

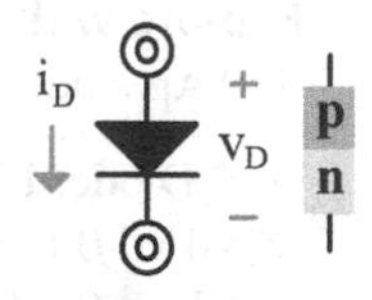

Fig. 1.14 Zener diode symbol

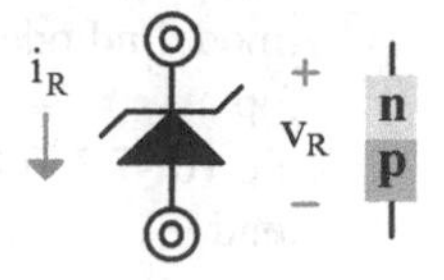

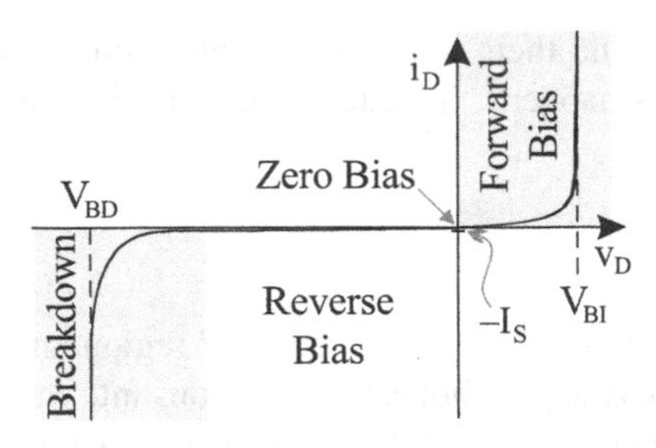

Fig. 1.15 Diode's current–voltage translation

The diode is practically a short in breakdown and at V_{BI} and an open circuit otherwise. This is why engineers often use them as on–off switches. When used this way, the diode switch closes and drops close to V_{BI} with forward current and opens whenever current reverses.

Example 5: Determine i_D when v_D is 650 mV, I_S is 50 fA, n_I is 1, and V_{BD} is −6.8 V at 27 °C.

Solution:

$$V_t = \frac{K_B T_J}{q_E} = \frac{(1.38 \times 10^{-23})(300)}{(1.60 \times 10^{-19})} = 25.9 \text{ mV}$$

$$i_D = I_S\left[\exp\left(\frac{v_D}{n_I V_t}\right) - 1\right] = (50\text{f})\left\{\exp\left[\frac{650\text{m}}{(1)(25.9\text{mV})}\right] - 1\right\} = 4.0 \text{ mA}$$

Explore with SPICE:
See Appendix A for notes on SPICE simulations.

```
* Diode: I-V Curve
vd vd 0 dc=650m
d1 vd 0 ndiode 1
.temp 27
.model ndiode d is=50f n=1 bv=6.8
.op
.dc vd -7.3 700m 10m
.end
```

Tip: Plot i(D1), comment or remove the .dc line, re-run the simulation, and view the output/log file.

C. Dynamic Response

Small Variations Shifting the operating point of a diode requires charge flow. Raising the barrier voltage, for example, repels recombined carriers back to their home regions. The number of carriers that diffuse across the junction also decreases. Moving these carriers changes the charge concentration across the junction.

This process requires time because i_D carries a finite amount of charge per second. So the current and charge needed Δi_D and Δq_D to vary the voltage Δv_D dictate the response time Δt_R of the diode. *Junction capacitance* C_J, which is the charge held across the junction with one volt, relates these parameters:

$$C_J = \frac{\Delta q_D}{\Delta v_D} = \frac{\Delta i_D \Delta t_R}{\Delta v_D} = C_{DEP} + C_{DIF}. \tag{1.12}$$

Depletion capacitance C_{DEP} is the component that the depletion region holds. *Diffusion capacitance* C_{DIF} is the diffusion charge held in-transit.

The depletion region is void of charge carriers and non-conducting, like an insulator, with the P and N regions as Ohmic contacts. This parallel-plate structure is what establishes C_{DEP}. C_{DEP} therefore increases with junction area and field intensity, and as a result, with A_J/d_W:

$$C_{DEP} = \frac{C_{J0}}{\sqrt{\frac{V_{BI} - v_D}{V_{BI}}}} = \frac{A_J C_{J0}''}{\sqrt{1 - \frac{v_D}{V_{BI}}}} = \frac{A_J C_{J0}''}{\sqrt{1 + \frac{v_R}{V_{BI}}}} \propto \frac{A_J}{d_W}. \tag{1.13}$$

Since a positive diode voltage narrows d_W, C_{DEP} also increases with higher v_D.

Doping concentration determines how many charge carriers are available and the barrier voltage $V_{BI} - v_D$ how many of those vanish in the depletion region. So C_{DEP} not only rises with v_D but also with N_A and N_D. Since diffusion current fades with zero bias, measuring C_J when v_D is zero to determine the *zero-bias junction capacitance* C_{J0} and using C_{J0} to extrapolate the effect of v_D in $V_{BI} - v_D$ on C_J is a practical way of quantifying C_{DEP}. Normalizing C_{J0} to area with C_{J0}'' is even better because A_J is a design variable.

Forward-biased carriers diffuse across the junction to become minority carriers. If the doped regions are short, these carriers in Fig. 1.16 can reach the metallic contacts without recombining. Irrespective, diffusing carriers require *forward transit time* τ_F to feed i_D. The voltage that sets this i_D dictates the number of in-transit charge q_{DIF} "held" by this mechanism.

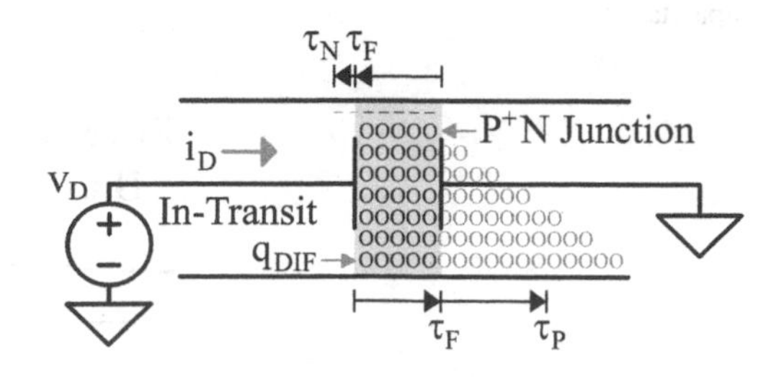

Fig. 1.16 In-transit diffusion charge when forward-biased

C_{DIF} is the charge needed Δq_{DIF} for the voltage to vary Δv_D:

$$C_{DIF} = \frac{\Delta q_{DIF}}{\Delta v_D} = \frac{\Delta i_D \tau_F}{\Delta v_D}\bigg|_{\text{Small Variations}} \approx \left(\frac{\partial i_D}{\partial v_D}\right)\tau_F \approx \left(\frac{I_D}{V_t}\right)\tau_F \equiv g_m \tau_F. \quad (1.14)$$

Since Δi_D avails Δq_{DIF} after τ_F, C_{DIF} is sensitive to i_D and the v_D that sets i_D. C_{DIF} is nonlinear because i_D is an exponential translation of v_D. For small variations, however, $\Delta i_D/\Delta v_D$ is roughly the linear translation that i_D's partial derivative $\partial i_D/\partial v_D$ or I_D/V_t sets. This derivative is the *small-signal transconductance* g_m of the diode, where I_D is the static (non-varying) component of i_D.

Since all carriers ultimately recombine in long diodes, τ_F is their average lifetime. τ_F in asymmetrically doped junctions is the average lifetime of the dominant carrier. So τ_F is nearly τ_P in P^+N junctions and τ_N in PN^+ junctions. τ_F, however, is a fraction of that in short diodes, in which case C_{DIF} is lower. In other words, short diodes are fast.

C_{DIF} climbs exponentially with v_D because diffused carriers in i_D increase exponentially with v_D. So C_{DIF} in Fig. 1.17 overwhelms C_{DEP} 200–400 mV before v_D reaches V_{BI}. In other words, C_J is practically C_{DEP} in reverse bias and light forward bias and C_{DIF} when v_D is within 100 mV or so of V_{BI}.

Example 6: Determine C_D for the diode in Example 5 when C_{JCO} is 100 fF, V_{BI} is 650 mV, and τ_F is 1 ns.

Solution:

$$v_D = 650\text{ mV} = V_{BI} = 650\text{ mV} \quad \therefore \quad C_D = C_{DEP} + C_{DIF} \approx C_{DIF}$$

$$C_{DIF} \approx \left(\frac{I_D}{V_t}\right)\tau_F = \left(\frac{4.0\text{m}}{25.9\text{m}}\right)(\text{1n}) = 154\text{ pF}$$

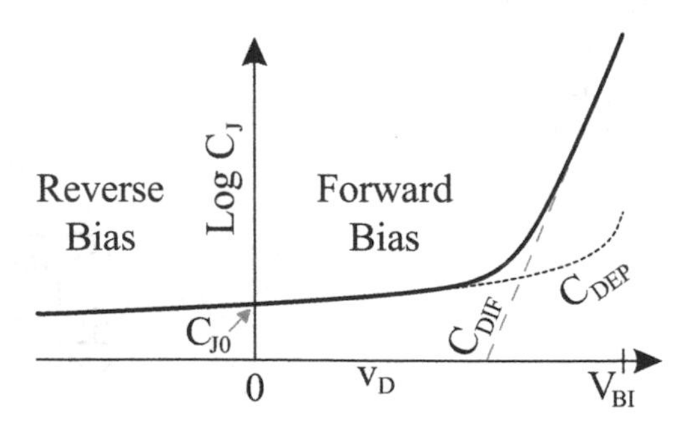

Fig. 1.17 Junction capacitance

Example 7: Determine C_D for the diode in Examples 5 and 6 when v_D is -2 V.

Solution:

$$v_D = -2\text{ V} < 0 \quad \therefore \quad C_D = C_{DEP} + C_{DIF} \approx C_{DEP}$$

$$C_{DEP} = \frac{C_{J0}}{\sqrt{1 - \frac{v_D}{V_{BI}}}} = \frac{100f}{\sqrt{1 - \left(\frac{-2}{650m}\right)}} = 50\text{ fF}$$

Note: Reverse-bias capacitance is usually much lower than forward-bias capacitance.

Explore with SPICE:
See Appendix A for notes on SPICE simulations.

```
* Diode: Capacitance
vd vd 0 dc=650m
d1 vd 0 ndiode 1
.temp 27
.model ndiode d is=50f n=1 bv=6.8 cjo=100f vj=650m tt=1n
.op
.end
```

Tip: View the output/log file.

Large Transitions When used as switches, diodes transition between the on and off states that forward- and reverse-bias conditions set. *Reverse-recovery time* t_{RR} refers to the time needed to reverse-bias the junction from a forward-bias state. t_{RR} is therefore the time needed to pull in-transit diffusion carriers q_{DIF} back to their home regions and to recombine carriers q_{DEP} in the depletion region. But of these, q_{DIF} is usually much greater than q_{DEP}.

t_{RR} is largely the time that reverse current i_R requires to collect the *reverse-recovery charge* q_{RR} that q_{DIF} sets:

$$t_{RR} = \frac{q_{RR}}{i_R} = \frac{q_{DIF} + q_{DEP}}{i_R} \approx \frac{q_{DIF}}{i_R} \approx \left(\frac{i_F}{i_R}\right)\tau_F, \tag{1.15}$$

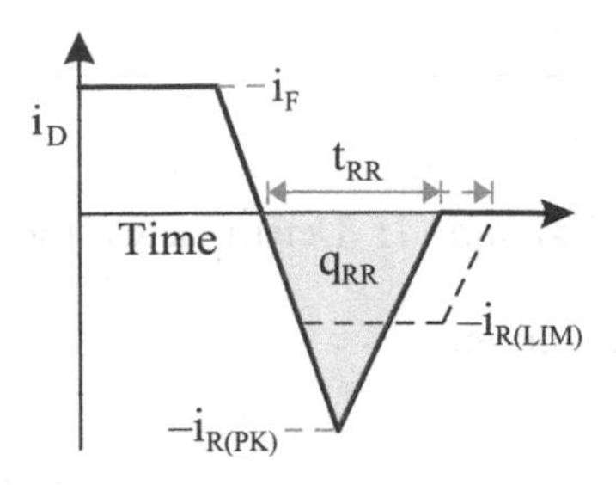

Fig. 1.18 Reverse-recovery current

where q_{DIF} is the charge that forward-bias current i_F produces with τ_F and i_R can vary with time. So t_{RR} ultimately hinges on i_F and the i_R that circuit conditions avail. When unchecked, i_R can peak to a level $i_{R(PK)}$ in Fig. 1.18 that is comparable to and possibly higher than $-i_F$. When limited to $i_{R(LIM)}$, $i_{R(LIM)}$ extends t_{RR}. This is unfortunate either way because the diode should be off, not conducting this much reverse current.

Forward-recovery time t_{FR} refers to the time needed to forward-bias a reverse-biased junction. t_{FR} is therefore the time needed to supply in-transit diffusion and depletion carriers. Since q_{DIF} is much greater than q_{DEP}, *forward-recovery charge* q_{FR} is nearly q_{DIF}.

This transition is more benign than reverse recovery in two ways. First, the forward current i_F needed to supply q_{DIF} flows from anode to cathode, as a diode should. Second, the circuit avails the i_F that sets q_{DIF} in the first place:

$$t_{FR} = \frac{q_{FR}}{i_F} = \frac{q_{DIF} + q_{DEP}}{i_F} \approx \frac{q_{DIF}}{i_F} \approx \frac{i_F \tau_F}{i_F} = \tau_F. \tag{1.16}$$

So t_{FR} is nearly the forward transit time of the diode.

Example 8: Determine t_{RR} for the diode in Examples 5 and 6 when recovering the reverse state in Example 7 and the resistance that limits this v_D variation is 10 kΩ.

Solution:

$$i_R = \frac{\Delta v_D}{R_R} = \frac{650\text{m} - (-2)}{10\text{k}} = 260\ \mu\text{A}$$

$$t_{RR} \approx \frac{q_{DIF}}{i_R} \approx \left(\frac{i_F}{i_R}\right)\tau_F = \left(\frac{4.0\text{m}}{260\mu}\right)(\text{ln}) = 15\ \text{ns}$$

Explore with SPICE:
See Appendix A for notes on SPICE simulations.

```
* Diode: Reverse Recovery
vd vx 0 dc=-2
rs vx vd 10k
d1 vd 0 ndiode 1
.ic v(vd)=650m
.model ndiode d is=50f n=1 bv=6.8 cjo=100f vj=600m tt=1n
.tran 10n
.end
```

Tip: Plot v(vd) and i(D1).

1.3 Metal–Semiconductor (Schottky) Diodes

Electrons in metal are so weakly bound that they are practically free and available for conduction. Still, the probability of finding electrons above the Fermi level of metal E_{FM} decreases exponentially with electron energy E_E. The Fermi level E_{FS} of a semiconductor is greater than E_{FM} when the semiconductor has more high-energy electrons, like the N-type semiconductor in Fig. 1.19.

1.3.1 Zero Bias

Since carrier concentration at higher energy levels is higher in the semiconductor, electrons diffuse into the metal when connected together. Diffusing electrons deplete and ionize the semiconductor region near the junction X_J. Electrons continue to diffuse until the growing electric field is high enough to repel further action.

Since current cannot flow under zero-bias conditions, the probability of finding charge carriers (i.e., E_{FM} and E_{FS}) in Fig. 1.20 is uniform across the junction. Higher electron density in the semiconductor $n_{E(S)}$ induces more diffusion. As a result, the semiconductor ionizes more, and the built-in voltage V_{BI} that the barrier establishes

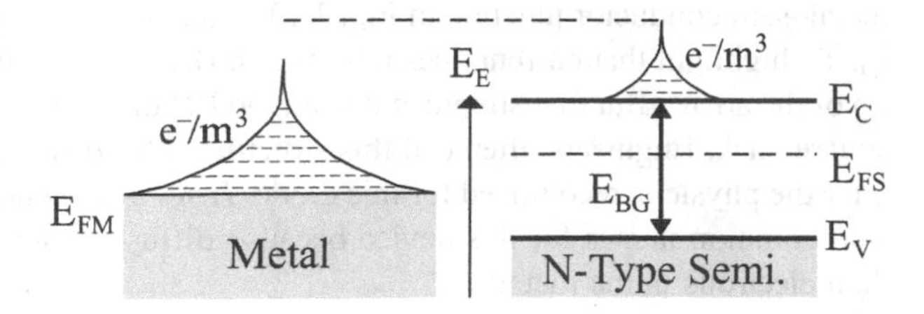

Fig. 1.19 Band diagram of separate metal and N-type semiconductor solids

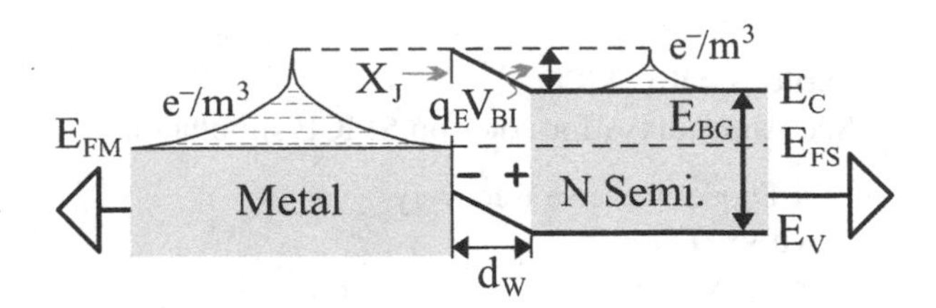

Fig. 1.20 Band diagram of zero-bias N-type metal–semiconductor junction

is higher. The depletion width is narrower when $n_{E(S)}$ is higher because depleting a region that is more denser with electrons is more difficult.

1.3.2 Reverse Bias

Applying a positive voltage v_R to the semiconductor raises the barrier voltage $V_{BI} + v_R$ that electrons in the semiconductor need to overcome and diffuse into the metal. So no electrons diffuse. Still, v_R pulls thermionic electrons in the metal into the semiconductor. But since thermionic electron density is low, reverse current saturates with little v_R to the reverse saturation current of the junction.

1.3.3 Forward Bias

Applying a positive voltage v_D to the metal does the opposite: reduces the barrier voltage $V_{BI} - v_D$ that electrons in the semiconductor need to overcome and diffuse into the metal. So electrons diffuse and establish current flow. Since the number of high-energy electrons climbs exponentially with a lower barrier voltage, i_D increases exponentially with v_D.

1.3.4 Model

A. Symbol

The metal–semiconductor junction conducts substantial i_D when forward-biased and hardly any i_R when reversed. So like the PN diode, the symbol that represents the metal–semiconductor junction in Fig. 1.21 is an arrow that points in the direction of i_D. To highlight that current does not reverse (by much), a blocking line crosses the tip of the arrow. But to distinguish it from the PN diode, the ends of the blocking line square back. Engineers often call this structure a *Schottky* or *Schottky barrier diode* after the physicist recognized for this diode. *Hot-electron* and *hot-carrier diodes* are also common names for this device because diffusing electrons carry more energy than electrons in the metal.

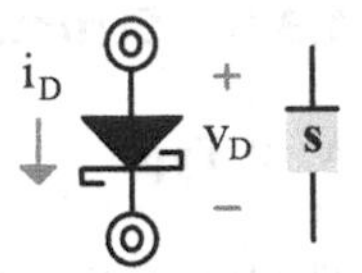

Fig. 1.21 Metal–semiconductor (Schottky) diode symbol

B. I–V Translation

i_D rises so much when v_D reaches V_{BI} that the junction practically shorts like Fig. 1.15 shows. Reinforcing the barrier with a negative v_D induces so little i_R that the device opens like a switch. When the reverse voltage is high enough, however, the junction "breaks down" and shorts to conduct substantial i_R. In short, this diode behaves very much like the PN diode.

This diode, however, normally diffuses fewer carriers with zero bias than the PN junction. So the V_{BI} that these diffusing electrons establish is usually lower, about 150–300 mV. The metal also avails more thermionic electrons, so reverse current is higher when v_D reverses. Since diffusing electrons into the metal do not recombine like they would in a P region, the ideality factor is closer to one.

C. Dynamic Response

Carriers that diffuse away and vanish from the depletion region ionize the junction. Since the barrier voltage $V_{BI} - v_D$ keeps more carriers from diffusing, reducing v_D reduces the number of carriers that vanish and the charge they establish. Depletion capacitance C_{DEP} therefore holds less charge when v_D falls and reverses like in the PN diode. C_{DEP} also scales with doping concentration because more carriers diffuse when carrier density is higher.

Unlike PN diodes, however, forward-bias electrons do not traverse a P region before reaching a metallic contact. And no holes diffuse into the semiconductor. So carriers do not require forward transit time τ_F to feed i_D. This means that the process of supplying and retrieving electrons is nearly instantaneous. And in-transit charge and diffusion capacitance are nil. With no q_{DIF} to recover, reverse-recovery time is very short.

D. Diode Distinctions

Metal–semiconductor diodes are faster than PN diodes. Plus, they drop lower voltages, so they burn less power. The drawback is, they also leak more reverse current. All this is because diffused electrons are majority carriers in metal and minority carriers in a P region. So forward i_D is the result of majority carrier conduction in Schottkys and minority carrier conduction in PN diodes. In a way, Schottkys behave like P^-N junctions with very short P regions because diffusing electrons alone establish V_{BI} and feed i_D, and τ_F in the P region is very short.

1.3.5 Structural Variations

A. P Type

The P-type Schottky diode is the complement of the N type. When E_{FS} in Fig. 1.22 is below E_{FM} and the two materials connect, semiconductor holes need less energy than metal electrons to diffuse. So metal electrons populate and pull semiconductor holes into the metal, ionizing the semiconductor until the field that is growing is high enough to keep other holes from diffusing.

V_{BI} in Fig. 1.23 is the barrier that keeps other holes from diffusing. The number of holes that diffuse and vanish from the depletion region is usually so low that V_{BI} is very low. So the reverse leakage current that results (with such a low V_{BI}) is correspondingly high. This is why N-type Schottkys are more prevalent in practice, because they are easier to optimize (for lower leakage).

B. Contacts

The depletion region is so narrow when the N semiconductor is highly doped that electrons can tunnel easily across. As a result, both positive and negative voltages across the junction induce substantial forward and reverse currents. This way, the junction forms a *Schottky contact*.

When high-energy electron density in the metal is higher than in the N semiconductor (i.e., E_{FM} is higher than E_{FS}) is higher than, metal electrons diffuse into the semiconductor with zero bias. When this happens, electrons *accumulate* near the junction. So electrons can flow easily in both directions. This way, the junction is an *Ohmic contact*.

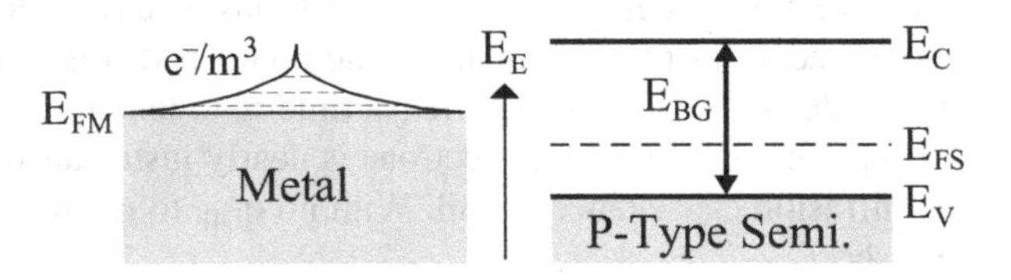

Fig. 1.22 Band diagram of separate metal and P-type semiconductor solids

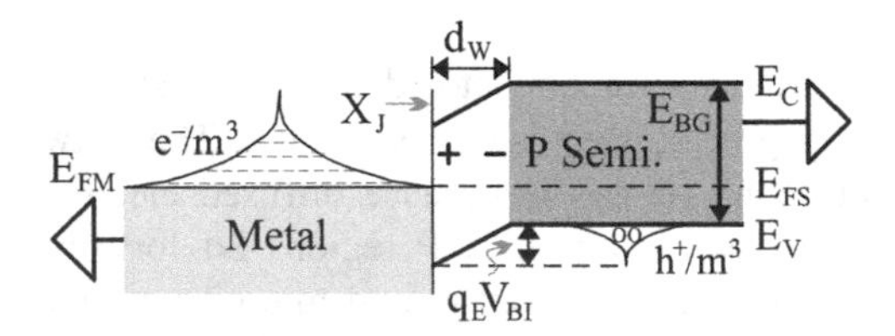

Fig. 1.23 Band diagram of zero-bias P-type metal–semiconductor junction

1.4 Bipolar-Junction Transistors

The *bipolar-junction transistor* (BJT) is basically two diodes head-to-head or back-to-back where the sandwiched "head" or sandwiched "back" is narrow. The transistor is "bipolar" because the structure is symmetrical, so the transistor can steer current in both directions. Engineers call the middle region the *base* because, when first built, it served as the mechanical base and support for the structure.

1.4.1 NPN

In an NPN, head-to-head diodes sandwich a thin P-type base like Fig. 1.24 shows. The short distance between the junctions is the *metallurgical base width* W_B. The *effective base width* w_B is shorter because base holes near the junctions diffuse away into the N regions. So the depletion regions effectively squeeze the base to w_B.

A. **Active**

Bias With a short base, the BJT *activates* when one diode forward-biases and the other diode reverses. In Fig. 1.25, the *base voltage* v_B forward-biases the junction on the left with respect to v_E and reverse-biases the junction on the right with respect to v_C. So v_C is greater than v_B, which is in turn greater than v_E. As a result, the field barrier and depletion region on the left decrease and the counterparts on the right increase.

Electrostatics Electrons and holes therefore diffuse across the junction on the left. w_B is so short that diffused electrons reach the opposite end of the base without recombining. Since v_C is greater than v_B, v_C pulls these base electrons across the depletion region into the N region on the right.

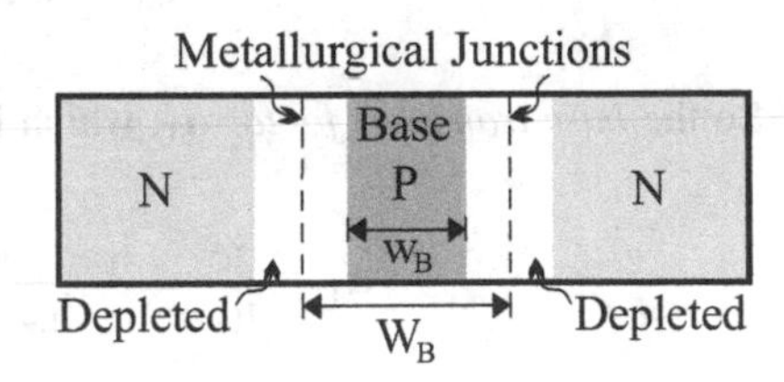

Fig. 1.24 NPN BJT structure

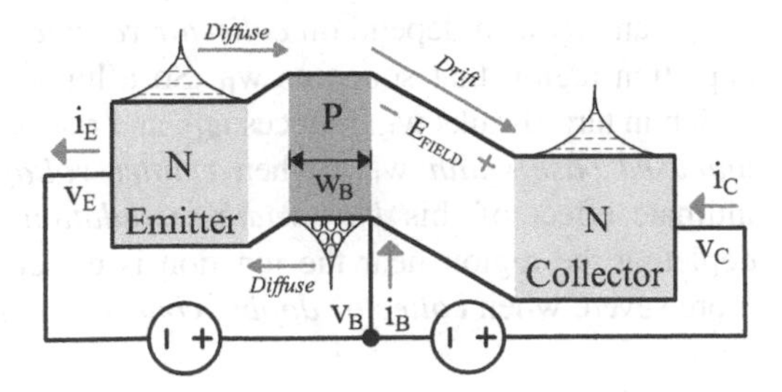

Fig. 1.25 Band diagram of NPN when activated

This way, the N-type region on the left "emits" the electrons that the N-type region on the right "collects." And as this happens, the P-type base injects holes into the N-type *emitter*. So the emitter receives the electron and hole currents i_{E-} and i_{E+} that the *collector* and base supply with *collector* and *base currents* i_C and i_B:

$$i_E = i_{E-} + i_{E+} = i_C + i_B. \tag{1.17}$$

Determinants Of the *emitter current* i_E, i_C loses i_{E+} to i_B. This i_{E+} is largely the fraction of i_{E-} that the *base* and *emitter doping concentrations* N_B and N_E and effective diffusion distance of electrons in the base w_B (because w_B is shorter than the average diffusion length of electrons across a long base L_{B-}) and average diffusion length L_{E+} of holes in the emitter set:

$$\frac{i_{E+}}{i_{E-}} = \left(\frac{N_B D_{E+}}{L_{E+}}\right)\left(\frac{w_B}{N_E D_{B-}}\right) \propto \left(\frac{N_B}{N_E}\right)\left(\frac{w_B}{L_{E+}}\right), \tag{1.18}$$

along with diffusivity of holes in the emitter D_{E+} and diffusivity of electrons in the base D_{B-}. So *emitter injection efficiency* γ_E, which is the fraction of i_E that i_{E-} avails, is mostly an $N_E L_{E+}$ fraction of $N_B w_B$ and $N_E L_{E+}$:

$$\gamma_E \equiv \frac{i_{E-}}{i_E} = \frac{i_{E-}}{i_{E+} + i_{E-}} \propto \frac{N_E L_{E+}}{N_B w_B + N_E L_{E+}}. \tag{1.19}$$

But not all electrons in i_{E-} reach the collector. Some of them recombine with holes in the base. And more recombine when w_B is a larger fraction of L_{B-}. The *recombination current* i_{RC} that results is a quadratic w_B/L_{B-} fraction of half i_{E-}:

$$i_{RC} \approx 0.5 i_{E-}\left(\frac{w_B}{L_{B-}}\right)^2. \tag{1.20}$$

So the *base transport factor* α_T, which is the fraction of i_{E-} that feeds i_C, is

$$\alpha_T \equiv \frac{i_C}{i_{E-}} = \frac{i_{E-} - i_{RC}}{i_{E-}} \approx 1 - 0.5\left(\frac{w_B}{L_{B-}}\right)^2. \tag{1.21}$$

Notice that α_T is nearly 90% when w_B is 45% of L_{B-}.

i_{RC} and α_T also depend on *collector voltage* v_C because v_C sets the width of the depletion region that squeezes w_B. So a higher v_C expands the depletion region, which in turn shrinks w_B, reduces i_{RC}, and raises α_T. w_B is therefore shorter than the *unbiased base width* w_{B0} (when *emitter voltage* v_E, v_B, and v_C are zero). The ultimate effect of this *base-width modulation* is a linear variation in i_C. Since depleting the region near the junction is easier when lightly doped, this effect is more severe when *collector doping concentration* N_C and N_B are lower.

Current Translations γ_E and α_T set the BJT's overall *baseline transport factor* α_0:

$$\alpha_0 \equiv \frac{i_C}{i_E} = \gamma_E \alpha_{T0} \propto \left(\frac{N_E L_{E+}}{N_B w_B + N_E L_{E+}}\right)\left[1 - 0.5\left(\frac{w_{B0}}{L_{B-}}\right)^2\right]. \quad (1.22)$$

This α_0 is nearly 100% when N_E is much greater than N_B and w_{B0} is much shorter than L_{E+} and L_{B-}. Note that α_0 is the emitter-to-collector current gain translation.

α_0 is close to 100% when i_C loses little to i_B. This is another way of saying that the base-to-collector *baseline current gain* β_0 is high. Since N_B and N_E dictate how i_E splits into i_{E-} and i_{E+} and w_B/L_{B-} determines how much of i_{E-} is lost to i_{RC}, w_{B0}/L_{B-} limits the gain that N_E/N_B sets:

$$\begin{aligned}\beta_0 &\equiv \frac{i_C}{i_B} = \frac{i_{E-} - i_{RC}}{i_{E+} + i_{RC}} = \frac{i_{E-} - 0.5 i_{E-}(w_{B0}/L_{B-})^2}{i_{E+} + 0.5 i_{E-}(w_{B0}/L_{B-})^2} \\ &= \left(\frac{N_E}{N_B}\right)\left\{\frac{1 - 0.5(w_{B0}/L_{B-})^2}{[(D_{E+} w_B)/(L_{E+} D_{B-})] + 0.5(N_E/N_B)(w_{B0}/L_{B-})^2}\right\}.\end{aligned} \quad (1.23)$$

α_0 and β_0 relate because i_E in α_0 feeds i_B in β_0 and i_C in α_0 and β_0:

$$\alpha_0 \equiv \frac{i_C}{i_E} = \frac{i_C}{i_B + i_C} = \frac{1}{i_B/i_C + 1} = \frac{\beta_0}{1 + \beta_0}. \quad (1.24)$$

α_0 and β_0 are ultimately measures of quality in a BJT that improve when w_B is shorter (i.e., when W_B is shorter, N_B is lower, and v_C is higher).

Collector Current The i_{E-} that i_C collects is the forward-bias diode current that v_B and v_E establish with *base–emitter voltage* v_{BE} or $v_B - v_E$:

$$i_C \approx i_{E-}\left(1 + \frac{v_{CE}}{V_A}\right) \approx I_S\left[\exp\left(\frac{v_{BE}}{n_I V_t}\right) - 1\right]\left(1 + \frac{v_{CE}}{V_A}\right) \propto A_{JBE}. \quad (1.25)$$

i_C is therefore proportional to I_S and, in consequence, to the cross-sectional area of the base–emitter junction A_{JBE}. But since raising v_C narrows w_B, which in turn increases the fraction of i_{E-} that feeds i_C, i_C climbs with *collector–emitter voltage* v_{CE}. V_A is the process-dependent constant that models this linear effect that base-width modulation produces. Engineers call this parameter *Early voltage* after the scientist that first observed and modeled this behavior.

B. Saturation

The BJT *saturates* when v_B forward-biases both junctions. As a result, both barrier voltages and depletion regions decrease and charge carriers in Fig. 1.26 diffuse across both junctions. Electron densities at the edges of the base match

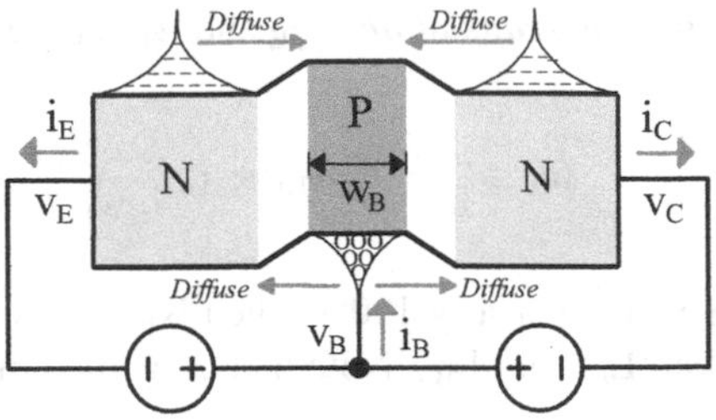

Fig. 1.26 Band diagram of symmetrically biased NPN in saturation

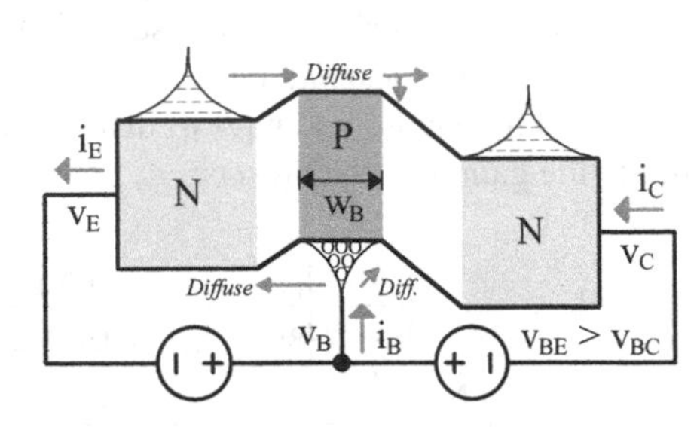

Fig. 1.27 Band diagram of asymmetrically biased NPN in saturation

when the junctions forward-bias equally. So instead of diffusing across the base, all emitter and collector electrons recombine with base holes. As a result, v_B feeds current to both N regions.

When one junction is less forward-biased than the other, fewer electrons diffuse across this junction. So the excess difference diffuses across the base to the least forward-biased end (to the right in Fig. 1.27). A fraction of these electrons recombines with the few base holes that diffuse in the same direction. So when v_{BE} exceeds the *base–collector voltage* v_{BC}, the electrons that survive establish an i_C that flows into v_C and, together with i_B, flow out of v_E as i_E.

Current Translations Collector electrons that diffuse into the base diminish electron diffusion across the base, so i_C collects fewer electrons. Decreasing N_C reduces this effect. i_C also loses electrons to forward-biased base holes, so i_{RC} falls when N_B is lower. i_C loses so much to these effects when deeply saturated that i_C can reverse direction. But even if only lightly saturated, α_0 and β_0 are still lower in saturation than in the active region.

C. **Optimal BJT**

The emitter injects more diffusion current that the collector receives when N_E is higher than N_B. Plus, the collector loses less diffusion current to the base when N_C is lower than N_B. So α_0 and β_0 are greater when $N_E > N_B > N_C$. But reducing N_B and N_C also increases base-width modulation, which sensitizes i_C to v_C with a lower V_A. The optimal BJT therefore reduces N_B and N_C only to the extent that an acceptably high V_A allows. Typical values for α_0, β_0, and V_A of an optimized N^+PN^- BJT are 98–99% A/A, 50–70 A/A, and 50–100 V.

Fig. 1.28 NPN BJT symbol

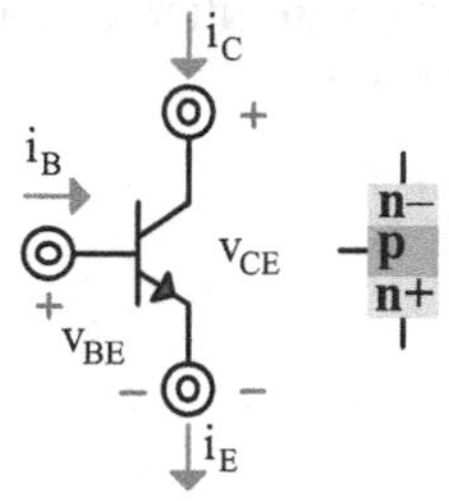

D. **Symbol**

Engineers design NPNs so collectors receive most of the electrons that emitters inject. This way, when active, i_B is very low. The orthogonal wall-like line into which i_B in Fig. 1.28 flows represents the base of the NPN for this reason, because a short base effectively blocks i_B.

In this mode, only the base–emitter junction forward-biases. So v_{BE} in an NPN is positive and i_B flows into the base and out of the emitter. To illustrate this diode-like behavior, an arrow between the base and emitter terminals points in the direction of i_B: toward the emitter.

E. **Modes**

Direction Given their symmetry, BJTs are bidirectional. As long as one diode forward-biases and the other reverses, the BJT is active. When saturated, the BJT favors the junction with the highest forward-bias current.

Orientation When asymmetrically doped, the end with the highest doping concentration can inject more carriers into the base that the other side can collect. This highly doped terminal is therefore more optimal as an emitter than a collector. So by convention, the "emitter" usually refers to the highest doped end. In an N^+PN^-, for example, the N^+ terminal is normally the emitter and the N^- terminal is the collector.

Forward Active The BJT is forward active when the base–emitter junction forward-biases and the base–collector junction reverses. So in the NPN, v_{BE} is positive and v_{BC} is negative. In other words, v_{CE} is higher than v_{BE} and i_C in Fig. 1.29 is exponential with v_{BE} and linear with v_{CE} when the BJT is forward active.

Forward Saturation The BJT saturates when both junctions forward-bias. So the forward-active NPN saturates when v_{CE} falls below v_{BE}. Since the transition happens when v_{CE} matches v_{BE}, i_C along the active–saturation boundary climbs exponentially with the v_{BE} that v_{CE}'s saturation point sets.

When the base–collector junction forward-biases by less than 200 mV or so, base–emitter diffusion is so much greater that the effects of base–collector diffusion

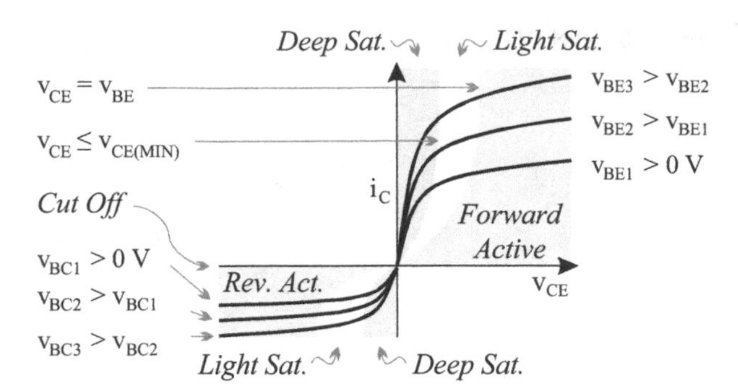

Fig. 1.29 N^+PN^- collector current

are negligible. So i_C remains exponential with v_{BE} and linear with v_{CE}. In other words, *light saturation* is largely an extension of the active region.

Forward-biasing the base–collector junction by more than 200 mV or so diffuses so many electrons and holes that their effects are no longer negligible. Fewer electrons reach the collector, and of these, a larger fraction recombines with the base holes that diffuse in the same direction. So i_C in *deep saturation* falls appreciably with v_{CE} when v_{CE} falls below the *minimum collector–emitter voltage* $v_{CE(MIN)}$ that doping concentrations and parasitic resistances ultimately dictate. $v_{CE(MIN)}$ is oftentimes 200–400 mV.

Reverse Active The junctions reverse roles in reverse modes. So the base–collector junction forward-biases and the base–emitter junction reverses in reverse active. In the NPN, v_{BC} is positive and v_{BE} is negative, and as a result, v_{EC} is higher than v_{BC} and i_C is exponential with v_{BC} and linear with v_{EC}.

Reverse Saturation The reverse-active NPN saturates when the base–emitter junction forward-biases, which happens when v_{EC} falls below v_{BC}. Forward-biasing the base–emitter junction by less than 200 mV or so, however, diffuses so few carriers that their effects are negligible. Current falls appreciably with v_{EC} when v_{EC} falls below the $v_{EC(MIN)}$ that process parameters and resistances set.

Optimal Behavior Since forward modes forward-bias the highly doped N region, forward injection efficiency is usually higher than in reverse. α_0 and β_0 are therefore greater in forward bias than in reverse under similar bias voltages. This is why the magnitude of i_C is usually higher in forward modes.

Cut off BJT currents are ultimately the result of carrier diffusion. So when both junctions reverse and diffusion stops, currents fade. This is the *cut off* region, which happens in the NPN when v_{BE} and v_{BC} are both zero and negative. The x axis in Fig. 1.29 marks this mode because i_C is zero along that line.

Example 9: Determine i_C and i_B when v_{BE} is 650 mV, v_C is 4 V, v_E is 0, β_F is 50 A/A, I_S is 50 fA, n_I is 1, and V_A is 50 V at 27 °C.

Solution:

$$i_C \approx I_S\left[\exp\left(\frac{v_{BE}}{n_I V_t}\right) - 1\right]\left(1 + \frac{v_{CE}}{V_A}\right) = (50f)\left\{\exp\left[\frac{650m}{(1)(25.9m)}\right] - 1\right\}\left(1 + \frac{4}{50}\right)$$

$$= 4.3 \text{ mA}$$

$$i_B \approx \frac{i_C}{\beta_F} = \frac{4.3m}{50} = 86 \text{ µA}$$

Explore with SPICE:

See Appendix A for notes on SPICE simulations.

```
* NPN BJT: I-V Curves
vc vc 0 dc=4
vb vb 0 dc=650m
q1 vc vb 0 npnbjt 1
.model npnbjt npn is=50f va=50 bf=50
.op
.dc vc 20m 5 10m vb 650m 600m 10m
.end
```

Tip: Plot ic(Q1), comment or remove .dc line, re-run the simulation, and view the output/log file.

1.4.2 PNP

The PNP is the NPN's complement. In a PNP, tail-to-tail diodes sandwich the thin N base in Fig. 1.30. W_B is the short metallurgical distance between the junctions. w_B is

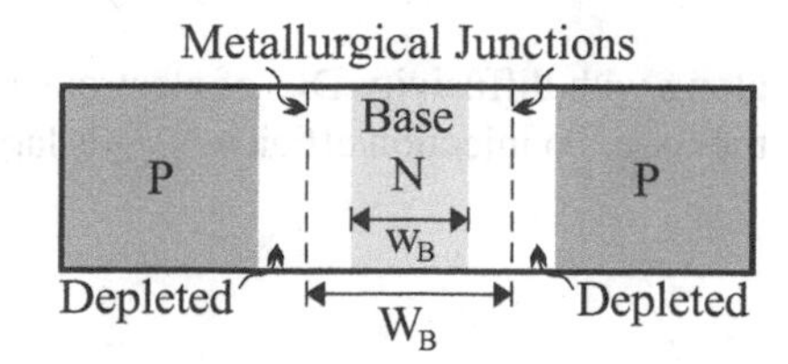

Fig. 1.30 PNP BJT structure

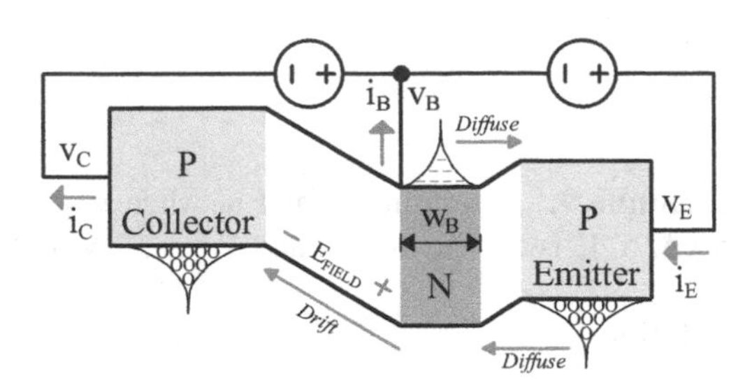

Fig. 1.31 Band diagram of PNP when activated

the distance between the depletion regions that base electrons near the junctions leave behind when diffusing into the P regions.

A. **Active**

Bias With a short base, the PNP activates when one diode forward-biases and the other diode reverses. The base voltage v_B in Fig. 1.31 forward-biases the junction on the right with respect to v_E and reverse-biases the junction on the left with respect to v_C. So v_E is greater than v_B, which is in turn greater than v_C. As a result, the field barrier and depletion region on the right decrease and the counterparts on the left increase.

Electrostatics With that junction forward-biased, electrons and holes diffuse across the junction on the right. w_B is so short that almost all diffused holes reach the other end of the base without recombining. Since v_C is lower than v_B, v_C pulls these base holes across the depletion region into the P-type region on the left.

In other words, the P-type region on the right "emits" the holes that the N-type region on the left "collects." And as this happens, the N-type base injects electrons into the emitter. So the emitter supplies the hole and electron currents i_{E+} and i_{E-} that the collector and base terminals receive with i_C and i_B:

$$i_E = i_{E+} + i_{E-} = i_C + i_B. \tag{1.26}$$

Determinants Of i_E, i_C loses i_{E-} to i_B. This i_{E-} is the fraction of i_{E+} that N_B and N_E and effective diffusion distance of holes in the base w_B (because w_B is shorter than the average diffusion length of holes across a long base L_{B+}) and average diffusion length of electrons in the emitter L_{E-} set:

$$\frac{i_{E-}}{i_{E+}} = \left(\frac{N_B D_{E-}}{L_{E-}}\right)\left(\frac{w_B}{N_E D_{B+}}\right) \propto \left(\frac{N_B}{N_E}\right)\left(\frac{w_B}{L_{E-}}\right), \tag{1.27}$$

along with diffusivity D_{E-} of electrons in the emitter and diffusivity of holes D_{B+} in the base. So injection efficiency γ_E is largely an $N_E L_{E-}$ fraction of $N_E L_{E-}$ and $N_B w_B$:

$$\gamma_E \equiv \frac{i_{E+}}{i_E} = \frac{i_{E+}}{i_{E+} + i_{E-}} \propto \frac{N_E L_{E-}}{N_E L_{E-} + N_B w_B}. \quad (1.28)$$

Some holes in i_{E+} recombine with base electrons. And more recombine when w_B is a larger fraction of L_{B+}. i_{RC} is a quadratic w_B/L_{B+} fraction of half i_{E+}:

$$i_{RC} \approx 0.5 i_{E+} \left(\frac{w_B}{L_{B+}}\right)^2. \quad (1.29)$$

So the fraction of i_{E-} that feeds i_C is

$$\alpha_T \equiv \frac{i_C}{i_{E+}} = \frac{i_{E+} - i_{RC}}{i_{E+}} \approx 1 - 0.5\left(\frac{w_B}{L_{B+}}\right)^2. \quad (1.30)$$

Notice that this base transport factor α_T is nearly 90% when w_B is 45% of L_{B+}.

i_{RC} and α_T also depend on v_C because v_C sets the width of the depletion region that squeezes w_B. So a lower v_C expands the depletion region, which in turn shrinks w_B, reduces i_{RC}, and raises α_T. w_B is therefore shorter than the unbiased base width w_{B0} (when v_E, v_B, and v_C match). This base-width modulation produces a linear variation in i_C that is more severe when the region near the junction is easier to deplete, which happens when N_B and N_C are lower.

Current Translations γ_E and α_T set the overall baseline transport factor α_0 of the PNP:

$$\alpha_0 \equiv \frac{i_C}{i_E} = \gamma_E \alpha_{T0} \propto \left(\frac{N_E L_{E-}}{N_E L_{E-} + N_B w_B}\right)\left[1 - 0.5\left(\frac{w_{B0}}{L_{B+}}\right)^2\right]. \quad (1.31)$$

α_0 nears 100% when N_E is much greater than N_B and w_{B0} is much shorter than L_{E-} and L_{B+}.

α_0 is close to 100% when i_C loses little to i_B, which happens when baseline β_0 is high. Since N_E and N_B dictate how i_E splits into i_{E+} and i_{E-} and w_B/L_{B+} determines how much i_{E+} is lost to i_R, w_{B0}/L_{B+} limits the gain that N_E/N_B sets:

$$\begin{aligned} \beta_0 &\equiv \frac{i_C}{i_B} = \frac{i_{E+} - i_{RC}}{i_{E-} + i_{RC}} = \frac{i_{E+} - 0.5 i_{E+}(w_B/L_{B+})^2}{i_{E-} + 0.5 i_{E+}(w_B/L_{B+})^2} \\ &= \left(\frac{N_E}{N_B}\right)\left\{\frac{1 - 0.5(w_B/L_{B+})^2}{[(D_{E-}w_B)/(L_{E-}D_{B+})] + 0.5(N_E/N_B)(w_B/L_{B+})^2}\right\}. \end{aligned} \quad (1.32)$$

α_0 and β_0 are higher when w_B is shorter (i.e., when W_B is shorter, N_B is lower, and v_C is higher).

Collector Current The i_{E+} that i_C receives is the forward-bias diode current that v_{EB} sets:

$$i_C \approx i_{E+}\left(1+\frac{v_{EC}}{V_A}\right) \approx I_S\left[\exp\left(\frac{v_{EB}}{V_t}\right)-1\right]\left(1+\frac{v_{EC}}{V_A}\right) \propto A_{JBE}. \tag{1.33}$$

i_C is therefore proportional to I_S and, in consequence, to A_{JBE}. But since raising v_{EC} narrows the base, which in turn increases the fraction of i_{E+} that feeds i_C, i_C climbs with v_{EC}. V_A models this base-width modulation effect in i_C.

B. Saturation

The PNP saturates when both junctions forward-bias. Charge carriers diffuse across both junctions when this happens. When symmetrically forward-biased, hole densities at the edges of the base in Fig. 1.32 match. So instead of diffusing across the base, holes recombine with base electrons. As a result, the base pulls current from both P regions.

When one junction forward-biases less than the other, fewer holes diffuse across that junction. So the excess difference diffuses across the base to the least forward-biased side (to the left in Fig. 1.33). A fraction of these holes recombines with the few base electrons that diffuse in the same direction. So when v_{EB} is greater than v_{CB}, the holes that survive establish an i_C that flows out of v_C, which is what remains of the i_E that flows into v_E after v_B pulls i_B.

Translations Collector holes that diffuse into the base diminish hole diffusion across the base, so i_C collects fewer holes. Decreasing N_C reduces this effect. i_C

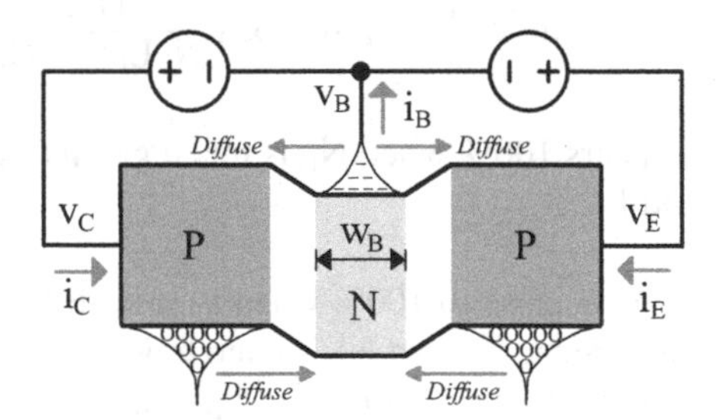

Fig. 1.32 Band diagram of symmetrically biased PNP in saturation

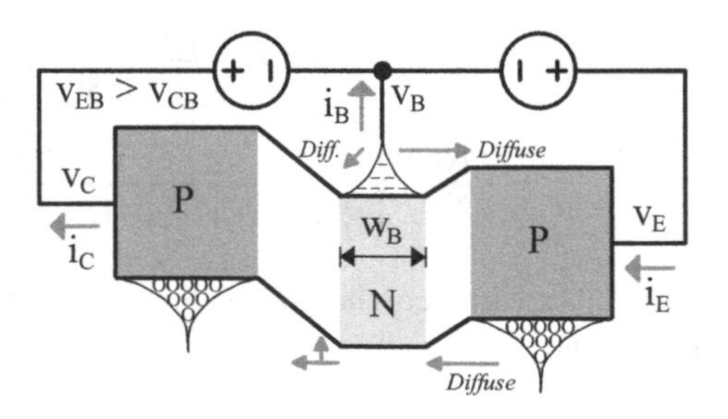

Fig. 1.33 Band diagram of asymmetrically biased PNP in saturation

loses holes to forward-biased electrons, so i_{RC} falls with lower N_B. i_C loses so much to these effects when deeply saturated that i_C can reverse. But even if only lightly saturated, α_0 and β_0 are still lower in saturation than in the active region.

C. Symbol

Engineers design PNPs so collectors receive most of the holes that emitters inject. This way, when active, i_B is very low. The orthogonal wall-like line out of which i_B in Fig. 1.34 flows represents the base of the PNP for this reason, because a short base effectively blocks this i_B.

In this mode, only the base–emitter junction forward-biases. So in a PNP, the emitter–base voltage v_{EB} is positive and i_B flows into the emitter and out of the base. To illustrate this diode-like behavior, an arrow between the emitter and base terminals points in the direction of i_B: toward the base.

D. Modes

Forward Active In forward active, the base–emitter junction forward-biases and the base–collector junction reverses. So v_{EB} is positive and v_{CB} is negative. v_{EC} is therefore higher than v_{EB} and i_C in Fig. 1.35 is exponential with v_{EB} and linear with v_{EC} when the PNP is forward active.

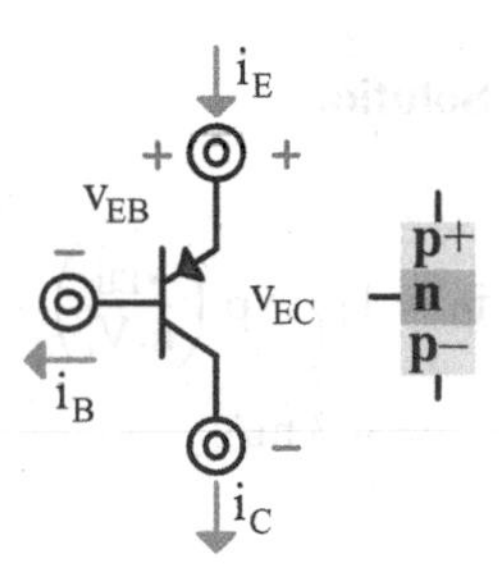

Fig. 1.34 PNP BJT symbol

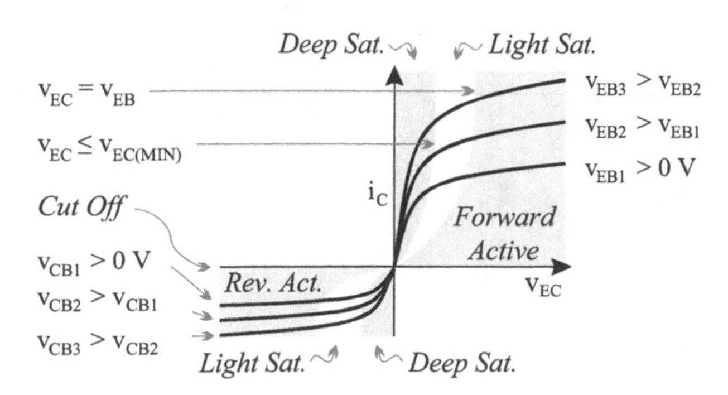

Fig. 1.35 P$^+$NP$^-$ collector current

Forward Saturation The forward-active PNP saturates when the base–collector junction forward-biases. This happens when v_{EC} falls below v_{EB}. i_C along this active–saturation boundary rises exponentially with the v_{EB} that v_{EC}'s saturation point sets. But when the base–collector junction forward-biases by less than 200 mV or so, the effects of saturation are negligible. i_C falls noticeably when this junction forward-biases by more, which happens when v_{EC} falls below $v_{EC(MIN)}$.

Reverse Modes Due to its symmetry, the PNP is reversible. So reverse modes behave the same way. But when asymmetrically doped, the higher-doped end injects more charge carriers into the base than the lighter-doped side under similar conditions. So the higher-doped terminal is, by convention, the emitter. This is why α_0, β_0, and i_C are higher in forward active and saturation (when the forward-biased base–emitter junction dominates) than in reverse active and saturation (when the forward-biased base–collector junction dominates).

Cut Off All currents fade when both junctions zero-bias or reverse. So i_C is zero when v_{EB} and v_{CB} are both zero and negative. The x axis in Fig. 1.35 marks this mode because i_C is zero along that line.

Example 10: Determine i_C and i_B when v_{EB} is 650 mV, v_E is 4 V, v_C is 0 V, β_F is 50 A/A, I_S is 50 fA, n_I is 1, and V_A is 50 V at 27 °C.

Solution:

$$i_C \approx I_S\left[\exp\left(\frac{v_{EB}}{n_I V_t}\right) - 1\right]\left(1 + \frac{v_{EC}}{V_A}\right) = (50\text{f})\left\{\exp\left[\frac{650\text{m}}{(1)(25.9\text{m})}\right] - 1\right\}\left(1 + \frac{4}{50}\right)$$
$$= 4.3\ \text{mA}$$

$$i_B \approx \frac{i_C}{\beta_F} = \frac{4.3\text{m}}{50} = 86\ \mu\text{A}$$

Explore with SPICE:
See Appendix A for notes on SPICE simulations.

```
* PNP BJT: I-V Curves
ve ve 0 dc=5
vb vb 0 dc=4.35
vc vc 0 dc=1
q1 vc vb ve pnpbjt 1
.model pnpbjt pnp bf=50 va=50 is=50f
.op
.dc vc 4.98 0 10m vb 4.35 4.40 10m
.end
```

Tip: Plot –ic(Q1), comment or remove .dc line, re-run the simulation, and view the output/log file.

1.4.3 Dynamic Response

A. Small Variations

Shifting the operating point of the diodes in the BJT requires charge q_D. And supplying or removing this q_D requires time. In the BJT, i_B and i_C supply or remove the q_D that base–emitter and base–collector junction capacitances C_{BE} and C_{BC} need. Engineers often use variables C_π and C_μ to refer to C_{BE} and C_{BC}.

Base–Emitter Capacitance Junction capacitance C_J includes two components: the charge held across the depletion region in C_{DEP} and the charge held in-transit in C_{DIF}. Since the base–emitter junction is zero- or reverse-biased in cut off, $C_{BE(DIF)}$ does not hold charge in those regions. So C_{BE} reduces to $C_{BE(DEP)}$ when v_{BE} is low, zero, or negative:

$$C_\pi\Big|_{v_{BE}\le 0}^{v_{BE}<300-500\text{mV}} \equiv C_{BE}\Big|_{v_{BE}\le 0}^{v_{BE}<300-500\text{mV}} \approx C_{BE(DEP)} = \frac{C_{JBE0}}{\sqrt{\dfrac{V_{BI}-v_{BE}}{V_{BI}}}} = \frac{A_{JBE}C_{JBE0}''}{\sqrt{1-\dfrac{v_{BE}}{V_{BI}}}}, \tag{1.34}$$

where C_{JBE0} is C_{BE}'s zero-bias capacitance, A_{JBE} is the cross-sectional area of the base–emitter junction, and C_{JBE0}'' is C_{JBE0} per unit area.

Forward-biasing the base–emitter junction increases the charge that $C_{BE(DIF)}$ holds exponentially. So $C_{BE(DIF)}$ surpasses $C_{BE(DEP)}$ when v_{BE} overcomes 300–500 mV. Since v_{BE} is normally 550–700 mV in active and saturation, C_{BE} is largely the in-transit charge that $C_{BE(DIF)}$ holds and i_C supplies with v_{BE} and forward transit time τ_F across the base sets:

$$C_\pi|_{v_{BE}>300-500mV} \equiv C_{BE}|_{v_{BE}>300-500mV} \approx C_{BE(DIF)} \approx \left(\frac{\partial i_C}{\partial v_{BE}}\right)\tau_F \approx \left(\frac{I_C}{V_t}\right)\tau_F \equiv g_m\tau_F, \tag{1.35}$$

Since small variations in i_C/v_{BE} are roughly the linear translation that i_C's partial derivative $\partial i_C/\partial v_{BE}$ or I_C/V_t or small-signal transconductance g_m sets, C_{BE} reduces to $C_{BE(DIF)}$'s $g_m\tau_F$.

Base–Collector Capacitance Since the base–collector junction only forward-biases in saturation, $C_{BC(DIF)}$ does not hold charge when the BJT activates or cuts off. So C_{BC} in Fig. 1.36 reduces to $C_{BC(DEP)}$ in the active and cut-off regions when v_{CE} nears, matches, or surpasses v_{BE}:

$$C_\mu\Big|_{v_{CE}\geq v_{BE}}^{v_{CE}\geq v_{CE(MIN)}} \equiv C_{BC}\Big|_{v_{CE}\geq v_{BE}}^{v_{CE}\geq v_{CE(MIN)}} \approx C_{BC(DEP)} = \frac{C_{JBC0}}{\sqrt{\frac{V_{BI}-v_{BC}}{V_{BI}}}} = \frac{A_{JBC}C_{JBC0}''}{\sqrt{1+\frac{v_{CB}}{V_{BI}}}} = \frac{A_{JBC}C_{JBC0}''}{\sqrt{1+\left(\frac{v_{CE}-v_{BE}}{V_{BI}}\right)}}. \tag{1.36}$$

C_{JBC0} here is C_{BC}'s zero-bias capacitance, A_{JBC} is the cross-sectional area of the base–collector junction, and C_{JBC0}'' is C_{JBC0} per unit area.

Forward-biasing the base–collector junction increases the charge that $C_{BC(DIF)}$ holds exponentially. But when lightly saturated, the effect is so minimal that $C_{BC(DEP)}$ dominates. So C_{BC} is nearly $C_{BC(DEP)}$ when v_{CE} matches or surpasses $v_{CE(MIN)}$. In deep saturation, however, $C_{BC(DIF)}$ holds more charge than $C_{BC(DEP)}$. So below $v_{CE(MIN)}$, C_{BC} is largely the in-transit charge that $C_{BC(DIF)}$ holds with v_{BC}. Note that C_{BC} is much lower than C_{BE} when lightly saturated and activated.

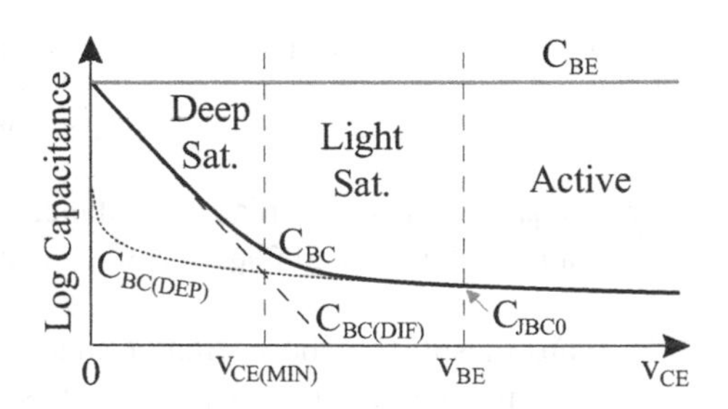

Fig. 1.36 BJT capacitances

Example 11: Determine C_{BE} and C_{BC} for the NPN BJT in Example 9 when τ_F is 100 ps, C_{JBC0} is 100 fF, and V_{BI} is 650 mV.

Solution:

$$v_{BE} = 650\text{ mV} = V_{BI} = 650\text{ mV}$$

$$\therefore \quad C_{BE} \approx C_{DIF} \approx \left(\frac{I_C}{V_t}\right)\tau_F = \left(\frac{4.3m}{25.9m}\right)(100p) = 17\text{ pF}$$

$$v_{BC} = v_{BE} - v_{CE} = 650m - 4 = -3.35\text{ V} < 0\text{ V}$$

$$\therefore \quad C_{BC} \approx C_{DEP} = \frac{C_{JBC0}}{\sqrt{1 - \frac{v_{BC}}{V_{BI}}}} = \frac{100f}{\sqrt{1 - \left(\frac{-3.35}{650m}\right)}} = 40\text{ fF}$$

Note: C_{BE} is usually much greater than C_{BC} in forward active.

Explore with SPICE:
See Appendix A for notes on SPICE simulations.

```
* NPN BJT: Capacitance
vc vc 0 dc=4
vb vb 0 dc=650m
q1 vc vb 0 npnbjt 1
.model npnbjt npn is=50f va=50 bf=50 tf=100p cjc=100f vjc=650m
+mjc=0.5
.op
.end
```

Tip: View the output/log file.

B. Large Transitions

BJTs activate after i_B and i_C supply the depletion and in-transit charge $q_{BE(DEP)}$ and $q_{BE(DIF)}$ that the base–emitter junction requires. To saturate, i_B and i_C must

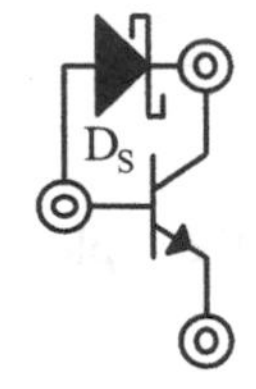

Fig. 1.37 Schottky-clamped BJT

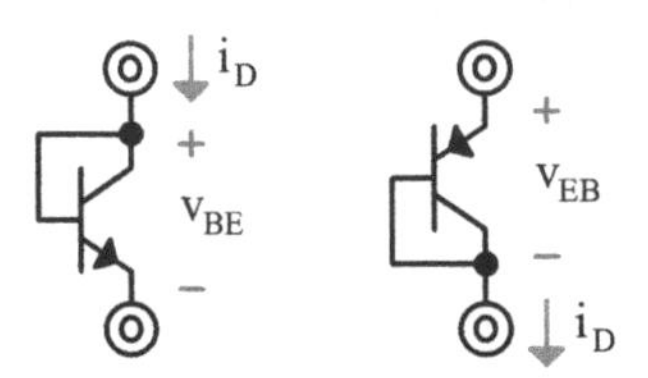

Fig. 1.38 Diode-connected NPN and PNP BJTs

similarly supply the $q_{BC(DEP)}$ and $q_{BE(DIF)}$ that the base–collector junction requires. So when used as a switch, the BJT activates and saturates after i_B and i_C supply this charge. The BJT cuts off after i_B and i_C reverse this charge.

These large transitions require substantial i_B and, when i_B is low, substantial *recovery time* t_R because $q_{BE(DIF)}$ and $q_{BC(DIF)}$ are high. Keeping the base–collector junction from entering deep saturation reduces $q_{BC(DIF)}$ to $q_{BC(DEP)}$ levels. This way, t_R can be shorter.

A Schottky diode across the base and collector (like D_S in Fig. 1.37) achieves this by shunting current away from the base–collector junction. Since D_S drops less voltage than the PN junction, D_S "clamps" v_{BC} to a level that keeps the PN junction from forward-biasing too much. This way, deep saturation does not occur, so $q_{BC(DIF)}$ is always as low as or lower than $q_{BC(DEP)}$. And D_S does not cancel the resulting reduction in t_R because D_S does not hold in-transit charge.

1.4.4 Diode-Connected BJT

The basic difference between PN diodes and BJTs is that BJTs split the diode current into its constituent electron and hole parts. In the NPN, for example, the forward-biased base–emitter junction produces a diode current i_D that flows entirely out of the emitter. Of i_D, the collector supplies the electron component i_{D-} the base supplies the hole component i_{D+}. Similarly, the PNP steers i_{D+} of the diffused emitter current i_D into the collector and i_{D-} into the base.

The base-to-collector connection in Fig. 1.38 combines and forces both parts to flow through one collector–base terminal. This connection combines the diode current that BJTs normally split. This way, BJTs behave like diodes, inducing a diode current i_D that climbs exponentially with v_{BE} in the NPN and v_{EB} in the PNP.

Engineers call this a *diode connection*. This connection essentially shorts and deactivates the base–collector junction. In other words, diode connections reduce BJTs to their base–emitter junctions. Sometimes this connection is *implicit*, like when other transistors or components connect these base and collector terminals together.

Example 12: Determine v_{BE} when i_C is 1 mA for the NPN BJT in Example 9.

Solution:

$$i_C = 1\text{ mA} \approx I_S\left[\exp\left(\frac{v_{BE}}{n_I V_t}\right) - 1\right]\left(1 + \frac{v_{BE}}{V_A}\right)$$

$$= (50\text{f})\left\{\exp\left[\frac{v_{BE}}{(1)(25.9\text{m})}\right] - 1\right\}\left(1 + \frac{v_{BE}}{50}\right)$$

$$\therefore \quad v_{BE} = 620\text{ mV}$$

Note: The effect of base-width modulation (v_{CE}) is minimal when V_A is 10 V or greater.

Explore with SPICE:
See Appendix A for notes on SPICE simulations.

```
* NPN BJT: Diode-Connected
ic 0 vbe dc=1m
q1 vbe vbe 0 npnbjt 1
.model npnbjt npn is=50f va=50 bf=50
.op
.end
```

Tip: View the output/log file.

1.5 Summary

Thermal energy can liberate and avail loosely bound electrons. When exposed to an electric field, these electrons and the holes they leave behind drift in opposite directions. Conductors, semiconductors, and insulators conduct these charge carriers easily, moderately well, and poorly, respectively. Dopant atoms add electrons or holes to semiconductors, so the resulting N- or P-type material avails more of one than the other. Between these, electrons flow more easily than holes.

The carrier concentration difference across a PN junction is so high that electrons and holes diffuse across, leaving behind a depleted carrier-free region. This process continues until the electric field they establish keeps other carriers from diffusing. Reinforcing the field with a reverse voltage keeps carriers from flowing. Opposing the field with a forward voltage, on the other hand, avails an exponentially increasing number of carriers for conduction. Only a negative diode current can reverse and drain these in-transit carriers.

Schottky diodes behave essentially the same way, except the metallic side is so full of electrons and void of holes that the metal does not deplete and holes do not diffuse into the semiconductor. Since diffusing carriers are immediately available for conduction, Schottkys are faster than PN diodes. Plus, their built-in field is usually lower, so they drop a lower voltage when they conduct.

BJTs are two head-to-head or back-to-back PN diodes with a short P- or N-type base. They activate when one diode forward-biases and the other reverses. The base is so short that, biased this way, diffused minority carriers reach the other end of the base, where the field of the reversed diode pulls them to the collector. When the emitter's doping concentration is much higher than that of the base, the majority carriers that the base feeds are much lower than the minority carriers that feed the collector. In other words, collector current is a much greater fraction of emitter current than base current is.

Forward-biasing the collector junction steers majority carriers away from the emitter. So emitter–collector and base–collector current translations fall. The reduction is lower when the collector is lightly doped and when the forward-biasing voltage across the base–collector junction is low. Reversing and draining forward-bias charges with limited base current requires considerable time. Schottky-clamped BJTs need less time because Schottkys do not hold in-transit carriers.

Switched-inductor power supplies need transistors to energize the inductor. Although they are not always BJTs, all MOSFETs incorporate unintended BJT structures, so understanding BJTs is essential. Plus, BJTs usually cost less money. Diodes are more basic and vital because they do not require a synchronizing signal to switch. So they can not only drain the inductor automatically but also steer current when transistors are busy transitioning between states.

Field-Effect Transistors

2

Abbreviations

BJT	Bipolar-junction transistor
CMOS	Complementary MOS
DIBL	Drain-induced barrier lowering
DMOS	Diffused-channel or double-diffused MOS
EHP	Electron–hole pair
FET	Field-effect transistor
JFET	Junction FET
LDMOS	Lateral DMOS
LDD	Lightly doped drain
MOS	Metal–oxide–semiconductor
NBTI	Negative bias temperature instability
RSS	Root sum of squares
SNR	Signal-to-noise ratio
SiO_2	Silicon dioxide
VCO	Voltage-controlled oscillator
VDMOS	Vertical DMOS
A_J	Junction area
β_0	Base–collector current gain
C_{CH}	Channel capacitance
C_{DB}	Drain–body capacitance
C_{DEP}	Depletion capacitance
C_{GB}	Gate–body capacitance
C_{GD}	Gate–drain capacitance
C_{GS}	Gate–source capacitance
C_J	Junction capacitance
C_{J0}	Zero-bias junction capacitance
C_{OL}	Overlap capacitance

G. A. Rincón-Mora, *Switched Inductor Power IC Design*,
https://doi.org/10.1007/978-3-030-95899-2_2

C_{OX}	Oxide capacitance
C_{SB}	Source–body capacitance
d_W	Depletion width
E_{BG}	Band-gap energy
$E_{CN/P}$	Critical electric field
E_F	Fermi energy level
E_K	Kinetic energy
ε_0	Permittivity in vacuum
ε_{OX}	Relative permittivity of SiO_2
f_C	Noise-corner frequency
f_O	Operating frequency
f_{SW}	Switching frequency
Δf_{BW}	Frequency bandwidth
g_m	Small-signal transconductance
$\gamma_{N/P}$	Body-effect parameter
i_B	Body current
i_D	Drain current
i_{DIF}	Diffusion current
i_{FLD}	Drift current
i_G	Gate current
i_{IN}	Input current
i_{nc}	Coupled noise current
i_{nf}	Flicker noise current
i_{ns}	Shot noise current
i_{nt}	Thermal noise current
i_S	Source current
i_{SUB}	Substrate current
$I_{SN/P}$	Saturation current
K_B	Boltzmann's constant
K_F	Flicker noise coefficient
$K_{N/P}$	Baseline conductivity
$K_{N/P}'$	Transconductance parameter
L_{CH}	Channel length
L_{MIN}	Minimum oxide length
L_{OL}	Overlap length
L_{OX} or L	Oxide length
$\lambda_{N/P}$	Channel-length modulation parameter
μ_N	Electron mobility
μ_P	Hole mobility
n_d	Spectral noise density
n_I	Ideality factor
q_E	Electronic charge
R_{CH}	Channel resistance
R_{ON}	On resistance
R_{DS}	Drain–source resistance

R_{SH}	Sheet resistivity
t_{OX}	Oxide thickness
T_J	Junction temperature
v_B	Body terminal/voltage
v_{BS}	Body–source voltage
v_D	Drain terminal/voltage
v_{DD}	Positive power supply
v_{DS}	Drain–source voltage
$v_{DS(SAT)}$	Drain–source pinch-off saturation voltage
$v_{DS(SAT)}'$	Drain–source sub-threshold saturation voltage
$v_{DS(SAT)}''$	Drain–source velocity-saturation voltage
v_E	Electron velocity
v_G	Gate terminal/voltage
v_{GS}	Gate–source voltage
v_{GSP}, v_{GST}, and v_{SGT}	Gate drive
v_H	Hole velocity
v_{JR}	Reverse junction voltage
v_{nt}	Thermal noise voltage
v_S	Source terminal/voltage
$v_{TN/P}$	Threshold voltage
V_A	Early voltage
V_{BI}	Built-in (potential) voltage
V_P	Pinch-off voltage
V_t	Thermal voltage
$V_{TN/P0}$	Zero-bias $v_{TN/P}$
ψ_B	Surface–body barrier potential
ψ_S	Surface potential
w_B	Effective base width
W_{CH} or W	Channel width

Power supplies use switches to steer and feed current into batteries and microelectronic systems. *Metal–oxide–semiconductor* (MOS) *field-effect transistors* (FETs) are popular in this space because they drop millivolts and do not require static gate current to close. Although *bipolar-junction transistors* (BJTs) usually cost less to fabricate, they need static base current to close and saturate when they close, so they require substantial reverse current to open. Diodes do not need this static current, but they drop 400–700 mV and only close when terminal voltages allow. Still, MOSFETs incorporate substrate diodes and BJTs that can help and at times also hurt.

The fundamental mechanism that establishes conductivity in FETs is an electric field. In the case of MOSFETs, parallel-plate MOS capacitors establish this field. The underlying purpose of the capacitor is to form a conducting *N-type channel* in NFETs and a *P-type channel* in PFETs. Although poly-silicon is nowadays more popular than metal as the upper plate, engineers still use MOS to refer to these and other oxide-sandwiched structures on semiconductors.

2.1 Junction FETs

The *junction FET* (JFET) is a simpler junction-based realization of the MOSFET. Although not as pervasive, JFETs are useful as resistors and low-noise transistors. *Electronic noise* in JFETs is low because carriers flow well below (and away from) the uneven silicon surface.

2.1.1 N Channel

N-channel JFETs are N-doped semiconductor strips sandwiched between P-type regions. In the case of Fig. 2.1, *top* and *bottom* P^+ and P *gates* sandwich an N channel contacted by highly doped N^+ regions. *Channel length* L_{CH} or L and *width* W_{CH} or W are the longitudinal length and transverse width of the overlapping P^+–N channel–P gate layers. The Ohmic surface contact of the bottom gate (on the left in Fig. 2.1) is another highly doped P^+ region.

A. Triode

The NJFET is basically an N-type resistor compressed by P-type regions. The geometry and doping concentration of the channel set the baseline *channel resistance* R_{CH}. R_{CH} increases as the depletion space against the top and bottom P regions in Fig. 2.2 expands to squeeze the channel. These P regions are the *gates* v_G of the JFET because their voltages adjust R_{CH}.

R_{CH} is high when L is long, W is narrow, and *baseline conductivity* K_N is low. The channel dematerializes (and opens) when the gate–channel voltage reverses

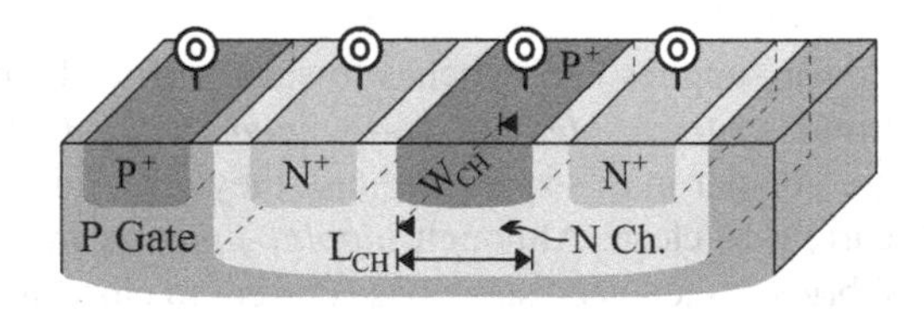

Fig. 2.1 N-channel JFET structure

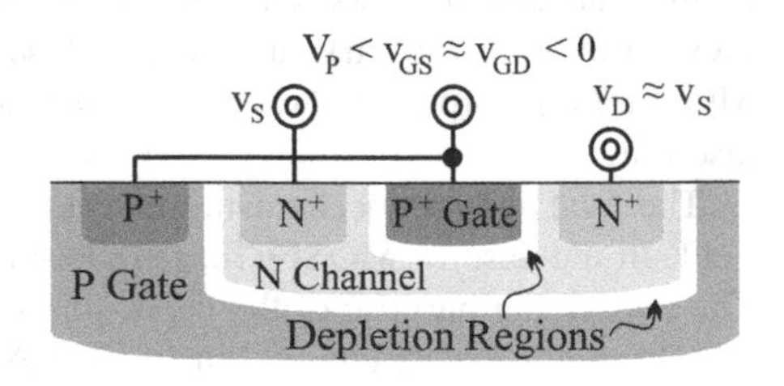

Fig. 2.2 Uniformly biased N-channel JFET in triode

enough to pinch the entire channel. This negative v_G is the *pinch-off voltage* V_P. So R_{CH} spikes sharply when v_{GS} and v_{GD} reach this V_P:

$$R_{CH}\Big|_{v_{DS}<<v_{GSP}}^{v_{GS}>V_P} \approx \left(\frac{L}{W}\right)\left[\frac{1}{K_N(v_{GS}-V_P)}\right] \equiv \left(\frac{L}{W}\right)\left(\frac{1}{K_N v_{GSP}}\right) \equiv \left(\frac{L}{W}\right)R_{SH}. \quad (2.1)$$

This R_{CH} is more accurate when v_{GS} is uniform across the channel and above the pinch-off point. This happens when v_{GS} and v_{GD} or $v_{GS} - v_{DS}$ are higher than V_P and v_{DS} is low. This means that *gate drive* $v_{GS} - V_P$ or v_{GSP} is positive and v_{DS} is much lower than v_{GSP}.

K_N and v_{GSP} set the *sheet resistivity* R_{SH} of R_{CH} (in Ohms per square). R_{SH} is the resistance of each W × W square of the channel. So when combined, R_{CH} is L/W squares of R_{SH}.

A voltage v_{DS} across R_{CH} induces channel current i_D. v_{DS}, however, expands the depletion space that squeezes half of the channel in Fig. 2.3. So half of v_{DS} raises and reduces the linear effects of v_{GS} on R_{CH} and i_D:

$$i_D\Big|_{v_{DS}<v_{GSP}}^{v_{GS}>V_P} = \frac{v_{DS}}{R_{CH}} \approx v_{DS}\left(\frac{W}{L}\right)K_N\left(v_{GS} - V_P - \frac{v_{DS}}{2}\right). \quad (2.2)$$

This i_D corresponds to *triode* because i_D is sensitive to v_{SG} and v_{SD}.

With v_{DS}, the negative N^+ terminal supplies the N-channel electrons that the positive N^+ terminal outputs. So the N^+ region on the left is the *source* v_S and the one on the right is the *drain* v_D. But when v_G connects to v_S or to a fixed negative voltage, the NJFET is just a *pinched resistor*.

B. Saturation

The channel pinches and charge disappears near v_D in Fig. 2.4 when v_{GD} reverses to V_P. Since v_{GD} is equivalent to $v_{GS} - v_{DS}$, this happens when v_{DS} reaches $v_{GS} - V_P$ or v_{GSP}. This is the *saturation voltage* $v_{DS(SAT)}$.

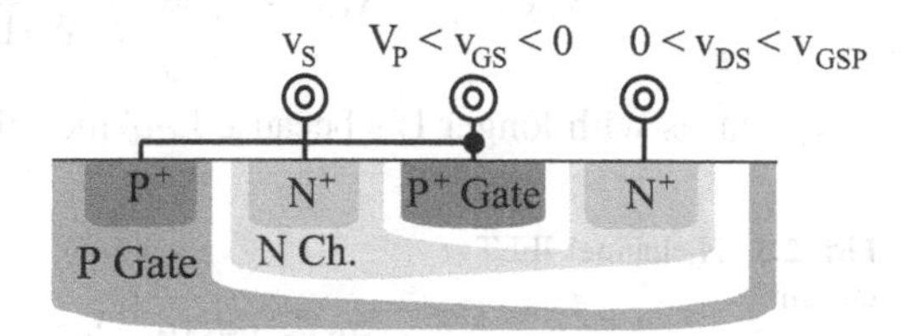

Fig. 2.3 Asymmetrically biased N-channel JFET in triode

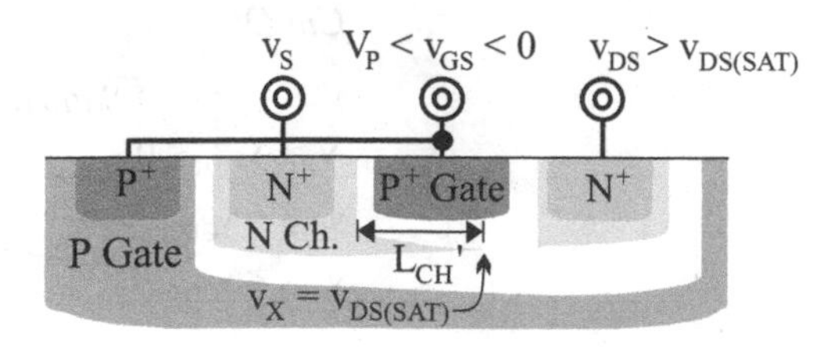

Fig. 2.4 N-channel JFET in saturation

When v_{DS} overcomes this $v_{DS(SAT)}$, the channel extends to the fraction of L that drops $v_{DS(SAT)}$: L_{CH}'. The depleted portion of L drops what remains of v_{DS}: $v_{DS} - v_{DS(SAT)}$. i_D saturates because the voltage across the channel that L_{CH}' establishes is $v_{DS(SAT)}$, which is independent of v_{DS}:

$$\begin{aligned} i_D\big|_{v_{DS}>v_{GSP}}^{v_{GS}>V_P} &= \frac{v_{DS(SAT)}}{R_{CH}} \approx v_{DS(SAT)}\left(\frac{W}{L_{CH}'}\right)K_N\left(v_{GS} - V_P - \frac{v_{DS(SAT)}}{2}\right) \\ &\approx \left(\frac{W}{L}\right)\left(\frac{K_N}{2}\right)(v_{GS} - V_P)^2(1 + \lambda_N v_{DS}). \end{aligned} \tag{2.3}$$

This i_D corresponds to *saturation* because i_D is sensitive to v_{GS} and largely insensitive to v_{DS}. This is another way of saying i_D saturates with respect to v_{DS}.

Since a higher v_D depletes more of the channel, L_{CH}' shrinks as v_{DS} increases, which means R_{CH} falls and i_D increases. This is *channel-length modulation. Channel-length modulation parameter* λ_N models this effect with respect to the L that circuit designers define.

Interestingly, the effects of v_{DS} fade as L lengthens. This is because the variation in L_{CH}' becomes a smaller fraction of L. So *drain current* i_D becomes less sensitive to v_{DS}, which is another way of saying λ_N is lower.

C. **I–V Translation**

i_D is sensitive to v_{DS} in triode and largely insensitive to v_{DS} in saturation like Fig. 2.5 shows. Since N^+ regions can reverse roles, negative v_{GD} and v_{DS} establish a negative i_D that mirrors the i_D that a negative v_{GS} and a positive v_{DS} produce. i_D is zero (in *cut off*) when v_{GS} and v_{GD} reach V_P. The x axis represents this mode because i_D is zero along that line.

i_D saturates when v_{DS} overcomes $v_{DS(SAT)}$'s v_{GSP}. Since i_D is a quadratic translation of v_{GSP} in saturation, the $v_{DS(SAT)}$ boundary in Fig. 2.5 is a squared-root reflection of i_D:

$$v_{DS(SAT)} = v_{GS} - V_P \approx \sqrt{\frac{2i_D}{(W/L)K_N(1 + \lambda_N v_{DS})}}. \tag{2.4}$$

$\lambda_N v_{DS}$ fades with longer L's because L_{CH}' modulation diminishes.

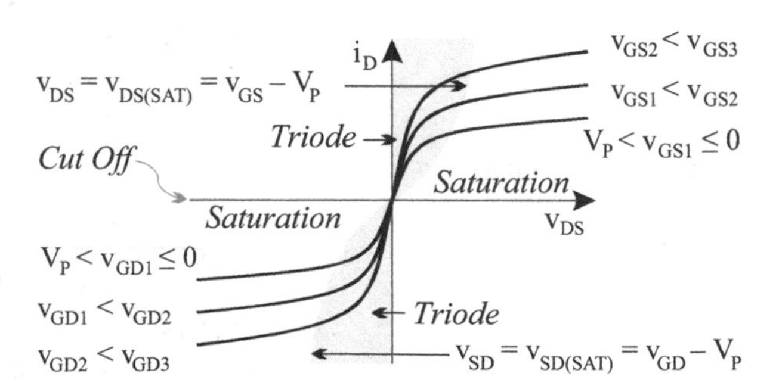

Fig. 2.5 N-channel JFET current

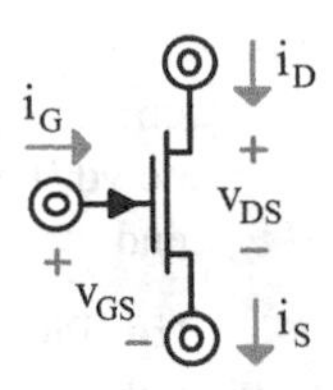

Fig. 2.6 N-channel JFET symbol

D. **Symbol**

JFETs are three-terminal devices with interchangeable v_S and v_D terminals that conduct i_D in Fig. 2.6 when v_{GS} is zero or negative and v_{DS} is positive. The two vertical lines into which v_G connects symbolize the PN depletion capacitance that pinches the channel. *Gate current* i_G is close to zero because v_G reverse-biases the gate–channel junction. So *source current* i_S outputs almost all the i_D that v_D receives. The arrow indicates the P gate and N channel form a PN junction.

Example 1: Determine i_D when v_D is 4 V, v_G and v_S are 0 V, W/L is 1, K_N is 100 μA/V², V_P is −2 V, and λ_N is 5%.

Solution:

$$v_{DS} = v_D - v_S = 4 - 0 = 4\ \text{V} \quad \text{and} \quad v_{GS} = v_G - v_S = 0 - 0 = 0\ \text{V}$$

$$v_{DS} = 4\ \text{V} > v_{DS(SAT)} = v_{GS} - V_P = 0 - (-2) = 2\ \text{V} \quad \therefore \quad \text{Saturated}$$

$$\begin{aligned} i_D &\approx \left(\frac{W}{L}\right)\left(\frac{K_N}{2}\right)(v_{GS} - V_P)^2(1 + \lambda_N v_{DS}) \\ &= (1)\left(\frac{100\mu}{2}\right)[0 - (-2)]^2[1 + 5\%(4)] \\ &= 240\ \mu\text{A} \end{aligned}$$

Explore with SPICE:
See Appendix A for notes on SPICE simulations.

```
NJFET: I-V Curves
vd vd 0 dc=4
vg vg 0 dc=0
vs vs 0 dc=0
j1 vd vg vs njfet 1
.model njfet njf beta=50u vto=-2 lambda=50m
```

(continued)

```
.op
.dc vd 0 5 10m vg 0 -1 100m
.end
```

Tip: Plot id(J1), comment or remove the .dc line, re-run simulation, and view the output/log file.

2.1.2 P Channel

PJFETs are and operate exactly the same way as NJFETs, except with a P channel. So top and bottom N^+ and N gates in Fig. 2.7 sandwich a P channel contacted by highly doped P^+ regions. L_{CH} or L and W_{CH} or W are the longitudinal length and transverse width of the overlapping N^+–P channel–N gate layers. A highly doped N^+ region (on the left in Fig. 2.7) contacts the bottom gate.

A. **Triode**

The PJFET is basically a P-type resistor compressed by N-type regions. R_{CH} is high when L is long, W is narrow, and K_P is low. The channel dematerializes (opens) when the gate–channel voltage reverses enough to pinch the entire channel. So R_{CH} spikes sharply when v_{SG} and v_{DG} reach V_P:

$$R_{CH}\Big|_{v_{SD}<<v_{SGP}}^{v_{SG}>V_P} \approx \left(\frac{L}{W}\right)\left[\frac{1}{K_P(v_{SG}-V_P)}\right] \equiv \left(\frac{L}{W}\right)\left(\frac{1}{K_P v_{SGP}}\right) \equiv \left(\frac{L}{W}\right)R_{SH}. \quad (2.5)$$

R_{CH} is more accurate when v_{SG} is uniform across the channel and above the pinch-off point. This happens when v_{SG} and v_{DG} or $v_{SG} - v_{SD}$ are higher than V_P and v_{SD} is low. This means that $v_{SG} - V_P$ or v_{SGP} is positive and v_{SD} is much lower than v_{SGP}.

A voltage v_{SD} across R_{CH} induces i_D. v_{SD} expands, however, the depletion space that squeezes half of the channel in Fig. 2.8. So half of v_{DS} raises and reduces the linear effects of v_{SG} on R_{CH} and i_D:

$$i_D\Big|_{v_{SD}<v_{SGP}}^{v_{SG}>V_P} = \frac{v_{SD}}{R_{CH}} \approx v_{SD}\left(\frac{W}{L}\right)K_P\left(v_{SG}-V_P-\frac{v_{SD}}{2}\right). \quad (2.6)$$

Since the positive P^+ supplies the P-channel holes that the negative P^+ outputs, the P^+ on the left is the source v_S and the one on the right is the drain v_D.

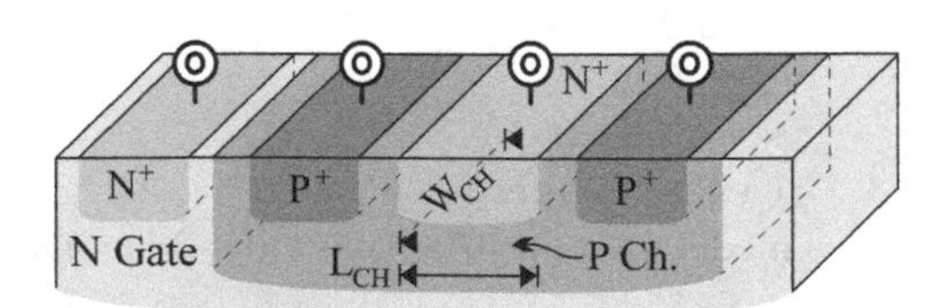

Fig. 2.7 P-channel JFET structure

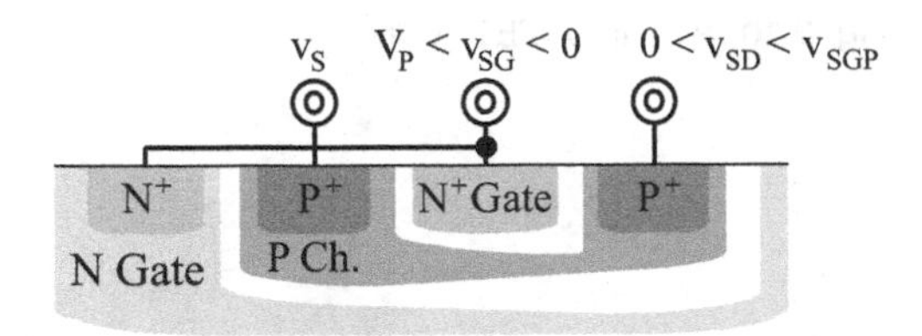

Fig. 2.8 P-channel JFET in triode

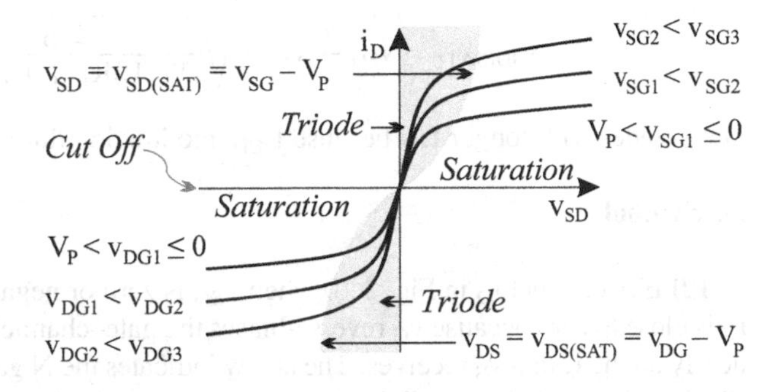

Fig. 2.9 P-channel JFET current

B. Saturation

The channel pinches and charge disappears near v_D when v_{DG} reverses to V_P. Since v_{DG} is equivalent to $v_{SG} - v_{SD}$, this happens when v_{SD} reaches the $v_{SD(SAT)}$ that $v_{SG} - V_P$ or v_{SGP} set. When v_{SD} surpasses this $v_{SD(SAT)}$, the channel extends to the fraction of L that drops $v_{SD(SAT)}$: L_{CH}'. The depleted portion of L drops the remainder: $v_{SD} - v_{SD(SAT)}$. So i_D saturates to the level that $v_{SD(SAT)}$ across L_{CH}' dictates:

$$
\begin{aligned}
i_D\Big|_{v_{SD}>v_{SGP}}^{v_{SG}>V_P} &= \frac{v_{SD(SAT)}}{R_{CH}} \approx v_{SD(SAT)}\left(\frac{W}{L_{CH}'}\right)K_P\left(v_{SG} - V_P - \frac{v_{SD(SAT)}}{2}\right) \\
&\approx \left(\frac{W}{L}\right)\left(\frac{K_P}{2}\right)(v_{SG} - V_P)^2(1 + \lambda_P v_{SD}).
\end{aligned}
\tag{2.7}
$$

L_{CH}' shrinks as v_{SD} climbs because a lower v_D depletes more of the channel. This means that R_{CH} falls and i_D increases. Lengthening L reduces λ_P because variations in L_{CH}' become a smaller fraction of L.

C. I–V Translation

i_D is sensitive to v_{SD} in triode and largely insensitive to v_{SD} in saturation like Fig. 2.9 shows. Since P^+ regions can reverse roles, negative v_{DG} and v_{SD} establish a negative i_D that mirrors the i_D that a negative v_{SG} and a positive v_{SD} produce. i_D is zero in cut off when v_{SG} and v_{DG} reach V_P.

i_D saturates when v_{SD} overcomes $v_{SD(SAT)}$'s v_{SGP}. Since i_D is a quadratic translation of v_{SGP} in saturation, the $v_{SD(SAT)}$ boundary in Fig. 2.9 is a squared-root reflection of i_D:

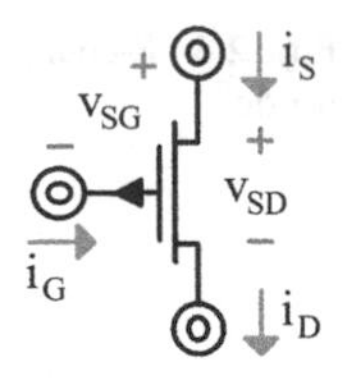

Fig. 2.10 P-channel JFET symbol

$$v_{SD(SAT)} = v_{SG} - V_P \approx \sqrt{\frac{2i_D}{(W/L)K_P(1 + \lambda_P v_{SD})}}. \quad (2.8)$$

$\lambda_P v_{SD}$ fades with longer L's because L_{CH}' modulation diminishes.

D. **Symbol**

PJFETs conduct i_D in Fig. 2.10 when v_{SG} is zero or negative and v_{SD} is positive. i_G is close to zero because v_G reverse-biases the gate–channel junction. So i_D outputs nearly all the i_S that v_S receives. The arrow indicates the N gate and P channel form a PN junction and the parallel lines next to it symbolize the capacitance that the junction establishes.

Example 2: Determine i_D when v_S and v_G are 5 V, v_D is 1 V, W/L is 1, K_P is 50 μA/V^2, V_P is −2 V, and λ_P is 5% V^{-1}.

Solution:

$$v_{SD} = v_S - v_D = 5 - 1 = 4\ \text{V} \quad \text{and} \quad v_{SG} = v_S - v_G = 5 - 5 = 0\ \text{V}$$

$$v_{SD} = 4\ \text{V} > v_{SD(SAT)} = v_{SG} - V_P = 0 - (-2) = 2\ \text{V} \quad \therefore \quad \text{Saturated}$$

$$\begin{aligned} i_D &\approx \left(\frac{W}{L}\right)\left(\frac{K_P}{2}\right)(v_{SG} - V_P)^2(1 + \lambda_P v_{SD}) \\ &= (1)\left(\frac{50\mu}{2}\right)[0 - (-2)]^2[1 + 5\%(4)] \\ &= 120\ \mu\text{A} \end{aligned}$$

Explore with SPICE:
See Appendix A for notes on SPICE simulations.

```
* PJFET: I-V Curves
vs vs 0 dc=5
vg vg 0 dc=5
vd vd 0 dc=1
j1 vd vg vs pjfet 1
.model pjfet pjf beta=25u vto=-2 lambda=50m
.op
.dc vd 5 0 10m vg 5 6 100m
.end
```

Tip: Plot is(J1), comment or remove the .dc line, re-run the simulation, and view the output/log file.

2.2 N-Channel MOSFETs

The N-channel MOSFET is a smaller and slightly more involved NJFET. Structurally, a MOS capacitor is on P-type material that constitutes the *body* or *bulk* of the transistor. The *thin oxide* SiO_2 in the MOS capacitor in Fig. 2.11 sets the *oxide length* L_{OX} or L of the FET and, together with the N^+ regions, the width of the channel W_{CH} or W to be established under this oxide. L_{CH} is the distance between these highly doped N^+ regions. The N^+ terminals extend into the oxide region (across *overlap length* L_{OL}) to ensure they connect to the N channel that the oxide field forms. The highly doped P^+ region (on the left of the structure in Fig. 2.11) is an Ohmic surface contact to the body.

2.2.1 Accumulation: Cut-Off

The P body usually connects to the most negative potential (ground in Fig. 2.12) to keep body–N^+ junctions from forward-biasing. So the electrons and holes that diffuse recombine and deplete the regions near those junctions. Applying a negative

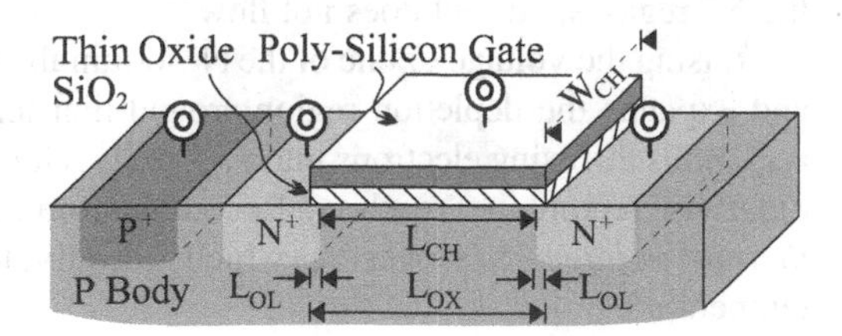

Fig. 2.11 N-channel MOSFET structure

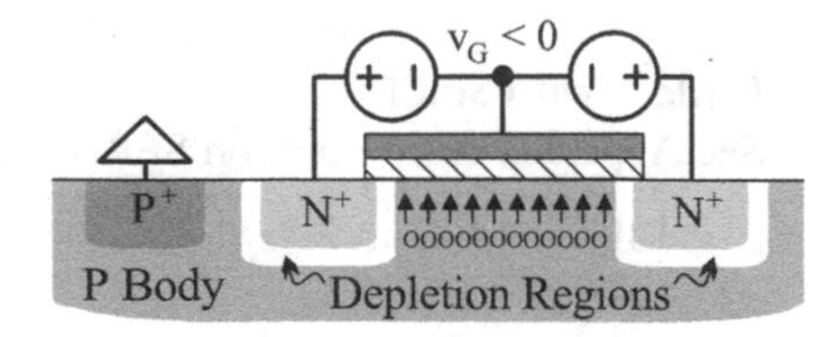

Fig. 2.12 N-channel MOSFET in accumulation and cutoff

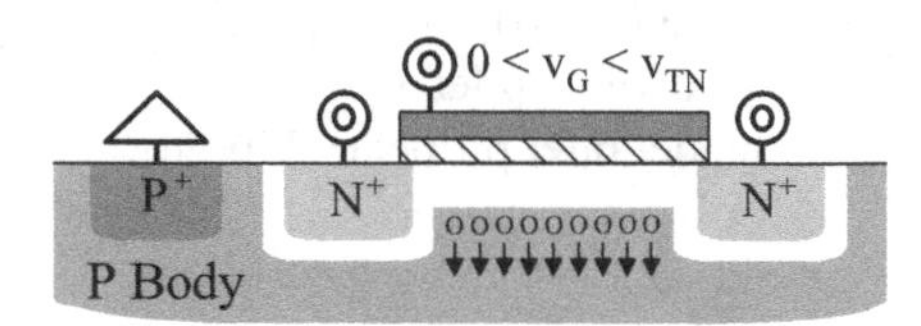

Fig. 2.13 N-channel MOSFET in depletion

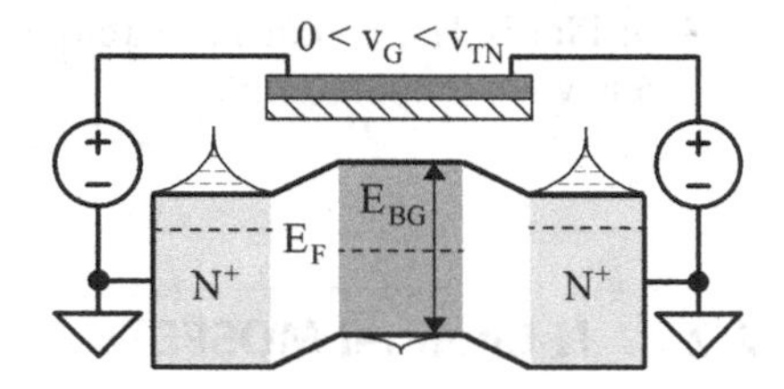

Fig. 2.14 Band diagram of N-channel MOSFET in depletion

v_G to the poly-silicon gate pulls holes in the body toward the oxide. Holes therefore accumulate near the surface of the semiconductor. This way, in *accumulation*, current cannot flow. So the NFET is in cut off.

2.2.2 Depletion: Sub-threshold

Applying a positive v_G does the opposite: pushes holes away from the semiconductor surface in Fig. 2.13. This depletes the region under the oxide of holes. With fewer holes with which to recombine, N^+ electrons diffuse farther before recombining. A positive v_G also pulls and loosens electrons from their N^+ home sites. So v_G reduces the barrier voltage that keeps N^+ electrons from diffusing and extends their diffusion length.

In depletion, the carrier density near the surface under the oxide falls to the point of becoming nearly intrinsic. This is why the *Fermi energy level* E_F in this region in Fig. 2.14 is nearly halfway across the *band gap* E_{BG}. Without any voltage between the N^+ regions, current does not flow.

Raising the voltage of one of the N^+ terminals (v_D in Fig. 2.15) elevates the barrier and expands the depletion region around that terminal. The resulting electric field ε_{FLD} pulls diffusing electrons into v_D. As ε_{FLD} intensifies, more electrons (that would otherwise recombine) reach v_D. Recombination nearly stops when v_D is three to four *thermal voltages* V_t's over v_S, which is equivalent to 75–100 mV or so at room temperature.

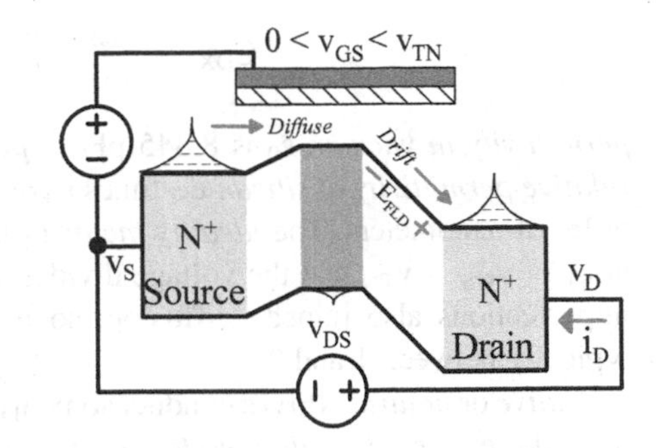

Fig. 2.15 Band diagram of N-channel MOSFET in sub-threshold

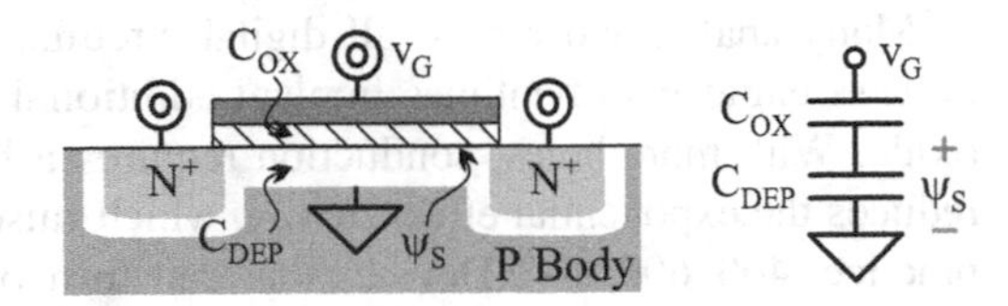

Fig. 2.16 Voltage divider across gate oxide and surface–body terminals

v_D is the drain because the N^+ terminal with the higher potential drains these diffusing electrons. And v_S is the source because the terminal with the lower potential supplies these electrons. The resulting flow of electrons establishes an i_D that flows into v_D and out of v_S.

A. Triode

i_D rises with v_{GS} because a positive gate–source voltage reduces the gate–source barrier. This rise is exponential because v_{GS} avails exponentially more electrons than *junction temperature* T_J avails with V_t:

$$\begin{aligned} i_D\big|^{0<v_{GS}<v_{TN}} &= \left(\frac{W}{L}\right) I_{SN} \exp\left[\left(\frac{C_{OX}''}{C_{OX}''+C_{DEP}''}\right)\left(\frac{v_{GST}}{V_t}\right)\right]\left[1-\exp\left(\frac{-v_{DS}}{V_t}\right)\right] \\ &= \left(\frac{W}{L}\right) C_{DEP}''\mu_N {V_t}^2 \exp\left(\frac{v_{GS}-v_{TN}}{n_I V_t}\right)\left[1-\frac{1}{\exp\left(v_{DS}/V_t\right)}\right]. \end{aligned} \tag{2.9}$$

i_D also increases with v_{DS} because v_{DS} intensifies the field that pulls electrons to v_D. This i_D corresponds to *triode* because i_D is sensitive to v_{GS} and v_{DS}.

i_D is high when W is wide, L is short, and the charge held in the *channel–body depletion capacitance* (per unit area) C_{DEP}'' is high. These, *electron mobility* μ_N and V_t, set the baseline conductivity of i_D. i_D's *saturation current* I_{SN} combines the effects of C_{DEP}'', μ_N, and V_t.

The surface-to-body *surface potential* ψ_S in Fig. 2.16 is ultimately the voltage-divided fraction that v_G couples through the *oxide capacitance* (per unit area) C_{OX}'' into C_{DEP}'', where

$$C_{OX}'' = \frac{\varepsilon_{OX}}{t_{OX}} = \frac{\varepsilon_{Si}\varepsilon_0}{t_{OX}} = \frac{3.9\varepsilon_0}{t_{OX}}, \tag{2.10}$$

permittivity in vacuum ε_0 is 8.845 pF/m, *permittivity of silicon dioxide* ε_{OX} is the *relative permittivity of silicon* ε_{Si} times ε_0 or $3.9\varepsilon_0$, and *oxide thickness* t_{OX} is on the order of nanometers. The *ideality factor* n_I in i_D models the reduction in gate drive v_{GST} or $v_{GS} - v_{TN}$ that the voltage divider across C_{OX}'' and C_{DEP}'' causes. Surface imperfections also impede diffusion, so n_I also accounts for these defects. n_I is typically between 1 and 2.

Native or *natural* NFETs conduct some i_D when v_{GS} is zero. Normally-on devices like these are *depletion-mode transistors.* They are useful in low-voltage applications that benefit from low v_{GS} values.

Many analog and almost all digital circuits, however, require an off state. So process engineers oftentimes implant additional acceptor dopant atoms below the oxide. With more holes, conduction requires a higher v_{GS}. This implant therefore reduces the exponential effect of v_{GS}, which raises the *threshold voltage* v_{TN} to, in practice, 400–600 mV. Devices that can turn on and off are *enhancement-mode transistors.*

B. Saturation

i_D climbs with v_{DS} because v_{DS} intensifies the field that reduces recombination. Diffusion length is so high in this quasi-intrinsic region that nearly 95% of the electrons reach the other end when v_{DS} is $3V_t$. Past this point, i_D saturates with respect to v_{DS} to

$$i_D\Big|_{v_{DS}>3V_t}^{0<v_{GS}<v_{TN}} \approx \left(\frac{W}{L}\right) I_{SN} \exp\left(\frac{v_{GS} - v_{TN}}{n_I V_t}\right), \tag{2.11}$$

where $v_{DS(SAT)}'$ is the *sub-threshold saturation voltage* that $3V_t$ sets. This i_D refers to the saturation region because i_D is insensitive to v_{DS}.

C. I–V Translation

i_D in Fig. 2.17 is sensitive to v_{DS} in triode and insensitive to v_{DS} in saturation, when v_{DS} is greater than $3V_t$ or so. N^+ regions can change roles and conduct reverse

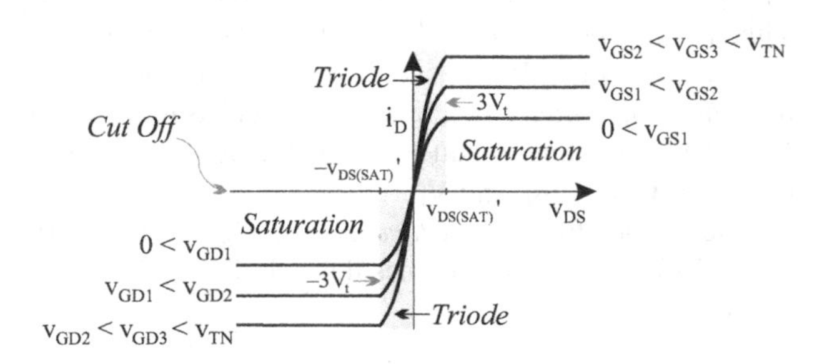

Fig. 2.17 Sub-threshold N-channel MOSFET current

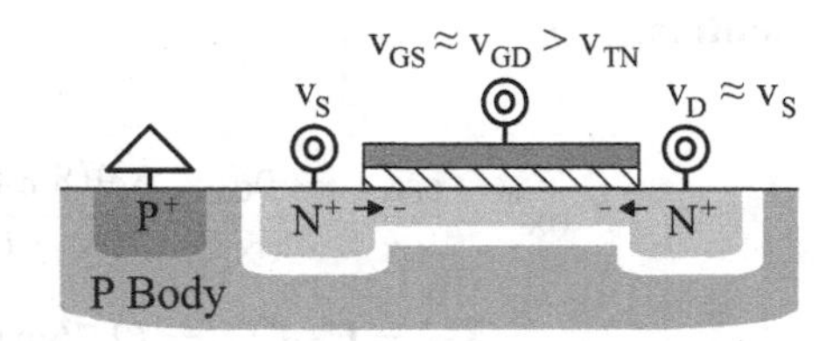

Fig. 2.18 Symmetrically inverted N-channel MOSFET in triode

current because the structure is symmetrical. So a positive v_{GD} and a negative v_{DS} establish a negative i_D that mirrors the i_D that positive v_{GS} and v_{DS} produce. i_D is zero in accumulation and cut off when v_{GS} and v_{GD} are both negative.

2.2.3 Inversion

A. **Triode**

Raising v_{GS} and v_{GD} first eliminates the barrier that keeps electrons from flowing. Past this barrier point, the electric field across the oxide is high enough to pull electrons from both N^+ terminals in Fig. 2.18 toward the region below the gate. When v_{GS} and v_{GD} reach v_{TN}, these electrons establish a conducting N channel with a carrier density that matches that of the original zero-bias P body. So overcoming this *threshold voltage* v_{TN} inverts this medium. This is the process to which *inversion* refers.

R_{CH} is high when L is long, W is narrow, and the conductivity that μ_N and C_{OX}'' in the *transconductance parameter* K_N' set is low. The channel dematerializes when v_{GS} and v_{GD} drop below v_{TN}. So R_{CH} spikes sharply when v_{GS} and v_{GD} reach v_{TN}:

$$\begin{aligned} R_{CH}\Big|_{v_{DS}<<v_{GST}}^{v_{GS}>v_{TN}} &\approx \left(\frac{L}{W}\right)\left[\frac{1}{\mu_N C_{OX}''(v_{GS}-v_{TN})}\right] \\ &\equiv \left(\frac{L}{W}\right)\left(\frac{1}{K_N' v_{GST}}\right) \equiv \left(\frac{L}{W}\right)R_{SH}. \end{aligned} \tag{2.12}$$

R_{CH} is more accurate when v_{GS} and v_{GD} match and overcome v_{TN}. So v_{GS} and v_{GD} or $v_{GS} - v_{DS}$ are higher than v_{TN} and v_{DS} is low. This means that gate drive $v_{GS} - v_{TN}$ or v_{GST} is positive and v_{DS} is much lower than v_{GST}. K_N' and v_{GST} set the sheet resistivity R_{SH} of R_{CH}.

Example 3: Determine R_{CH} when W is 10 μm, L is 180 nm, L_{OL} is 30 nm, v_{GS} is 1.8 V, v_{DS} is 50 mV, μ_N is 72k $mm^2/V{\cdot}s$, t_{OX} is 12.5 nm, and v_{TN} is 400 mV.

Solution:

$$C_{OX}'' = \frac{\varepsilon_{OX}}{t_{OX}} = \frac{3.9\varepsilon_0}{t_{OX}} = \frac{3.9(8.845 \times 10^{-12})}{12.5n} = 2.76\ \text{fF}/\mu\text{m}^2$$

$$K_N' = C_{OX}''\mu_N = (2.76m)(72m) = 200\ \mu\text{A}/\text{V}^2$$

$$v_{GS} = 1.8\ \text{V} > v_{TN} = 400\ \text{mV} \quad \therefore \quad \text{Inverted}$$

$$v_{DS} = 50\ \text{mV} << v_{GS} - v_{TN} = 1.8 - 400\ \text{m} = 1.4\ \text{V} \quad \therefore \quad \text{Deep in Triode}$$

$$R_{CH} \approx \left(\frac{L - 2L_{OL}}{W_{CH}}\right)\left[\frac{1}{K_N'(v_{GS} - v_{TN})}\right]$$

$$= \left[\frac{180n - 2(30n)}{10\mu}\right]\left[\frac{1}{(200\mu)(1.8 - 400m)}\right] = 43\ \Omega$$

Explore with SPICE:
See Appendix A for notes on SPICE simulations.

```
* NMOSFET: N-Channel Resistor
vgs vgs 0 dc=1.8
vds vds 0 dc=50m
mr vds vgs 0 0 nmosfet w=10u l=180nm
.model nmosfet nmos vto=400m kp=200u ld=30n
.op
.end
```

Tip: View the output/log file.

v_{DS} across this R_{CH} induces i_D. But since v_D opposes the inversion action of v_{GD} across half the channel, half of v_{DS} in Fig. 2.19 opposes the linear effects of v_{GS}. So R_{CH} rises and i_D falls with $0.5v_{DS}$:

$$i_D\Big|_{v_{DS}<v_{GST}}^{v_{GS}>v_{TN}} = \frac{v_{DS}}{R_{CH}} = v_{DS}\left(\frac{W}{L}\right)K_N'\left(v_{GS} - v_{TN} - \frac{v_{DS}}{2}\right). \tag{2.13}$$

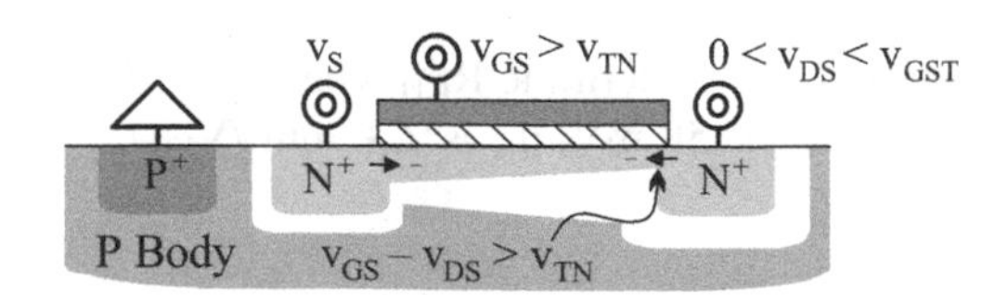

Fig. 2.19 Asymmetrically inverted N-channel MOSFET in triode

i_D climbs almost linearly with v_{GS} because an electric field (not diffusion) alters conductivity. This i_D corresponds to triode inversion because i_D is sensitive to v_{GS} and v_{DS}.

B. Saturation

Charge q_D near the edge of v_D's N^+ region in Fig. 2.20 disappears when v_{GD} falls to v_{TN}. Since v_{GD} is v_{TN} when this happens and v_{GD} is equivalent to $v_{GS} - v_{DS}$, the channel pinches when v_{DS} reaches $v_{GS} - v_{TN}$. So gate drive v_{GST} sets the saturation voltage $v_{DS(SAT)}$ of the NFET in inversion.

When v_{DS} overcomes $v_{DS(SAT)}$, the channel extends to the fraction of L_{CH} that drops $v_{DS(SAT)}$: L_{CH}'. The depleted portion of L_{CH} drops what remains of v_{DS}: $v_{DS} - v_{DS(SAT)}$. i_D saturates because the voltage across the channel that L_{CH}' establishes is $v_{DS(SAT)}$, which is independent of v_{DS}:

$$i_D\Big|_{v_{DS}>v_{GST}}^{v_{GS}>v_{TN}} = \frac{v_{DS(SAT)}}{R_{CH}} = v_{DS(SAT)}\left(\frac{W}{L_{CH}'}\right)K_N'\left(v_{GS} - v_{TN} - \frac{v_{DS(SAT)}}{2}\right)$$
$$\approx \left(\frac{W}{L}\right)\left(\frac{K_N'}{2}\right)(v_{GS} - v_{TN})^2(1 + \lambda_N v_{DS}) \equiv \left(\frac{W}{L}\right)\left(\frac{K_N'}{2}\right)v_{GST}^2\left(1 + \frac{v_{DS}}{V_{AN}}\right). \tag{2.14}$$

In other words, i_D is sensitive to v_{GS} and largely insensitive to v_{DS}.

Since a higher v_D depletes more of the channel, L_{CH}' shrinks as v_{DS} rises. So R_{CH} falls and i_D increases with v_{DS} according to λ_N with respect to the L that circuit designers define. v_{DS} effects fade (and λ_N falls) as L lengthens because the variation in L_{CH}' becomes a smaller fraction of L.

Although different in nature, the effect of λ_N mirrors *base-width modulation* in BJTs. This is why engineers often use *Early voltage* V_A to quantify this effect in FETs. So when applied to NFETs, V_{AN} is $1/\lambda_N$.

Since C_{GS} falls with L_{CH}', v_{GS} pulls fewer source electrons into the N channel when L is shorter, which means R_{CH} rises. Although a shorter channel also increases conduction, the field effect of C_{GS} on R_{CH} dominates. So R_{CH} is roughly 50% higher in saturation than deep in triode:

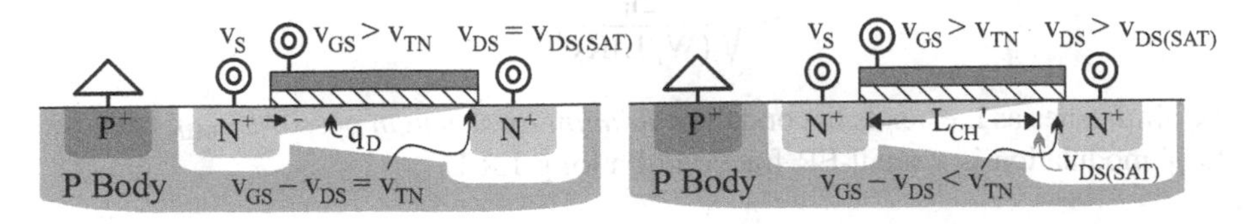

Fig. 2.20 Inverted N-channel MOSFET in saturation

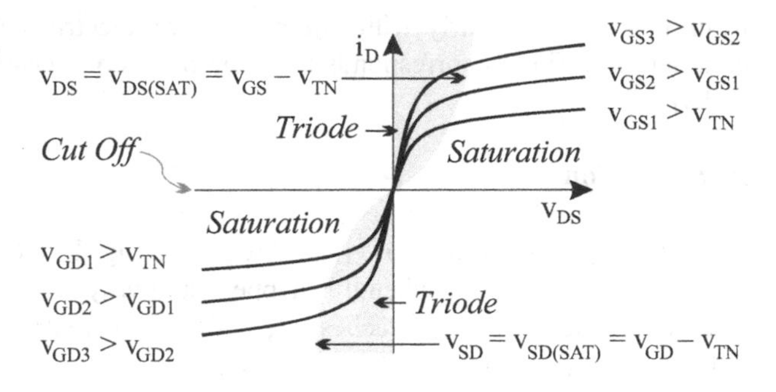

Fig. 2.21 Inverted N-channel MOSFET current

$$R_{CH}\Big|_{v_{DS}>v_{GST}}^{v_{GS}>v_{TN}} = \frac{v_{DS(SAT)}}{i_D} = \left(\frac{L_{CH}'}{W}\right)\left[\frac{v_{GS} - v_{TN}}{0.5K_N'(v_{GS} - v_{TN})^2}\right]$$
$$\approx \left(\frac{3}{2}\right)\left(\frac{L}{W}\right)\left[\frac{1}{0.5K_N'(v_{GS} - v_{TN})}\right] = 1.5R_{CH}\Big|_{v_{DS}<<v_{GST}}^{v_{GS}>v_{TN}}. \quad (2.15)$$

Note R_{CH} is only the resistance of the channel, which does not extend to v_D in saturation. The total *drain–source resistance* R_{DS} or $\partial v_{DS}/\partial i_D$ is usually much greater than R_{CH} because i_D is largely insensitive to v_{DS} in saturation.

C. I–V Translation

Like in sub-threshold, i_D in inversion is sensitive to v_{DS} in triode and largely insensitive to v_{DS} in saturation like Fig. 2.21 shows. N^+ regions can still reverse roles, so a positive v_{GD} and a negative v_{DS} establish a negative i_D that mirrors the i_D that positive v_{GS} and v_{DS} produce. i_D is zero when v_{GS} and v_{GD} are both negative.

The basic difference in inversion is that i_D is a squared translation of v_{GST}. The other difference is that i_D saturates when v_{DS} surpasses v_{GST}. So the $v_{DS(SAT)}$ boundary in Fig. 2.21 is a squared-root reflection of i_D:

$$v_{DS(SAT)} = v_{GS} - v_{TN} \approx \sqrt{\frac{2i_D}{(W/L)K_N'(1 + \lambda_N v_{DS})}}\Bigg|_{L>>L_{MIN}} \approx \sqrt{\frac{2i_D}{(W/L)K_N'}}. \quad (2.16)$$

λ_N diminishes as L extends beyond the *minimum oxide length* possible L_{MIN} because L_{CH}' modulation is a small-ER fraction of a long-ER L.

Example 4: Determine $v_{DS(SAT)}$ when i_D is 100 μA, v_{DS} is 1 V, W is 10 μm, L is 180 nm, L_{OL} is 30 nm, K_N' is 200 μA/V^2, and λ_N is 2%.

Solution:

$$v_{DS(SAT)} \approx \sqrt{\frac{2i_D}{(W_{CH}/L_{CH})K_N'(1+\lambda_N v_{DS})}} = \sqrt{\frac{2(100\mu)[180n-2(30)]}{(10\mu)(200\mu)[1+(2\%)(1)]}}$$
$$= 110 \text{ mV}$$

2.2.4 Body Effect

v_{GS} induces a field through C_{OX} that avails electrons for v_{DS} to pull. The body–source or bulk–source voltage v_{BS} also avails electrons the same way via C_{DEP}. So the P^+ body terminal in Fig. 2.11 is a bottom gate.

A positive v_{BS} in NFETs pulls electrons into the channel and a negative v_{BS} repels some of the electrons that v_{GS} avails. So v_{BS} effectively reduces v_{TN}:

$$v_{TN} = V_{TN0} + \gamma_N\left(\sqrt{2\psi_B - v_{BS}} - \sqrt{2\psi_B}\right). \tag{2.17}$$

v_{TN} is the baseline *zero-bias threshold* V_{TN0} when v_{BS} is zero.

v_{TN} is the voltage that v_{GS} overcomes to invert a channel. To invert in equal proportion, v_{GS} should not only negate the *surface–body barrier potential* ψ_B but also re-assert another ψ_B in the opposite direction. v_{BS} reduces this $2\psi_B$ translation. So v_{BS} alters v_{TN} by the amount that the *body-effect parameter* γ_N allows. This is the *body* or *bulk effect*, where ψ_S is $2\psi_B$ in inversion and γ_N and ψ_B can be 600 m$\sqrt{\text{V}}$ and 300 mV, respectively.

Example 5: Determine v_{TN} when V_{TN0} is 400 mV, v_{BS} is −100 mV, γ_N is 600 m$\sqrt{\text{V}}$, and ψ_B is 300 mV.

Solution:

$$v_{TN} = V_{TN0} + \gamma_N\left(\sqrt{2\psi_B - v_{BS}} - \sqrt{2\psi_B}\right)$$
$$= 400\text{m} + (600\text{m})\left(\sqrt{2(300\text{m}) + 100\text{m}} - \sqrt{2(300\text{m})}\right) = 440 \text{ mV}$$

Note: A negative v_B repels channel electrons back to v_S, so v_{GS} needs to overcome a higher v_{TN} to induce conduction.

Example 6: Determine v_{TN} when V_{TN0} is 400 mV, v_{BS} is 100 mV, γ_N is 600 m$\sqrt{V}$, and ψ_B is 300 mV.

Solution:

$$v_{TN} = V_{TN0} + \gamma_N\left(\sqrt{2\psi_B - v_{BS}} - \sqrt{2\psi_B}\right)$$
$$= 400\text{m} + (600\text{m})\left(\sqrt{2(300\text{m}) - 100\text{m}} - \sqrt{2(300\text{m})}\right) = 360 \text{ mV}$$

Note: A positive v_B pulls v_S electrons into the channel region, so v_{GS} induces conduction more easily (with a lower v_{TN}).

Explore with SPICE:
See Appendix A for notes on SPICE simulations.

```
* NMOSFET: Body Effect
vbs vbs 0 dc=100m
m1 0 0 0 vbs nmosfet w=10u l=180nm
.model nmosfet nmos vto=400m kp=200u ld=30n lambda=100m
+gamma=600m phi=600m
.op
.end
```

Tip: View the output/log file.

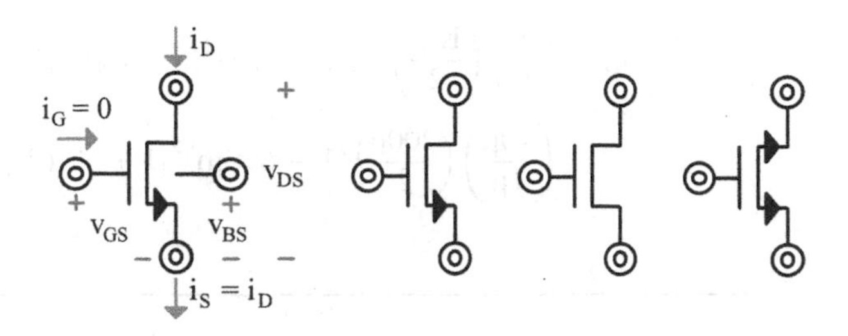

Fig. 2.22 N-channel MOSFET symbols

2.2.5 Symbols

The NMOS is a four-terminal device with interchangeable v_S and v_D terminals that conduct i_D in Fig. 2.22 when v_{GS} and v_{DS} are positive. The two vertical lines at the gate symbolize the oxide capacitance that induces an N-type channel. Static gate current i_G is zero because dc current into this C_{OX} is zero. So i_D is also the i_S that flows out of v_S. The arrow attaches to the v_S terminal that sets v_{GS} and points in the direction of i_S.

The symbol sometimes excludes the body terminal to indicate other transistors share the same body. In these cases, independent access to the body is not possible. The arrow is also sometimes absent to show that source and drain terminals can reverse roles – this is typical in digital circuits. Although less of a convention, some switching power supplies add arrows to both terminals to indicate i_D can reverse direction.

Example 7: Determine i_D when v_D is 3 V, v_G is 2 V, v_S is 1 V, v_B is 0 V, W is 10 μm, L is 1 μm, K_N' is 200 μA/V^2, V_{TN0} is 400 mV, γ_N is 600 m$\sqrt{\text{V}}$, ψ_B is 300 mV, and λ_N is 5%.

Solution:

$$v_{DS} = v_D - v_S = 3 - 1 = 2\text{ V}$$
$$v_{GS} = v_G - v_S = 2 - 1 = 1\text{ V}$$
$$v_{BS} = v_B - v_S = 0 - 1 = -1\text{ V}$$

$$\begin{aligned} v_{TN} &= V_{TN0} + \gamma_N\left(\sqrt{2\psi_B - v_{BS}} - \sqrt{2\psi_B}\right) \\ &= 400\text{m} + (600\text{m})\left[\sqrt{2(300\text{m}) - (-1)} - \sqrt{2(300\text{m})}\right] = 690\text{ mV} \end{aligned}$$

$$v_{DS} = 2\text{ V} > v_{DS(SAT)} = v_{GS} - v_{TN} = 1 - 690\text{m} = 310\text{ mV} \quad \therefore \quad \text{Saturation}$$

$$i_D \approx \left(\frac{W}{L}\right)\left(\frac{K_N'}{2}\right)(v_{GS} - v_{TN})^2(1 + \lambda_N v_{DS})$$

$$= \left(\frac{10\mu}{1\mu}\right)\left(\frac{200\mu}{2}\right)(1 - 690m)^2[1 + 5\%(2)] = 110\ \mu A$$

Explore with SPICE:
See Appendix A for notes on SPICE simulations.

```
* NMOSFET: I-V Curves
vd vd 0 dc=3
vg vg 0 dc=2
vs vs 0 dc=1
vb vb 0 dc=0
m1 vd vg vs vb nmosfet w=10u l=1u
.model nmosfet nmos vto=400m kp=200u lambda=50m gamma=600m
+phi=600m
.op
.dc vd 1 5 10m vg 2 3 200m
.end
```

Tip: Plot id(M1), comment or remove the .dc line, re-run the simulation, and view the output/log file.

2.3 P-Channel MOSFETs

PFETs and NFETs operate the same way, except PFETs rely on holes for conduction. So the PMOS structure is the complement of the NMOS. In PFETs, the $W_{CH} \times L_{OX}$ MOS capacitor in Fig. 2.23 hangs over N material and overlaps highly doped P^+ regions (across L_{OL}) that are L_{CH} apart. The P^+ terminals extend into the oxide region to connect them to the P channel that the oxide field forms. The highly doped N^+ region (on the left in Fig. 2.23) is an Ohmic surface contact to the body.

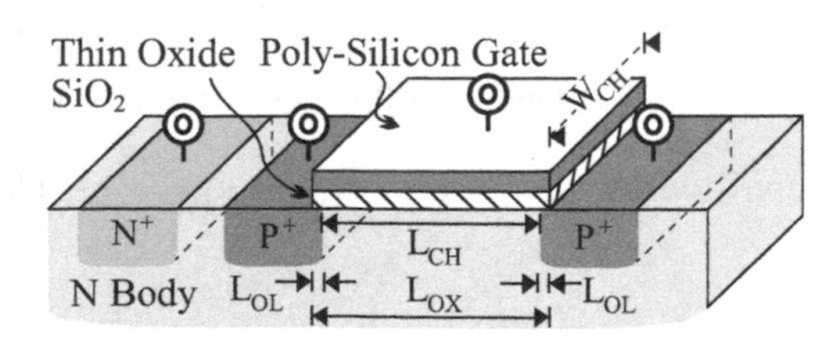

Fig. 2.23 P-channel MOSFET structure

2.3.1 Accumulation: Cut Off

The N body usually connects to the most positive potential (*positive power supply* v_{DD} in Fig. 2.24) to keep P^+N junctions from forward-biasing. So the electrons and holes that diffuse recombine and deplete the regions near those junctions. Applying a positive v_G to the gate pulls electrons in the body toward the oxide. As a result, electrons accumulate near the semiconductor surface. This way, in accumulation, current cannot flow, so the PFET is in cut off.

2.3.2 Depletion: Sub-threshold

Applying a negative v_G does the opposite: pushes electrons away from the semiconductor surface in Fig. 2.25, which depletes the region under the oxide of electrons. With fewer electrons with which to recombine, P^+ holes diffuse farther before recombining. A negative v_G also presses bound electrons into their home sites, which eases hole movement. So v_G reduces the barrier voltage that keeps P^+ holes from diffusing and extends their diffusion length.

The carrier density near the surface under the oxide falls to the point of becoming nearly intrinsic. E_F in this region in Fig. 2.26 is therefore nearly halfway across E_{BG}. But without any voltage between the P^+ regions, current does not flow.

Decreasing the voltage of one of the P^+ terminals elevates the barrier and expands the depletion region around that terminal (v_D in Fig. 2.27). The resulting field pulls diffusing holes into v_D. As ε_{FLD} intensifies, more diffusing holes that would

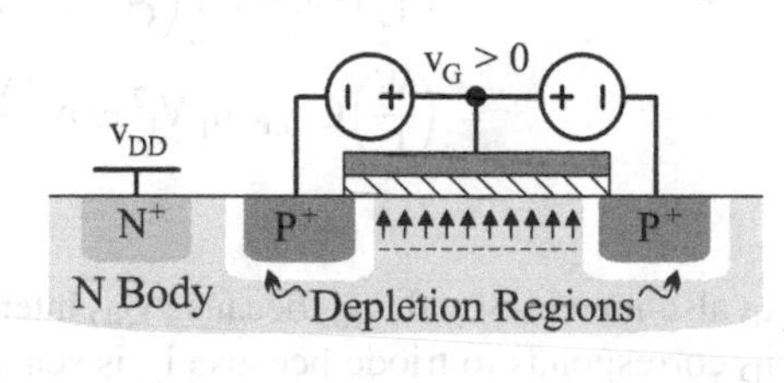

Fig. 2.24 P-channel MOSFET in accumulation and cutoff

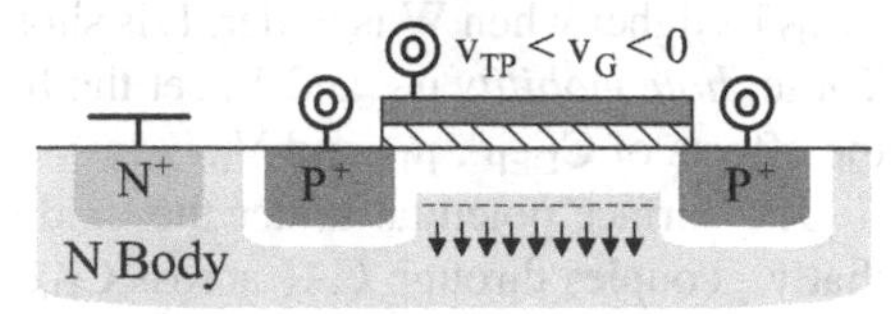

Fig. 2.25 P-channel MOSFET in depletion

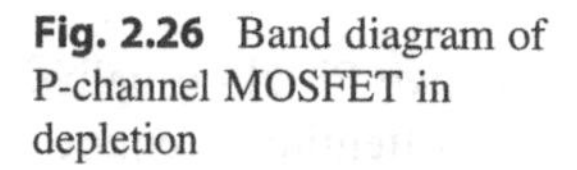

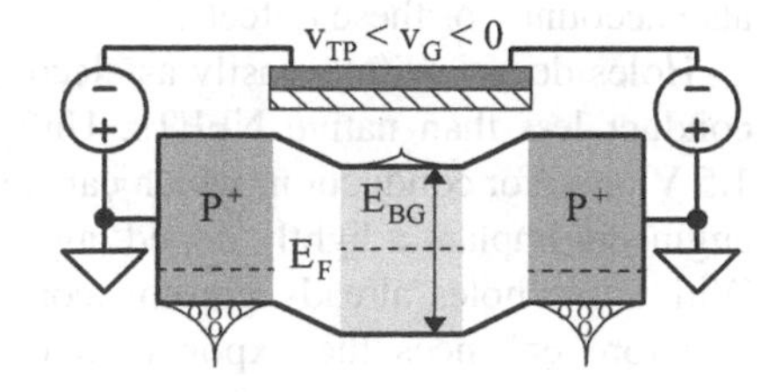

Fig. 2.26 Band diagram of P-channel MOSFET in depletion

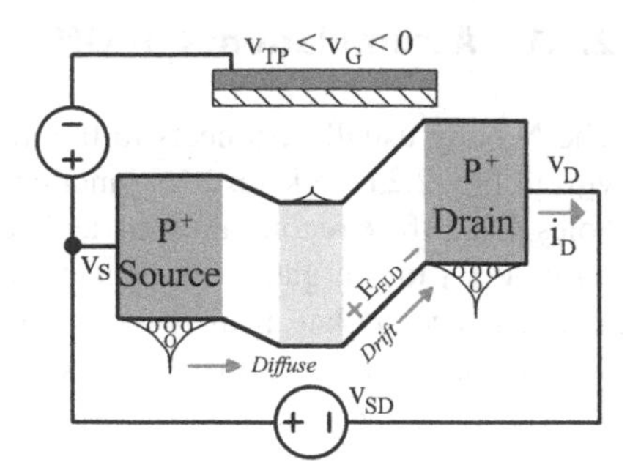

Fig. 2.27 Band diagram of P-channel MOSFET in sub-threshold

otherwise recombine reach v_D. Recombination nearly stops when v_D is 3 or 4 V_ts below v_S.

The negative terminal is the drain because v_D outputs diffusing holes. The positive terminal is the source because v_S supplies these holes. The resulting flow of holes establishes an i_D that flows into v_S and out of v_D.

A. Triode

i_D increases with v_{SG} because a positive v_{SG} reduces the gate–source barrier. This rise is exponential because v_{SG} avails exponentially more holes than T_J avails with V_t:

$$i_D\Big|^{0<v_{SG}<|v_{TP}|} = \left(\frac{W}{L}\right) I_{SP} \exp\left[\left(\frac{C_{OX}''}{C_{OX}'' + C_{DEP}''}\right)\left(\frac{v_{SGT}}{V_t}\right)\right]\left[1 - \exp\left(\frac{-v_{SD}}{V_t}\right)\right]$$
$$= \left(\frac{W}{L}\right) C_{DEP}'' \mu_P V_t^2 \exp\left(\frac{v_{SG} - |v_{TP}|}{n_I V_t}\right)\left[1 - \frac{1}{\exp\left(v_{SD}/V_t\right)}\right]. \tag{2.18}$$

i_D also increases with v_{SD} because v_{SD} intensifies the field that pulls them to v_D. This i_D corresponds to triode because i_D is sensitive to v_{SG} and v_{SD}.

i_D is higher when W is wider, L is shorter, and the charge held in C_{DEP}'' is higher. These, *hole mobility* μ_P, and V_t set the baseline conductivity of i_D. I_{SP} incorporates the effects of C_{DEP}'', μ_P, and V_t.

The surface potential under the oxide is ultimately the voltage-divided fraction that v_G couples through C_{OX} across C_{DEP}. n_I models this voltage-divided reduction in gate drive v_{SGT} or $v_{SG} - |v_{TP}|$. Surface imperfections also impede diffusion, so n_I also accounts for these defects.

Holes do not drift as easily as electrons. Not surprisingly, native PFETs usually conduct less than native NFETs. Unfortunately, these PFETs oftentimes require 1.5 V or so for conduction, which can be too high for many applications. So process engineers implant a lightly doped layer of acceptor dopant atoms below the oxide. With some holes already present, conduction requires a lower v_{SG}. This implant therefore enhances the exponential effect of v_{SG}, which reduces the threshold

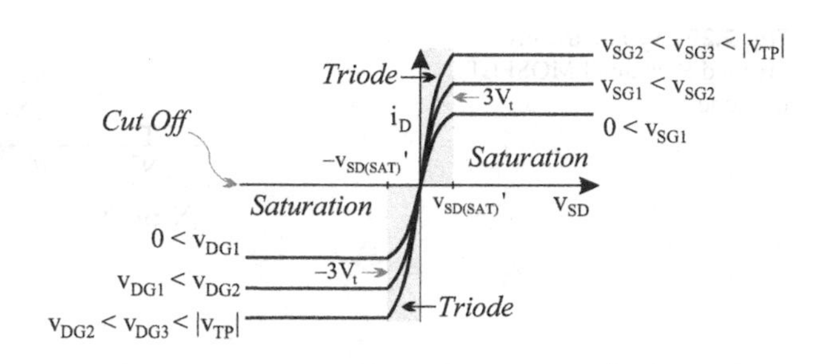

Fig. 2.28 Sub-threshold P-channel MOSFET current

voltage. Engineers define v_{TP} as a negative value because v_{GS} and v_{DS} in PFETs are correspondingly negative. But when using v_{SG} and v_{SD}, thinking of $|v_{TP}|$ is more insightful.

B. Saturation

i_D climbs with v_{SD} because v_{SD} intensifies the longitudinal field, which reduces recombination. Diffusion length is so high in this quasi-intrinsic region that nearly 95% of the holes reach the other end when v_{SD} is $3V_t$. i_D becomes largely insensitive to v_{SD} past this point, saturating to

$$i_D\Big|_{v_{SD}>3V_t}^{0<v_{SG}<|v_{TP}|} \approx \left(\frac{W}{L}\right) I_{SP} \exp\left(\frac{v_{SG} - |v_{TP}|}{n_I V_t}\right). \tag{2.19}$$

C. I–V Translation

i_D in Fig. 2.28 is sensitive to v_{SD} in triode and insensitive to v_{SD} in saturation, when v_{SD} is greater than $3V_t$ or so. P^+ regions can change roles and conduct reverse current because the structure is symmetrical. So a positive v_{DG} and a negative v_{SD} establish a negative i_D that mirrors the i_D that positive v_{SG} and v_{SD} produce. i_D is zero in accumulation, when v_{SG} and v_{DG} are both negative.

2.3.3 Inversion

A. Triode

Reducing v_G (raising v_{SG} and v_{DG}) first eliminates the barrier that keeps holes from flowing. Past this barrier point, the electric field across the oxide is high enough to pull holes from both P^+ regions in Fig. 2.29 toward the region below the gate. When v_{SG} and v_{DG} reach $|v_{TP}|$, these holes form a P channel with a carrier density that matches that of the original zero-bias N body. In other words, v_{SG} and v_{DG} "invert" a channel.

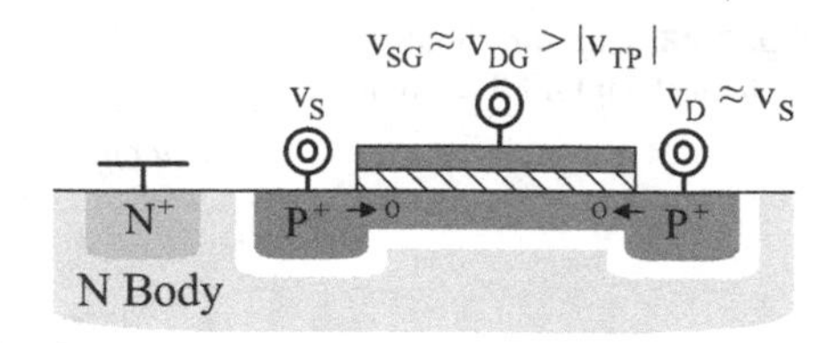

Fig. 2.29 Symmetrically inverted P-channel MOSFET in triode

R_{CH} is high when L is long, W is narrow, and the conductivity that μ_P and C_{OX}'' in K_P' set is low. The channel dematerializes when v_{SG} and v_{DG} fall below $|v_{TP}|$. So R_{CH} spikes sharply when v_{SG} and v_{DG} reach $|v_{TP}|$:

$$R_{CH}\Big|_{v_{SD}<<v_{SGT}}^{v_{SG}>|v_{TP}|} \approx \left(\frac{L}{W}\right)\left[\frac{1}{\mu_P C_{OX}''(v_{SG}-|v_{TP}|)}\right] \equiv \left(\frac{L}{W}\right)\left(\frac{1}{K_P' v_{SGT}}\right) \equiv \left(\frac{L}{W}\right)R_{SH}. \quad (2.20)$$

R_{CH} is more accurate when v_{SG} and v_{DG} match and overcome $|v_{TP}|$. So for this, v_{SG} and v_{DG} or $v_{SG} - v_{SD}$ are higher than $|v_{TP}|$ and v_{SD} is low. This means that gate drive $v_{SG} - |v_{TP}|$ or v_{SGT} is positive and v_{SD} is much lower than v_{SGT}. K_P' and v_{SGT} set the sheet resistivity R_{SH} of R_{CH}.

Example 8: Determine W so R_{CH} is no greater than 45 Ω when L is 180 nm, v_{SG} is 1.8 V, v_{SD} is 50 mV, μ_P is 14.4k mm²/V·s, t_{OX} is 12.5 nm, L_{OL} is 30 nm, and v_{TP} is −400 mV.

Solution:

$$C_{OX}'' = \frac{\varepsilon_{OX}}{t_{OX}} = \frac{3.9\varepsilon_0}{t_{OX}} = \frac{3.9(8.845 \times 10^{-12})}{12.5\text{n}} = 2.76\ \text{fF}/\mu\text{m}^2$$

$$K_P' = C_{OX}''\mu_P = (2.76\text{m})(14.4\text{m}) = 40\ \mu\text{A}/\text{V}^2$$

$$v_{SD} = 50\,\text{mV} \ll v_{SG} - |v_{TP}| = 1.8 - 400\text{m} = 1.4\ \text{V}$$

$$\therefore R_{CH} \approx \left(\frac{L_{CH}}{W_{CH}}\right)\left[\frac{1}{K_P'(v_{SG}-|v_{TP}|)}\right] = \left[\frac{180\text{n} - 2(30\text{n})}{W_{CH}}\right]\left[\frac{1}{(40\mu)(1.8-400\text{m})}\right] \leq 45\ \Omega$$

$$\rightarrow \quad W \equiv W_{CH} \geq 48\ \mu\text{m}$$

Note: W for the PFET is 5× higher than for the NFET in Example 3 because μ_P is that much lower than μ_N.

Explore with SPICE:
See Appendix A for notes on SPICE simulations.

```
* PMOSFET: P-Channel Resistor
vs vs 0 dc=1.8
vg vg 0 dc=0
vd vd 0 dc=1.75
mr vd vg vs vs pmosfet w=48u l=180nm
.model pmosfet pmos vto=-400m kp=40u ld=30n
.op
.end
```

Tip: View the output/log file.

v_{SD} across R_{CH} induces i_D. But since v_D opposes the inversion action of v_{DG} across half the channel, half of v_{SD} in Fig. 2.30 opposes the linear effects of v_{SG}. So R_{CH} rises and i_D falls with half v_{SD}:

$$i_D\Big|_{v_{SD}<v_{SGT}}^{v_{SG}>|v_{TP}|} = \frac{v_{SD}}{R_{CH}} = v_{SD}\left(\frac{W}{L}\right)K_P{}'\left(v_{SG} - |v_{TP}| - \frac{v_{SD}}{2}\right). \tag{2.21}$$

i_D climbs almost linearly with v_{SG} because an electric field (not diffusion) alters conductivity. This i_D corresponds to triode because i_D is sensitive to v_{SG} and v_{SD}.

B. Saturation

q_D near the edge of v_D's P^+ region in Fig. 2.31 disappears when v_{DG} falls to $|v_{TP}|$. Since v_{DG} is $|v_{TP}|$ when this happens and v_{DG} is equivalent to $v_{SG} - v_{SD}$, the channel pinches when v_{SD} reaches $v_{SG} - |v_{TP}|$. So gate drive v_{SGT} sets this saturation voltage $v_{SD(SAT)}$.

When v_{SD} overcomes this $v_{SD(SAT)}$, the channel extends to the fraction of L_{CH} that drops $v_{SD(SAT)}$: across L_{CH}'. The depleted portion of L_{CH} drops what remains of

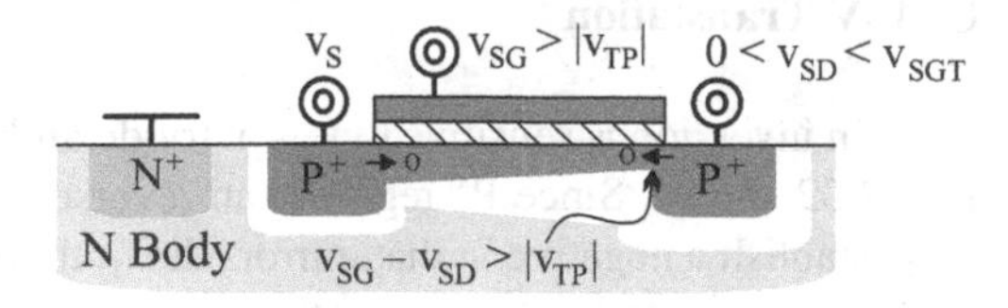

Fig. 2.30 Asymmetrically inverted P-channel MOSFET in triode

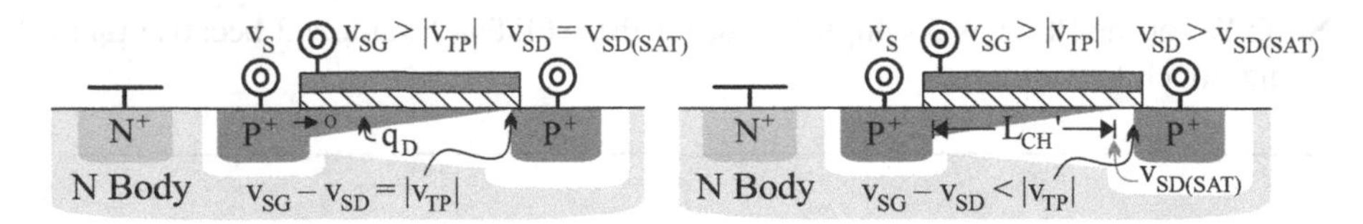

Fig. 2.31 Inverted P-channel MOSFET in saturation

v_{SD}: $v_{SD} - v_{SD(SAT)}$. i_D saturates because the voltage across the channel that L_{CH}' establishes is $v_{SD(SAT)}$, which is independent of v_{SD}:

$$\begin{aligned} i_D\Big|_{v_{SD}>v_{SGT}}^{v_{SG}>|v_{TP}|} &= \frac{v_{SD(SAT)}}{R_{CH}} = v_{SD(SAT)}\left(\frac{W}{L_{CH}'}\right)K_P'\left(v_{SG} - |v_{TP}| - \frac{v_{SD(SAT)}}{2}\right) \\ &\approx \left(\frac{W}{L}\right)\left(\frac{K_P'}{2}\right)(v_{SG} - |v_{TP}|)^2(1 + \lambda_P v_{SD}) \\ &\equiv \left(\frac{W}{L}\right)\left(\frac{K_P'}{2}\right)v_{SGT}{}^2\left(1 + \frac{v_{SD}}{V_{AP}}\right). \end{aligned} \tag{2.22}$$

So i_D is sensitive to v_{SG} and largely insensitive to v_{SD}.

Since a lower v_D depletes more of the channel, L_{CH}' shrinks as v_{SD} increases. So R_{CH} falls and i_D rises with v_{SD}. λ_P models this L_{CH}' modulation with respect to the L that circuit designers define. The effects of v_{DS} fade as L lengthens because the variation becomes a smaller fraction of L, which is equivalent to saying λ_P decreases and V_{AP} increases.

Since C_{GS} falls with L_{CH}', v_{SG} pulls fewer source holes into the P channel when L is shorter, which means R_{CH} is higher. Although a shorter L_{CH}' also increases conduction, the field effect of C_{GS} is dominant. So R_{CH} is roughly 50% higher in saturation than in triode:

$$\begin{aligned} R_{CH}\Big|_{v_{SD}>v_{SGT}}^{v_{SG}>|v_{TP}|} &= \frac{v_{SD(SAT)}}{i_D} \\ &= \left(\frac{L_{CH}'}{W}\right)\left[\frac{v_{SG} - |v_{TP}|}{0.5K_P'(v_{SG} - |v_{TP}|)^2}\right] \\ &\approx \left(\frac{3}{2}\right)\left(\frac{L}{W}\right)\left[\frac{1}{0.5K_P'(v_{SG} - |v_{TP}|)}\right] = 1.5R_{CH}\Big|_{v_{SD}>v_{SGT}}^{v_{SG}>|v_{TP}|}. \end{aligned} \tag{2.23}$$

R_{CH}, however, is only a fraction of R_{DS} in saturation. R_{DS} is usually much higher than R_{CH} because i_D is largely insensitive to v_{DS}.

C. I–V Translation

i_D in inversion is sensitive to v_{SD} in triode and insensitive to v_{SD} in saturation like Fig. 2.32 shows. Since P^+ regions can reverse roles, a positive v_{DG} and a negative v_{SD} establish a negative i_D that mirrors the i_D that positive v_{SG} and v_{SD} produce. i_D is zero when v_{SG} and v_{DG} are negative.

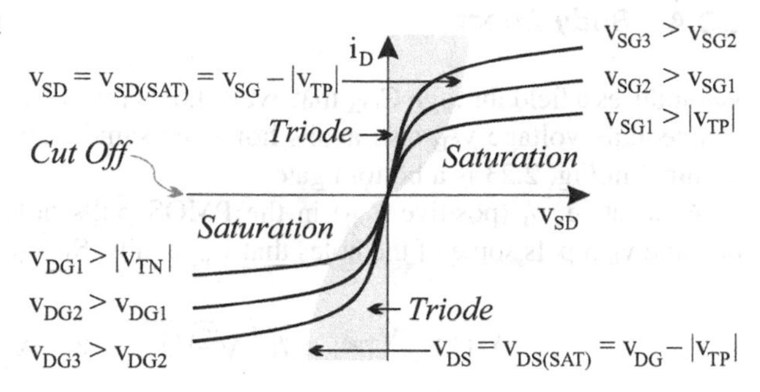

Fig. 2.32 Inverted P-channel MOSFET current

In inversion, i_D is a squared translation of v_{SGT} and i_D saturates when v_{SD} overcomes v_{SGT}. So the $v_{SD(SAT)}$ boundary in Fig. 2.32 is a squared-root reflection of i_D:

$$v_{SD(SAT)} = v_{SG} - |v_{TP}| \approx \sqrt{\frac{2i_D}{(W/L)K_P'(1+\lambda_P v_{SD})}}\Bigg|_{L>>L_{MIN}} \approx \sqrt{\frac{2i_D}{(W/L)K_P'}}. \tag{2.24}$$

$\lambda_P v_{SD}$ fades as L extends beyond L_{MIN} because L_{CH}' modulation becomes a smaller fraction of L.

Example 9: Determine W so $v_{SD(SAT)}$ is no greater than 110 mV, i_D is 100 μA, v_{SD} is 1 V, L is 180 nm, L_{OL} is 30 nm, K_P' is 40 μA/V^2, and λ_P is 2%.

Solution:

$$v_{SD(SAT)} \approx \sqrt{\frac{2i_D}{(W_{CH}/L_{CH})K_P'(1+\lambda_P v_{SD})}} = \sqrt{\frac{2(100\mu)[180n - 2(30n)]}{W_{CH}(40\mu)[1+(2\%)(1)]}}$$
$$\leq 110\,\text{mV}$$

$$\therefore W \equiv W_{CH} \geq 49\ \mu\text{m}$$

Note: W is 5× wider than the NMOS in Example 4 because μ_P in K_P' is that much lower than μ_N in K_N'.

2.3.4 Body Effect

v_{SG} induces a field through C_{OX} that avails holes for v_{SD} to pull. The source–body or source–bulk voltage v_{SB} also avails holes the same way via C_{DEP}. So the N body terminal in Fig. 2.23 is a bottom gate.

A negative v_B (positive v_{SB}) in the PMOS pulls holes into the channel and a positive v_B repels some of the holes that v_{SG} avails. So v_{SB} effectively reduces $|v_{TP}|$:

$$|v_{TP}| = |V_{TP0}| + \gamma_P\left(\sqrt{2\psi_B - v_{SB}} - \sqrt{2\psi_B}\right). \tag{2.25}$$

$|v_{TP}|$ is the baseline V_{TP0} when v_{SB} is zero.

$|v_{TP}|$ is the voltage that v_{SG} overcomes when inverting a channel. To invert in equal proportion, v_{SG} should not only negate the barrier ψ_B but also re-assert another ψ_B in the opposite direction. v_{SB} reduces this $2\psi_B$ translation and alters $|v_{TP}|$ by the amount that γ_P allows.

2.3.5 Symbols

The PMOS is a four-terminal device with interchangeable v_S and v_D terminals that conduct i_D in Fig. 2.33 when v_{SG} and v_{SD} are positive. The two vertical lines at the gate symbolize the C_{OX} that induces a P-type channel. i_G is zero because dc current into C_{OX} is zero. So i_D is also the i_S that flows into v_S. The arrow attaches to the v_S that sets v_{SG} and points in the direction of i_S.

The symbol sometimes excludes the body terminal to indicate other transistors share the same body. In these cases, independent access to the body is not possible. The arrow is also sometimes absent in digital circuits to show that source and drain terminals can reverse roles or on both terminals in switching power supplies to show they can reverse roles. A "bubble" next to the gate distinguishes arrowless PFETs from NFETs. This indicates, like in a digital inverter, that PFETs "invert" the action of NFETs.

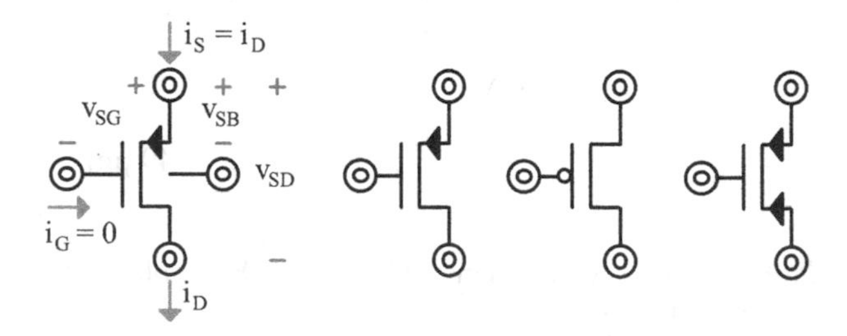

Fig. 2.33 P-channel MOSFET symbols

Example 10: Determine i_D when v_B is 5 V, v_S is 4 V, v_G is 3 V, v_D is 2 V, W is 10 μm, L is 1 μm, K_P' is 40 μA/V^2, V_{TP0} is −400 mV, γ_P is 600 m√V, ψ_B is 300 mV, and λ_P is 5%.

Solution:

$$v_{SD} = v_S - v_D = 4 - 2 = 2\text{ V}$$
$$v_{SG} = v_S - v_G = 4 - 3 = 1\text{ V}$$
$$v_{SB} = v_S - v_B = 4 - 5 = -1\text{ V}$$

$$|v_{TP}| = |V_{TP0}| + \gamma_P\left(\sqrt{2\psi_B - v_{SB}} - \sqrt{2\psi_B}\right)$$
$$= |-400\text{m}| + (600\text{m})\left(\sqrt{2(300\text{m}) - (-1)} - \sqrt{2(300\text{m})}\right) = 690\text{ mV}$$

$$v_{SD} = 2\text{ V} > v_{SD(SAT)} = v_{SG} - |v_{TP}| = 1 - 690\text{m} = 310\text{ mV} \quad \therefore \text{ Saturated}$$

$$i_D \approx \left(\frac{W}{L}\right)\left(\frac{K_P'}{2}\right)(v_{SG} - |v_{TP}|)^2(1 + \lambda_P v_{SD})$$
$$= \left(\frac{10\mu}{1\mu}\right)\left(\frac{40\mu}{2}\right)(1 - 690\text{m})^2[1 + 5\%(2)] = 23\text{ μA}$$

Explore with SPICE:
See Appendix A for notes on SPICE simulations.

```
* PMOSFET: I-V Curves
vb vb 0 dc=5
vs vs 0 dc=4
vg vg 0 dc=3
vd vd 0 dc=2
m1 vd vg vs vb pmosfet w=10u l=1u
.model pmosfet pmos vto=-400m kp=40u lambda=50m gamma=600m
+phi=600m
.op
.dc vd 4 0 10m vg 3 2 200m
.end
```

Tip: Plot is(M1), comment or remove the .dc line, re-run the simulation, and view the output/log file.

2.3.6 Unifying Convention

PFETs and NFETs function the same way. Accumulation, depletion, and inversion result, respectively, when v_{GS} in NFETs and v_{SG} in PFETs are negative, positive and below v_{TN} and $|v_{TP}|$, and positive and above v_{TN} and $|v_{TP}|$. i_D saturates when v_{DS} in NFETs and v_{SD} in PFETs reach $v_{DS(SAT)}'$ and $v_{SD(SAT)}'$ in sub-threshold and $v_{DS(SAT)}$ and $v_{SD(SAT)}$ in inversion. v_{TN} and $|v_{TP}|$ decrease when v_{BS} in NFETs and v_{SB} in PFETs are positive.

Everything that is positive in one is negative in the other. In PFETs, v_{GS}, v_{DS}, and v_{TP} are negative, and a negative v_{BS} reduces v_{TP}. So NFETs and PFETs deplete when v_{GS} reverses, invert when v_{GS} overcomes v_T's v_{TN} or v_{TP}, and saturate when v_{DS} overcomes the $v_{DS(SAT)}'$ or $v_{DS(SAT)}$ that $3V_t$ or $v_{GS} - v_T$ sets. So general v_{GS}, v_{DS}, v_{BS}, and v_T discussions apply to both: NFETs and PFETs.

2.4 Capacitances

The terminals of the MOSFET require time to charge and discharge. The most noticeable of these is the gate because t_{OX} is very thin (on the order of nanometers), which means the resulting C_{OX}'' is substantial. Still, PN junction capacitances at the source and drain also require charge and time to transition.

2.4.1 PN Junction Capacitances

Body terminals normally connect to voltages that zero- or reverse-bias their PN junctions to sources and drains. A *reverse junction voltage* v_{JR} reinforces the diminishing effect that the *built-in potential* V_{BI} induces on *zero-bias junction capacitance* (per unit area) C_{J0}''. So *junction capacitance* C_J peaks to the C_{J0} that C_{J0}'' and *junction area* A_J set and v_{JR} reduces C_J to

$$C_J = \frac{C_{J0}''A_J}{\sqrt{\frac{V_{BI}+v_{JR}}{V_{BI}}}} = \frac{C_{J0}}{\sqrt{1 + \frac{v_{JR}}{V_{BI}}}}, \tag{2.26}$$

where C_{J0}'' and V_{BI} depend strongly on the body's doping concentration.

Although source and drain geometries do not always match, A_J's are usually the same or comparable. v_{SB}, however, is usually lower than v_{DB}. So *source–body capacitance* C_{SB} is normally higher than *drain–body capacitance* C_{DB}.

v_{JR}'s in NFETs are v_{SB} and v_{DB} and in PFETs are v_{BS} and v_{BD}. Although not necessarily so, doping densities in NFETs and PFETs usually differ. C_{J0}'' in NFETs is therefore different in PFETs.

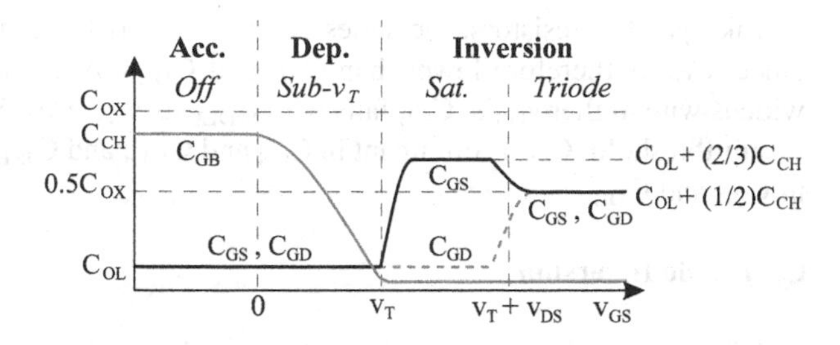

Fig. 2.34 Gate-oxide capacitances

2.4.2 Gate-Oxide Capacitances

Oxide capacitance decomposes into overlap and channel components. *Overlap capacitance* C_{OL} is the C_{OX} fraction that hangs over source and drain diffusions (along W_{CH} and across L_{OL} in Figs. 2.11 and 2.23):

$$C_{OL} = C_{OX}''W_{CH}L_{OL}. \quad (2.27)$$

Channel capacitance C_{CH} is the L_{CH} fraction of L_{OX} along W_{CH} that L_{OL}'s over source and drain diffusions exclude:

$$C_{CH} = C_{OX}''W_{CH}L_{CH} = C_{OX}''W_{CH}(L_{OX} - 2L_{OL}). \quad (2.28)$$

But since the channel does not always extend across L_{CH}, C_{OX} decomposes into gate components differently across regions in Fig. 2.34.

A. **Cut-Off**

In accumulation, the region under the oxide in Figs. 2.12 and 2.24 is a good Ohmic contact to the body. This means that C_{CH} connects the gate to the body, so *gate–body capacitance* C_{GB} comprehends all of C_{CH} in Fig. 2.34. Since the gate overlaps the source and drain diffusions across L_{OL}, *gate–source* and *gate–drain capacitances* C_{GS} and C_{GD} only incorporate C_{OL}. So in cut off, the largest fraction of C_{OX} is in C_{GB}. C_{SB} and C_{DB} are the C_J's that their A_J's and v_{JR}'s set.

B. **Sub-threshold**

When depleted, a depletion region separates the semiconductor surface (in Figs. 2.13 and 2.25) from the body. So C_{CH} stacks over C_{DEP} and C_{GB} is the series combination that results:

$$C_{GB} = C_{CH} \oplus C_{DEP} = \left(\frac{1}{C_{CH}} + \frac{1}{C_{DEP}}\right)^{-1} = \frac{C_{CH}C_{DEP}}{C_{CH} + C_{DEP}}$$
$$\leq \text{Min}\{C_{CH}, C_{DEP}\}. \quad (2.29)$$

Like parallel resistors, the series combination is lower than the smaller capacitance. C_{GB} is therefore lower than C_{CH} and C_{DEP}. And since the depletion region widens with higher v_{GS}'s, C_{GB} falls with C_{DEP} as v_{GS} rises. With so little conduction in sub-threshold, C_{OL} is dominant in C_{GS} and C_{GD}, and C_{JSB} and C_{JDB} are dominant in C_{SB} and C_{DB}.

C. **Triode Inversion**

When inverted in triode, a channel across the semiconductor surface connects v_S and v_D. C_{CH} therefore connects to v_G and the channel and C_{DEP} to the channel and v_B. Since v_S and v_D both connect to the channel, C_{GS} and C_{GD} share C_{CH}, and C_{SB} and C_{DB} share C_{DEP}. So combined, C_{GS} and C_{GD} incorporate their C_{OL}'s plus matching C_{CH} halves:

$$C_{GS/GD(TRI)} = C_{OL} + 0.5C_{CH}, \tag{2.30}$$

and C_{SB} and C_{DB} incorporate their C_J's plus matching C_{DEP} halves:

$$C_{SB/DB(TRI)} = C_{JSB/DB} + 0.5C_{DEP}. \tag{2.31}$$

In other words, C_{GS} and C_{GD} carry equal C_{OX} fractions.

Note that FETs invert into triode when v_{GS} overcomes $v_T + v_{DS}$ because v_{DS} overcomes $v_{DS(SAT)}$'s $v_{GS} - v_T$ when this happens. Also notice that C_{GS}, C_{GD}, C_{SB}, and C_{DB} model all capacitances present in triode inversion. C_{GB} is the series combination of C_{DEP} (which is very low in inversion) and the C_{CH} that C_{GS} and C_{GD} model (and v_S and v_D affect).

D. **Saturated Inversion**

Inverted MOSFETs saturate when v_{GD} or $v_{GS} - v_{DS}$ drops below v_T, which happens when v_{GS} is between v_T and $v_T + v_{DS}$. In this mode, the channel (in Figs. 2.20 and 2.31) disconnects from v_D and shortens to L_{CH}'. So C_{DB} is C_{JDB} and C_{GD} is only the C_{OL} that L_{OL} sets. Since the channel still connects to v_S, C_{SB} carries C_{JSB} plus the C_{DEP} fraction that L_{CH}' sets, and C_{GS} carries C_{OL} plus a similar fraction of C_{CH}:

$$C_{SB(SAT)} \approx C_{JSB} + (2/3)C_{DEP} \tag{2.32}$$

$$C_{GS(SAT)} \approx C_{OL} + (2/3)C_{CH}, \tag{2.33}$$

where this fraction is roughly two-thirds.

The C_{CH} and C_{DEP} fractions near v_D that L_{CH}' excludes are no longer in C_{GD} and C_{DB}. C_{GB} carries these fractions, but since C_{GD} is the series combination of these fractions and C_{DEP} and C_{DEP} is much lower, C_{GB} is usually negligible. In this mode, the largest fraction of C_{OX} is in C_{GS}.

Example 11: Determine C_{GS} and C_{GD} in saturation when W is 10 μm, L is 180 nm, L_{OL} is 30 nm, and C_{OX}" is 2.76 fF/μm^2.

Solution:

$$C_{OL} = C_{OX}''W_{CH}L_{OL} = (2.76m)(10\mu)(30n) = 0.83 \text{ fF}$$

$$C_{CH} = C_{OX}''W_{CH}L_{CH} = C_{OX}''W_{CH}(L_{OX} - 2L_{OL})$$
$$= (2.76m)(10\mu)[180n - 2(30n)] = 3.3 \text{ fF}$$

$$C_{GS} = C_{OL} + (2/3)C_{CH} = 0.83f + (2/3)(3.3f) = 3.0 \text{ fF}$$

$$C_{GD} = C_{OL} = 0.83 \text{ fF}$$

Explore with SPICE:
See Appendix A for notes on SPICE simulations.

```
* NMOSFET: Capacitance in Saturated Inversion
vd vd 0 dc=600m
m1 vd vd 0 0 nmosfet w=10u l=180n
.model nmosfet nmos vto=400m kp=200u tox=12.5n ld=30n cgso=83p
+cgdo=83p
.op
.end
```

Tip: View the output/log file.

E. **Transition**

C_{GS} and C_{GD} share C_{CH} equally when inverted in triode and v_{DS} is zero. As v_{DS} rises, the charge in the channel shifts toward the source. So C_{GS} acquires the corresponding C_{CH} fraction that C_{GD} in Fig. 2.35 loses. This continues until the channel pinches, when v_{DS} reaches $v_{DS(SAT)}$'s $v_{GS} - v_T$.

As the channel recedes from the drain past $v_{DS(SAT)}$, the source and drain lose a small but growing fraction of C_{CH} to the body. As a result, C_{GS} acquires less of C_{CH} than C_{GD} loses as v_{DS} rises past $v_{GS} - v_T$. This continues until C_{GD}'s fraction fades and C_{GS}'s share maxes to two-thirds. C_{GS} and C_{GD} reach their saturation limits when v_{DS} matches v_{GS}.

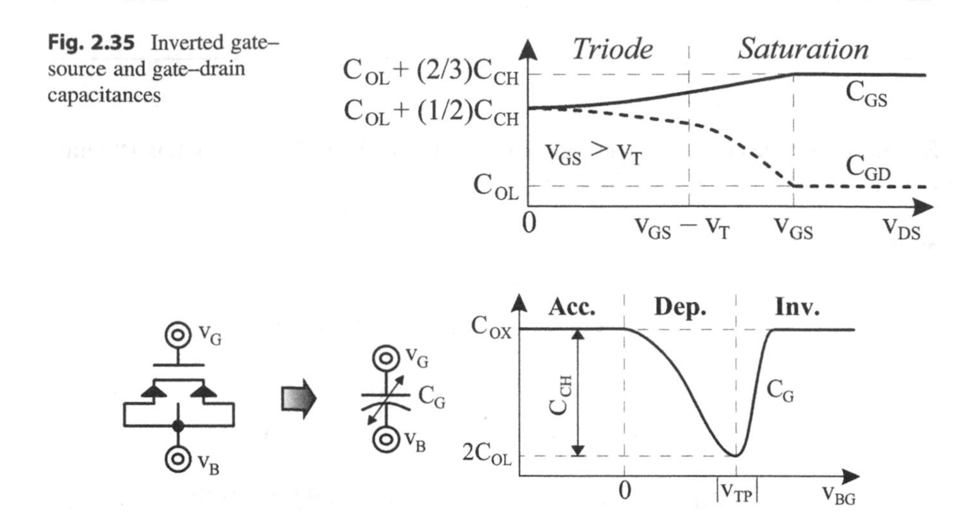

Fig. 2.35 Inverted gate–source and gate–drain capacitances

Fig. 2.36 Bi-modal P-channel MOSFET varactor

2.4.3 MOS Varactors

A. **Bi-modal**

The MOS structure is fundamentally a parallel-plate capacitor with a thin dielectric. When used as a capacitor, paralleling all oxide components yields the highest capacitance. The PMOS in Fig. 2.36 combines all capacitive components by shorting v_S, v_B, and v_D terminals. This way, gate capacitance C_G incorporates C_{GS}, C_{GB}, and C_{GD}:

$$C_G = C_{GS} + C_{GB} + C_{GD} \leq 2C_{OL} + C_{CH} = C_{OX} = C_{OX}"W_{CH}L_{OX}. \quad (2.34)$$

When a positive v_G (negative v_{BG}) accumulates electrons in the channel region, C_{GB} is C_{CH}, so C_G includes C_{GS} and C_{GD}'s $2C_{OL}$ and C_{GB}'s C_{CH}. When a v_{BG} that is greater than $|v_{TP}|$ inverts the channel that connects v_S and v_D, C_{GS} and C_{GD} each carry C_{OL} and half of C_{CH}. So even though C_{GB} is very low, C_{GS} and C_{GD} in C_G still carry $2C_{OL}$ and C_{CH}.

When a negative v_G (positive v_{BG}) depletes the channel region without inverting it, C_{GB} becomes the series combination of C_{CH} and C_{DEP}. As v_{BG} climbs, v_G depletes more of the body, so C_{DEP} decreases, and with it, C_{GB}. So as v_{BG} rises above zero toward $|v_{TP}|$, C_G loses C_{GB}'s C_{CH}, falling from C_{OX} to the $2C_{OL}$ that C_{GS} and C_{GD} carry.

This structure is useful as a capacitor because C_G is high at C_{OX} when v_{BG} is negative and greater than $|v_{TP}|$. C_G is not a perfect variable capacitor or *varactor* because C_G is not monotonic with v_{BG}. A rise in v_{BG} does not always raise C_G because C_G is *bi-modal*. (The PMOS symbol in Fig. 2.36 has two "source" arrows to indicate both P^+ diffusions supply holes when inverting the channel.)

B. Inversion Mode

Disconnecting v_B from v_S and v_D removes C_{GB} from C_G. This way, the C_{CH} and C_{DEP} that C_{GB} in Fig. 2.34 adds in accumulation and depletion disappear from C_G in Fig. 2.37. So C_G's transition between $2C_{OL}$ and C_{OX} is now monotonic with v_{SG}. The drawback to this *inversion-mode* varactor is that the v_{SG} range that changes C_G is usually narrow. (v_B connects to the highest potential to reverse-bias the PN junctions that connect v_S and v_D to v_B.)

C. Accumulation Mode

C_{GB}'s transition in depletion in Fig. 2.34 is more gradual than C_{GS} and C_{GD}'s in inversion. A gradual transition is appealing because extending the voltage range that transitions capacitance is usually desirable in a varactor. So the purpose of the *accumulation-mode* structure in Fig. 2.38 is to eliminate the inversion mode from the bi-modal case.

The fundamental difference here is that the body is the same type of material as the source and drain. So the body connects the two N^+ terminals when v_{GS} is zero. The line across the source/drain terminals of the NMOS in Fig. 2.38 represents this connection.

A positive v_{GS} reinforces this connection because it pulls and accumulates electrons under the oxide. A negative v_G, however, repels electrons and depletes the channel. This way, depletion reduces C_G from C_{OX} to $2C_{OL}$. Since no part of the structure can avail holes, the channel never inverts.

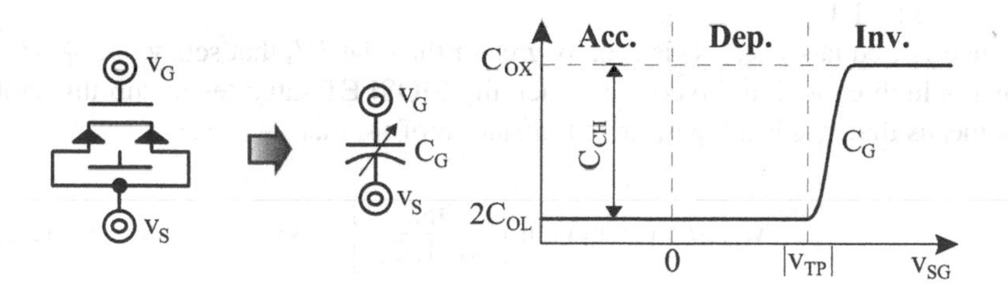

Fig. 2.37 Inversion-mode P-channel MOSFET varactor

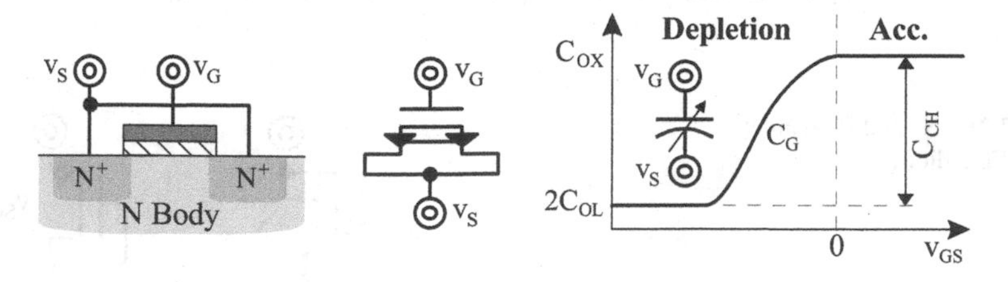

Fig. 2.38 Accumulation-mode N-channel MOSFET varactor

D. **Variations**

The varactors in Figs. 2.36, 2.37, and 2.38 are P-, P-, and N-channel FETs because most fabrication technologies stack NFETs on P substrates. This means that independent P-type body terminals are not accessible. If they were, N-, N-, and P-channel devices would also be possible. These varactors are useful in *voltage-controlled oscillators* (VCO) because their voltages can adjust the frequency that their capacitances set.

2.4.4 MOS Diodes

Another useful application of MOSFETs in power supplies and other analog circuits are as diodes. Key to their realization is C_{GS}, first as a stand-alone component and then as the agent that induces conduction. The other critical element is an inverting feedback loop.

A. **Diode Connection**

In Fig. 2.39, the drain–gate connections close a feedback loop around C_{GS} and the channel that induce the MOSFETs to behave like diodes. The NMOS, for example, is off when v_S is grounded and v_G is low. But when a circuit feeds an *input current* i_{IN} into the drain–gate node, i_{IN} charges C_{GS}. When v_{GS} overcomes v_T, channel current i_D begins to sink some of i_{IN}. The rest of i_{IN} continues to charge C_{GS} until i_D is able to sink all of i_{IN}, at which point C_{GS} stops charging. So in effect, i_{IN} "forward-biases" the NFET.

Since v_{DS} equals v_{GS}, v_{DS} is usually greater than the $3V_t$ that sets $v_{DS(SAT)}'$. So if i_{IN} is not high enough to invert a channel, the MOSFET saturates in sub-threshold. This means that v_{GS} is a logarithmic translation of i_{IN} that v_T offsets:

$$v_{GS} \approx v_T + n_I V_t \ln\left[\frac{i_{IN}}{(W/L)I_{SN}}\right] < v_T. \tag{2.35}$$

When i_{IN} inverts a channel, v_{DS} is also greater than the $v_{GS} - v_T$ that sets $v_{DS(SAT)}$ because v_{DS} matches v_{GS}. So the MOSFET inverts into saturation. v_{GS} is therefore not only a $v_{DS(SAT)}$ reflection but also a squared-root translation of i_{IN} that v_T offsets:

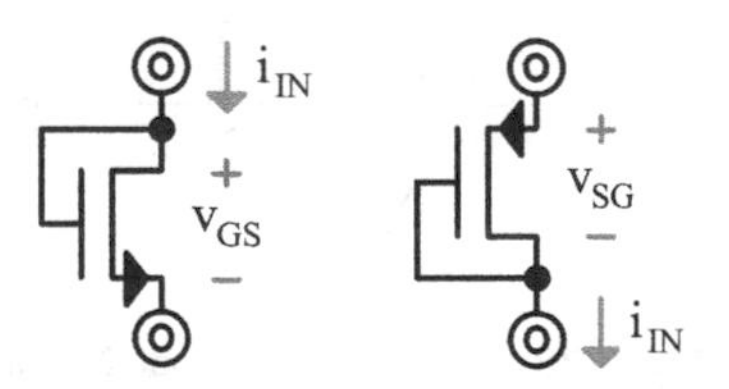

Fig. 2.39 N- and P-channel MOSFET diodes

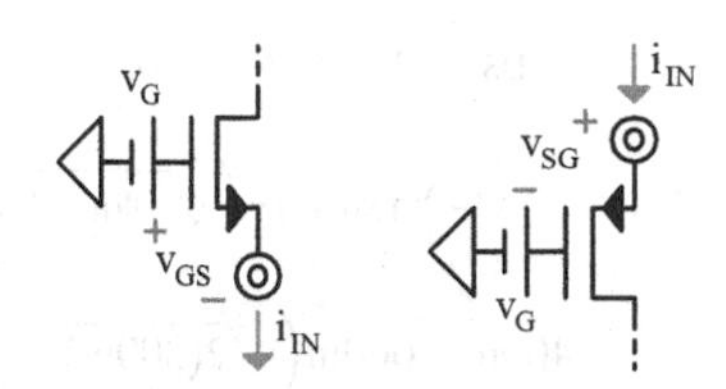

Fig. 2.40 Implicit N- and P-channel MOS diode action

$$v_{GS} = v_T + v_{DS(SAT)}$$

$$\approx V_{TN0} + \gamma_N\left(\sqrt{2\psi_B - v_{BS}} - \sqrt{2\psi_B}\right) + \sqrt{\frac{2i_{IN}}{(W/L)K'(1+\lambda v_{DS})}}. \quad (2.36)$$

Note that body effect alters v_T, L_{CH}' modulation alters $v_{DS(SAT)}$, and L_{CH}' modulation is higher when L is shorter.

When i_{IN} is no longer present, i_D discharges C_{GS} to v_T in inversion and below v_T in sub-threshold. So i_{IN} activates and cuts off *diode-connected* MOSFETs like i_{IN} would a junction diode. The only difference is that a diode drops a logarithmic translation of i_{IN} and MOSFETs drop a squared-root translation offset by v_T. *MOS diodes* that drop 300–500 mV, however, are often preferable because i_{IN} burns less power across 300–500 mV than across the 600–800 mV that diodes usually establish.

Pulling i_{IN} from v_S when the gate–drain terminals connect to a voltage that is above ground similarly charges C_{GS} until i_D conducts i_{IN}. The PMOS does the same when a circuit feeds or pulls i_{IN} to v_S or from the gate–drain node. The only difference is their v_{GS} because v_{TN} and μ_N do not match $|v_{TP}|$ and μ_P.

B. Diode Action

Sometimes this *diode action* results from an *implicit* connection. One example is when other circuit components connect gate and drain terminals together. Another example is pulling i_{IN} from v_S in Fig. 2.40 when a voltage or a large capacitor holds v_G. Here, i_{IN} charges C_{GS} until i_S (and whatever connects to v_D) supplies i_{IN}. The PMOS does the same when a circuit feeds i_{IN} to v_S: i_{IN} charges C_{GS} until i_S (and whatever connects to v_D) sinks i_{IN}.

Example 12: Determine v_S for an NMOSFET when i_{IN} pulls 100 μA from v_S, W is 10 μm, L is 180 nm, L_{OL} is 30 nm, K_N' is 200 μA/V^2, V_{TN0} is 400 mV, γ_N is 600 m$\sqrt{V}$, ψ_B is 300 mV, λ_N is 10%, and v_B, v_G, and v_D are 0 V.

Solution:

$$L_{CH} = L - 2L_{OL} = 180n - 2(30n) = 120 \text{ nm}$$

$$v_{DS} = v_D - v_S = v_{GS} = v_G - v_S = v_{BS} = v_B - v_S = 0 - v_S = -v_S$$

$$v_{GS} = -v_S \approx V_{TN0} + \gamma_N\left(\sqrt{2\psi_B - v_{BS}} - \sqrt{2\psi_B}\right) + \sqrt{\frac{2i_{IN}}{(W_{CH}/L_{CH})K_N{}'(1+\lambda_N v_{DS})}}$$

$$= 400m + 600m\left(\sqrt{2(300m) + v_S} - \sqrt{2(300m)}\right) + \sqrt{\frac{2(100\mu)}{(10\mu/120n)(200\mu)(1 - 10\% v_S)}}$$

$$= 340 \text{ mV} \quad \rightarrow \quad v_S = -340 \text{ mV}$$

Note: v_{BS} reduces v_T more than $V_{DS(SAT)}$ raises v_{GS}. λ_N suppresses v_{DS} more than γ_N suppresses v_{BS}, so neglecting L_{CH} modulation yields a similar v_{GS}. These implicit diode conditions are typical for ground NMOSFETs in switched-inductor power supplies. (Example 15 shows why the FET is in inversion.)

Explore with SPICE:
See Appendix A for notes on SPICE simulations.

```
* NMOSFET: Diode Action
m1 0 0 vs 0 nmosfet w=10u l=180n
iin vs 0 dc=100u
.model nmosfet nmos vto=400m kp=200u ld=30n lambda=100m
+gamma=600m phi=600m
.op
.end
```

Tip: View the output/log file.

2.5 Short Channels

Smaller geometries are appealing in three basic ways. First, they occupy less *silicon area*, so FETs cost less and microchips can fit more circuits. Second, capacitances are lower, so FETs require less charge power and less time to transition. And third, electric fields are stronger, so conduction requires less voltage, and as a result, less power.

Unfortunately, geometric reductions and stronger electric fields also produce unappealing effects in conductivity and noise that are not always easy to model or counter. These usually surface when L_{CH} is comparable to the *depletion widths* d_W

around the source and drain regions. This is why sub-micron devices suffer from *short-channel effects* that fade and disappear in longer-channel devices.

2.5.1 Drain-Induced Barrier Lowering

In NFETs, increasing v_D in Fig. 2.41 pulls N^+ electrons away from the PN junction toward v_D's contact and pushes P-body holes away into the body. So v_D's depletion region expands in all directions. When L_{CH} is comparable to d_W's, v_D's depletion region can extend and merge into v_S's. The merging of two depletion regions this way is *punch-through.*

v_D's field is so close to v_S that it pushes holes away from the channel region near v_S and loosens nearby electrons. So raising v_{DS} helps deplete and invert the channel. This way, a lower v_{GS} can more easily induce conduction. This means, v_{DS} reduces the barrier voltage v_B in Fig. 2.42 that keeps electrons from diffusing. This *drain-induced barrier lowering* (DIBL) effectively reduces v_T.

This is a problem because v_{DS} can induce current flow with zero v_{GS}. So v_G may not be able to cut the NFET off, which is another way of saying v_G can lose control of the NFET. Unfortunately, this effect is not static because v_D changes with time.

In PFETs, v_D's field is so close to v_S that it pushes electrons away from the oxide region near v_S and presses nearby valence electrons into their home sites. Holes can therefore drift more easily. So PFETs also suffer a dynamic reduction in v_T when v_D falls.

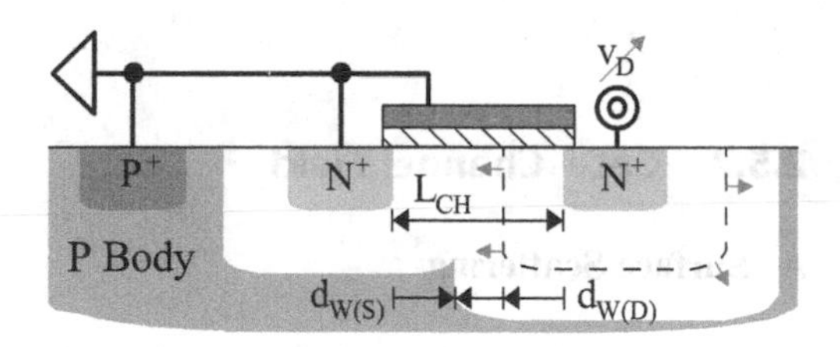

Fig. 2.41 Drain-induced punch-through

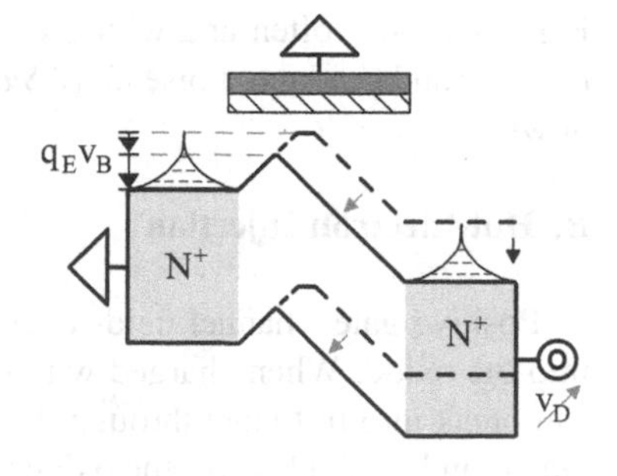

Fig. 2.42 Drain-induced barrier lowering in an N-channel MOSFET

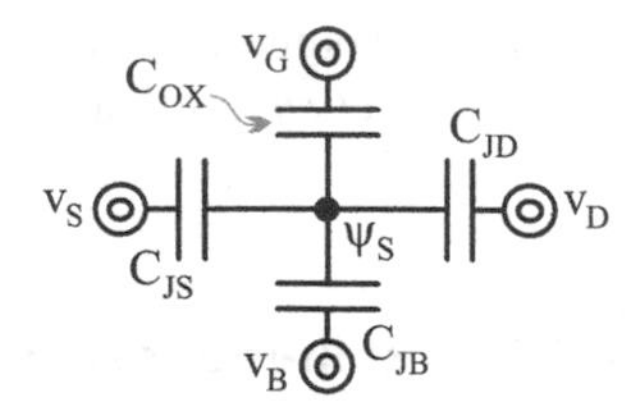

Fig. 2.43 Channel coupling components

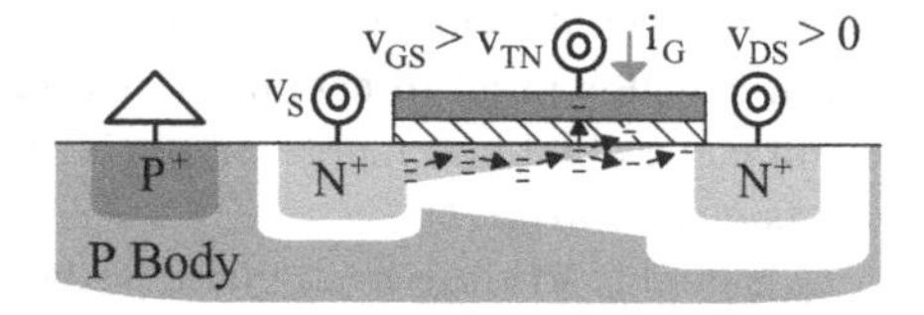

Fig. 2.44 Surface scattering and hot-electron injection in the NMOS

A. **Thinner Oxide**

The surface potential is ultimately the result of capacitor coupling from v_G, v_B (via the body effect), and v_D (with DIBL). So in the absence of v_G and v_B, ψ_S in Fig. 2.43 is the voltage-divided fraction that v_D's depletion capacitance C_{JD} to the channel couples across v_G's C_{OX}, v_S's C_{JS}, and v_B's C_{JB}. DIBL is noticeable because short channels increase C_{JD}'s coupling.

The effect of C_{OX}, C_{JS}, and C_{JB} is to shunt C_{JD}'s coupling. So raising C_{OX} reduces DIBL. This is one of the driving reasons why engineers scale t_{OX} with L_{OX}. Another reason is higher current density i_D/W_{CH} and v_{GS}-to-i_D gain because C_{OX}'' in K' climbs with reductions in t_{OX}. Reducing t_{OX} from 25 to 5 nm, for example, can suppress the 250-mV reduction in v_T that 100 mV across v_{DS} can produce when L_{OX} is 40 nm.

2.5.2 Gate–Channel Field

A. **Surface Scattering**

Thinner t_{OX}'s intensify vertical gate–channel fields. So on their way to the drain, carriers accelerate and collide with the oxide on the surface of the semiconductor in Fig. 2.44 more often and with greater force. This scattering effect reduces *surface mobility* and produces noise in i_D. *Surface scattering* intensifies as L_{CH} and t_{OX} scale down.

B. **Hot-Electron Injection**

Positive gate–channel fields energize, accelerate, and direct N-channel electrons into the oxide. When charged with sufficient *kinetic energy* E_K, these *hot electrons* can break into or tunnel through the oxide. Electrons that break into and stay in the lattice (in Fig. 2.44) leave the oxide negatively charged. So over time, a higher v_{GS} is

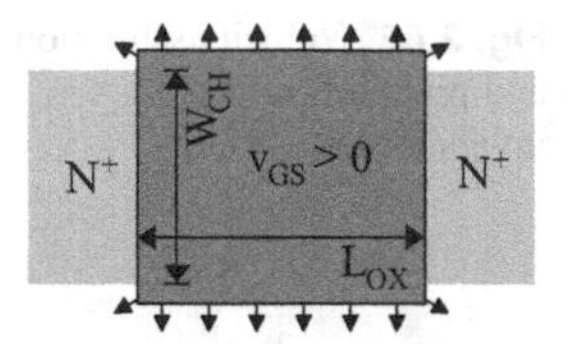

Fig. 2.45 Fringing electric field lines around an N-channel MOSFET

necessary to deplete and invert the channel, which means v_{TN} increases. And electrons that tunnel through the thin oxide establish a gate current i_G. PFETs are largely immune to *hot-electron injection* because their v_{GS}'s are usually negative, whose effect is to repel electrons.

C. Oxide-Surface Ejections

In repelling electrons, negative gate–channel fields weaken electron bonds along the silicon–oxide interface. When sustained and at elevated temperatures, they break silicon–hydrogen (Si–H) bonds and dispel negatively charged H atoms into the body. These atoms leave behind positively charged "hole" traps that counter the action of negative v_{GS}'s.

So PFETs need a higher v_{SG} to deplete and invert the channel, which means $|v_{TP}|$ increases. But as v_{SG} weakens the field, electrons repopulate holes, so v_{TP} recovers. v_{TP} therefore fluctuates as v_{SG} stresses and relaxes the oxide. This *negative bias temperature instability* (NBTI) is less prevalent in NFETs because they are less prone to negative v_{GS}'s.

D. Fringing Fields

Shrinking planar dimensions enhance the effects of fringing fields along the periphery of the gate-oxide region in Fig. 2.45 on the channel. These field lines deplete and invert space outside the W_{CH} that source and drain diffusions define, extending the width of the channel. This W_{CH} variation ΔW_{CH} is negligible in larger FETs, but substantial in sub-micron devices.

2.5.3 Source–Drain Field

A. Velocity Saturation

Electron velocity v_E in the *conduction band* scales linearly with voltage up to 100 km/s or so. *Hole velocity* v_H scales similarly, but saturates at 60 km/s or so. $v_{H(SAT)}$ is lower than $v_{E(SAT)}$ because the *valence electrons* that shift holes are more tightly bound to their home sites than electrons in the conduction band. Their mobility ultimately determines the *critical electric fields* ε_C that accelerate them to these levels:

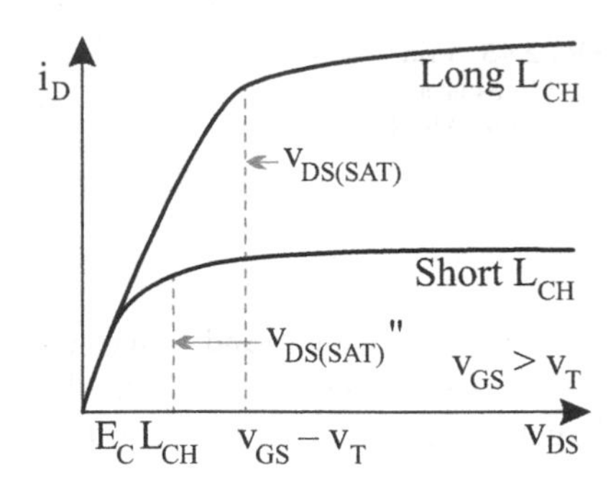

Fig. 2.46 Velocity saturation and pinch-off effects in inversion

$$\varepsilon_{CN} = \frac{v_{E(SAT)}}{\mu_N} \tag{2.37}$$

and

$$\varepsilon_{CP} = \frac{v_{H(SAT)}}{\mu_P}. \tag{2.38}$$

ε_{CN} and ε_{CP} in silicon are 1.4 and 4.2 V/μm at room temperature. ε_{CP} is higher because μ_P is lower than μ_N more than $v_{H(SAT)}$ is lower than $v_{E(SAT)}$.

In triode inversion, i_D scales with v_{DS} until v_{DS} saturates v_E or v_H. The *velocity saturation voltage* $v_{DS(SAT)}''$ that ε_C across L_{CH} sets is

$$v_{DS(SAT)}'' = \varepsilon_C L_{CH}. \tag{2.39}$$

v_E and v_H therefore saturate when v_{DS} across a 1-μm channel reaches 1.4 and 4.2 V. But if the $v_{DS(SAT)}$ that $v_{GS} - v_T$ sets is less than 1.4 V in 1-μm NFETs and 4.2 V in 1-μm PFETs, v_E and v_H do not saturate.

$v_{DS(SAT)}''$ for sub-micron channels can be so low that i_D can saturate before v_{GD} pinches the channel at $v_{DS(SAT)}$. i_D in Fig. 2.46, for example, scales with v_{DS} in triode inversion until v_{DS} reaches $v_{GS} - v_T$ when L_{CH} is long and $\varepsilon_C L_{CH}$ when L_{CH} is short. Normally, MOSFETs begin to suffer from velocity saturation when L_{CH} is less than 1 μm.

Example 13: Determine $v_{DS(SAT)}''$ and $v_{SD(SAT)}''$ when L_{CH} is 180 nm.

Solution:

$$v_{DS(SAT)}'' = \varepsilon_{CN} L_{CH} = (1.4/\mu)(180n) = 250 \text{ mV}$$

$$v_{SD(SAT)}'' = \varepsilon_{CP} L_{CH} = (4.2/\mu)(180n) = 760 \text{ mV}$$

Note: $v_{DS(SAT)}$'' and $v_{SD(SAT)}$'' are over the 130-mV $v_{DS(SAT)}$ and $v_{SD(SAT)}$ that 10- and 50-μm-wide and 180-nm-long N- and P-channel MOSFETs set with 100 μA (from previous examples), so v_{GD} pinches their channels before carrier velocities saturate.

B. Impact Ionization

Electrons gain speed and ε_K when source–drain fields intensify. ε_K can be so great that these *hot electrons* can collide and liberate otherwise immobile electrons from their home sites. Engineers call this process *impact ionization* because atoms ionize on impact.

Each energized electron can free an electron that avails a hole. Two electrons can then gain enough ε_K to liberate another two *electron–hole pairs* (EHP), like Fig. 2.47 shows. This v_{DS}-induced process repeats, multiplies, and grows in *avalanche* fashion.

Impact ionization is a problem in MOSFETs because, while electrons in NFETs scatter toward the positively charged drain, holes flow into the negatively charged body. Since the body is moderately doped, body resistance drops a voltage that forward-biases v_{BS}'s PN junction. So v_S's N^+ electrons also diffuse across the junction into the body.

Forward-biasing this body–source junction activates the lateral NPN across the source–body–drain regions. When L_{CH} is very short, the N^+ drain "collects" most of the electrons that the N^+ source "emits" into the channel. So the NPN draws *body current* i_B from v_B and conducts i_D. In other words, the NPN can short the NFET with zero v_{GS}.

PFETs suffer the same effect, except holes scatter toward the negatively charged drain and electrons flow into the positively charged body. The resulting i_B drops a voltage that forward-biases v_{SB}'s PN junction and activates the lateral PNP across the source–body–drain regions. So even with zero v_{SG}, the PNP can short the PFET.

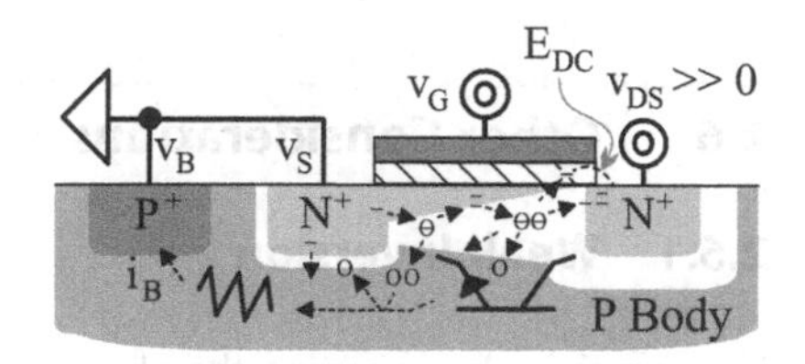

Fig. 2.47 Impact ionization, avalanche, and hot-electron injection

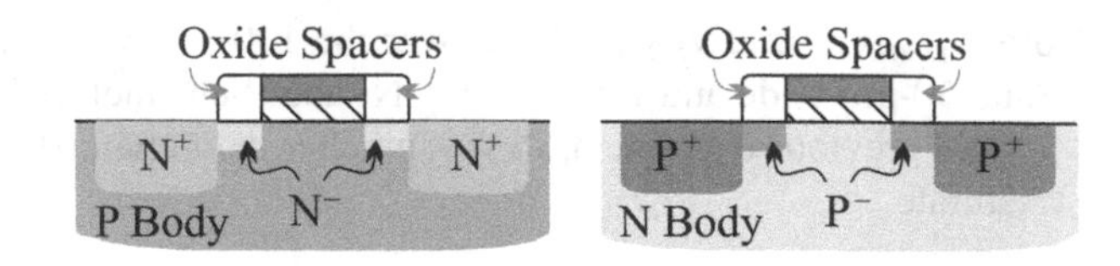

Fig. 2.48 Lightly doped drain MOSFETs

C. **Arcing Field**

Shrinking planar dimensions enhance the effects of arcing channel–drain field ε_{DC} lines that pass through the gate oxide near the drain in Fig. 2.47. This ε_{DC} intensifies as L_{OX} shortens, v_D rises, and the $v_{DS(SAT)}$ across the channel that $v_{GS} - v_T$ sets falls. L_{OX} in sub-micron devices is so short and lateral and arcing fields are so intense as a result that N-channel electrons can gain enough ε_K to break into the oxide and stay there. These hot electrons charge the oxide, increasing v_{TN}.

D. **Lightly Doped Drain**

In triode inversion, the channel drops v_{DS} across L_{CH}. When pinched, the channel drops the $v_{DS(SAT)}$ that $v_{GS} - v_T$ sets across L_{CH}'. So the depletion region between the drain and channel drops the remainder $v_{DS} - v_{DS(SAT)}$ or $v_{DG} + v_T$ across $L_{CH} - L_{CH}'$. Of these distances, $L_{CH} - L_{CH}'$ is the shorter. So the most intense field usually results across this short drift space. This field intensifies when L_{CH} shortens and v_{DG} rises.

Outside of lengthening L_{CH} and reducing v_{DG}, the only other way of weakening this field is by extending v_D's depletion length between the drain and the channel. The *lightly doped drain* (LDD) regions in Fig. 2.48 do this by depleting farther into the drain. Because with fewer carriers, v_D depletes farther into the LDD region. This way, the resulting drift length is longer than $L_{CH} - L_{CH}'$.

For this, engineers first implant LDD dopants into the silicon without the oxide spacers shown. Then, with the spacers, implanting more dopants forms the highly doped regions without altering the LDDs. Both source and drain regions receive LDDs because they swap roles when i_D reverses. Only the voltage at the terminal that acts as the drain can deplete, leaving the source largely intact. This practice reduces impact ionization, avalanche, and electron injection. And since LDDs are shallow and lightly doped, v_D depletes less channel space, so DIBL is also lower.

2.6 Other Considerations

2.6.1 Weak Inversion

Not surprisingly, inverting the channel does not keep carriers from diffusing across source–drain terminals. In fact, v_{DS} induces carriers to both diffuse and drift. And v_{GS} determines to what extent.

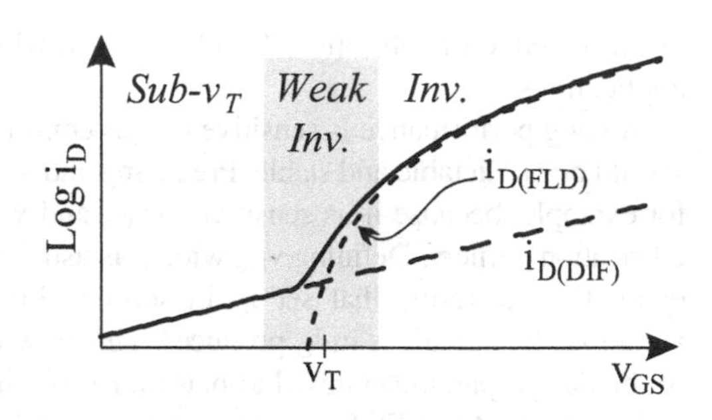

Fig. 2.49 Drain current as MOSFET channel forms

Deep in sub-threshold, when v_{GS} is well below v_T in Fig. 2.49, *drift current* i_{FLD} is so low that *diffusion current* i_{DIF} dominates i_D. And i_{FLD} is so high in *strong inversion*, when v_{GS} is well above v_T, that i_{FLD} dwarfs i_{DIF}. Near v_T, i_{DIF} and i_{FLD} are comparable. In this context, v_T is the v_{GS} that produces matching i_{DIF} and i_{FLD} components. So in *weak inversion*, as the channel forms, i_D reflects both conduction mechanisms.

Solving accurate sub-threshold and inversion expressions for every single transistor across time in a system that incorporates thousands if not billions of transistors can be time-consuming for a computer. To save computing time, computer algorithms model one region well and estimate the other. Or if they model both regions well, they approximate weak inversion. So for more predictable and reliable operation, engineers often design MOSFETs to operate deep in sub-threshold or in strong inversion.

A. **Voltage Bias**

Digital circuits and switching power supplies normally use MOSFETs as switches. These FETs close and open into the on and off states that v_{GS} determines. The v_{GS} that closes FETs is usually much greater than v_{DS} because v_{DS} is only millivolts after FETs close. FETs open when v_{GS} is zero. So these FETs switch between triode and cut off and saturate only during transitions. When closed, they are in sub-threshold when v_{GS} cannot overcome v_T and in inversion when v_{GS} can.

To assert the least and most resistive on and off states, circuit designers normally apply the highest and lowest v_{GS}'s possible. When voltages higher than v_T are not available, MOSFETs cannot invert a channel when they close. So they switch between sub-threshold and cut off. Inversion is only possible when voltages higher than v_T are available, in which case *on resistance* R_{ON} is R_{CH} in triode inversion.

B. **Current Bias**

Amplifiers and linear power supplies normally bias FETs at particular v_{GS}–v_{BS}–v_{DS}–i_D settings. Then, they vary one or two of these variables and use variations in one or two of the others to drive other circuits into action. v_{GS} is normally 0.5–2 V and v_{DS} is higher than 300 mV. So for the most part, these FETs operate in saturation or on the edge of saturation. In other words, triode operation is less likely in these applications.

Analog performance is sensitive to bias conditions, so current and voltage settings should be predictable and stable. Predicting and stabilizing a v_{GS}-defined i_D is difficult, for example, because i_D is sensitive to v_{GS} and v_T, v_{GS} is noisy, and v_T varies across fabrication corners. Defining v_{GS} with i_D is usually better because the exponential and quadratic v_{GS} terms that set i_D in sub-threshold and inversion suppress the v_{GS} variations that changes in i_D produces. This is why i_D (instead of v_{GS}) is usually one of the design parameters used to bias transistors in analog circuits.

Inverting a MOSFET, however, is not an option when voltages higher than v_T are not available. But it is otherwise. Still, *analog designers* often prefer sub-threshold for low-power consumption because both voltages and currents are low. They like inversion for high speed because the higher currents that inversion induces charge and discharge capacitances faster.

Since v_{GS} is an indirect logarithmic or square-root translation of i_D that v_T shifts, ensuring i_D-biased MOSFETs are in sub-threshold or inversion is not straightforward. Luckily, what usually matters most to analog design engineers is predictable *small-signal performance*. And this hinges on *small-signal transconductance* g_m' because g_m translates small-signal variations in v_{GS} on i_D and *vice versa*.

Small v_{GS} signals are so much smaller than v_{GS} that a linear slope translation can approximate their effect on i_D fairly well. This g_m slope is i_D's first partial derivative $\partial i_D/\partial v_{GS}$ with respect to v_{GS}. Since v_{GS} is within an exponential term in i_D in sub-threshold, $\partial i_D/\partial v_{GS}$ matches i_D, but with v_{GS}'s $1/n_I V_t$ coefficient as a multiplier:

$$g_m\Big|_{v_{DS}>3V_t}^{0<v_{GS}<v_T} \equiv \frac{\partial i_D}{\partial v_{GS}} \approx \left(\frac{1}{n_I V_t}\right)\left(\frac{W}{L}\right) I_S \exp\left(\frac{v_{GS}-v_T}{n_I V_t}\right) \approx \frac{i_D}{n_I V_t} \approx \frac{I_D}{n_I V_t}, \quad (2.40)$$

where variations in i_D are so much smaller than i_D's static (non-varying) component I_D that i_D reduces to I_D. In inversion, $\partial i_D/\partial v_{GS}$ loses i_D's quadratic effect, so g_m is a square-root translation of I_D:

$$\begin{aligned} g_m\Big|_{v_{DS}>v_{GST}}^{v_{GS}>v_T} &\equiv \frac{\partial i_D}{\partial v_{GS}} = \left(\frac{W}{L}\right) K'(v_{GS}-v_T)(1+\lambda v_{DS}) \\ &= \sqrt{2 i_D K'(W/L)(1+\lambda v_{DS})} \approx \sqrt{2 I_D K'(W/L)}, \end{aligned} \quad (2.41)$$

where λv_{DS}'s channel-length modulation effect fades when L is long.

Note that g_m rises with W/L in inversion, but not in sub-threshold. And a square-root suppresses i_D in inversion, but not in sub-threshold. So with the same i_D, g_m in Fig. 2.50 climbs with W/L until g_m maxes in sub-threshold. From this perspective, v_T is the v_{GS} that maxes g_m in inversion to the $i_D/n_I V_t$ that sets g_m in sub-threshold:

$$g_{m(MAX)}\Big|_{v_{DS}>v_{GST}}^{v_{GS}>v_T} = \sqrt{2 i_D K'(W/L)_X(1+\lambda v_{DS})} = g_m\Big|_{v_{DS}>3V_t}^{0<v_{GS}<v_T} \approx \frac{i_D}{n_I V_t}. \quad (2.42)$$

This happens when W/L is $(W/L)_X$ and $v_{DS(SAT)}$ is $2n_I V_t$:

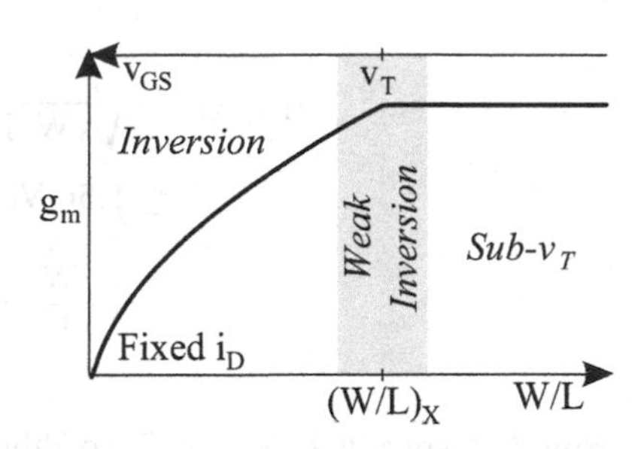

Fig. 2.50 Small-signal transconductance across regions

$$\left(\frac{W}{L}\right)_X \approx \frac{i_D}{2n_I^2V_t^2K'(1+\lambda v_{DS})} \tag{2.43}$$

and

$$v_{DS(SAT)}\Big|_{(W/L)_X}^{v_{GS}>V_T} = \sqrt{\frac{2i_D}{(W/L)_X K'(1+\lambda v_{DS})}} = 2n_IV_t. \tag{2.44}$$

In other words, MOSFETs invert when the $v_{DS(SAT)}$ that W/L sets at a particular i_D is greater than $2n_IV_t$.

This is handy because i_D, W, and L are the variables that engineers use to design circuits. But since MOSFETs invert weakly near $2n_IV_t$ and weak-inversion models are not always accurate, adding a ±25% guard band is prudent. So $v_{DS(SAT)}$ should be less than $1.5n_IV_t$ for deep sub-threshold and greater than $2.5n_IV_t$ for strong inversion.

Example 14: Determine the W/L that ensures a PMOSFET is in sub-threshold when i_D is 1 μA, T_J is 27 °C, K_P' is 40 μA/V² at 27 °C, n_I is 1.75, *Boltzmann's constant* K_B is 1.38×10^{-23} J/K, *electronic charge* q_E is 1.60×10^{-19}, and L is $10L_{MIN}$.

Solution:

$$V_t = \frac{K_BT_J}{q_E} = \frac{(1.38)(10^{-23})(300)}{(1.60)(10^{-19})} = 26\text{ mV}$$

$$L >> L_{MIN}$$

$$\therefore \quad v_{DS(SAT)} \approx \sqrt{\frac{2i_D}{(W/L)K_P'}} = \sqrt{\frac{2(1\mu)}{(W/L)(40\mu)}}$$
$$\leq 1.5n_IV_t = 1.5(1.75)(26m) = 68 \text{ mV}$$
$$\rightarrow \quad \frac{W}{L} \geq 11 \quad \therefore \quad \frac{W}{L} \equiv 15$$

Note: K_P' typically falls with T_J, so although not to the same extent, $v_{DS(SAT)}$ tends to track V_t's rise with T_J. So this W/L may keep the PFET in sub-threshold across temperature.

Example 15: Determine if the NMOSFET in Example 12 is inverted when n_I is 1.75.

Solution:

$$v_{DS(SAT)} = \sqrt{\frac{2i_{IN}}{(W_{CH}/L_{CH})K_N'(1+\lambda_N v_{DS})}}$$
$$\approx \sqrt{\frac{2i_{IN}}{(W_{CH}/L_{CH})K_N'}} = \sqrt{\frac{2(100\mu)}{(10\mu/120n)(200\mu)}}$$
$$= 110 \text{ mV} > 2n_IV_t = 2(1.75)25.6m) = 90 \text{ mV}$$
$$\therefore \quad \text{Inverted}$$

2.6.2 Junction Isolation

Basic *complementary MOS* (CMOS) technologies integrate N- and P-channel FETs into one silicon substrate. More sophisticated technologies integrate and optimize other components. But they cost more because they require additional fabrication steps. Still, the benefits of an expanded list of optimized components can outweigh the added cost.

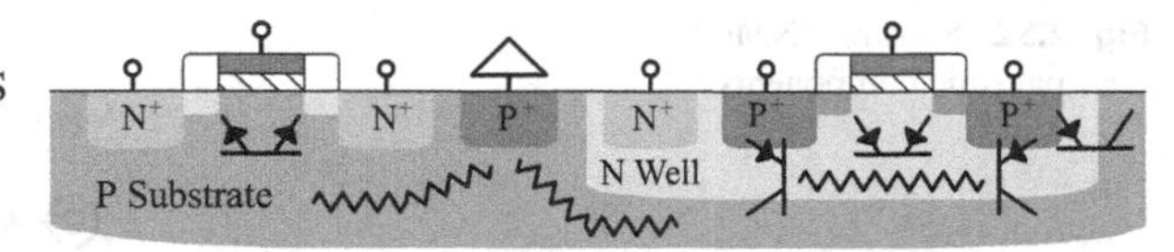

Fig. 2.51 Junction-isolated single-well P-substrate CMOS FETs

Sharing a common silicon substrate requires electronic isolation from the substrate. This means that no circuit component should inject or draw *substrate current* i_{SUB}. Luckily, reverse-biasing all PN junctions to the substrate does this. So when ground is the most negative potential, the P substrate in Fig. 2.51 should connect to ground. N substrates should similarly connect to the most positive potential, which is often v_{DD}.

Junction isolation is not perfect, however. For one, every component in the microchip incorporates junction capacitance C_J to the substrate, which requires current and time to charge and discharge. C_J's also couple noise into the substrate and, through the substrate, into other devices. Plus, unintended breakdown and transient excursions can forward-bias substrate PN junctions and inject i_{SUB} into the shared substrate. Still, junction isolation is straightforward and cost-effective.

A. **Channel BJTs**

Oppositely doped source–body–drain diffusions also realize source–drain BJTs. Source, body, and drain terminals are the emitter, base, and collector, respectively, because sources supply the electrons in NFETs and holes in PFETs that drains collect. *Base–collector current gain* β_0 can be high because the *effective base width* w_B across sub-micron channels is short.

Channel BJTs in the substrate are usually in cut off because substrate PN junctions to source and drain diffusions are usually reverse-biased. Source–drain BJTs in wells also cut off when their N bodies (in the case of PFETs) connect to the highest potential possible. But since β_0 can be high, engineers sometimes activate them on purpose.

B. **Substrate BJTs**

Oppositely doped source–body–substrate and drain–body–substrate diffusions in wells also realize vertical and lateral *substrate BJTs*. In these cases, sources and drains supply the carriers that the substrate collects, so sources and drains are emitters, wells are bases, and the substrate is the collector. β_0 can be high because moderately doped and relatively shallow wells can establish short w_B's.

Connecting N-well bodies (in the case of PFETs) to the highest potential keeps these substrate BJTs off. Analog engineers sometimes activate them on purpose for their β_0. But this is not a common practice because they inject i_{SUB} into the shared substrate.

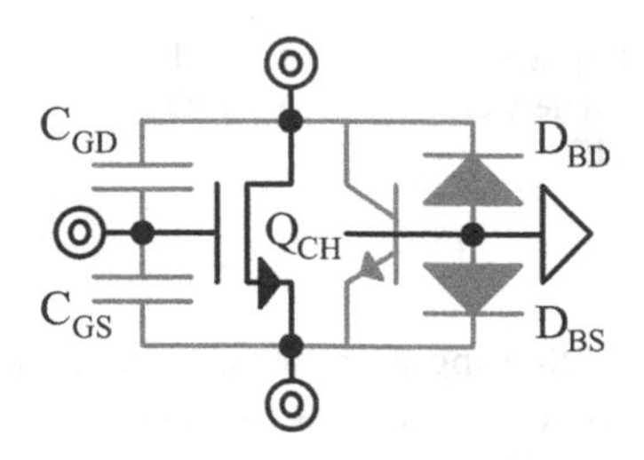

Fig. 2.52 Substrate NMOS with parasitic components

C. **Substrate MOSFETs**

NFETs built directly over a P substrate (like in Fig. 2.51) share their body with the rest of the die. So independent access to their body terminals is not possible. Engineers sometimes use three-terminal symbols (like in Fig. 2.22) to indicate this. In the case of substrate NFETs, the P body's connection to ground or to the most negative potential is implied. *Substrate MOSFETs* incorporate the channel BJT and substrate diodes that Figs. 2.51 and 2.52 show.

D. **Welled MOSFETs**

The body of PFETs built in N wells over a P substrate (like in Fig. 2.51) is the N well. So independent access to their bodies is possible. The four-terminal symbol (in Fig. 2.33) is therefore more appropriate and almost always used. On occasion, engineers use three-terminal symbols to indicate a pool of welled PFETs share one well and one body connection. *Welled MOSFETs* incorporate the channel and substrate BJTs that Figs. 2.51 and 2.53 show. Connecting the N body to the highest potential deactivates these BJTs.

E. **Process Variants**

Although less popular, integrating N- and P-channel MOSFETs in N substrates is also possible. In these cases, PFETs sit directly over an N substrate, NFETs lie in P wells, and the N substrate connects to the most positive potential. PFET body terminals are not available.

Independent access to the body offers a degree of design flexibility that can help optimize and improve circuit performance. Substrate MOSFETs in "vanilla" *single-well* technologies do not offer this option. *Twin-well* or *dual-well* process technologies do because they can embed NFETs and PFETs in their own independent wells. The drawback is the additional expense of more fabrication steps.

2.6.3 Diffused-Channel MOSFETs

Automotive, laptop, and other higher-voltage consumer products call for v_{DS} breakdown voltages that exceed the rating of 1.8–5.5-V LDD MOSFETs. *Diffused-*

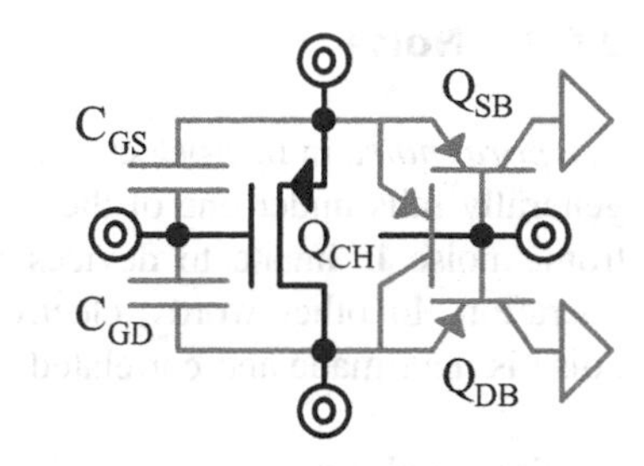

Fig. 2.53 Welled PMOS with parasitic components

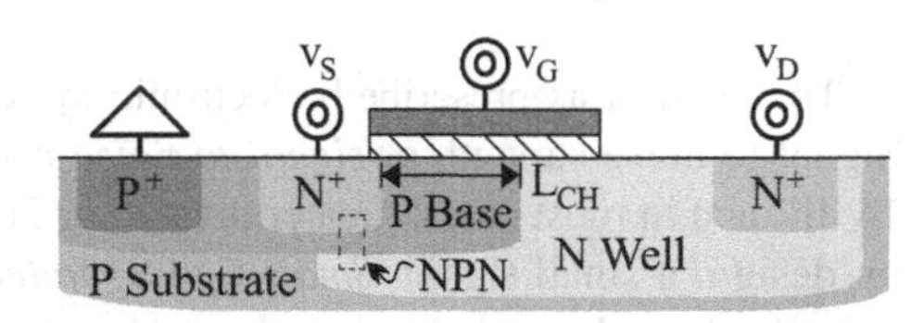

Fig. 2.54 Lateral diffused-channel N-channel MOSFET

channel MOS (DMOS) transistors are popular in this space because they can break at 12–100 V. v_{DS} breakdown is higher because their channel–drain drift regions are longer. The drawback is the cost of the additional fabrication steps needed to build them. Still, many power supplies cannot survive without these *double-diffused* structures.

Adding an NPN P base layer to the CMOS process illustrated in Fig. 2.51 not only avails the vertical N^+–P base–N well BJT structure in Fig. 2.54 but also the *lateral DMOS* (LDMOS) under the oxide. Here, the P base is the body of the NMOS that v_G inverts. The N well extends the N^+ drain to the edge of the P-base body. So L_{CH} is the distance across the P base between the edges of the N^+ source and the N-well drain.

Operationally, v_G inverts the P-base region under the oxide to form an N channel that connects the N^+ source to the N-well drain. When v_D reduces v_{GD} below v_T, the channel pinches and the region near the P base–N well junction depletes. So the saturated channel drops $v_{DS(SAT)}$'s $v_{GS} - v_T$, the non-depleted part of the N well drops an Ohmic portion of v_{DS}, and the depleted region drops the rest.

The N well is so lightly doped that it can drop 10–50 V, depending on doping concentration and t_{OX}. Unfortunately, the non-depleted portion of the N well adds on resistance to R_{CH}. So the lateral distance between the P base and N^+ drain should be as short as limitations allow.

This *drain-extended* NMOS is not the only way to realize a DMOS. Depending on the fabrication technology, *vertical DMOS* (VDMOS) transistors are also possible. Thicker oxides and V-shaped gate structures are other ways of extending the breakdown limits of the MOSFET.

2.6.4 Noise

Temporal noise in microelectronics refers to unintended variations in conduction. It generally falls under one of the two categories: electronic or *systemic noise*. Electronic noise is innate to devices and systemic noise is the byproduct of circuit operation. In other words, electronic noise is natural and random and systemic noise is man-made and correlated.

A. Terminology

To function as prescribed, electronic systems must discern signals from noise. Signal-to-noise strength or *signal-to-noise ratio* (SNR) should be no less than 5 or 14 dB, and in most cases, greater than 10 or 20 dB. This means that noise should be, by design, a small signal. *Small-signal models* can therefore predict the effects of noise fairly well.

Noise manifests over time. Although hardly ever a single sinusoid, a combination of sinusoids can replicate any behavior. So decomposing noise into frequencies is convenient. *Noise spectrum* is a popular term in this respect because it refers to the frequency content of noise.

Spectral noise density n_d refers to noise power at one of these tones in Watts per Hertz W/Hz, amps per root Hertz A/√Hz, or volts per root Hertz V/√Hz. Total noise n_t is the summation of these n_d's across *operating frequencies* f_O. But since zero crossings (phase) seldom align, n_t is the statistical *root sum of squares* (RSS):

$$n_t = \sqrt{\int_{f_{LOW}}^{f_{HIGH}} {n_d}^2 df_o}. \quad (2.45)$$

Capacitances require current and time to charge and discharge. Since their voltages cannot rise or fall instantly, circuits cannot track infinitely fast signals. Capacitances also block static low-frequency signals. So circuits only process signals and noise within the *frequency band* or *bandwidth* $f_{HIGH} - f_{LOW}$ or Δf_{BW} that these capacitances ultimately allow. n_t is therefore the statistical sum of noise components across this *noise bandwidth*.

B. Electronic Noise

Thermal Temperature energizes electrons into random mobile states that cause them to collide. These collisions produce small-signal variations in current i_{nt} that intensify with temperature. Since resistivity impedes motion, the resistance R_X of the material suppresses this *thermal noise current* i_{nt}:

$$\frac{\overline{i_{nt}^2}}{\int df_O} = \frac{\overline{i_{nt}^2}}{\Delta f_{BW}} = \frac{4k_B T_J}{R_X}. \tag{2.46}$$

R_X therefore drops a *thermal noise voltage* v_{nt} that scales with T_J and R_X:

$$\frac{\overline{v_{nt}^2}}{\int df_O} = \frac{(i_{nt}R_X)^2}{\Delta f_{BW}} = 4k_B T_J R_X. \tag{2.47}$$

These collisions are so random in nature that i_{nt} and v_{nt} decompose, like white light, into all frequencies equally. *Thermal noise* is therefore a form of *white noise*. As such, frequency strength is uniform across the spectrum and total noise is the statistical sum of strengths across Δf_{BW}.

Holes suffer from the same effect because they drift as valence electrons shift positions. The only difference is that bound valence electrons resist motion more than loosely bound electrons in the conduction band. So i_{nt} is lower and v_{nt} is higher in P material.

In MOSFETs, R_{CH} generates i_{nt} and v_{nt}. Interestingly, R_{CH} in triode inversion is equivalent to $1/g_m$ for long channels in saturated inversion:

$$R_{CH}\Big|_{v_{DS}<<v_{GST}}^{v_{GS}>v_T} \approx \left(\frac{L}{W}\right)\left[\frac{1}{K'(v_{GS}-v_T)}\right] \approx \frac{1}{g_m}\Big|_{v_{DS}>v_{DS(SAT)}}^{v_{GS}>v_T}. \tag{2.48}$$

So R_{CH} is $1/g_m$ in triode and $1.5/g_m$ in saturation. Since t_{OX} is thinner with shorter L_{CH}'s, surface scattering in short-channel devices raises R_{CH} in saturation to $2/g_m$ or $3/g_m$. So R_{CH}'s $1/g_m$ to $3/g_m$ generates thermal noise.

Shot Electrons "shoot" through gaps randomly. Diodes, BJTs, JFETs, and MOSFETs all suffer from this *shot noise* because their electrons cross the drift space that their respective depletion regions establish. The resulting *shot noise current* i_{ns} is proportional to *electronic charge* q_E and intensifies with higher conduction i_D:

$$\frac{\overline{i_{ns}^2}}{\int df_O} = \frac{\overline{i_{ns}^2}}{\Delta f_{BW}} = 2q_E i_D. \tag{2.49}$$

Shot noise is so random in nature that it spreads evenly across frequency. So like thermal noise, shot noise is also a form of white noise. i_{ns} is therefore the statistical sum of individual strengths across Δf_{BW}.

Flicker Like the "flicker" of a flame, *flicker noise* is mostly a low-frequency phenomenon. In electronics, it refers to *1/f noise* because noise power falls with f_O

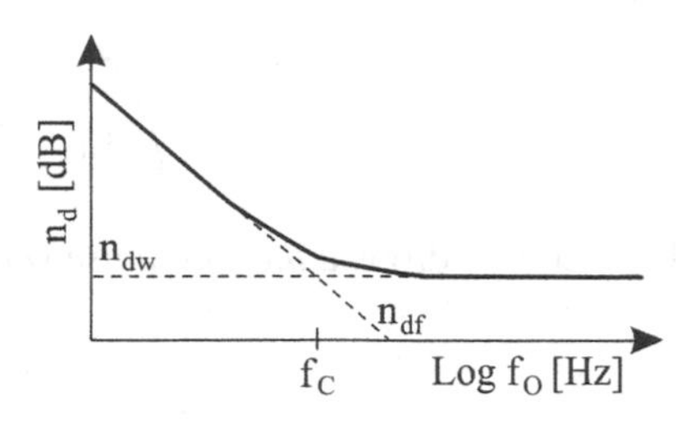

Fig. 2.55 Spectrum of flicker and thermal (white) noise

at 20 dB per decade. It is a form of *pink noise* (from audio engineering) for this reason.

Since flicker noise n_{df} fades with f_O, white noise n_{dw} in Fig. 2.55 overpowers n_{df} past the *noise corner frequency* f_C, where n_{df} and n_{dw} cross. f_C is an indirect measure of noise content n_d because f_C increases with higher n_{df}. In other words, n_{df} is more powerful when f_C is higher.

Slow carriers have more time to recombine (when crossing PN junctions and BJT bases) and to fall into surface oxide traps (when crossing MOS channels) than fast carriers. Slow carriers also have more time to scatter (along the oxide surface when crossing MOS channels) than fast carriers. So the noise that these random mechanisms produce fades with f_O.

Flicker noise is usually worse in MOSFETs than in BJTs and JFETs because the effects of random recombination on conduction are less severe than oxide irregularities and scattering along the oxide surface. The NMOS is worse in this respect because their channel electrons are more loosely bound than the valence electrons that shift holes in the PMOS. BJTs and JFETs generate less 1/f noise because they conduct well beneath the surface of the semiconductor, where the silicon structure is less imperfect.

Surface effects in MOSFETs are more profound across shorter channels because surface defects are a larger fraction of the channel. More intense v_{GS} fields and higher conduction increases their effects. So *flicker noise current* i_{nf} climbs with lower L_{CH} and C_{OX}'' and higher i_D:

$$\frac{i_{nf}^2}{\int df_O} = \frac{K_F i_D}{C_{OX}'' L_{CH}^2 f_O}. \tag{2.50}$$

Traps and their effects on scattering are so dependent on the fabrication process that engineers normally derive the *flicker noise coefficient* K_F empirically from measurements.

Side Note: *Brown noise* falls quadratically with f_O. This noise is relevant in audio engineering because low-frequency $1/f^2$ noise can overpower 1/f noise. In electronics, pink 1/f noise is usually more powerful, so brown $1/f^2$ noise is less discernable and, as a result, less relevant.

C. Systemic Noise

Coupling As capacitances charge and discharge, they draw and supply *displacement currents*. When capacitances are the unintended byproducts of devices, these currents become and produce *coupled noise current* i_{nc}. i_{nc} appears everywhere because all components incorporate capacitances.

i_{nc} is systemic because the capacitances and voltages that produce i_{nc} depend entirely on the circuit. i_{nc} is therefore consistent and predictable over time. Substrate capacitances that result from junction isolation are especially problematic because they couple noise into the shared substrate. So the substrate collects, integrates, and spreads this noise to every component embedded in the substrate. Large digital blocks and switching power supplies, for example, generate *switching noise* at their *switching frequency* f_{SW} that normally appears almost everywhere in the system.

Injection The effects of forward-biasing substrate junctions are more severe. This is because i_{SUB} can be higher and more sustained. Substrate and well resistances can therefore drop voltages that *de-bias* parasitic substrate diodes and lateral and vertical BJTs into action.

Audio amplifiers, power supplies, and power amplifiers are more prone to *injection* because they conduct lots of power at near-breakdown voltages. So impact ionization and avalanche currents are more likely to flow into and across the substrate and wells. Switched inductors in power supplies are also more likely to feed and forward-bias these junctions.

2.7 Summary

MOSFETs are the evolutionary offspring of JFETs. And JFETs are nothing but resistors that gate fields pinch with their depletion regions. What is perhaps most interesting is that current saturates when the channel voltage saturates. Past this $v_{DS(SAT)}$, i_D is only sensitive to v_{GS}.

MOSFETs similarly use fields to alter channel resistance. But to invert channels, v_G should overcome v_T. Their channel regions deplete below v_T and accumulate opposite charge carriers when v_G reverses.

v_G in sub-threshold reduces the barrier that carriers must overcome to diffuse, v_G in inversion pulls carriers into the channel, and v_{DS} propels all these carriers across the channel. v_D, however, opposes v_{GD}'s barrier reduction and charge formation until i_D saturates. In inversion, v_D pinches and saturates the channel voltage like JFETs. Also like JFETs, v_B is a bottom gate that can reinforce or counter the action of v_G.

Without a channel, C_{GB} incorporates all the oxide capacitance to the channel region. C_{GS} and C_{GD} are low in cut off because gates overlap sources and drains across a very short L_{OL}. C_{GB}, however, loses C_{CH} to C_{GS} and C_{GD} when the channel forms. And when v_D pinches the channel, C_{GB} loses its share of C_{CH} to C_{GS} and

C_{GD}, but mostly to C_{GS} because the inverted channel is a large fraction of L_{CH} that still connects to v_S.

When paralleled, these capacitances become a bi-modal varactor. Disconnecting or reversing the semiconductor type of the body eliminates the non-monotonicity that bi-modal behavior engenders. And connecting the gate and source closes a feedback loop around C_{GS} that converts the MOSFET into a diode.

Short channels are desirable in microelectronics because microchips can fit more transistors and perform more functions. Unfortunately, the depletion region that v_D induces when L_{CH} is short reaches so far into the channel that it reinforces the action of v_G. Thinning the oxide helps shunt the effect of v_D on the channel. But the stronger gate–channel and source–drain fields that result induce surface scattering, hot-electron injection, oxide-surface ejections, velocity saturation, and impact ionization. Reducing v_D's doping concentration helps because it extends the short drain–channel drift region, which weakens the lateral field.

As MOSFETs invert, drift and diffusion currents compete. Luckily, switching applications switch between cut off and triode. So predicting i_D accurately across short-lived transitions is not critical. To avoid this relatively unpredictable region of operation, analog designs often bias MOSFETs deep in sub-threshold or in strong inversion. Noting that MOSFETs effectively "invert" their channels when $v_{DS(SAT)}$ surpasses $2n_IV_t$ is helpful in this respect.

When integrated into the same substrate, substrate and welled MOSFETs incorporate substrate diodes and source–drain and substrate BJTs. Engineers zero- or reverse-bias all substrate PN junctions to deactivate unintended components and isolate designed-in devices. Diffusing a body region under the source that extends to a lightly doped N-well drain extends v_{DS}'s breakdown limit. These diffused-channel MOSFETs cost more because they require additional fabrication steps.

Thermal energy, conduction across depleted spaces, and silicon-surface imperfections produce random electronic noise in i_D. Substrate capacitances also couple and spread circuit-generated noise. The effects of near-breakdown operation are worse because impact-ionization currents de-bias substrate diodes and BJTs into action. Systems with large switching transistors, like switched-inductor power supplies, usually suffer the most from noise coupling and injection.

Switched Inductors

3

Abbreviations

BJT	Bipolar-junction transistor
FET	Field-effect transistor
LED	Light-emitting diode
MOS	Metal–oxide–semiconductor
RMS	Root-mean-squared
CCM	Continuous-conduction mode
DCM	Discontinuous-conduction mode
A_{Si}	Silicon area
d_D	Drain duty cycle
D_{DG}	Ground drain diode
D_{DO}	Output drain diode
d_E	Energize duty cycle
d_{IN}	Input duty cycle
d_O	Output duty cycle
d_X	Coil distance
E_M	Magnetic energy
E_L	Inductor energy
f_{LC}	LC resonant frequency
f_O	Operating frequency
f_{SW}	Switching frequency
i_{IN}	Input current
i_L	Inductor current
$i_{L(HI)}$	Inductor current's peak in CCM
$i_{L(LO)}$	Inductor current's valley in CCM
$i_{L(MIN)}$	Minimum inductor current
$i_{L(PK)}$	Inductor current's peak in DCM/peak variation in PDCM
i_O	Output current

G. A. Rincón-Mora, *Switched Inductor Power IC Design*,
https://doi.org/10.1007/978-3-030-95899-2_3

$i_{XI/O}$	Input/output-referred transformer current
Δi_L	Inductor CCM ripple current
Δi_{LD}	Load dump
k_C	Coupling factor/coefficient
k_L	Transformer translation
$K_{N/P}'$	MOS transconductance parameter
L_{CH}	MOS channel length
L_X	Switched transfer inductor
M_{DG}	Ground drain MOSFET
M_{DO}	Output drain MOSFET
M_{EG}	Ground energize MOSFET
M_{EI}	Input energize MOSFET
P_{DD}	Diode drain power
P_{DT}	Dead-time power
P_{IN}	Input power
P_L	Inductor power
P_{LOSS}	Power losses
P_O	Output power
P_R	Ohmic power
R_{CH}	MOS channel resistance
R_D	Drain resistance
R_E	Energize resistance
R_L	Inductor resistance
$R_{L(AC)}$	Inductor ac resistance
$R_{L(DC)}$	Inductor dc resistance
R_{ESR}	Equivalent series resistance
R_{SER}	Series resistance
S_{DG}	Ground drain switch
S_{DO}	Output drain switch
S_{EG}	Ground energize switch
S_{EI}	Input energize switch
t_C	Conduction time
t_D	Drain time
t_{DT}	Dead time
t_E	Energize time
t_{SW}	Switching period
t_{LC}	LC resonant period
τ_{LC}	LC time constant
σ_{LOSS}	Fractional losses
v_B	Body voltage
v_D	Drain voltage
v_{DI}	Ideal drain voltage
v_E	Energize voltage
v_{EI}	Ideal energize voltage
v_G	Gate voltage

v_{IN}	Input node/voltage
v_L	Intrinsic inductor voltage
v_L'	Extrinsic inductor voltage
v_O	Output node/voltage
v_S	Source voltage
v_{SW}	Switching node/voltage
v_{SWI}	Input switching node/voltage
v_{SWO}	Output switching node/voltage
$v_{TN/P}$	N/P-channel MOS threshold voltage
Δv_O	Output voltage ripple
W_{CH}	MOS channel width

The fundamental purpose of power supplies is to transfer power. *Switched inductors* (SLs) are pervasive in this space because they output a large fraction of the power they draw from an input source. The fundamental reason for this is low Ohmic losses, and that's because switches in the network only drop millivolts. So the *Ohmic power* P_R these switches burn when they conduct current is low.

This means that *power-conversion efficiency* η_C, which is the fraction of *input power* P_{IN} delivered to the output, is usually high, between 85% and 95% when P_{IN} is moderate to high. This is because *output power* P_O is the P_{IN} that *power losses* P_{LOSS} avail. So *fractional losses* σ_{LOSS}, which is the fraction of P_{IN} lost to P_{LOSS}, are what determine and limit η_C:

$$\eta_C \equiv \frac{P_O}{P_{IN}} = \frac{P_{IN} - P_{LOSS}}{P_{IN}} = 1 - \frac{P_{LOSS}}{P_{IN}} = 1 - \sigma_{LOSS}. \tag{3.1}$$

In the absence of these losses, switched inductors deliver all the P_{IN} they draw. Diode switches, however, consume power and drop voltages that alter some of the switching characteristics of the circuit. Resistors produce similar effects, but to a lesser extent because resistances are (by design) very low. Either way, the basic mechanics are the same.

3.1 Transfer Media

3.1.1 Inductor

A. Ideal Inductor

Inductors magnetize and demagnetize with voltages of opposing polarity. In other words, they energize and drain their magnetic fields with positive and negative voltages. So as the alternating *inductor voltage* $\pm v_L$ in Fig. 3.1 raises and lowers *inductor current* i_L across time t_X, the *magnetic* or *inductor energy* E_M or E_L that the *transfer inductor* L_X holds rises and falls quadratically with i_L:

$$i_L = \left(\frac{v_L}{L_X}\right)t_X \quad (3.2)$$

$$E_M \equiv E_L = 0.5L_Xi_L{}^2. \quad (3.3)$$

$+v_L$ is the *energize voltage* $+v_E$ that magnetizes L_X and $-v_L$ is the *drain voltage* $-v_D$ that demagnetizes L_X. But since i_L can also flow in the opposite direction, $-v_L$ can be the $+v_E$ that energizes L_X and $+v_L$ the $-v_D$ that drains L_X. Notice $+v_L$ in Fig. 3.1 supplies power when i_L flows to the right and sinks power when i_L flows to the left. $-v_L$ similarly supplies power when i_L flows to the left and sinks power when i_L flows to the right.

L_X's inductance is a measure of how much energy i_L can hold. So a higher L_X holds more energy with the same current than a lower L_X. Since more magnetic space stores more E_M, L_X scales with the *cross-sectional area* A_L of the loop across the coil that implements the inductor and the *number of turns* N_L in the coil. L_X scales more with N_L when the loops align because the magnetic fields of the loops reinforce one another to establish an even stronger magnetic field.

B. **Actual Inductor**

In practice, the coil is resistive. This means that inductors also burn Ohmic power. The *equivalent series resistance* R_{ESR} in inductors is the *inductor resistance* R_L in Fig. 3.2. This R_L drops a voltage that effectively reduces the voltage across the intrinsic L_X. As a result, the *intrinsic* v_L is lower than the *extrinsic inductor voltage* v_L' that a circuit applies.

R_L scales with N_L because the length of the wire is longer with more turns. R_L is also inversely proportional to the cross-sectional area of the wire, so R_L is higher with thinner coils. Plus, nearby coils produce an alternating magnetic field that induces local eddy currents that effectively push current away from the edges.

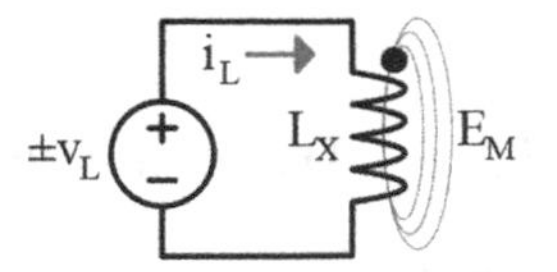

Fig. 3.1 Magnetizing inductor

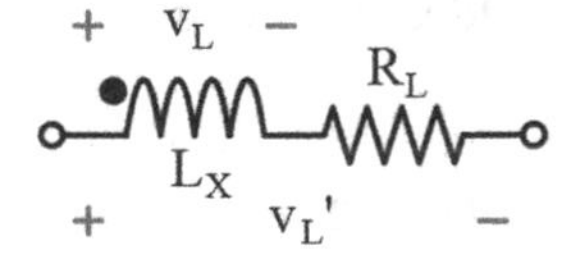

Fig. 3.2 Actual inductor

This means R_L also scales with the number of nearby coils and the alternating frequency of the current flowing through the coils. This is the *proximity effect*.

Fast-moving charges also tend to flow along the outer edges of the coil. This is why R_L also scales with *operating frequency* f_O, which is the *skin effect*. So from a circuit's perspective, P_R in R_L decomposes into the low- and high-frequency components that *dc* and *ac inductor resistances* $R_{L(DC)}$ and $R_{L(AC)}$ consume:

$$P_{RL} = i_{L(AVG)}{}^2 R_{L(DC)} + \Delta i_{L(RMS)}{}^2 R_{L(AC)}. \quad (3.4)$$

Here, $i_{L(AVG)}$ is i_L's average dc current and $\Delta i_{L(RMS)}$ is the *root-mean-squared* (RMS) equivalent of the alternating ac ripple.

C. **Optimal Inductor**

The optimal L_X carries lots of E_L and burns little P_{RL}. L_X should therefore deliver high E_L with low i_L. This happens when *magnetic permeability*, which is the ability to form a magnetic field, is high, for which a large magnetic core is usually necessary. R_L should also be low, which means the coil should be thick. Although higher N_L raises L_X, lengthening the coil and the proximity effect of more nearby turns also increase R_L. So the optimal (high-inductance and low-resistance) L_X is large. Confining L_X to smaller dimensions and cheaper materials ultimately sacrifices power for space and cost savings.

3.1.2 Transformer

A. **Ideal Transformer**

A transformer is nothing but two inductor coils that share the same magnetic space and have access to the same magnetic field. So when neglecting unintended parasitic effects, L_I and L_O in Fig. 3.3 hold the same E_M:

$$E_M = 0.5 L_I i_{LI}{}^2 = 0.5 L_O i_{LO}{}^2. \quad (3.5)$$

This means that the coil with the higher inductance conducts lower current. This is why L_I's current i_{LI} is the *transformer translation* k_L of L_O's current i_{LO} that L_O/L_I establish. When L_O is higher than L_I, i_{LO} is a reverse k_L fraction of i_{LI}:

$$\frac{i_{LI}}{i_{LO}} = \sqrt{\frac{L_O}{L_I}} \equiv k_L. \quad (3.6)$$

The ideal transformer has no resistive components. So P_R is negligible, which means the output receives the power that the input supplies:

Fig. 3.3 Ideal transformer

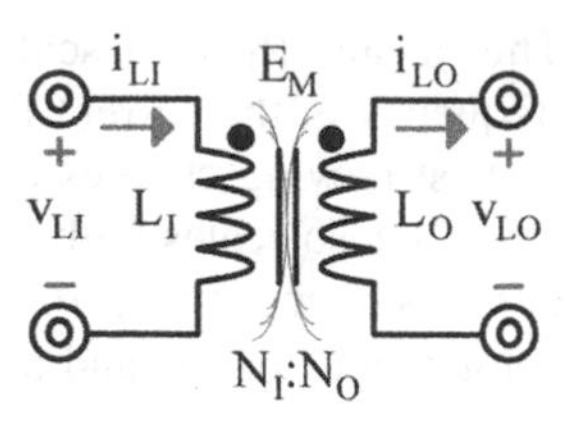

$$P_{IN} = i_{LI}v_{LI} = P_O = i_{LO}v_{LO}. \tag{3.7}$$

L_O's voltage v_{LO} is therefore the k_L translation that i_{LI}/i_{LO} and, ultimately, L_O/L_I determine. In short, i_{LI} is a k_L translation of i_{LO} and v_{LO} is a k_L translation of v_{LI}:

$$\frac{v_{LO}}{v_{LI}} = \frac{i_{LI}}{i_{LO}} = k_L = \sqrt{\frac{L_O}{L_I}} \approx \frac{N_O}{N_I}. \tag{3.8}$$

When L_I's and L_O's geometries match, k_L reduces to the ratio of the number of loops in L_O to those in L_I. This is the *turns ratio* to which literature refers when using N_O/N_I to quantify k_L.

Example 1: Derive an expression for P_{IN} when v_{IN} supplies L_I and R_{LD} loads L_O.

Solution:

$$P_{IN} = i_{LI}v_{IN} = i_{LO}k_Lv_{IN} = \left(\frac{v_{LO}}{R_{LD}}\right)k_Lv_{IN} = \left(\frac{v_{IN}k_L}{R_{LD}}\right)k_Lv_{IN}$$

$$= \frac{(v_{IN}k_L)^2}{R_{LD}} = \frac{v_{LO}{}^2}{R_{LD}} = P_{LD} = P_O$$

B. **Actual Transformer**

In practice, coupled inductors access a fraction of the magnetic field they share. When decomposed into the pieces that actually couple (L_I and L_O) and the ones that do not (L_I' and L_O') like Fig. 3.4 shows, only a k_{CI} fraction of the input couples to a k_{CO} fraction of the output:

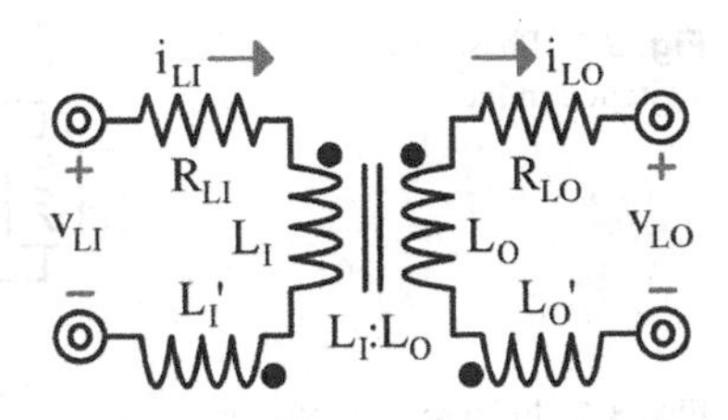

Fig. 3.4 Actual transformer

$$k_{CI} = \frac{L_I}{L_I + L_I'} \tag{3.9}$$

$$k_{CO} = \frac{L_O}{L_O + L_O'}. \tag{3.10}$$

This means that L_O avails a k_{CI} fraction of the energy that v_{LI} supplies. So L_I' reduces k_L to

$$k_L' \equiv \frac{v_{LO}}{v_{LI}} = \frac{v_{LI}k_{CI}k_L}{v_{LI}} = \left(\frac{L_I}{L_I + L_I'}\right)\sqrt{\frac{L_O}{L_I}}. \tag{3.11}$$

This also means that L_O' is a load to L_O. k_{CI} and k_{CO}, which set the *coupling factor* or *coupling coefficient* k_C of the transformer, are one in ideal transformers and less than one in practical realizations.

Since separating the coils decreases the fraction of inductance that couples, L_I and L_O and their resulting k_C fall with increasing *coil distance* d_X. Misalignment between the coils also reduces k_C. Coupling also depends on the geometry of the coils, so variations in d_X manifest in k_C in a variety of ways.

L_I and L_O are also resistive and, as a result, not lossless. So input and output resistances R_{LI} and R_{LO} further alter the k_C fraction that couples and reaches v_{LO}. Needless to say, better transformers *couple* more and *resist* less.

3.2 Switched Inductors

Switched inductors energize and drain in alternating phases of a switching cycle. An input source at v_{IN} first energizes a *switched inductor* L_X (in Fig. 3.5) across the *energize time* t_E. The energize voltage v_E should always include elements of the *input voltage* v_{IN} to receive P_{IN}. The load that the output v_O feeds then drains L_X across the *drain time* t_D. For this, the switching network connects L_X in such a way that elements of the *output voltage* v_O apply an opposing v_D across L_X. This way, L_X drains into v_O.

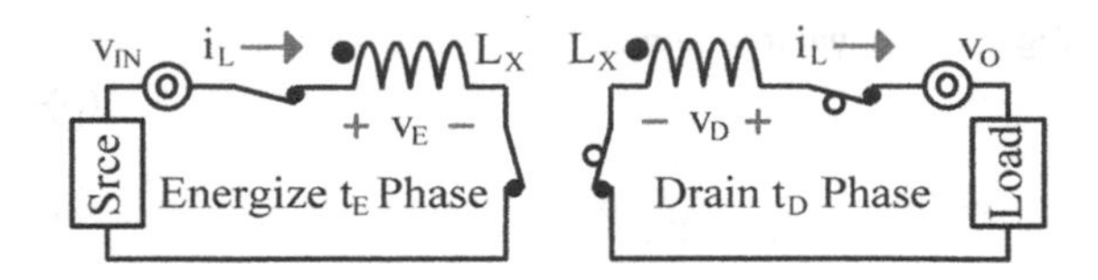

Fig. 3.5 Phases of the switched inductor

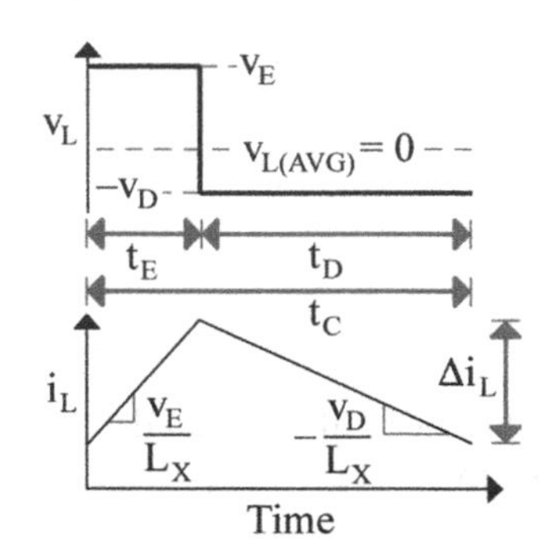

Fig. 3.6 Inductor waveforms in dc-supplied switched inductors

3.2.1 DC–DC Applications

Many consumer applications transfer power from static dc sources to loads that impose or require steady voltages. Conventional sources include *lithium-ion* (Li-ion), *nickel* (Ni), and lead-acid batteries and ac–dc rectifiers that convert dynamic ac sources to static dc outputs. Chargers recharge Li-ion and nickel batteries, voltage regulators feed systems that require steady supplies, and *light-emitting diode* (LED) drivers feed diodes whose voltages are, for the most part, steady.

All these outputs are essentially dc because the capacity (or equivalent capacitance) of batteries is usually high and feedback controllers in regulators and LED drivers steady their outputs. Although controllers cannot respond instantly, designers normally add capacitance to their outputs. So switched-inductor inputs and outputs in all these applications are, for all intents and purposes, nearly dc.

3.2.2 Inductor Current

Since v_{IN} and v_O in dc–dc applications are largely static, the i_L that v_E and v_D establish is a steady linear ramp di_L/dt_X:

$$\frac{di_L}{dt_X} = \frac{v_L}{L_X}. \tag{3.12}$$

So like Fig. 3.6 illustrates, v_E ramps i_L up linearly across t_E. The opposing voltage that v_D applies similarly ramps i_L down across t_D. t_E and t_D are opposite phases of the *conduction time* t_C. Energizing and draining L_X this way across t_C produces the triangular *current ripple* Δi_L shown.

3.2.3 Duty Cycle

The *energize duty cycle* d_E is the t_E fraction of t_C across which v_E energizes L_X:

$$d_E \equiv \frac{t_E}{t_C}. \tag{3.13}$$

The *drain duty cycle* d_D is the opposite: the t_D fraction of t_C across which v_D drains L_X:

$$d_D \equiv \frac{t_D}{t_C} = \frac{t_C - t_E}{t_C} = 1 - d_E. \tag{3.14}$$

But since t_D is the time that remains across t_C after t_E elapses, t_D is also $t_C - t_E$ and d_D is, in consequence, the complement that $1 - d_E$ sets.

Under static steady-state conditions, i_L in one cycle should rise as much as it falls. v_E across t_E therefore increases i_L by the same Δi_L that v_D across t_D decreases i_L:

$$\Delta i_L = \left(\frac{v_E}{L_X}\right) t_E = \left(\frac{v_D}{L_X}\right) t_D. \tag{3.15}$$

This means (*i*) $v_E t_E$'s and $v_D t_D$'s *volts–seconds products* match; (*ii*) time, duty-cycle, and reciprocal voltage ratios t_E/t_D, d_E/d_D, and v_D/v_E match:

$$\frac{t_E}{t_D} = \frac{d_E}{d_D} = \frac{d_E}{1 - d_E} = \frac{v_D}{v_E}; \tag{3.16}$$

and (*iii*) d_E is a v_D fraction of the combined v_E and v_D applied:

$$d_E = \frac{v_D}{v_E + v_D}. \tag{3.17}$$

A. **Ohmic Loss**

A lossless L_X delivers all the input energy it receives. In practice, however, resistances in the network burn energy E_R that L_X does not receive or deliver. So L_X requires longer t_E and higher d_E to energize.

R_L and *energize resistances* R_E drop voltages v_{RL} and v_{RE} that decrease v_E below its ideal level v_{EI}. Since i_L flows to v_O when L_X drains, R_L and *drain resistances* R_D drop voltages v_{RL} and v_{RD} that increase v_D over its ideal level v_{DI}. Reducing v_E extends the t_E that L_X needs to energize across Δi_L in Fig. 3.7 and raising v_D reduces the t_D that L_X needs to drain E_L. This rise in t_E and fall in t_D increase d_E to the extent that i_L, R_L, R_E, and R_D dictate:

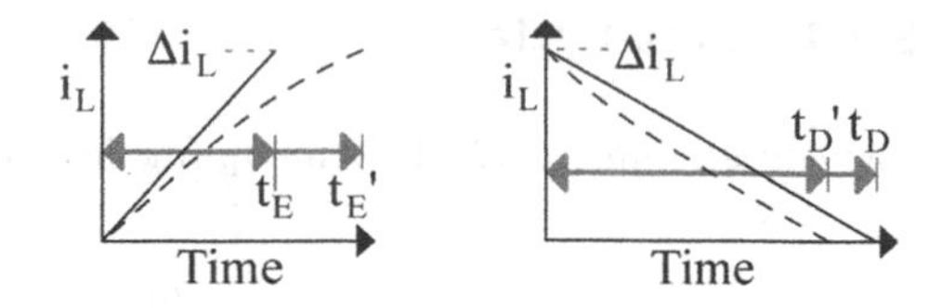

Fig. 3.7 Inductor current with Ohmic losses

$$\begin{aligned} d_E' &= \frac{v_D}{v_E + v_D} \\ &= \frac{v_{DI} + v_{RL} + v_{RD}}{(v_{EI} - v_{RL} - v_{RE}) + (v_{DI} + v_{RL} + v_{RD})} \\ &= \frac{v_{DI} + v_{RL} + v_{RD}}{(v_{EI} - v_{RE}) + (v_{DI} + v_{RD})} \\ &\approx \frac{v_{DI} + v_{RL} + v_{RD}}{v_{EI} + v_{DI}} > d_E. \end{aligned} \tag{3.18}$$

On average, R_L subtracts from v_E the same v_{RL} that R_L adds to v_D. Similarly, R_E and R_D subtract from v_E and add to v_D similar voltages when R_E and R_D are similar, which is not unlikely. But since d_E is a v_D fraction of v_E and v_D, v_{RL} and v_{RD} in v_D invariably raise d_E' over its ideal counterpart d_E.

In all, v_R's scale with i_L, oppose v_E, and reinforce v_D. i_L is increasingly less linear (and more parabolic) across t_E and t_D. In short, resistances distort i_L and increase d_E.

So when a feedback controller adjusts d_E so v_O nears a target, the resulting rise in d_E is a reflection of the Ohmic power lost in the switching network. Since v_{IN} is usually steady and v_{IN}-to-v_O translations need higher d_E's when resistances are present, resistances shift v_O from their ideal targets when controllers do not adjust d_E. In these cases, which are less typical, the shift in v_O reflects the power lost. v_O in simulations that exclude feedback controllers, for example, varies with R_L, R_E, and R_D and the i_L that sets their voltages.

3.2.4 Continuous Conduction

In *continuous-conduction mode* (CCM), L_X conducts continuously across time. In this mode, L_X's conduction period extends across the entire *switching period* t_{SW}. This way, i_L is never static, which is to say di_L/dt is never zero. In steady state, i_L's *peaks* and *valleys* $i_{L(HI)}$'s and $i_{L(LO)}$'s do not vary, so i_L ripples periodically like Fig. 3.8 shows. Since t_C is t_{SW} in CCM, d_E and d_D become

$$d_E\big|_{CCM} = \frac{t_E}{t_C}\bigg|_{CCM} = \frac{t_E}{t_{SW}} \tag{3.19}$$

and

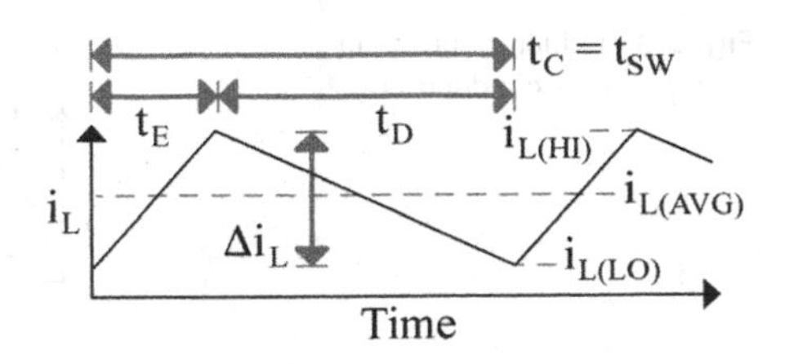

Fig. 3.8 Inductor current in continuous conduction

$$d_D\big|_{CCM} = \frac{t_D}{t_C}\bigg|_{CCM} = \frac{t_D}{t_{SW}} = 1 - d_E. \tag{3.20}$$

In this mode, i_L ripples about i_L's average $i_{L(AVG)}$, $i_{L(AVG)}$ is i_L's low $i_{L(LO)}$ plus half i_L's *CCM ripple* Δi_L, and $i_{L(AVG)}$ can be positive or negative, which happens when i_L reverses direction:

$$i_{L(AVG)}\big|_{CCM} = i_{L(LO)} + \Delta i_{L(AVG)}\big|_{CCM} = i_{L(LO)} + \frac{\Delta i_L}{2}. \tag{3.21}$$

Example 2: Determine t_E, t_D, and Δi_L in CCM when v_E is 2 V, v_D is 1 V, t_{SW} is 1 μs, and L_X is 10 μH.

Solution:

$$d_E = \frac{v_D}{v_E + v_D} = \frac{1}{2+1} = 33\%$$

$$\therefore \quad t_E = d_E t_C = d_E t_{SW} = (33\%)(1\mu) = 330 \text{ ns}$$

$$t_D = t_C - t_E = t_{SW} - t_E = 1\mu - 330\text{n} = 670 \text{ ns}$$

$$\Delta i_L = \left(\frac{v_E}{L_X}\right) t_E = \left(\frac{2}{10\mu}\right)(330\text{n}) = 66 \text{ mA}$$

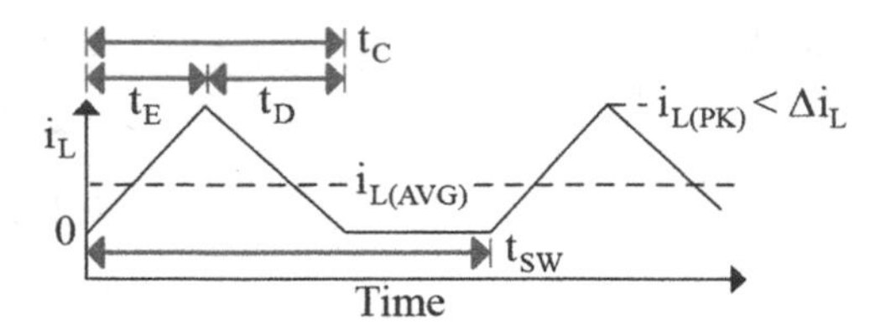

Fig. 3.9 Inductor current in discontinuous conduction

3.2.5 Discontinuous Conduction

In *discontinuous-conduction mode* (DCM), L_X energizes and depletes before t_{SW} ends. This way, like Fig. 3.9 shows, i_L climbs across t_E to *peak inductor current* $i_{L(PK)}$, falls across t_D to zero, and remains zero until another cycle begins. The dead period between t_C's is the discontinuity in conduction that characterizes DCM.

In other words, L_X's conduction period does not extend across the switching cycle. Since t_C is not t_{SW}, d_E and d_D do not relate to t_{SW} like they do in CCM. d_E and d_D should therefore remain in their more primitive forms: t_E/t_C and t_D/t_C.

CCM borders DCM when i_L reaches zero at the end of the switching cycle. When this happens, $i_{L(LO)}$ is zero, $i_{L(HI)}$ is Δi_L, and $i_{L(AVG)}$ is half Δi_L. L_X enters DCM just below this level, when $i_{L(AVG)}$ is less than half the CCM ripple Δi_L:

$$i_{L(AVG)}\big|_{DCM} = \Delta i_{L(AVG)}\big|_{DCM} = \left(\frac{i_{L(PK)}}{2}\right)\left(\frac{t_C}{t_{SW}}\right) < \frac{\Delta i_L}{2}. \tag{3.22}$$

In DCM, $i_{L(PK)}$ is lower than Δi_L, and $i_{L(AVG)}$ across t_C is half $i_{L(PK)}$ and across t_{SW} is a t_C fraction of t_{SW} lower.

Example 3: Determine t_E, t_D, and Δi_L when v_E is 2 V, v_D is 1 V, $i_{L(AVG)}$ is 25 mA, i_L valleys to 0 mA, t_{SW} is 1 μs, and L_X is 10 μH.

Solution:

$$d_E = 33\% \text{ and } \Delta i_L = 66 \text{ mA in CCM from previous example}$$

$$i_{L(AVG)} = 25\,\text{mA} < i_{L(LO)} + \frac{\Delta i_L}{2}\bigg|_{CCM} = 0 + \frac{66\text{m}}{2} = 33\,\text{mA} \quad \therefore \quad \text{DCM}$$

$$i_{L(AVG)} = \left(\frac{i_{L(PK)}}{2}\right)\left(\frac{t_C}{t_{SW}}\right)$$

$$i_{L(PK)} = \left(\frac{v_E}{L_X}\right)t_E = \left(\frac{v_E}{L_X}\right)d_E t_C$$

$$\rightarrow t_C = \sqrt{2i_{L(AVG)}\left(\frac{L_X}{v_E}\right)\left(\frac{t_{SW}}{d_E}\right)} = \sqrt{2(25m)\left(\frac{10\mu}{2}\right)\left(\frac{1\mu}{33\%}\right)} = 870 \text{ ns}$$

$$t_E = d_E t_C = (33\%)(870n) = 290 \text{ ns}$$

$$t_D = t_C - t_E = 870n - 290n = 580 \text{ ns}$$

$$i_{L(PK)} = \left(\frac{v_E}{L_X}\right)t_E = \left(\frac{2}{10\mu}\right)(290n) = 58 \text{ mA}$$

Pseudo-discontinuous-conduction mode (PDCM) mimics DCM. L_X in these cases energizes and drains to a minimum static level before t_{SW} ends. So like in DCM, i_L rises and falls across t_C before t_{SW} lapses. But unlike DCM, i_L falls to a *minimum inductor current* $i_{L(MIN)}$ in Fig. 3.10 that is not zero. This way, L_X energizes, drains, and "holds" until the next cycle begins.

This means that di_L/dt is greater than, less than, and equal to zero across t_E, t_D, and the dead period between t_C's, respectively. For i_L to remain static between conduction periods this way, v_L across L_X must be zero. L_X enters this mode of operation when $i_{L(AVG)}$ is over this $i_{L(MIN)}$ by less than half the CCM ripple Δi_L:

$$\begin{aligned} i_{L(AVG)}\big|_{PDCM} &= i_{L(MIN)} + \Delta i_{L(AVG)}\big|_{DCM} \\ &= i_{L(MIN)} + \left(\frac{i_{L(PK)}}{2}\right)\left(\frac{t_C}{t_{SW}}\right) < i_{L(MIN)} + \frac{\Delta i_L}{2}. \end{aligned} \tag{3.23}$$

$i_{L(AVG)}$ in PDCM is over $i_{L(MIN)}$ by a t_C/t_{SW} fraction of half i_L's *peak variation* $i_{L(PK)}$.

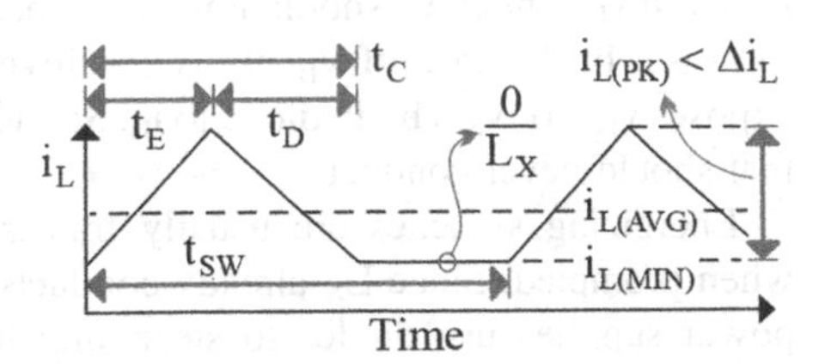

Fig. 3.10 Inductor current in pseudo-discontinuous-conduction mode

3.2.6 CMOS Implementations

Switches in a *complementary metal–oxide–semiconductor* (CMOS) implementation are *MOS field-effect transistors* (MOSFETs). Replacing each switch with the parallel combination of complementary N- and P-channel MOSFETs is the most straightforward translation, though not always the most effective one. Available *gate drive* v_{GST} or $v_{GS} - v_T$ and the *sheet resistivity* R_{SH} that v_{GST} establishes can dictate which type of transistor is more efficient. This is why switches in many applications are N- *or* P-channel transistors, not the parallel combination of N- *and* P-channel devices.

Although *electron mobility* μ_N is usually two to three times greater than *hole mobility* μ_P, v_{GST} can outweigh that difference in R_{SH}:

$$R_{SH} = R_{CH}\left(\frac{W_{CH}}{L_{CH}}\right) = \frac{v_{DS}}{i_{D(TRI)}}\left(\frac{W_{CH}}{L_{CH}}\right)\Bigg|_{v_{DS}<<v_{DS(SAT)}} \approx \frac{1}{K_{N/P}'|v_{GST}|} = \frac{1}{C_{OX}''\mu_{N/P}|v_{GST}|}, \quad (3.24)$$

where R_{CH} is the *channel resistance*, W_{CH} and L_{CH} are the *width* and *length* of the *channel*, $K_{N/P}'$ is the *transconductance parameter* that accounts for $\mu_{N/P}$, and $i_{D(TRI)}$ is the drain current in *triode*, when v_{DS} is much lower than the saturation level $v_{DS(SAT)}$. Still, NFETs are naturally less resistive than PFETs under equivalent v_{GST}'s and *silicon areas* A_{Si}. A PFET is therefore less resistive only when its v_{SGT} is more than μ_N/μ_P times (or 2× to 3×) greater than the v_{GST} of an NFET.

The *source voltage* v_S that the *input* and *output switching nodes* v_{SWI} and v_{SWO} set is key to determining v_{GS} in v_{GST}. This v_S is also the voltage that the *drain voltage* v_D approaches (within 10–200 mV) after the MOSFET closes. v_{GS} is usually higher than v_{SG} when v_S is low because a higher *gate voltage* v_G is more likely than a lower v_G. v_{SGT} is similarly higher than v_{GST} when v_S is high because a lower v_G is more likely than a higher v_G. So generally, NFETs are good *low-side switches* and PFETs are good *high-side switches*. When v_S is neither high nor low, *threshold voltage* v_T plays a more decisive role in v_{GST} and v_{SGT}.

After using v_{GST} and v_{SGT} to determine which FET is least resistive, the *body terminal* v_B is next. v_B should not "float" because disconnected nodes are vulnerable to noise. In the case of v_B, v_T is sensitive to v_B. But connecting v_B to v_S or v_D exposes v_D's or v_S's body diode to i_L. So the connection should short the body diode that should never conduct.

Energizing switches are usually transistors because L_X should energize only when prompted. Since L_X already conducts current when energized, *asynchronous* power supplies use diodes to steer and drain i_L into v_O. The advantage of this approach is that diodes do not require a synchronizing control signal to switch. *Synchronous* power supplies, however, use transistors only, so they need a synchronizing signal to close and open all MOSFETs, including the ones that drain L_X.

Since energizing events should be synchronized to commands, body diodes should not energize L_X asynchronously. So the body connections of energizing switches should short the body diodes that would conduct *input current* i_{IN}. Drain switches, on the other hand, should block reverse *output current* i_O. Their body connections should therefore short the body diodes that would conduct reverse i_O.

Synchronous converters are popular in many applications because diodes often consume more power with the 600–800 mV that they drop than MOSFETs with the 10–200 mV that MOSFETs drop. But since gate signals can easily crisscross, adjacent MOSFETs can inadvertently short their inputs together. Controllers must therefore insert *dead time* t_{DT} between the conduction periods of adjacent switches. Ironically, diodes must conduct L_X's current across these dead times because all MOSFETs are off across these times.

3.2.7 Design Limits

i_L is linear across t_E and t_D when P_R is very low. This happens when *series resistances* R_{SER} drop a small fraction of v_E and v_D. So the $v_{R(MAX)}$ that $i_{L(HI)}$ into R_{SER} produces should be less than a small fraction of the lowest v_L:

$$v_{R(MAX)} = i_{L(HI)} R_{SER} = \left(i_{L(AVG)} + \frac{\Delta i_L}{2} \right) R_{SER} << v_{L(MIN)}. \tag{3.25}$$

Since R_{SER} includes the resistances of one or two switches and L_X, this $v_{R(MAX)}$ limits the R_{CH}'s of the switches and the maximum R_L allowed.

Feeding L_X's triangular i_L into C_O produces an *output ripple voltage* Δv_O that also deviates v_O in v_D from its intended setting. C_O should therefore be high enough to keep Δv_O within a small fraction of v_O:

$$\Delta v_{O(MAX)} << v_O. \tag{3.26}$$

C_O in voltage regulators is on the order of microfarads primarily for this reason: to keep i_L and wide and sudden *load dumps* $\Delta i_{LD(MAX)}$ from pulling v_O beyond this $\Delta v_{O(MAX)}$ limit. C_O for batteries is so high (often on the order of millifarads) that Δv_O in chargers is normally very low.

When interconnected, inductors and capacitors exchange energy every quarter cycle of their *LC resonant period* t_{LC}, which is 2π radians per cycle times their corresponding *LC time constant* τ_{LC}:

$$t_{LC} = 2\pi \tau_{LC} = 2\pi \sqrt{L_X C_O} = \frac{1}{f_{LC}}. \tag{3.27}$$

So their voltages and currents oscillate at the *resonant frequency* f_{LC} that t_{LC} sets. Since L_X in switched inductors drains into v_O, L_X connects to C_O no less than a t_D fraction of t_{SW}. If t_{LC} is shorter than t_{SW}, L_X and C_O can exchange energy a few times before t_{SW} ends. To avoid these oscillations, t_{SW} should be much shorter than t_{LC}'s quarter cycle:

$$t_{SW} = \frac{1}{f_{SW}} << \frac{t_{LC}}{4} = 25\%t_{LC} = \frac{1}{4f_{LC}}. \tag{3.28}$$

This means the *switching frequency* f_{SW} should be much higher than four times f_{LC}.

Example 4: Determine 10% limits for R_{CH} and f_{SW} when two switches conduct i_L, v_D is 1 V, v_E is 2 V, L_X is 10 μH, R_L is 200 mΩ, C_O is 10 μF, $i_{L(AVG)}$ is 70 mA, and Δi_L is 30 mA.

Solution:

$$R_{SER} \le \frac{10\%v_{L(MIN)}}{i_{L(HI)}} = \frac{10\%v_D}{i_{L(AVG)} + 0.5\Delta i_L} = \frac{10\%(1)}{70m + 0.5(30m)} = 1.2\ \Omega$$

$$\rightarrow \quad R_{CH} \le \frac{R_{SER} - R_L}{2} = \frac{1.2 - 200m}{2} = 500\ m\Omega$$

$$t_{SW} \le 2.5\%t_{LC} = 2.5\%(2\pi)\sqrt{L_X C_O} = 2.5\%(2\pi)\sqrt{(10\mu)(10\mu)} = 1.6\ \mu s$$

$$\rightarrow \quad f_{SW} \ge 620\ kHz$$

3.3 Buck–Boost

3.3.1 Ideal Buck–Boost

A. Power Stage

As the name implies, *buck–boost* switched inductors *buck* or *boost* v_{IN} to a *lower* or *higher* v_O. *Input* and *ground energize switches* S_{EI} and S_{EG} in the buck–boost of Fig. 3.11 draw input power by connecting L_X across v_{IN} and ground. This way, with a positive v_E that is equal to v_{IN}, L_X energizes.

After energizing, S_{EI} and S_{EG} open and *ground* and *output drain switches* S_{DG} and S_{DO} close to connect L_X to v_O. v_L's polarity reverses this way to $-v_O$, so L_X drains into v_O. v_L in Fig. 3.6 therefore swings between $+v_E$'s v_{IN} and $-v_D$'s $-v_O$ and i_L ramps up with $+v_E$'s v_{IN} and down with $-v_D$'s $-v_O$. In short, v_E is v_{IN} and v_D is v_O.

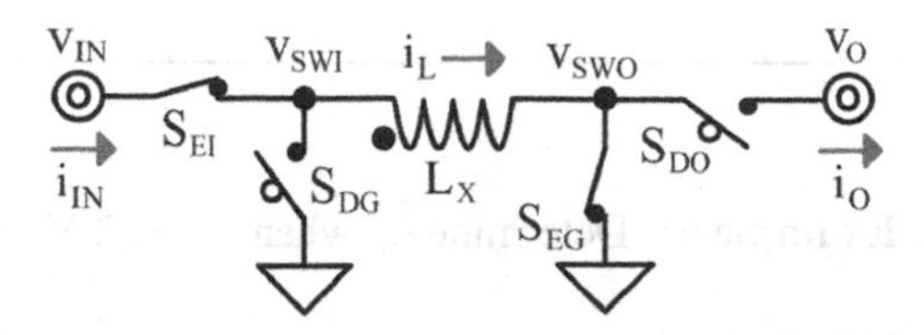

Fig. 3.11 Ideal buck–boost

B. **Duty-Cycle Translation**

In steady state, L_X's average voltage is zero, which is why engineers often say inductors are "dc shorts." The *input* and *output switching voltages* v_{SWI} and v_{SWO} in Fig. 3.11 are therefore, on average, equal. Since v_{IN} connects to v_{SWI} a t_E fraction of t_C and v_O connects to v_{SWO} a t_D fraction of t_C, their averages are matching duty-cycled fractions of v_{IN} and v_O:

$$\begin{aligned} v_{SW(AVG)} &= v_{SWI(AVG)} = v_E\left(\frac{t_E}{t_C}\right) = v_{IN}d_E \\ &= v_{SWO(AVG)} = v_D\left(\frac{t_D}{t_C}\right) = v_Od_D = v_O(1 - d_E). \end{aligned} \tag{3.29}$$

This means that v_O is a duty-cycled scalar translation of v_{IN}:

$$v_O = v_{IN}\left(\frac{d_E}{d_D}\right) = v_{IN}\left(\frac{d_E}{1 - d_E}\right). \tag{3.30}$$

And since d_D is $1 - d_E$, d_E is a v_O fraction of $v_{IN} + v_O$:

$$d_E = \frac{v_D}{v_E + v_D} = \frac{v_O}{v_{IN} + v_O}. \tag{3.31}$$

When d_E is greater than 50%, the average *switching voltage* $v_{SW(AVG)}$ is more than half of v_{IN}. But since d_D is d_E's opposite fraction (less than 50%), $v_{SW(AVG)}$ is also less than half of v_O. With this d_E, only a v_O that is higher than v_{IN} can establish a $v_{SWO(AVG)}$ that matches $v_{SWI(AVG)}$. The opposite is true when d_E is less than 50%: $v_{SW(AVG)}$ is less than half of v_{IN} and more than half of v_O, so v_O is lower than v_{IN}. In other words, L_X bucks when d_E is less than 50% (and d_D is greater than 50%) and boosts when d_E is greater than 50% (and d_D is less than 50%).

Example 5: Determine d_E when v_{IN} is 2 V and v_O is 1 V.

Solution:

$$d_E = \frac{v_D}{v_E + v_D} = \frac{v_O}{v_{IN} + v_O} = \frac{1}{2+1} = 33\%$$

Example 6: Determine d_E when v_{IN} is 2 V and v_O is 4 V.

Solution:

$$d_E = \frac{v_D}{v_E + v_D} = \frac{v_O}{v_{IN} + v_O} = \frac{4}{2+4} = 67\%$$

Explore with SPICE:
See Appendix A for notes on SPICE simulations.

```
* Ideal Buck-Boost in CCM
vde de 0 dc=0 pulse 0 1 50n 1n 1n 670n 1u
vin vin 0 dc=2
sei vin vswi de 0 sw1v
ddg 0 vswi idiode
lx vswi vswo 10u
seg vswo 0 de 0 sw1v
ddo vswo vo idiode
co vo 0 5u
ro vo 0 10
.lib lib.txt
.tran 700u
.end
```

Tip: Plot v(vo), i(Lx), v(vswi), and v(vswo) and view across 700 μs and from 695 to 700 μs.

C. Power

Since v_{IN} connects to L_X a t_E fraction of t_C, v_{IN} supplies a d_E fraction of $i_{L(AVG)}$. Similarly, v_O receives a d_D fraction of $i_{L(AVG)}$ because buck–boosts duty-cycle L_X into v_O across the opposing t_D fraction of t_C. $i_{IN(AVG)}$ and $i_{O(AVG)}$ in P_{IN} and P_O are therefore d_E and d_D fractions of $i_{L(AVG)}$, which also means $i_{L(AVG)}$ is a reverse d_D translation of $i_{O(AVG)}$:

$$\begin{aligned} P_{IN} = v_{IN} i_{IN(AVG)} &= v_{IN} i_{L(AVG)} d_{IN} = v_{IN} i_{L(AVG)} d_E \\ &= v_{IN} \left(\frac{i_{O(AVG)}}{d_O} \right) d_E = v_{IN} \left(\frac{i_{O(AVG)}}{d_D} \right) d_E \end{aligned} \tag{3.32}$$

$$P_O = v_O i_{O(AVG)} = v_O i_{L(AVG)} d_O = v_O i_{L(AVG)} d_D. \tag{3.33}$$

The *input* and *output duty cycles* d_{IN} and d_O that d_E and d_D set split opposite fractions of $i_{L(AVG)}$ between $i_{IN(AVG)}$ and $i_{O(AVG)}$. In the ideal buck–boost, P_O equals P_{IN} because ideal switches do not burn power (and $v_E d_E$'s $v_{IN} d_E$ matches $v_D d_D$'s $v_O d_D$). In other words, v_O receives all the energy v_{IN} supplies with L_X.

3.3.2 Asynchronous Buck–Boost

A. Power Stage

L_X in the asynchronous buck–boost energizes with transistors and drains with diodes. The *input* and *ground* energize switches S_{EI} and S_{EG} in Fig. 3.11 are therefore *MOSFETS* M_{EI} and M_{EG} in the asynchronous buck–boost in Fig. 3.12. The connectivity of the switches and the gate drive that the circuit can establish ultimately dictate which type of MOSFET is more effective for M_{EI} and M_{EG}.

When closed, M_{EI}'s source and drain voltages v_S and v_D are close to v_{IN}. Since v_O is not always greater than v_{IN}, M_{EI}'s gate voltage v_G can swing reliably only between v_{IN} and ground. Under these conditions, the maximum gate drive for an NFET would be v_{GST}, $v_G - v_S - v_{TN}$, $v_{IN} - v_{IN} - v_{TN}$, or just $-v_{TN}$, which is too low to close an NFET. The maximum gate drive for a PFET would be v_{SGT}, $v_S - v_G - |v_{TP}|$, $v_{IN} - 0 - |v_{TP}|$, or $v_{IN} - |v_{TP}|$, which is greater and more likely to close a PFET. This is why M_{EI} is usually a PFET, because a high v_{SWI} calls for a *high-side switch*.

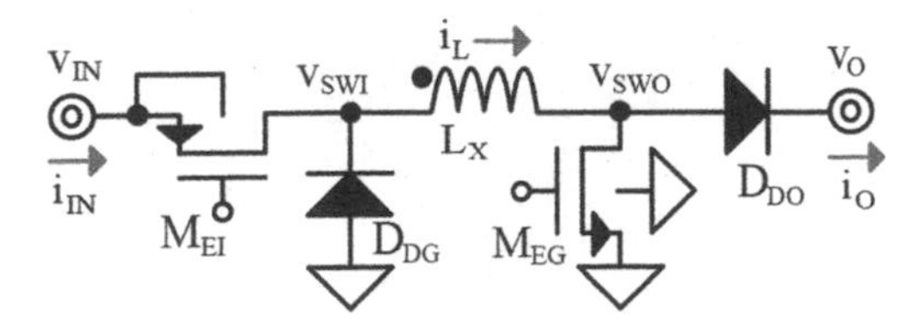

Fig. 3.12 Asynchronous buck–boost

Since PMOS source terminals receive current, v_{IN} connects to M_{EI}'s source and M_{EI}'s source arrow points into the transistor (into the channel through which i_L flows). To ensure M_{EI}'s body diode does not inadvertently energize L_X by conducting i_{IN} when v_{SWI} falls, v_B should short the body diode that conducts i_{IN}. v_{IN}, to which the P-type source connects, should therefore connect to M_{EI}'s N-type body.

When closed, M_{EG}'s source and drain voltages v_S and v_D are close to ground. The only type of MOSFET that can close when v_S is the most negative potential like this is an N-channel transistor because P-channel devices need a gate voltage that is lower than v_S to close. M_{EG} is therefore an NFET: a *low-side switch* with a source that connects and points to ground (away from the channel through which i_L flows) because N-channel source terminals output current.

The P-type body terminal connects to the N-type source (to ground) to block the diode that would otherwise conduct i_{IN} when M_{EG} is off. NFETs in P-type substrates normally sit on the substrate, so the substrate is M_{EG}'s body. To isolate all devices in the *integrated circuit* (IC), P-type substrates normally connect to the most negative potential, which is usually ground.

After M_{EI} and M_{EG} energize L_X, i_L flows to the right. The *diodes* that implement the *ground* and *output drain* switches S_{DG} and S_{DO} must therefore derive i_L from ground and direct it to v_O. This is why D_{DG} and D_{DO} in Fig. 3.12 point up and to the right: to draw and steer i_L from ground to v_O.

B. **Duty-Cycle Translation**

Since D_{DG} drops a diode voltage below ground and D_{DO} a diode voltage above v_O, L_X drains with v_O plus two diode voltages. This higher v_D drains L_X faster than in the ideal buck–boost, so t_D and t_C are shorter, and as a result, t_E is a larger fraction of t_C. In other words, d_E in the asynchronous buck–boost is higher than in the ideal buck–boost.

v_{SWI} swings between v_{IN} and $-v_{DG}$ and v_{SWO} between zero and $v_O + v_{DO}$, like Fig. 3.13 illustrates. v_{SWI} connects to v_{IN} a t_E fraction of t_C or d_E' and to $-v_{DG}$ a t_D fraction d_D', so $v_{SWI(AVG)}$ is

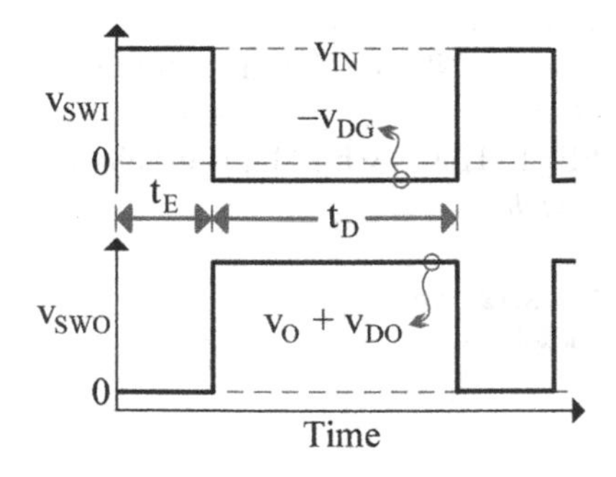

Fig. 3.13 Asynchronous buck–boost voltages

$$v_{SWI(AVG)} \approx v_{IN}d_E' - v_{DG}d_D' = v_{IN}d_E' - v_{DG}(1 - d_E'). \quad (3.34)$$

v_{SWO} connects to zero a t_E fraction d_E' and to $v_O + v_{DG}$ a t_D fraction d_D', so $v_{SWO(AVG)}$ is

$$v_{SWO(AVG)} \approx (0)d_E' + (v_O + v_{DO})d_D' = (v_O + v_{DO})(1 - d_E'). \quad (3.35)$$

Since these averages match, d_E'/d_D' is higher than the ideal d_E/d_D by the fraction of v_{IN} that two diode voltages represent:

$$\frac{d_E'}{d_D'} = \frac{v_D}{v_E} \approx \frac{v_O + v_{DG} + v_{DO}}{v_{IN}} = \frac{v_O}{v_{IN}} + \frac{v_{DG} + v_{DO}}{v_{IN}}. \quad (3.36)$$

The diodes raise the buck-to-boost transition point of d_E'/d_D' (when v_O equals v_{IN}) from v_O/v_{IN}'s one (when d_E and d_D in the ideal case are 50%) to one plus two diode fractions of v_{IN}. Since d_D' is $1 - d_E'$, the diodes raise v_D's v_O in the numerator of d_E' by a larger fraction than in the denominator that v_E's v_{IN} and v_D's v_O set:

$$d_E' = \frac{v_D}{v_E + v_D} \approx \frac{v_O + v_{DG} + v_{DO}}{v_{IN} + v_O + v_{DG} + v_{DO}} > d_E. \quad (3.37)$$

In short, the effect of the diodes is to increase d_E to d_E'.

Example 7: Determine d_E' when v_{IN} is 2 V, v_O is 4 V, and D_{DG} and D_{DO} drop 800 mV.

Solution:

$$d_E' = \frac{v_D}{v_E + v_D} \approx \frac{v_O + v_{DO} + v_{DG}}{v_{IN} + v_O + v_{DG} + v_{DO}} = \frac{4 + 800m + 800m}{2 + 4 + 800m + 800m} = 74\%$$

Note 1: d_E' is higher than d_E in the ideal example because D_{DG} and D_{DO} increase L_X's drain voltage. So t_D shortens and t_E's fraction of t_C's $t_D + t_E$ increases.

Explore with SPICE:
See Appendix A for notes on SPICE simulations.

```
* Asynchronous Buck-Boost in CCM
vge vge 0 dc=0 pulse 0 2 50n 1n 1n 740n 1u
vgeb vgeb 0 dc=2 pulse 2 0 50n 1n 1n 740n 1u
vin vin 0 dc=2
mei vswi vgeb vin vin pmos0 w=300m l=250n
ddg 0 vswi diode1
lx vswi vswo 10u
meg vswo vge 0 0 nmos0 w=100m l=250n
ddo vswo vo diode1
co vo 0 5u
ro vo 0 10
.lib lib.txt
.tran 700u
.end
```

Tip: Plot v(vo), i(Lx), v(vswi), and v(vswo) and view across 700 μs and from 695 to 700 μs.

C. Power

With a higher d_E and, as a result, a lower d_D, $i_{L(AVG)}$ is a higher reverse d_D translation of $i_{O(AVG)}$ and $i_{IN(AVG)}$ is a higher d_E fraction of $i_{L(AVG)}$. So for the same i_O and v_O, v_{IN} supplies more i_{IN}. In other words, the asynchronous buck–boost draws more P_{IN} than the ideal buck–boost when supplying the same P_O. In short, P_{IN} is higher than P_O.

But since $v_E d_E$ matches $v_D d_D$, v_E supplies $d_E i_{L(AVG)}$, and v_D receives $d_D i_{L(AVG)}$, v_E's v_{IN} still outputs the power v_D consumes. v_D, however, is v_{DG} plus v_{DO} over v_O, not v_O like in the ideal case. v_O therefore loses (to D_{DG} and D_{DO}) two diode fractions of the *inductor power* P_L that L_X and v_{IN} deliver:

$$\begin{aligned}
P_{DD} &= i_{L(AVG)}(v_{DG} + v_{DO})d_D \\
&= \left(\frac{i_{IN(AVG)}}{d_E}\right)(v_{DG} + v_{DO})\left(\frac{v_D}{v_O + v_{DG} + v_{DO}}\right)d_D \\
&= \left(\frac{i_{IN(AVG)}}{d_E}\right)\left(\frac{v_{DG} + v_{DO}}{v_O + v_{DG} + v_{DO}}\right)v_E d_E \\
&= \left(\frac{v_{DG} + v_{DO}}{v_O + v_{DG} + v_{DO}}\right)P_L \approx \left(\frac{v_{DG} + v_{DO}}{v_O + v_{DG} + v_{DO}}\right)P_{IN},
\end{aligned} \tag{3.38}$$

where $i_{L(AVG)}(v_{DG} + v_{DO})$ is the average power D_{DG} and D_{DO} consume across t_D, $i_{L(AVG)}(v_{DG} + v_{DO})d_D$ is the average they consume across t_C, and P_{IN} is nearly P_L when P_{DD} is a small fraction of P_{IN}, which happens when v_E and v_D are much higher than v_{DG} and v_{DO}.

This *diode drain power* P_{DD} is what the buck–boost sacrifices to S_{DG} and S_{DO} when using diodes. S_{EI} and S_{EG} also consume power, but much less than D_{DG} and D_{DO}. This is because the voltages M_{EI} and M_{DG} drop are usually much lower.

D. Conduction Modes

In continuous conduction, i_L rises and falls about the $i_{L(AVG)}$ that keeps i_L above zero, like the thinner traces of i_L in Fig. 3.14 illustrate. As the controller adjusts how much current v_{IN} supplies or v_O pulls, $i_{L(AVG)}$ shifts. i_L reaches zero at the end of t_{SW} when $i_{L(AVG)}$ is half the CCM ripple Δi_L and before t_{SW} ends when $i_{L(AVG)}$ is below that level. Since the diodes cannot conduct reverse current, i_L reaches and remains near zero until the next t_{SW}. L_X is in discontinuous conduction this way.

Just before the diodes stop conducting (toward the end of t_D), v_{SWI} is a diode voltage below ground and v_{SWO} is a diode voltage above v_O. Parasitic capacitances at the switching nodes (C_{SWI} and C_{SWO} in Fig. 3.15) therefore hold energy when L_X depletes: $0.5C_{SWI}v_{DG}^2$ and $0.5C_{SWO}(v_O + v_{DO})^2$. So when all switches open (at the end of t_D) before the t_{SW} cycle elapses, v_{SWI} and v_{SWO} impress $v_O + v_{DG} + v_{DO}$ across L_X. This first drains C_{SWO} and C_{SWI} into L_X, and after v_{SWI} reaches zero, drains C_{SWO} into L_X and C_{SWI}. v_{SWO} therefore drops as v_{SWI} climbs. v_L then reverses when v_{SWO} crosses and falls below v_{SWI}. So past this point, C_{SWO} and L_X drain into C_{SWI}.

After L_X depletes, v_{SWI} is greater than v_{SWO}. L_X therefore draws a current from C_{SWI} that drains C_{SWI} and energizes L_X and C_{SWO}. L_X stops energizing when v_{SWO} and v_{SWI} criss-cross. i_L nevertheless continues to flow to, this time, drain C_{SWI} and L_X into C_{SWO}. When L_X depletes, v_{SWO} is again greater than v_{SWI}, so the entire sequence repeats.

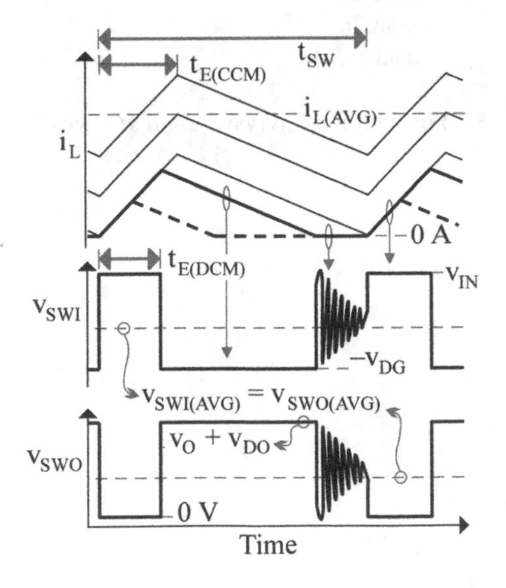

Fig. 3.14 Discontinuous-conduction waveforms

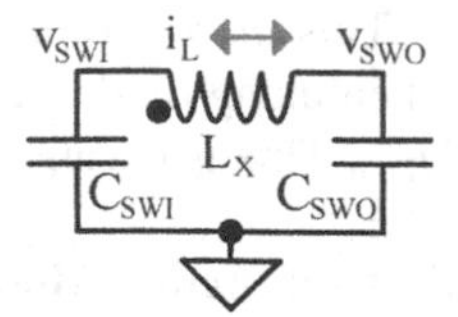

Fig. 3.15 Drained and disconnected asynchronous buck–boost inductor

L_X and the capacitors exchange energy this way until series resistances burn the energy or a new t_{SW} begins. This is why v_{SWI} and v_{SWO} in Fig. 3.14 oscillate about their averages when i_L is close to zero. These damped oscillations are characteristic of DCM operation.

Explore with SPICE:
See Appendix A for notes on SPICE simulations.

```
* Asynchronous Buck-Boost in DCM
vge vge 0 dc=0 pulse 0 2 0 1n 1n 100n 1u
vgeb vgeb 0 dc=2 pulse 2 0 0 1n 1n 100n 1u
vin vin 0 dc=2
mei vswi vgeb vin vin pmos0 w=300m l=250n
ddg 0 vswi diode1
lx vswi vl 10u
rl vl vswo 25
meg vswo vge 0 0 nmos0 w=100m l=250n
ddo vswo vo diode1
vo vo 0 dc=4
.lib lib.txt
.tran 2u
.end
```

Tip: Plot i(Lx), v(vswi), and v(vswo).

3.3.3 Synchronous Buck–Boost

A. **Power Stage**

The difference between asynchronous and synchronous power supplies is what drains L_X: diodes in one and transistors in the other. So the FETs that energize the asynchronous buck–boost and implement S_{EI} and S_{EG} in the ideal case also energize the synchronous counterparts in Fig. 3.16: M_{EI} and M_{EG}. The difference in Fig. 3.16 is that *ground* and *output drain MOSFETs* M_{DG} and M_{DO} drain L_X, which is the function of S_{DG} and S_{DO} and D_{DG} and D_{DO} in the ideal and asynchronous cases. The connectivity of the switches and the gate drive v_{GST} available ultimately dictate which type of MOSFET is more effective for M_{DG} and M_{DO}.

Since digital gate signals can crisscross, input and output power FETs can momentarily short v_{IN} and v_O to ground. To avoid this, the controller inserts dead times t_{DT} between the conduction periods of adjacent switches: between M_{EI}'s and M_{DG}'s and between M_{EG}'s and M_{DO}'s. Body diodes must therefore connect in such a way that they not only block reverse inductor current but, if needed, also steer i_L across these t_{DT}'s.

When closed, M_{DG}'s source and drain voltages v_S and v_D are close to ground. The only type of MOSFET that can close when v_S is the most negative potential like this is an N-channel transistor because P-channel devices need a gate voltage that is lower than v_S to close. M_{DG} is therefore a low-side NFET with a source that connects and points to v_{SWI} (away from the channel through which i_L flows) because N-channel source terminals output current.

The P-type body terminal connects to the N-type drain (to ground) to ensure M_{DG}'s body diode does not conduct reverse i_L. This terminal is also often the substrate because NFETs in P-type substrates normally sit directly over the substrate. This connection also ensures the body diode can steer i_L across t_{DT}'s.

When closed, M_{DO}'s source and drain voltages v_S and v_D are close to v_O. Since v_{IN} is not always greater than v_O, M_{DO}'s gate voltage v_G can swing reliably only between v_O and ground. Under these conditions, the maximum gate drive for an NFET would be v_{GST}, $v_G - v_S - v_{TN}$, $v_O - v_O - v_{TN}$, or just $-v_{TN}$, which is too low to close an NFET. The maximum gate drive for a PFET would be v_{SGT}, $v_S - v_G - |v_{TP}|$, $v_O - 0 - |v_{TP}|$, or $v_O - |v_{TP}|$, which is greater and more likely to close the switch. This is why M_{DO} is oftentimes a PFET, because a high v_{SWO} calls for a high-side switch.

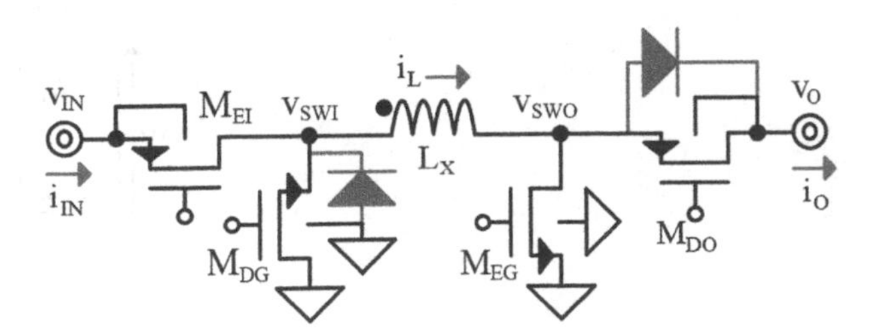

Fig. 3.16 Synchronous buck–boost

Since PMOS source terminals receive current, L_X feeds M_{DO}'s source and M_{DO}'s source arrow points into the transistor (into the channel through which i_L flows). To ensure M_{DO}'s body diode does not conduct reverse i_O when v_{SWO} falls, v_B should short the body diode that would conduct reverse i_O. v_O, to which the P-type drain connects, should therefore connect to M_{DO}'s N-type body. This connection also allows the surviving body diode to steer dead-time current into v_O.

B. Duty-Cycle Translation

The difference between asynchronous and synchronous operation across t_C is that M_{DG} and M_{DO} drop lower voltages than D_{DG} and D_{DO}. So v_{SWI} in Fig. 3.17 is close to zero and v_{SWO} close to v_O when M_{DG} and M_{DO} close across t_D. But since M_{DG} and M_{DO} open across the dead-time portions of t_D, D_{DG} drops v_{SWI} to $-v_{DG}$ and D_{DO} raises v_{SWO} to $v_O + v_{DO}$ across t_{DT}'s.

This means that L_X drains two t_{DT} fractions of t_C with v_O plus two diode voltages and the remainder of t_D's fraction with v_O. This higher v_D drains L_X faster than in the ideal buck–boost, but not as fast as in the asynchronous implementation (because v_D is higher only across t_{DT}'s). t_D and t_C are therefore shorter, but not as short as in the asynchronous buck–boost. t_E is similarly a longer fraction of t_C, but also not as long a fraction as the asynchronous d_E is.

v_{SWI} connects to v_{IN} a t_E fraction of the t_C that t_{SW} sets, $-v_{DG}$ two t_{DT} fractions, and zero for the fraction of t_D that remains. So v_{SWI}'s average is

$$v_{SWI(AVG)} \approx v_{IN}d_E'' - v_{DG}\left(\frac{2t_{DT}}{t_{SW}}\right) + (0)\left(\frac{t_D - 2t_{DT}}{t_{SW}}\right)$$
$$= v_{IN}d_E'' - v_{DG}\left(\frac{2t_{DT}}{t_{SW}}\right). \tag{3.39}$$

v_{SWO} connects to zero a t_E fraction of t_{SW}, $v_O + v_{DG}$ two t_{DT} fractions, and v_O across what remains of t_D:

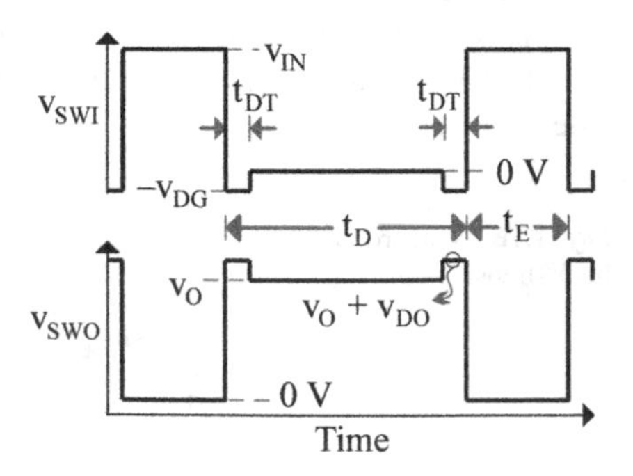

Fig. 3.17 Continuous conduction with non-reversing inductor current

$$\begin{aligned} v_{SWO(AVG)} &\approx (0)d_E'' + (v_O + v_{DO})\left(\frac{2t_{DT}}{t_{SW}}\right) + v_O\left(\frac{t_D - 2t_{DT}}{t_{SW}}\right) \\ &= v_O d_D'' + v_{DO}\left(\frac{2t_{DT}}{t_{SW}}\right) \\ &= v_O(1 - d_E'') + v_{DO}\left(\frac{2t_{DT}}{t_{SW}}\right). \end{aligned} \quad (3.40)$$

Since these averages match in steady state and d_D'' is $1 - d_E''$, the two dead-time fractions raise d_E by two diode fractions of $v_{IN} + v_O$:

$$\begin{aligned} d_E'' &= \frac{v_D}{v_E + v_D} \\ &\approx \frac{v_O}{v_{IN} + v_O} + \left(\frac{v_{DO} + v_{DG}}{v_{IN} + v_O}\right)\left(\frac{2t_{DT}}{t_{SW}}\right) \\ &= d_E + \left(\frac{v_{DO} + v_{DG}}{v_{IN} + v_O}\right)\left(\frac{2t_{DT}}{t_{SW}}\right) > d_E. \end{aligned} \quad (3.41)$$

Example 8: Determine d_E'' when v_{IN} is 2 V, v_O is 4 V, MOS diodes drop 800 mV, t_{DT} is 50 ns, and t_{SW} is 1 μs.

Solution:

$$\begin{aligned} d_E'' &= \frac{v_D}{v_E + v_D} \\ &\approx \frac{v_O}{v_{IN} + v_O} + \left(\frac{v_{DO} + v_{DG}}{v_{IN} + v_O}\right)\left(\frac{2t_{DT}}{t_{SW}}\right) \\ &= \frac{4}{2+4} + \left(\frac{800m + 800m}{2+4}\right)\left[\frac{2(50n)}{1\mu}\right] = 69\% \end{aligned}$$

Note: d_E'' is higher than d_E in the ideal example because D_{DG} and D_{DO} raise v_D. But since the diodes do so only two t_{DT} fractions of t_C, t_D shortens and t_E's fraction of t_C climbs less than d_E' in the asynchronous example.

Explore with SPICE:
See Appendix A for notes on SPICE simulations.

```
* Synchronous Buck-Boost in CCM
vge vge 0 dc=0 pulse 0 4 50n 1n 1n 690n 1u
vgeb vgeb 0 dc=2 pulse 2 0 50n 1n 1n 690n 1u
vgd vgd 0 dc=2 pulse 2 0 0n 1n 1n 790n 1u
vgdb vgdb 0 dc=0 pulse 0 4 0n 1n 1n 790n 1u
vin vin 0 dc=2
mei vswi vgeb vin vin pmos0 w=300m l=250n
mdg 0 vgd vswi 0 nmos0 w=100m l=250n
lx vswi vswo 10u
meg vswo vge 0 0 nmos0 w=100m l=250n
mdo vo vgdb vswo vo pmos0 w=300m l=250n
co vo 0 5u
ro vo 0 10
.lib lib.txt
.tran 700u
.end
```

Tip: Plot v(vo), i(Lx), v(vswi), and v(vswo) and view across 700 μs and from 695 to 700 μs.

C. **Power**

With a higher d_E, and as a result, a lower d_D, $i_{L(AVG)}$ is a higher reverse d_D translation of $i_{O(AVG)}$ and $i_{IN(AVG)}$ is a higher d_E fraction of $i_{L(AVG)}$. So for the same i_O and v_O, v_{IN} supplies more i_{IN} than in the ideal buck–boost. But since v_{DG} and v_{DO} raise v_D only across two t_{DT} fractions of t_{SW}, i_{IN} is lower than the asynchronous i_{IN}. In other words, P_{IN} is higher than P_O, but not as high as in the asynchronous case.

Since $v_E d_E$ matches $v_D d_D$, v_E supplies $d_E i_{L(AVG)}$, and v_D receives $d_D i_{L(AVG)}$, v_E's v_{IN} still delivers the power v_D consumes. v_D, however, is two diodes over v_O across two t_{DT}'s. v_O therefore loses about two t_{DT} fractions of the P_{DD} that the diodes consume across t_D in the asynchronous stage:

$$
\begin{aligned}
P_{DT} &= i_{L(AVG)}(v_{DG} + v_{DO})\left(\frac{2t_{DT}}{t_{SW}}\right) \\
&= i_{L(AVG)}(v_{DG} + v_{DO})d_D\left(\frac{2t_{DT}}{t_D}\right) \\
&= P_{DD}\left(\frac{2t_{DT}}{t_D}\right).
\end{aligned}
\tag{3.42}
$$

This *dead-time power* P_{DT} is what the buck–boost sacrifices to D_{DG} and D_{DO} across t_{DT}'s. The switches also consume power, but much less than D_{DG} and D_{DO}. This is because the voltages M_{EI}, M_{EG}, M_{DG}, and M_{DO} drop are usually much lower.

D. Conduction Modes

The core difference between asynchronous and synchronous operation hinges on the controller. If the controller opens and closes the synchronous drain transistors M_{DG} and M_{DO} when the asynchronous diodes D_{DG} and D_{DO} naturally would, the only difference is the voltage dropped across the switches: millivolts with FETs and 600–800 mV with PN junction diodes. But since M_{DG} and M_{DO} are off across dead-time periods, M_D's and M_{DO}'s body diodes in synchronous converters conduct across t_{DT}'s like their asynchronous counterparts.

If the controller does not open the drain FETs when i_L reaches zero, discontinuous conduction is not possible with the synchronous stage. The reason for this is FETs are bidirectional. So if i_L reaches zero before t_{SW} ends, the drain voltage that M_{DG} and M_{DO} in Fig. 3.18 impress across L_X induces i_L to fall below zero, like the thicker trace in the graph illustrates. In other words, i_L reverses and pulls current from v_O.

Since M_{DG}'s and M_{DO}'s body diodes are not bidirectional, reverse dead-time i_L does not flow through these diodes. Instead, negative i_L flows through M_{EG}'s and M_{EI}'s body diodes. So when M_{DG} and M_{DO} open before starting another t_{SW}, v_{SWO} falls to $-v_{EG}$ and v_{SWI} rises to $v_{IN} + v_{EI}$ across t_{DT} like Fig. 3.18 shows, instead of rising to $v_O + v_{DO}$ and falling to $-v_{DG}$ like a positive i_L would induce.

The worst part about this is that L_X pulls power from v_O and returns it to v_{IN} when i_L reverses, which is the opposite of what a power supply should do. This is why designers often open M_{DG} and M_{DO} when i_L reaches zero, like diodes would. By forcing discontinuous conduction this way, the system does not burn Ohmic power when unnecessarily delivering and returning energy that v_O does not ultimately receive.

Note one t_{DT} is in t_D and another in t_E when i_L reverses, whereas without negative conduction, t_D includes both t_{DT}'s. And since v_{SWI} and v_{SWO} rise and fall by a similar diode voltage across similar t_{DT}'s, diode effects on $v_{SWI(AVG)}$ and $v_{SWO(AVG)}$ tend to cancel. So with negative conduction, d_E is closer to the ideal case (lower than d_E'').

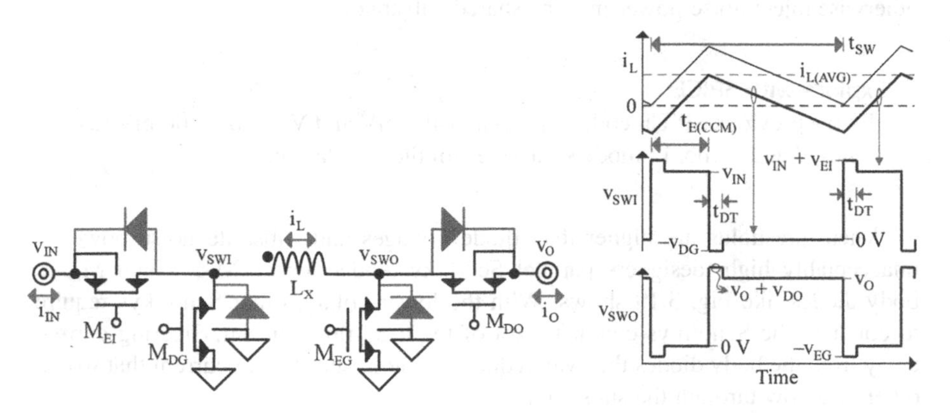

Fig. 3.18 Buck–boost in continuous conduction with reversible inductor current

Explore with SPICE:
See Appendix A for notes on SPICE simulations.

```
* Synchronous Buck-Boost with Negative Conduction
vge vge 0 dc=0 pulse 0 4 50n 1n 1n 616n 1u
vgeb vgeb 0 dc=2 pulse 2 0 50n 1n 1n 616n 1u
vgd vgd 0 dc=2 pulse 2 0 0n 1n 1n 716n 1u
vgdb vgdb 0 dc=0 pulse 0 4 0n 1n 1n 716n 1u
vin vin 0 dc=2
mei vswi vgeb vin vin pmos0 w=300m l=250n
mdg 0 vgd vswi 0 nmos0 w=100m l=250n
lx vswi vswo 10u
meg vswo vge 0 0 nmos0 w=100m l=250n
mdo vo vgdb vswo vo pmos0 w=300m l=250n
vo vo 0 dc=4
.ic i(lx)=-70m
.lib lib.txt
.tran 2u
.end
```

Tip: Plot i(Lx), v(vswi), and v(vswo).

E. **Diode Conduction**

M_{DG}'s and M_{DO}'s body diodes conduct dead-time i_L like D_{DG} and D_{DO} in Fig. 3.12. So the synchronous network operates like the asynchronous converter across t_{DT}'s. If MOSFET threshold voltages are lower than diode voltages, v_{SWI} falls below M_{DG}'s grounded gate voltage until v_{GS} is high enough to conduct i_L. v_{SWO} similarly climbs over M_{DO}'s v_O-supplied gate voltage until v_{SG} is high enough to conduct i_L. This way, M_{DG}'s and M_{DO}'s body diodes do not activate, which would otherwise inject noise power into the shared substrate.

Explore with SPICE:
Use the previous SPICE code, set V_{TN0} to 400 mV and V_{TP0} to −400 mV (use "nmos1" and "pmos1" models), and re-run the simulation.

When thresholds are higher than diode voltages and substrate noise power is unacceptably high, designers parallel Schottky diodes across M_{DG}'s and M_{DO}'s body diodes like Fig. 3.19 shows. With the lower voltages that Schottkys require to conduct, the Schottkys conduct most of the dead-time current. Steering current away from the body diodes this way reduces the noise-producing current that would otherwise flow through the substrate.

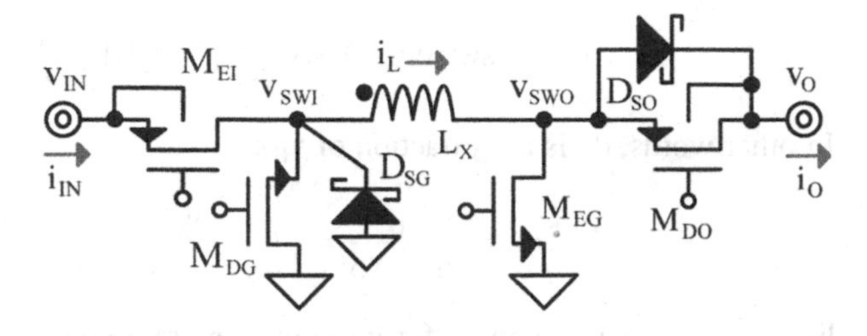

Fig. 3.19 Schottky-clamped synchronous buck–boost

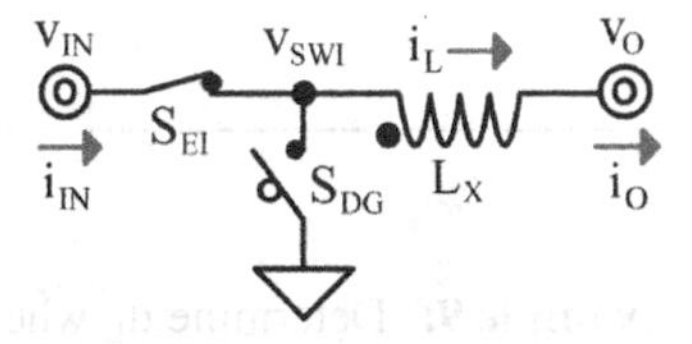

Fig. 3.20 Ideal buck

3.4 Buck

3.4.1 Ideal Buck

A. Power Stage

As the name implies, bucking switched inductors *buck* v_{IN} to a lower v_O. Since v_{IN} is always greater than v_O, $v_{IN} - v_O$ can be the v_E that energizes L_X. v_{IN} can energize L_X into v_O directly this way.

This is good because v_{IN} delivers energy with i_L into v_O that L_X does not need to transfer. As a result, i_L peaks to a lower value than in the buck–boost. And with a lower i_L, series resistances burn less power. In other words, P_O in a buck is usually a higher fraction of P_{IN} than P_O in a buck–boost.

The switches that connect L_X to v_O and ground in the buck–boost in Fig. 3.11 disappear in the buck of Fig. 3.20. This is because L_X energizes and drains to v_O. This way, the energize switch S_{EI} draws input power by connecting L_X across v_{IN} and v_O.

After L_X energizes, S_{EI} opens and the ground drain switch S_{DG} closes to connect L_X's switching terminal to ground. v_L's polarity reverses to $-v_O$ this way, draining L_X into v_O. v_L (in Fig. 3.6) therefore swings between $+v_E$'s $v_{IN} - v_O$ and $-v_D$'s $-v_O$ and i_L ramps up with $+v_E$'s $v_{IN} - v_O$ and down with $-v_D$'s $-v_O$. In effect, the buck is the part of the buck–boost that bucks: S_{EI}, S_{DG}, and L_X.

B. Duty-Cycle Translation

In steady state, v_L's average is zero, which means v_{SWI}'s average matches v_O. Since v_{IN} connects to v_{SWI} a t_E fraction of t_C, v_{SWI}'s average and v_O are a duty-cycled fraction of v_{IN}:

$$v_O = v_{SWI(AVG)} = v_{IN}\left(\frac{t_E}{t_C}\right) + (0)\left(\frac{t_D}{t_C}\right) = v_{IN}d_E. \quad (3.43)$$

In other words, d_E is a v_O fraction of v_{IN}:

$$d_E = \frac{v_D}{v_E + v_D} = \frac{v_O}{(v_{IN} - v_O) + v_O} = \frac{v_O}{v_{IN}}, \quad (3.44)$$

like $v_{IN} - v_O$ for v_E and v_O for v_D in the general expression for d_E also predict.

Example 9: Determine d_E when v_{IN} is 2 V and v_O is 1 V.

Solution:

$$d_E = \frac{v_D}{v_E + v_D} = \frac{v_O}{v_{IN}} = \frac{1}{2} = 50\%$$

Explore with SPICE:
See Appendix A for notes on SPICE simulations.

```
* Ideal Buck in CCM
vde de 0 dc=0 pulse 0 1 50n 1n 1n 500n 1u
vin vin 0 dc=2
sei vin vswi de 0 sw1v
ddg 0 vswi idiode
lx vswi vo 10u
co vo 0 5u
ro vo 0 10
.lib lib.txt
.tran 700u
.end
```

Tip: Plot v(vo), i(Lx), and v(vswi) and view across 700 μs and from 695 to 700 μs.

C. **Power**

Since v_{IN} connects to L_X a t_E fraction of t_C, v_{IN} supplies a d_E fraction of $i_{L(AVG)}$. v_O receives all of $i_{L(AVG)}$ because L_X connects directly to v_O. As a result, $i_{IN(AVG)}$ in P_{IN} is a d_E fraction of $i_{L(AVG)}$ and $i_{O(AVG)}$ in P_O is $i_{L(AVG)}$:

$$P_{IN} = v_{IN} i_{IN(AVG)} = v_{IN} i_{L(AVG)} d_{IN} = v_{IN} i_{L(AVG)} d_E = v_{IN} i_{O(AVG)} d_E \tag{3.45}$$

and

$$P_O = v_O i_{O(AVG)} = v_O i_{L(AVG)} d_O = v_O i_{L(AVG)}. \tag{3.46}$$

d_{IN} is d_E and d_O is one or 100% because d_D does not duty-cycle L_X into v_O. In ideal bucks, P_O also equals P_{IN} because ideal switches do not burn power (and $v_E d_E$'s $(v_{IN} - v_O)d_E$ matches $v_D d_D$'s $v_O(1 - d_E)$). But since v_O receives v_{IN} power when L_X energizes, v_O receives more energy across t_C than L_X delivers across t_D. In other words, v_O receives the energy v_{IN} supplies with L_X plus the energy v_{IN} supplies directly to v_O.

3.4.2 Asynchronous Buck

A. **Power Stage**

L_X in the asynchronous buck energizes with a transistor and drains with a diode. The energize switch S_{EI} in Fig. 3.20 is therefore a FET M_{EI} in Fig. 3.21 and the drain switch S_{DG} is a diode D_{DG}. Since i_L flows to the right after M_{EI} energizes L_X, D_{DG} must derive i_L from ground and direct it to v_O. This is why D_{DG}'s anode connects to ground and D_{DG}'s cathode to v_{SWI}. This circuit is essentially the part of the asynchronous buck–boost in Fig. 3.12 that bucks: M_{EI}, D_{DG}, and L_X. So the selection and connectivity of M_{EI} here matches that of M_{EI} in the buck–boost.

B. **Duty-Cycle Translation**

Since D_{DG} drops a diode voltage below ground, L_X drains with v_O plus a diode voltage. This higher v_D drains L_X faster than in the ideal buck, so t_D and t_C are shorter and t_E is a larger fraction of t_C. In other words, d_E in the asynchronous buck is higher than in the ideal buck.

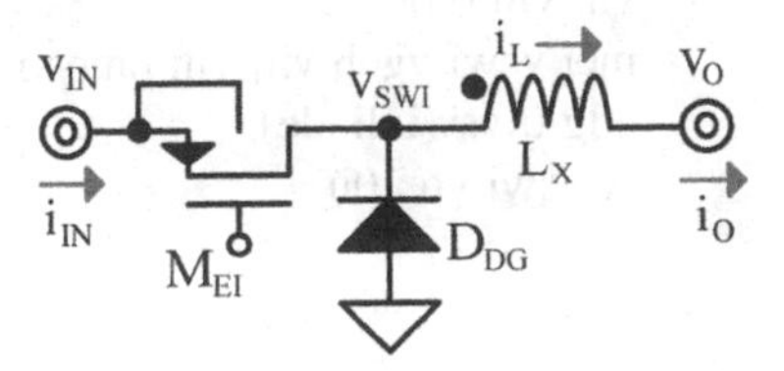

Fig. 3.21 Asynchronous buck

v_{SWI} swings between v_{IN} and $-v_{DG}$ like in Figs. 3.13 and 3.14. v_{SWI} therefore connects to v_{IN} a t_E fraction of t_C and to $-v_{DG}$ a t_D fraction d_D', so v_{SWI}'s average, which matches v_O, is

$$v_O = v_{SWI(AVG)} \approx v_{IN}d_E' - v_{DG}d_D' = v_{IN}d_E' - v_{DG}(1 - d_E'). \quad (3.47)$$

Since d_D' is $1 - d_E'$ and v_O is less than v_{IN} in a buck, the diode raises v_D's v_O in the numerator of d_E' by a larger fraction than in the denominator that v_E's $v_{IN} - v_O$ and v_D's v_O set:

$$d_E' = \frac{v_D}{v_E + v_D} \approx \frac{v_O + v_{DG}}{(v_{IN} - v_O) + (v_O + v_{DG})} \approx \frac{v_O + v_{DG}}{v_{IN} + v_{DG}} > d_E. \quad (3.48)$$

So the effect of the diode is to increase d_E to d_E'.

Example 10: Determine d_E' when v_{IN} is 2 V, v_O is 1 V, and D_{DG} drops 800 mV.

Solution:

$$d_E' = \frac{v_D}{v_E + v_D} \approx \frac{v_O + v_{DG}}{v_{IN} + v_{DG}} = \frac{1 + 800\text{m}}{2 + 800\text{m}} = 64\%$$

Note: d_E' is higher than d_E in the ideal example because D_{DG} increases L_X's drain voltage. So t_D shortens and t_E's fraction of t_C's t_D + t_E increases.

Explore with SPICE:
See Appendix A for notes on SPICE simulations.

```
* Asynchronous Buck in CCM
vgeb vgeb 0 dc=2 pulse 2 0 50n 1n 1n 640n 1u
vin vin 0 dc=2
mei vswi vgeb vin vin pmos0 w=300m l=250n
ddg 0 vswi diode1
lx vswi vo 10u
```

(continued)

```
co vo 0 5u
ro vo 0 10
.lib lib.txt
.tran 700u
.end
```

Tip: Plot v(vo), i(Lx), and v(vswi) and view across 700 μs and from 695 to 700 μs.

C. **Power**

With a higher d_E, $i_{IN(AVG)}$ is a higher d_E fraction of $i_{L(AVG)}$'s $i_{O(AVG)}$. So for the same i_O and v_O, v_{IN} supplies more i_{IN}. In other words, the asynchronous buck draws more P_{IN} than the ideal buck when supplying the same P_O. Or more to the point, P_{IN} is higher than P_O.

But since $v_E d_E$ matches $v_D d_D$, v_E supplies $d_E i_{L(AVG)}$, and v_D receives $d_D i_{L(AVG)}$, v_E still delivers the power v_D consumes. v_D, however, is v_{DG} over v_O, not v_O like in the ideal case. v_O therefore loses (to D_{DG}) a diode fraction of the P_L that L_X delivers after energizing with $i_{IN(AVG)}/d_E$ and v_E across t_E:

$$
\begin{aligned}
P_{DD} &= i_{L(AVG)} v_{DG} d_D \\
&= \left(\frac{i_{IN(AVG)}}{d_E}\right) v_{DG} \left(\frac{v_D}{v_O + v_{DG}}\right) d_D \\
&= \left(\frac{i_{IN(AVG)}}{d_E}\right) \left(\frac{v_{DG}}{v_O + v_{DG}}\right) v_E d_E \\
&= \left(\frac{v_{DG}}{v_O + v_{DG}}\right) P_L,
\end{aligned}
\tag{3.49}
$$

where $i_{L(AVG)} v_{DG}$ is the average power D_{DG} consumes across t_D and $i_{L(AVG)} v_{DG} d_D$ is the average across t_C.

This P_{DD} is what the buck sacrifices to S_{DG} when using a diode. S_{EI} also consumes power, but much less than D_{DG}. This is because the voltage M_{EI} drops is usually much lower.

D. **Conduction Modes**

In static applications, the output is either a battery or a load that requires a steady v_O or i_O. Designers normally add capacitance to the output of voltage regulators and LED drivers because controllers cannot respond instantly. So either way, C_O is inherently very high or intentional and therefore much higher than the parasitic capacitance C_{SWI} present at the switching node.

In continuous conduction, i_L climbs and falls about the $i_{L(AVG)}$ that keeps i_L above zero. As the controller adjusts how much current v_{IN} supplies or v_O pulls, $i_{L(AVG)}$

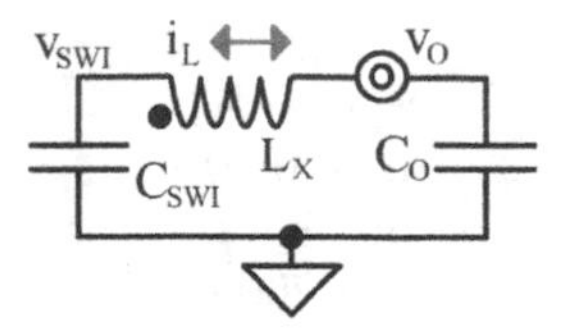

Fig. 3.22 Drained and disconnected asynchronous buck inductor

shifts. Like in the asynchronous buck–boost of Fig. 3.14, i_L reaches zero at the end of t_{SW} when $i_{L(AVG)}$ is half the CCM ripple Δi_L and before t_{SW} ends when $i_{L(AVG)}$ is below that level. Since D_{DG} cannot conduct reverse current, i_L reaches and remains near zero until the next t_{SW}. This is discontinuous conduction.

Just before the diode stops conducting (toward the end of t_D), v_{SWI} is a diode voltage below ground. C_{SWI} in Fig. 3.22 therefore holds energy $0.5C_{SWI}v_{DG}^2$ when L_X depletes. So when the switches open (at the end of t_D) before the t_{SW} cycle lapses, v_O and v_{SWI} impress $v_O + v_{DG}$ across L_X. This drains C_{SWI} and C_O into L_X. After v_{SWI} rises to ground, C_O drains into both L_X and C_{SWI}. L_X stops energizing when v_{SWI} reaches v_O.

Since L_X holds energy when v_{SWI} reaches v_O, i_L continues to charge C_{SWI} until L_X depletes. C_{SWI} therefore charges over v_O when L_X depletes. The voltage that v_{SWI} and v_O now impress across L_X draws a reverse current that first drains C_{SWI} into L_X and C_O and then (when v_{SWI} falls below v_O) drains both C_{SWI} and L_X into C_O.

When L_X depletes, v_O is again greater than v_{SWI}, so the entire sequence repeats. L_X and the capacitors exchange energy this way until series resistances burn the energy or a new t_{SW} begins. This is why v_{SWI} oscillates about the average v_O sets when i_L nears zero, like v_{SWI} in the asynchronous buck–boost from Fig. 3.14.

Explore with SPICE:
See Appendix A for notes on SPICE simulations.

```
* Asynchronous Buck in DCM
vgeb vgeb 0 dc=2 pulse 2 0 50n 1n 1n 100n 1u
vin vin 0 dc=2
mei vswi vgeb vin vin pmos0 w=300m l=250n
ddg 0 vswi diode1
lx vswi vl 10u
rl vl vo 25
vo vo 0 dc=1
.lib lib.txt
.tran 2u
.end
```

Tip: Plot i(Lx) and v(vswi).

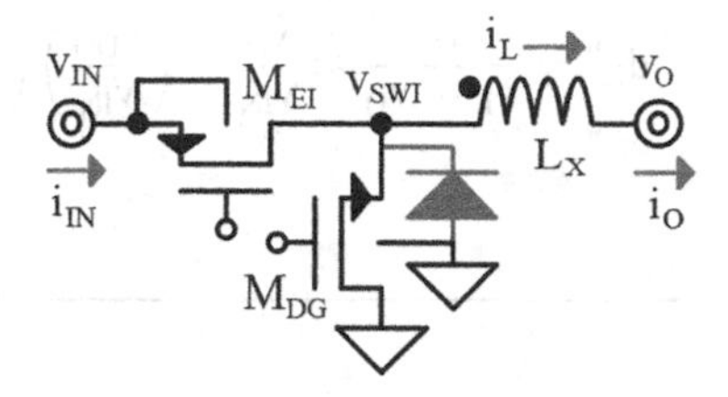

Fig. 3.23 Synchronous buck

3.4.3 Synchronous Buck

A. **Power Stage**

The difference between asynchronous and synchronous power supplies is what drains L_X: diodes in one and transistors in the other. So the FET M_{EI} that energizes the asynchronous buck and implements S_{EI} in the ideal buck also energizes the synchronous counterpart in Fig. 3.23. The difference in Fig. 3.23 is that M_{DG} drains L_X, which is the function of S_{DG} and D_{DG} in the ideal and asynchronous bucks. This circuit is the part of the synchronous buck–boost in Fig. 3.16 that bucks: M_{EI}, M_{DG}, and L_X. So the selection and connectivity of M_{DG} and behavior of v_{SWI} in continuous and discontinuous conduction match those of M_{DG} and v_{SWI} in the synchronous buck–boost.

B. **Duty-Cycle Translation**

The difference between asynchronous and synchronous operation across t_C is that M_{DG} drops a lower voltage than D_{DG}. So v_{SWI} is close to zero when M_{DG} closes across t_D. But since M_{DG} and M_{DO} open across the dead-time portions of t_D, v_{SWI} reaches $-v_{DG}$ across t_{DT}'s, like v_{SWI} in the synchronous buck–boost from Fig. 3.17.

This means that L_X drains two t_{DT} fractions of t_C with v_O plus one diode voltage and the remainder of t_D's fraction with v_O. This higher v_D drains L_X faster than in the ideal buck, but not as fast as in the asynchronous implementation because v_D is higher only across t_{DT}'s. t_D and t_C are therefore shorter, but not as short as in the asynchronous buck. t_E is similarly a longer fraction of t_C, but also not as long a fraction as the asynchronous d_E' is.

v_{SWI} connects to v_{IN} a t_E fraction of the t_C that t_{SW} sets, $-v_{DG}$ two t_{DT} fractions, and zero for the fraction of t_D that remains. So v_{SWI}'s average, which matches v_O, is

$$v_O = v_{SWI(AVG)} \approx v_{IN}d_E'' - v_{DG}\left(\frac{2t_{DT}}{t_{SW}}\right) + (0)\left(\frac{t_D - 2t_{DT}}{t_{SW}}\right). \tag{3.50}$$

Since v_O matches v_{SWI}'s average in steady state, the two dead-time fractions raise d_E by a diode fraction of v_{IN}:

$$d_E'' = \frac{v_D}{v_E + v_D} \approx \frac{v_O}{v_{IN}} + \left(\frac{v_{DG}}{v_{IN}}\right)\left(\frac{2t_{DT}}{t_{SW}}\right) = d_E + \left(\frac{v_{DG}}{v_{IN}}\right)\left(\frac{2t_{DT}}{t_{SW}}\right) > d_E. \quad (3.51)$$

Example 11: Determine d_E'' when v_{IN} is 2 V, v_O is 1 V, D_{DG} drops 800 mV, t_{DT} is 50 ns, and t_{SW} is 1 μs.

Solution:

$$d_E'' = \frac{v_D}{v_E + v_D} \approx \frac{v_O}{v_{IN}} + \left(\frac{v_{DG}}{v_{IN}}\right)\left(\frac{2t_{DT}}{t_{SW}}\right) = \frac{1}{2} + \left(\frac{800m}{2}\right)\left[\frac{2(50n)}{1\mu}\right] = 54\%$$

Note: d_E'' here is higher than d_E in the ideal example because D_{DG} raises v_D. But since it does so only two t_{DT} fractions of t_C, t_D and t_C shorten and t_E's fraction of t_C rises less than d_E' in the asynchronous example.

Explore with SPICE:
See Appendix A for notes on SPICE simulations.

```
* Synchronous Buck in CCM
vgeb vgeb 0 dc=2 pulse 2 0 50n 1n 1n 540n 1u
vgd vgd 0 dc=2 pulse 2 0 0n 1n 1n 640n 1u
vin vin 0 dc=2
mei vswi vgeb vin vin pmos0 w=300m l=250n
mdg 0 vgd vswi 0 nmos0 w=100m l=250n
lx vswi vo 10u
co vo 0 5u
ro vo 0 10
.lib lib.txt
.tran 500u
.end
```

Tip: Plot v(vo), i(Lx), and v(vswi) and view across 500 μs and from 495 to 500 μs.

C. Power

With a higher d_E, $i_{IN(AVG)}$ is a higher d_E fraction of $i_{L(AVG)}$'s $i_{O(AVG)}$. So for the same i_O and v_O, v_{IN} supplies more i_{IN} than in the ideal buck. But since v_{DG} raises v_D only across two t_{DT} fractions of t_{SW}, i_{IN} is lower than the asynchronous i_{IN}. In other words, P_{IN} is higher than P_O, but not as high as in the asynchronous case.

Since $v_E d_E$ matches $v_D d_D$, v_E supplies $d_E i_{L(AVG)}$, and v_D receives $d_D i_{L(AVG)}$, v_E still delivers the power v_D consumes. v_D, however, is a diode over v_O across two t_{DT}'s. v_O therefore loses about two t_{DT} fractions of the P_{DD} that D_{DG} consumes across t_D in the asynchronous stage:

$$P_{DT} = i_{L(AVG)} v_{DG} \left(\frac{2t_{DT}}{t_{SW}}\right) = i_{L(AVG)} v_{DG} d_D \left(\frac{2t_{DT}}{t_D}\right) = P_{DD} \left(\frac{2t_{DT}}{t_D}\right). \quad (3.52)$$

This P_{DT} is what the buck sacrifices to D_{DG} across t_{DT}'s. S_{EI} and S_{DG} also consume power, but much less than D_{DG}. This is because the voltages M_{EI} and M_{EG} drop are usually much lower.

D. Conduction Modes

Since the buck is the part of the buck–boost that bucks, M_{EI} and M_{DG} switch the same way and produce the same v_{SWI}. So like in the buck–boost, the synchronous buck operates like the asynchronous counterpart when the controller opens and closes M_{DG} like D_{DG} naturally would, except M_{DG} drops millivolts and D_{DG} drops 600–800 mV. But since M_{DG} is off across t_{DT}'s, M_{DG}'s body diode plays the role of D_{DG} across t_{DT}'s.

If the controller does not open M_{DG} when i_L falls to zero before t_{SW} lapses, M_{DG} lets i_L reverse direction. Since M_{DG}'s body diode cannot conduct reverse current, M_{EI}'s body diode steers this negative i_L to v_{IN} across the t_{DT} that follows. So v_{SWI} climbs to $v_{IN} + v_{EI}$ like Fig. 3.18 shows. Returning energy to v_{IN} this way means L_X transfers and burns more power than necessary.

Note one t_{DT} is in t_D and another in t_E when i_L reverses, whereas without negative conduction, t_D includes both t_{DT}'s. And since v_{SWI} rises and falls by a similar diode voltage across similar t_{DT}'s, diode effects on $v_{SWI(AVG)}$ tend to cancel. So with negative conduction, d_E is close to the ideal case (lower than d_E").

Explore with SPICE:
See Appendix A for notes on SPICE simulations.

```
* Synchronous Buck with Negative Conduction
vgeb vgeb 0 dc=2 pulse 2 0 50n 1n 1n 450n 1u
vgd vgd 0 dc=2 pulse 2 0 0n 1n 1n 550n 1u
vin vin 0 dc=2
mei vswi vgeb vin vin pmos0 w=300m l=250n
```

(continued)

```
mdg 0 vgd vswi 0 nmos0 w=100m l=250n
lx vswi vo 10u
vo vo 0 dc=1
.ic i(lx)=-25m
.lib lib.txt
.tran 2u
.end
```

Tip: Plot i(Lx) and v(vswi).

E. **Diode Conduction**

If M_{DG}'s threshold voltage is lower than a diode voltage, M_{DG} conducts across t_{DT}'s when v_{SWI} falls a v_{TN} below M_{DG}'s grounded gate voltage. M_{DG}'s body diode does not steer current into the substrate when this happens. A Schottky diode across M_{DG} similarly channels dead-time current away from the body diode and the substrate into which the body diode conducts.

Explore with SPICE:
Use the previous SPICE code, set V_{TN0} to 400 mV and V_{TP0} to −400 mV (use "nmos1" and "pmos1" models), and re-run the simulation.

3.5 Boost

3.5.1 Ideal Boost

A. **Power Stage**

As the name implies, boosting switched inductors *boost* v_{IN} to a higher v_O. Since v_O is always greater than v_{IN}, $v_O - v_{IN}$ can be the v_D that drains L_X. This way, v_{IN} also supplies power as L_X drains.

So v_{IN} delivers energy with i_L into v_O that L_X does not need to transfer. i_L therefore peaks to a lower value than in the buck–boost, which means series resistances burn less power. P_O in a boost is, as a result, usually a higher fraction of P_{IN} than P_O is in a buck–boost.

The switches that connect v_{IN} to L_X in the buck–boost in Fig. 3.11 disappear in the boost of Fig. 3.24. This is because L_X energizes and drains from v_{IN}. Here, the ground energize switch S_{EG} draws input power by connecting L_X across v_{IN} and ground.

Fig. 3.24 Ideal boost

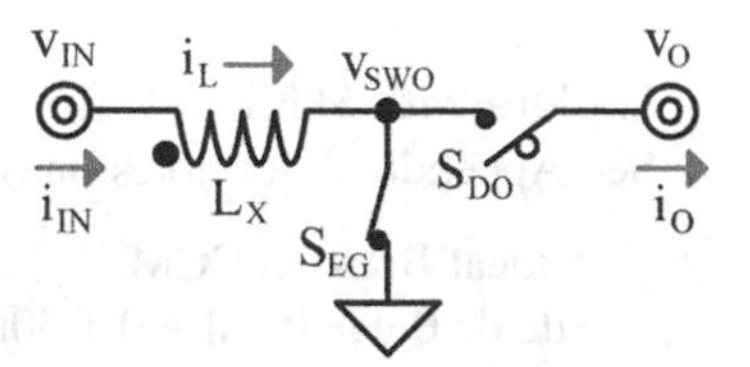

After energizing, S_{EG} opens and the output drain switch S_{DO} closes to connect L_X's switching terminal to v_O. This way, v_L's polarity is v_{IN} when energizing L_X and inverts to $v_{IN} - v_O$ when draining L_X into v_O. v_L (in Fig. 3.6) therefore swings between $+v_E$'s v_{IN} and $-v_D$'s $v_{IN} - v_O$ and i_L ramps up with $+v_E$'s v_{IN} and down with $-v_D$'s $v_{IN} - v_O$. The boost is basically the part of the buck–boost that boosts: L_X, S_{EG}, and S_{DO}.

B. **Duty-Cycle Translation**

In steady state, v_L's average is zero, which means the switching voltage v_{SWO} is, on average, equal to v_{IN}. Since v_O connects to v_{SWO} a t_D fraction of t_C, v_{SWO}'s average, which matches v_{IN}, is a duty-cycled fraction of v_O:

$$v_{IN} = v_{SWO(AVG)} = (0)\left(\frac{t_E}{t_C}\right) + v_O\left(\frac{t_D}{t_C}\right) = v_O d_D = v_O(1 - d_E). \tag{3.53}$$

This is why v_O is greater than v_{IN} by the multiplying scalar that $1/d_D$ sets:

$$v_O = \frac{v_{IN}}{d_D} = \frac{v_{IN}}{1 - d_E}. \tag{3.54}$$

And since d_D is $1 - d_E$, d_E is a $v_O - v_{IN}$ fraction of v_O:

$$d_E = \frac{v_D}{v_E + v_D} = \frac{v_O - v_{IN}}{v_{IN} + (v_O - v_{IN})} = \frac{v_O - v_{IN}}{v_O} = 1 - \frac{v_{IN}}{v_O} = 1 - d_D, \tag{3.55}$$

like v_{IN} for v_E and $v_O - v_{IN}$ for v_D in the general expression for d_E predict.

Example 12: Determine d_E when v_{IN} is 2 V and v_O is 4 V.

Solution:

$$d_E = \frac{v_D}{v_E + v_D} = \frac{v_O - v_{IN}}{v_O} = \frac{4 - 2}{4} = 50\%$$

Explore with SPICE:
See Appendix A for notes on SPICE simulations.

```
* Ideal Boost in CCM
vde de 0 dc=0 pulse 0 1 50n 1n 1n 500n 1u
vin vin 0 dc=2
lx vin vswo 10u
seg vswo 0 de 0 sw1v
ddo vswo vo idiode
co vo 0 5u
ro vo 0 10
.lib lib.txt
.tran 700u
.end
```

Tip: Plot v(vo), i(Lx), and v(vswo) and view across 700 μs and from 695 to 700 μs.

C. **Power**

Since boosts connect v_{IN} directly to L_X, v_{IN} supplies $i_{L(AVG)}$. v_O receives a d_E fraction of this $i_{L(AVG)}$ because L_X connects to v_O a t_D fraction of t_C. As a result, $i_{IN(AVG)}$ in P_{IN} is $i_{L(AVG)}$ and $i_{O(AVG)}$ in P_O is a d_D fraction of $i_{L(AVG)}$, which is to say, $i_{L(AVG)}$ is a reverse d_D translation of $i_{L(AVG)}$:

$$\begin{aligned} P_{IN} &= v_{IN} i_{IN(AVG)} = v_{IN} i_{L(AVG)} d_{IN} = v_{IN} i_{L(AVG)} \\ &= v_{IN}\left(\frac{i_{O(AVG)}}{d_O}\right) = v_{IN}\left(\frac{i_{O(AVG)}}{d_D}\right). \end{aligned} \tag{3.56}$$

$$P_O = v_O i_{O(AVG)} = v_O i_{L(AVG)} d_O = v_O i_{L(AVG)} d_D. \tag{3.57}$$

d_{IN} is one or 100% and d_O is d_D in boosts because d_E does not duty-cycle L_X to v_{IN}. In ideal boosts, P_O equals P_{IN} because ideal switches do not burn power (and $v_E d_E$'s $v_{IN} d_E$ matches $v_D d_D$'s $(v_O - v_{IN})(1 - d_E)$). But since v_{IN} also supplies power when L_X drains, v_O receives more energy than L_X transfers. In other words, v_O receives the energy v_{IN} supplies with L_X plus the energy v_{IN} supplies directly to v_O.

3.5.2 Asynchronous Boost

A. **Power Stage**

L_X in the asynchronous boost energizes with a transistor and drains with a diode. The ground energize switch S_{EG} in Fig. 3.24 is therefore a transistor M_{EG} in Fig. 3.25 and

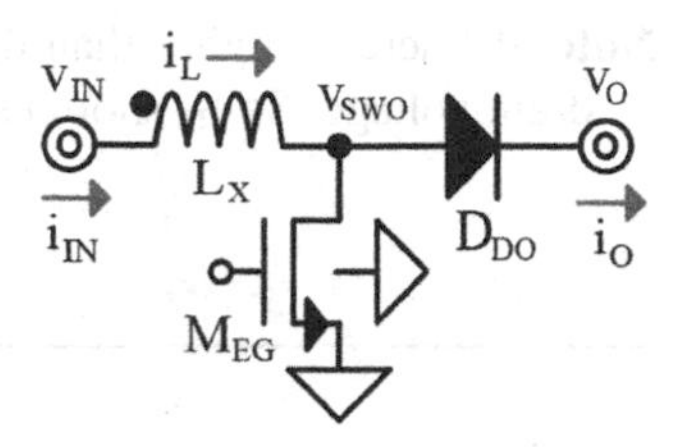

Fig. 3.25 Asynchronous boost

the output drain switch S_{DO} is a diode D_{DO}. Since i_L flows to the right after M_{EG} energizes L_X, D_{DO} steers i_L into v_O. This circuit is basically the part of the asynchronous buck–boost in Fig. 3.12 that boosts: L_X, M_{EG}, and D_{DO}. So the selection and connectivity of M_{EG} matches that of M_{EG} in the asynchronous buck–boost.

B. **Duty-Cycle Translation**

Since D_{DO} drops a diode voltage over v_O, L_X drains with $v_O - v_{IN}$ plus a diode voltage. This higher v_D drains L_X faster than in the ideal boost, so t_D and t_C are shorter and t_E is a larger fraction of t_C. In short, d_E is higher than in the ideal boost.

As in the buck–boost, v_{SWO} swings between ground and $v_O + v_{DO}$, like Fig. 3.13 shows. v_{SWO} therefore connects to ground a t_E fraction of t_C and to $v_O + v_{DO}$ a t_D fraction d_D'. So v_{SWO}'s average, which matches v_{IN}, is

$$v_{IN} = v_{SWO(AVG)} \approx (0)d_E' + (v_O + v_{DO})d_D' = (v_O + v_{DO})(1 - d_E'). \quad (3.58)$$

Since d_D' is $1 - d_E'$ and $v_O - v_{IN}$ is less than v_O, the diode raises v_D's $v_O - v_{IN}$ in the numerator of d_E' by a larger fraction than in the denominator that v_E's v_{IN} and v_D's $v_O - v_{IN}$ set:

$$d_E' = \frac{v_D}{v_E + v_D} \approx \frac{v_O + v_{DO} - v_{IN}}{v_{IN} + (v_O + v_{DO} - v_{IN})} \approx \frac{v_O + v_{DO} - v_{IN}}{v_O + v_{DO}} > d_E. \quad (3.59)$$

So the effect of the diodes is to increase d_E.

Example 13: Determine d_E when v_{IN} is 2 V, v_O is 4 V, and D_{DO} drops 800 mV.

Solution:

$$d_E' = \frac{v_D}{v_E + v_D} \approx \frac{v_O + v_{DO} - v_{IN}}{v_O + v_{DO}} = \frac{4 + 800m - 2}{4 + 800m} = 58\%$$

Note: d_E' here is higher than d_E in the ideal example because D_{DO} increases L_X's drain voltage. So t_D shortens and t_E's fraction of t_C's $t_D + t_E$ increases.

Explore with SPICE:
See Appendix A for notes on SPICE simulations.

```
* Asynchronous Boost in CCM
vge vge 0 dc=0 pulse 0 4 50n 1n 1n 580n 1u
vin vin 0 dc=2
lx vin vswo 10u
meg vswo vge 0 0 nmos0 w=100m l=250n
ddo vswo vo diode1
co vo 0 5u
ro vo 0 10
.lib lib.txt
.tran 700u
.end
```

Tip: Plot v(vo), i(Lx), and v(vswo) and view across 700 μs and from 695 to 700 μs.

C. Power

With a higher d_E, and as a result, a lower d_D, i_{IN}'s $i_{L(AVG)}$ is a higher reverse d_D translation of $i_{O(AVG)}$. So for the same i_O and v_O, v_{IN} supplies more i_{IN}. In other words, the asynchronous boost draws more P_{IN} than the ideal boost when supplying the same P_O. Or more to the point, P_{IN} is higher than P_O.

But since $v_E d_E$ matches $v_D d_D$, v_E supplies $d_E i_{L(AVG)}$, and v_D receives $d_D i_{L(AVG)}$, v_E still delivers the power v_D consumes. v_D, however, is v_{DO} over v_O, not v_O like in the ideal case. v_O therefore loses (to D_{DO}) a diode fraction of the P_L that L_X delivers after energizing with $i_{IN(AVG)}$ and v_E across t_E:

$$\begin{aligned} P_{DD} &= i_{L(AVG)} v_{DO} d_D \\ &= i_{L(AVG)} v_{DO} \left(\frac{v_D}{v_O + v_{DO}}\right) d_D \\ &= i_{IN(AVG)} \left(\frac{v_{DO}}{v_O + v_{DO}}\right) v_E d_E \\ &= \left(\frac{v_{DO}}{v_O + v_{DO}}\right) P_L \approx \left(\frac{v_{DO}}{v_O + v_{DO}}\right) P_{IN} d_E, \end{aligned} \tag{3.60}$$

where $i_{L(AVG)}v_{DO}$ is the average power D_{DO} consumes across t_D and $i_{L(AVG)}v_{DO}d_D$ is the average power across t_C, and P_{IN} nears P_O when power lost is a small fraction of P_{IN}.

This P_{DD} is what the boost sacrifices to S_{DO} when using a diode. S_{EG} also consumes power, but much less than D_{DO}. This is because the voltage M_{EG} drops is usually much lower.

D. **Conduction Modes**

In static applications, the input is either a dc source or the output of a voltage regulator that feeds a steady v_{IN}. Designers normally add capacitance to the output of voltage regulators because controllers cannot respond instantaneously. So either way, v_{IN}'s equivalent C_{IN} is inherently high or intentional and therefore much higher than the parasitic capacitance C_{SWO} present at the switching node.

In continuous conduction, i_L climbs and falls about the $i_{L(AVG)}$ that keeps i_L above zero. As the controller adjusts how much current v_{IN} supplies or v_O pulls, $i_{L(AVG)}$ shifts. Like in the asynchronous buck–boost from Fig. 3.14, i_L reaches zero at the end of t_{SW} when $i_{L(AVG)}$ is half the CCM ripple Δi_L and before t_{SW} ends when $i_{L(AVG)}$ is below that level. Since D_{DO} cannot conduct reverse current, i_L reaches and remains near zero until the next t_{SW}. L_X is in discontinuous conduction this way.

Just before the diode stops conducting (towards the end of t_D), v_{SWO} is a diode voltage over v_O. C_{SWO} (in Fig. 3.26) therefore holds energy $0.5C_{SWO}(v_O + v_{DO})^2$ when L_X depletes. So when the switches open (at the end of t_D) before the cycle lapses, v_{IN} and v_{SWO} impress $v_O + v_{DG} - v_{IN}$ across L_X, which drains C_{SWO} into L_X and v_{IN}. L_X stops energizing when v_{SWO} falls to v_{IN} (when v_L is zero). Past that point, i_L drains L_X and C_{SWO} into v_{IN}.

Since v_{IN} is greater than v_{SWO} when L_X depletes, L_X draws a reverse current that first drains v_{IN} into L_X and C_{SWO} and then (when v_{SWO} rises above v_{IN}) drains v_{IN} and L_X into C_{SWO}. v_{SWO} is again greater than v_{IN} when L_X depletes, so the entire sequence repeats. L_X and the capacitors exchange energy this way until series resistances burn the energy or a new t_{SW} begins. This is why v_{SWO} oscillates when i_L is zero like v_{SWO} in the asynchronous buck–boost from Fig. 3.14.

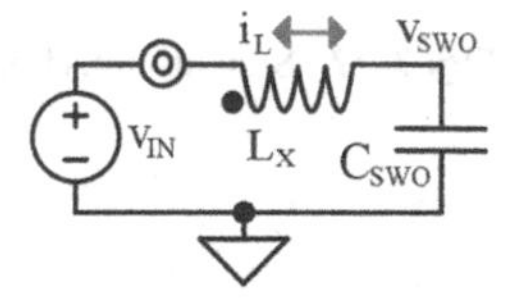

Fig. 3.26 Drained and disconnected asynchronous boost inductor

Explore with SPICE:
See Appendix A for notes on SPICE simulations.

```
* Asynchronous Boost in DCM
vge vge 0 dc=0 pulse 0 4 50n 1n 1n 100n 1u
vin vin 0 dc=2
lx vin vl 10u
rl vl vswo 25
meg vswo vge 0 0 nmos0 w=100m l=250n
ddo vswo vo diode1
vo vo 0 dc=4
.lib lib.txt
.tran 2u
.end
```

Tip: Plot i(Lx) and v(vswo).

3.5.3 Synchronous Boost

A. **Power Stage**

The difference between asynchronous and synchronous power supplies is what drains L_X: diodes in one and transistors in the other. So the ground FET M_{EG} that energizes the asynchronous boost and implements S_{EG} in the ideal case in Fig. 3.16 also energizes the synchronous counterpart in Fig. 3.27. The difference in Fig. 3.27 is that M_{DO} drains L_X, which is the function of S_{DO} and D_{DO} in the ideal and asynchronous boosts. This circuit is basically the part of the synchronous buck–boost that boosts: L_X, M_{EG}, and M_{DO}. So the selection and connectivity of M_{DO} and behavior of v_{SWO} in continuous and discontinuous conduction match those of M_{DO} and v_{SWO} in the synchronous buck–boost.

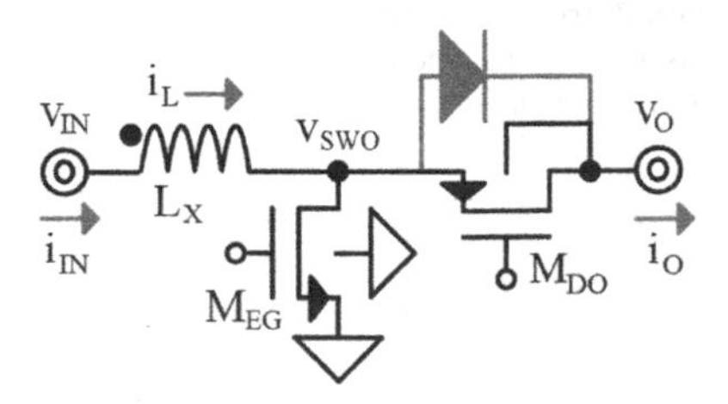

Fig. 3.27 Synchronous boost

B. Duty-Cycle Translation

The difference between asynchronous and synchronous operation across t_D is that M_{DO} drops a lower voltage than D_{DO}. So v_{SWO} is close to v_O when M_{DO} closes within t_D. But since M_{EG} and M_{DO} open across the dead-time portions of t_D, v_{SWO} is $v_O + v_{DG}$ across the t_{DT}'s that t_D incorporates, like v_{SWO} in the synchronous buck–boost from Fig. 3.17.

This means that L_X drains two t_{DT} fractions of t_C with v_O plus one diode voltage and the remainder of t_D's fraction with v_O. This higher v_D drains L_X faster than in the ideal boost, but not as fast as in the asynchronous implementation because v_D is higher only across t_{DT}'s. t_D and t_C are therefore shorter, but not as short as in the asynchronous boost. t_E is similarly a longer fraction of t_C, but also not as long a fraction as the asynchronous d_E' is.

v_{SWO} connects to zero a t_E fraction of the t_C that t_{SW} sets, $v_O + v_{DG}$ two t_{DT} fractions, and v_O for the fraction of t_D that remains. So $v_{SWO(AVG)}$ is

$$v_{IN} = v_{SWO(AVG)} \approx (0)d_E'' + (v_O + v_{DG})\left(\frac{2t_{DT}}{t_{SW}}\right) + v_O\left(\frac{t_D - 2t_{DT}}{t_{SW}}\right). \quad (3.61)$$

Since v_{IN} matches v_{SWO}'s average in steady state and t_D/t_C is d_D'' or $1 - d_E''$, the two dead-time fractions raise d_E by a diode fraction of v_O:

$$d_E'' = \frac{v_D}{v_E + v_D} \approx \frac{v_O - v_{IN}}{v_O} + \left(\frac{v_{DO}}{v_O}\right)\left(\frac{2t_{DT}}{t_{SW}}\right) = d_E + \left(\frac{v_{DO}}{v_O}\right)\left(\frac{2t_{DT}}{t_{SW}}\right) > d_E. \quad (3.62)$$

Example 14: Determine d_E'' when v_{IN} is 2 V, v_O is 4 V, D_{DO} drops 800 mV, t_{DT} is 50 ns, and t_{SW} is 1 μs.

Solution:

$$d_E'' = \frac{v_D}{v_E + v_D} \approx \frac{v_O - v_{IN}}{v_O} + \left(\frac{v_{DO}}{v_O}\right)\left(\frac{2t_{DT}}{t_{SW}}\right) = \frac{4-2}{4} + \left(\frac{800m}{4}\right)\left[\frac{2(50n)}{1\mu}\right] = 52\%$$

Note: d_E'' here is higher than d_E in the ideal example because D_{DO} raises v_D. But since the diode does so only two t_{DT} fractions of t_C, t_D shortens and t_E's fraction of t_C increases less than d_E' in the asynchronous example.

Explore with SPICE:
See Appendix A for notes on SPICE simulations.

```
* Synchronous Boost in CCM
vge vge 0 dc=0 pulse 0 4 50n 1n 1n 520n 1u
vgdb vgdb 0 dc=0 pulse 0 4 0n 1n 1n 620n 1u
vin vin 0 dc=2
lx vin vswo 10u
meg vswo vge 0 0 nmos0 w=100m l=250n
mdo vo vgdb vswo vo pmos0 w=300m l=250n
co vo 0 5u
ro vo 0 10
.lib lib.txt
.tran 700u
.end
```

Tip: Plot v(vo), i(Lx), and v(vswo) and view across 700 μs and from 695 to 700 μs.

C. **Power**

With a higher d_E, and as a result, a lower d_D, $i_{L(AVG)}$'s $i_{IN(AVG)}$ is a higher reverse d_D translation of $i_{O(AVG)}$. So for the same i_O and v_O, v_{IN} supplies more i_{IN} than in the ideal boost. But since v_{DO} raises v_D only across two t_{DT} fractions of t_{SW}, i_{IN} is lower than the asynchronous i_{IN}. In other words, P_{IN} is higher than P_O, but not as high as in the asynchronous case.

Since $v_E d_E$ matches $v_D d_D$, v_E supplies $d_E i_{L(AVG)}$, and v_D receives $d_D i_{L(AVG)}$, v_E still delivers the power v_D consumes. v_D, however, is a diode over v_O across two t_{DT}'s. v_O therefore loses about two t_{DT} fractions of the P_{DD} that D_{DO} consumes across t_D in the asynchronous stage:

$$P_{DT} = i_{L(AVG)} v_{DO} \left(\frac{2t_{DT}}{t_{SW}}\right) = i_{L(AVG)} v_{DO} d_D \left(\frac{2t_{DT}}{t_D}\right) = P_{DD} \left(\frac{2t_{DT}}{t_D}\right). \quad (3.63)$$

This P_{DT} is what the boost sacrifices to D_{DO} across t_{DT}'s. S_{EG} and S_{DO} also consume power, but much less than D_{DO}. This is because the voltages M_{EG} and M_{DO} drop are usually much lower.

D. **Conduction Modes**

Since the boost is the part of the buck–boost that boosts, M_{EG} and M_{DO} switch the same way and produce the same v_{SWO}. So like in the buck–boost, the synchronous boost operates like the asynchronous sibling when the controller opens and closes M_{DO} like D_{DO} naturally would, except M_{DO} drops millivolts and D_{DO} drops

600–800 mV. But since M_{DO} is off across t_{DT}'s, M_{DO}'s body diode plays the role of D_{DO} across t_{DT}'s.

If the controller does not open M_{DO} when i_L falls to zero before t_{SW} lapses, M_{DO} lets i_L reverse direction. Since M_{DO}'s body diode cannot conduct reverse current, M_{EG}'s body diode steers this negative i_L to v_{IN} across the t_{DT} that follows. v_{SWO} therefore falls to $-v_{EG}$ across t_{DT} like Fig. 3.18 shows. Returning energy to v_{IN} this way means L_X transfers and burns more power than necessary.

Note one t_{DT} is in t_D and another in t_E when i_L reverses, whereas without negative conduction, t_D includes both t_{DT}'s. And since v_{SWO} rises and falls by a similar diode voltage across similar t_{DT}'s, diode effects on $v_{SWI(AVG)}$ tend to cancel. So with negative conduction, d_E is close to the ideal case (lower than d_E'').

Explore with SPICE:
See Appendix A for notes on SPICE simulations.

```
* Synchronous Boost with Negative Conduction
vge vge 0 dc=0 pulse 0 4 50n 1n 1n 450n 1u
vgdb vgdb 0 dc=0 pulse 0 4 0n 1n 1n 550n 1u
vin vin 0 dc=2
lx vin vswo 10u
meg vswo vge 0 0 nmos0 w=100m l=250n
mdo vo vgdb vswo vo pmos0 w=300m l=250n
vo vo 0 dc=4
.ic i(lx)=-50m
.lib lib.txt
.tran 2u
.end
```

Tip: Plot i(Lx) and v(vswo).

E. **Diode Conduction**

If M_{DO}'s threshold voltage is lower than a diode voltage, M_{DO} conducts across t_{DT}'s when v_{SWO} climbs a $|v_{TP}|$ over M_{DO}'s v_O-supplied gate voltage. M_{DO}'s body diode does not inject substrate current when this happens because it does not conduct. A Schottky diode across M_{DO} similarly steers dead-time current away from the body diode and the substrate into which the parasitic *bipolar-junction transistors* (BJTs) present inject current.

Explore with SPICE:
Use the previous SPICE code, set V_{TN0} to 400 mV and V_{TP0} to −400 mV (use "nmos1" and "pmos1" models), and re-run the simulation.

3.6 Flyback

3.6.1 Ideal Flyback

A. **Power Stage**

The flyback is an interesting variation of the buck–boost. Like all switched inductors, v_{IN} magnetizes the core of an inductor L_I. v_O similarly demagnetizes the core, but with another inductor L_O. In other words, L_I draws power from v_{IN} that a coupled L_O delivers to v_O.

The advantage of this setup is separate grounds for v_{IN} and v_O for what engineers call *galvanic isolation*. This way, without a direct connection, stray noise currents do not couple. Ground levels can also be at different potentials. Galvanic isolation is ultimately a form of protection.

Aside from separate inductors, flybacks switch and operate like buck–boosts. L_I in Fig. 3.28 magnetizes the core when the input switch S_{EI} connects v_{IN} across L_I. L_O demagnetizes the core after S_{EI} opens, when the output drain switch S_{DO} connects v_O across L_O. v_O drains the core because L_I and L_O couple in opposite directions, so v_{LO} is $-v_O$.

L_I's i_{LI} in Fig. 3.29 therefore ramps up with the v_E that v_{LI}'s v_{IN} impresses across L_I and L_O's i_{LO} ramps down with the $-v_D$ that v_{LO}'s $-v_O$ impresses across L_O.

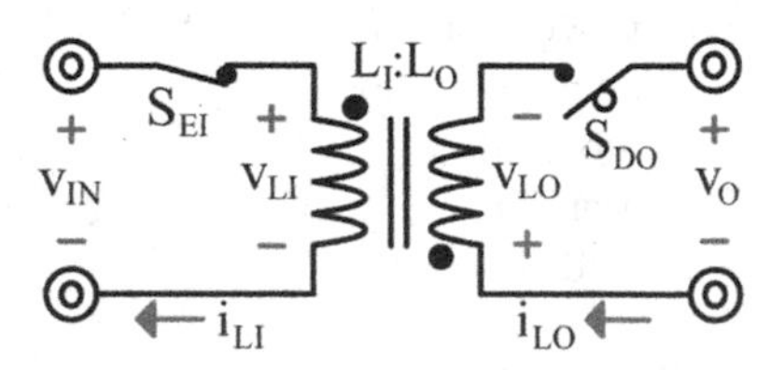

Fig. 3.28 Ideal (supply-switched) flyback

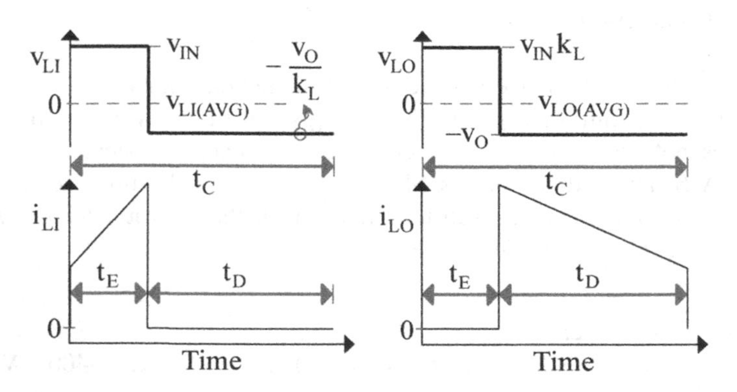

Fig. 3.29 Continuous-conduction waveforms in the flyback

Since S_{DO} opens across t_E and S_{EI} opens across t_D, i_{LO} is zero across t_E and i_{LI} is zero across t_D. As a result, v_{IN} couples a transformer translation of v_{IN} to L_O across t_E and v_O couples "back" a transformer translation of $-v_O$ to L_I across t_D. So when S_{EI} opens, v_{LI} practically "flies" from v_{IN} to $-v_O/k_L$ and v_{LO} from $v_{IN}k_L$ to $-v_O$. This "flyback" action on L_I is how this power converter derives its name.

As a whole, the coupled inductors L_I:L_O operate and behave like L_X in the buck–boost. They energize and drain with v_{IN} and v_O. And the combined current they produce i_L ripples about an average $i_{L(AVG)}$ that the controller adjusts like Fig. 3.6 shows.

B. Duty-Cycle Translation

In steady state, the average voltages across L_I and L_O are zero. Since v_{IN} is across L_I a t_E fraction of t_C and $-v_O/k_L$ couples back across L_I a t_D fraction, $v_{LI(AVG)}$ incorporates duty-cycled fractions of v_{IN} and $-v_O/k_L$:

$$v_{LI(AVG)} = v_{IN}\left(\frac{t_E}{t_C}\right) - \left(\frac{v_O}{k_L}\right)\left(\frac{t_D}{t_C}\right) = v_{IN}d_E - v_O\left(\sqrt{\frac{L_I}{L_O}}\right)d_D = 0. \tag{3.64}$$

Similarly, $v_{LO(AVG)}$ incorporates duty-cycled fractions of $v_{IN}k_L$ and $-v_O$ because $v_{IN}k_L$ couples across L_O a t_E fraction of t_C and $-v_O$ is across L_O a t_D fraction:

$$v_{LO(AVG)} = v_{IN}k_L\left(\frac{t_E}{t_C}\right) - v_O\left(\frac{t_D}{t_C}\right) = v_{IN}\left(\sqrt{\frac{L_O}{L_I}}\right)d_E - v_Od_D = 0. \tag{3.65}$$

Since d_D is $1 - d_E$, d_E is a v_O fraction of $v_{IN}k_L + v_O$:

$$d_E = \frac{v_D}{v_E + v_D} = \frac{v_O/k_L}{v_{IN} + (v_O/k_L)} = \frac{v_O}{v_{IN}k_L + v_O}, \tag{3.66}$$

like L_I's v_{IN} for v_E and v_O/k_L for v_D and L_O's $v_{IN}k_L$ for v_E and v_O for v_D in the general expression predict.

Notice that v_O is a duty-cycled scalar d_E/d_D of $v_{IN}k_L$. And together, v_O/v_{IN} scales with k_L and d_E/d_D:

$$\frac{v_O}{v_{IN}} = k_L\left(\frac{d_E}{d_D}\right) = \left(\sqrt{\frac{L_O}{L_I}}\right)\left(\frac{d_E}{1 - d_E}\right). \tag{3.67}$$

d_E/d_D is greater than one, and v_O is correspondingly greater than the transformer translation of v_{IN} when d_E is greater than 50% (and d_D is less than 50%). d_E/d_D is less than one and v_O is correspondingly less than the transformer translation of v_{IN} otherwise. So like the buck–boost, the flyback can buck and boost v_{IN}.

Example 15: Determine d_E when v_{IN} is 2 V, L_I is 5 μH, L_O is 20 μH, and v_O is 2 V.

Solution:

$$k_L = \sqrt{\frac{L_O}{L_I}} = \sqrt{\frac{20\mu}{5\mu}} = 2$$

$$d_E = \frac{v_D}{v_E + v_D} = \frac{v_O}{v_{IN}k_L + v_O} = \frac{2}{2(2) + 2} = 33\%$$

Example 16: Determine d_E for the transformer in Example 15 when v_{IN} is 2 V and v_O is 6 V.

Solution:

$$k_L = 2 \text{ from Example 15}$$

$$d_E = \frac{v_D}{v_E + v_D} = \frac{v_O}{v_{IN}k_L + v_O} = \frac{6}{2(2) + 6} = 60\%$$

Explore with SPICE:
See Appendix A for notes on SPICE simulations.

```
* Ideal Flyback in CCM
vde de 0 dc=0 pulse 0 1 50n 1n 1n 333n 1u
vin vin 0 dc=2
sei vin vswi de 0 sw1v
li vswi 0 5u
k1 li lo 1
lo 0 vswo 20u
```

(continued)

```
ddo vswo vo idiode
co vo 0 5u
ro vo 0 10
.lib lib.txt
.tran 700u
.end
```

Tip: Plot v(vo), i(Li), i(Lo), v(vswi), and v(vswo) and view across 700 μs and from 695 to 700 μs.

C. Power

Without a physical (galvanic) connection, v_{IN} cannot deliver power directly to v_O like the buck and boost can. L_I:L_O therefore carries all the energy v_{IN} delivers. This means that the two switches carry more current than their buck and boost counterparts. So like the buck–boost, the flyback usually burns more Ohmic power than the buck and boost.

Together, i_{LI} and i_{LO} carry the *transformer current* i_X that L_I:L_O's magnetic core carries. With respect to L_I, i_{XI} carries i_{LI} and a k_L translation of i_{LO}. i_{XO} similarly carries i_{LO} and a reverse k_L translation of i_{LI} with respect to L_O. So i_{XI} is $k_L i_{XO}$ and i_{XO} is i_{XI}/k_L:

$$i_{XI} = k_L i_{XO} = i_{LI} + k_L i_{LO} \tag{3.68}$$

$$i_{XO} = \frac{i_{XI}}{k_L} = \frac{i_{LI}}{k_L} + i_{LO}. \tag{3.69}$$

Like the buck–boost, v_O receives a d_D fraction of i_{XO}, so the output delivers $v_O i_{XO(AVG)} d_D$ with $v_O i_{O(AVG)}$:

$$P_O = v_O i_{O(AVG)} = v_O i_{XO(AVG)} d_O = v_O i_{XO(AVG)} d_D = v_O i_{XO(AVG)} (1 - d_E). \tag{3.70}$$

And because v_{IN} supplies a d_E fraction of i_{XI} and i_{XI} is a k_L translation of i_{XO}, v_{IN} supplies $v_{IN}(i_{O(AVG)}/d_D)k_L d_E$ with $v_{IN} i_{XI(AVG)} d_E$:

$$\begin{aligned} P_{IN} &= v_{IN} i_{IN(AVG)} = v_{IN} i_{XI(AVG)} d_{IN} = v_{IN} i_{XI(AVG)} d_E \\ &= v_{IN} i_{XO(AVG)} k_L d_E = v_{IN} \left(\frac{i_{O(AVG)}}{d_D} \right) k_L d_E. \end{aligned} \tag{3.71}$$

Since ideal switches do not burn power, this P_{IN} matches P_O (because $v_D d_D$'s $v_O d_D$ matches $v_E d_E$'s $v_{IN} d_E$, which is to say $i_{XI(AVG)}$ is $i_{XO(AVG)} k_L$, $i_{XO(AVG)}$ is $i_{O(AVG)}/d_D$, and v_O is $v_{IN} k_L d_E/d_D$). In short, v_O receives the energy v_{IN} supplies with L_I:L_O.

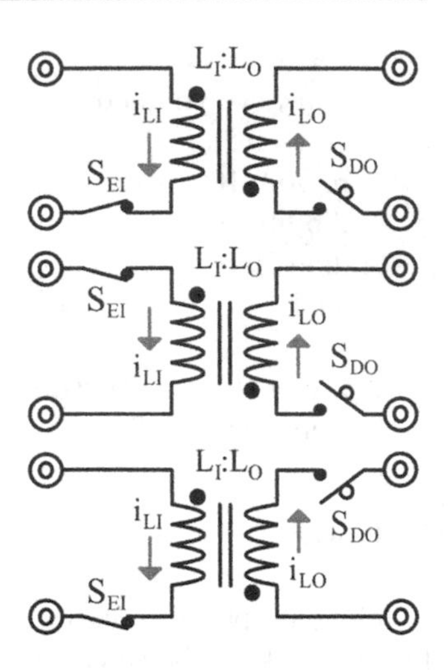

Fig. 3.30 Ground- and supply-switched flyback variations

D. **Variants**

Although ground and supply switches can (at the same time) connect and disconnect L_I from v_{IN} and L_O from v_O, only one switch per side is necessary like Figs. 3.28 and 3.30 show. A second switch would burn power, require space, and complicate the controller needlessly. Of these, the ground-switched input and supply-switched output variant in Fig. 3.30 is probably the most popular because a low-side switch is often less resistive and connecting L_O to v_O's ground plane produces less ground noise. Device availability, breakdown voltage, and conductivity ultimately dictate which switches are possible, more reliable, and less lossy.

E. **Snubbers**

In practice, parts of L_I and L_O do not couple. This means that L_O cannot drain the energy that v_{IN} injects across t_E into L_I's uncoupled fraction. So when S_{EI} opens, remnant i_{LI} charges the parasitic capacitances C_{SWI} that remain attached to S_{EI}'s switching node v_{SWI}. This is often a problem because v_{SWI} can swing above S_{EI}'s breakdown level. *Snubbers* protect switches from overvoltage conditions of this sort.

Without protection, L_I's uncoupled fraction L_I' drains into C_{SWI}, C_{SWI} drains back into L_I', and if S_{EI} does not break, C_{SWI} and L_I' exchange energy until parasitic resistances burn the energy or t_{SW} lapses. One way of limiting v_{SWI}'s swing is to dissipate some of this energy quickly. The purpose of R_{SI} in the *damper* that R_{SI} and C_{SI} implement in Fig. 3.31 is just this: to burn remnant energy in the core.

For this, C_{SI} should shunt and short below the resonant frequency f_{LC}. In other words, R_{SI} and C_{SI}'s combined impedance Z_{SI} at the resonant frequency f_{LC} should be lower than C_{SWI}'s Z_{SWI}. This way, Z_{SI} can steer remnant i_{LI} away from C_{SWI} into R_{SI} when S_{EI} opens. So R_{SI} burns energy, C_{SWI} peaks to a lower voltage, and L_I and C_{SI} together with C_{SWI} exchange energy across fewer cycles.

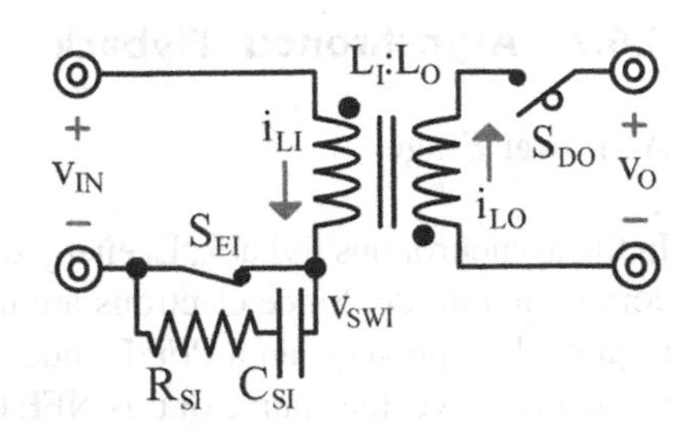

Fig. 3.31 Input-damped flyback

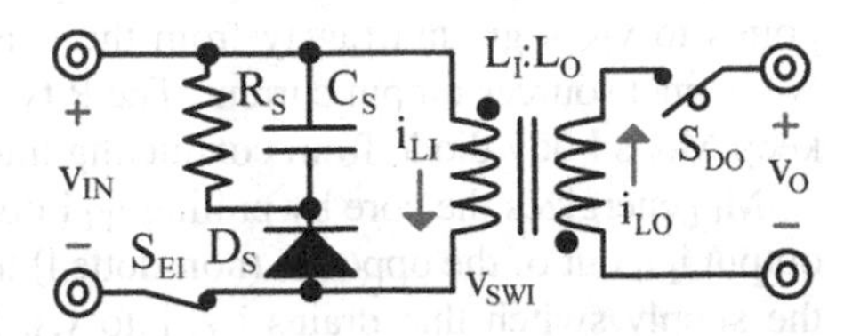

Fig. 3.32 Input-clamped flyback

Z_{SI}, however, should not load L_I across t_{SW} to the extent that C_{SWI} cannot "fly" to $v_{IN} + v_O/k_L$. In other words, C_{SI} should only shunt and short above f_{SW}, not below. More to the point, R_{SI} should current-limit C_{SI} at a frequency f_{SI} that is greater than f_{SW}:

$$\left.\frac{1}{sC_{SI}}\right|_{f_{SW}<f_{SI}=\frac{1}{2\pi R_{SI}C_{SI}}<f_{LC}} = R_{SI}. \tag{3.72}$$

This way, Z_{SI} is greater than R_{SI} below and near f_{SW}. But since R_{SI} should also burn LC energy, f_{SI} should also be lower than f_{LC}.

Another way to limit C_{SWI}'s swing is to clamp v_{SWI}. For this, D_S and C_S in Fig. 3.32 charge C_S to v_O's coupling target v_O/k_L minus D_S's diode voltage v_{DS}. If C_S is high enough to behave like a battery, C_S holds $v_O/k_L - v_{DS}$ and clamps v_{SWI} to $v_{IN} + v_O/k_L$ when S_{EI} opens. In other words, C_S absorbs remnant i_{LI} in L_I away from C_{SWI} without altering v_{SWI} too much.

But since excess i_{LI} is often systemic, energy in C_S accumulates and grows over time. The purpose of R_S is to leak and burn this excess energy before the next cycle begins. If R_S is too low, however, R_S leaks energy from C_S that the core needs to replenish with what could have been delivered to v_O. Still, R_S should be lower than necessary to ensure v_{SWI} does not drift above $v_{IN} + v_O/k_L$. This additional margin dictates how much power R_S burns.

The amount of L_I that does not couple depends on the manufacturing process. And, R_S, C_S, and D_S vary with technology, fabrication lots, and temperature. So engineers often resort to empirical methods (via simulations or experiments) when choosing R_S and C_S. Dampers are usually preferable over clampers because R_S in clampers burns not only excess L_I energy but also core energy meant for v_O. Although less frequently necessary, a damper across S_{DO} similarly protects S_{DO} from the effects of remnant i_{LO} in L_O.

3.6.2 Asynchronous Flyback

A. Power Stage

In the asynchronous flyback, L_I energizes the core with a transistor and L_O drains the core with a diode. Since electrons are more mobile than holes, an NFET burns and requires less power than a PFET under equivalent gate-drive conditions. For maximum gate drive, the source of this NFET should connect to the lowest potential. This is why M_{EI} in Fig. 3.33 is the N-channel ground switch that energizes L_I. The source points to v_{IN}'s ground (away from the channel through which i_{LI} flows) because N-channel sources output current. The P-type body also connects to v_{IN}'s ground to keep M_{EI}'s body diode from conducting undesirable i_{IN} when v_{LI} "flies" high.

M_{EI} energizes the core by pulling i_{LI} into L_I's dotted terminal. L_O must therefore output i_{LO} out of the opposite (non-dotted) terminal. This is why D_{DO} in Fig. 3.33 is the supply switch that drains i_{LO} into v_O. Since D_{DO} blocks reverse current, v_{LO} "flies" high uninterruptedly when M_{EI} energizes L_I. Although connecting D_{DO} in v_O's ground path also works, grounding L_O without a switch in series ties L_O to v_O's common ground plane with lower resistance. This way, ground noise is common to all v_O components, so its effects are minimal.

B. Duty-Cycle Translation

Since D_{DO} drops a diode voltage above v_O, L_O drains the core with v_O plus D_{DO}'s diode voltage v_{DO}. This higher v_D drains the core faster than in the ideal flyback, so t_D and t_C are shorter and t_E is a larger fraction of t_C. In other words, d_E is higher than in the ideal flyback.

v_{LI} therefore swings between v_{IN} and $-(v_O + v_{DO})/k_L$ and v_{LO} between $v_{IN}k_L$ and $-(v_O + v_{DO})$. v_{LI} is v_{IN} a d_E' fraction of the t_C, $-(v_O + v_{DO})/k_L$ a d_D' fraction, and zero on average:

$$\begin{aligned} v_{LI(AVG)} &\approx v_{IN}d_E' - \left(\frac{v_O + v_{DO}}{k_L}\right)d_D' \\ &= v_{IN}d_E' - (v_O + v_{DO})\left(\sqrt{\frac{L_I}{L_O}}\right)d_D' = 0. \end{aligned} \tag{3.73}$$

v_{LO} is similarly $v_{IN}k_L$ a d_E' fraction of t_C, $-(v_O + v_{DO})$ a d_D' fraction, and zero on average:

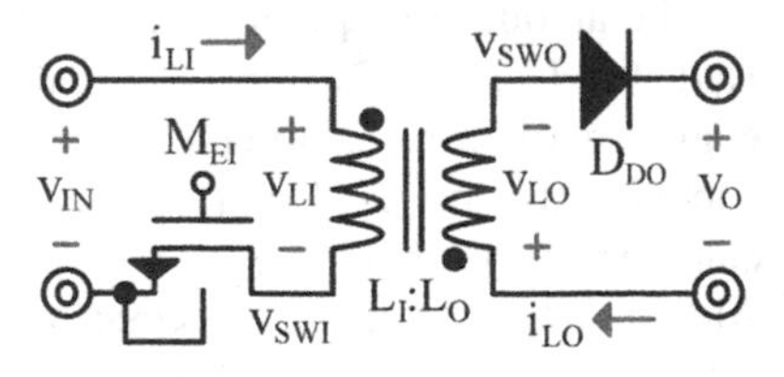

Fig. 3.33 Asynchronous flyback

$$v_{LO(AVG)} \approx v_{IN}k_L d_E' - (v_O + v_{DO})d_D' = v_{IN}\left(\sqrt{\frac{L_O}{L_I}}\right)d_E' - (v_O + v_{DO})d_D' = 0. \quad (3.74)$$

Since d_D' is $1 - d_E'$, the diode raises v_O in L_O's v_D in the numerator of d_E' by a larger fraction than in the denominator that v_E's $v_{IN}k_L$ and v_D's v_O set:

$$d_E' = \frac{v_D}{v_E + v_D} \approx \frac{(v_O + v_{DO})/k_L}{v_{IN} + (v_O + v_{DO})/k_L} = \frac{v_O + v_{DO}}{v_{IN}k_L + v_O + v_{DO}} > d_E. \quad (3.75)$$

So the effect of the diode is to increase d_E to d_E'.

Example 17: Determine d_E' for the transformer in Example 15 when v_{IN} is 1 V, v_O is 6 V, and D_{DO} drops 800 mV.

Solution:

$$k_L = 2 \text{ from Example 15}$$

$$d_E' = \frac{v_D}{v_E + v_D} \approx \frac{v_O + v_{DO}}{v_{IN}k_L + v_O + v_{DO}} = \frac{6 + 800\text{m}}{2(2) + 6 + 800\text{m}} = 63\%$$

Note: d_E' here is higher than d_E in the ideal example because D_{DO} increases L_O's drain voltage. So t_D shortens and t_E's fraction of t_C's t_D + t_E increases. d_E's variation (from ideal to asynchronous) is higher than in the buck–boost because only one diode raises v_D (two diodes raise v_D in the buck–boost).

Explore with SPICE:
See Appendix A for notes on SPICE simulations.

```
* Asynchronous Flyback in CCM
vge vge 0 dc=0 pulse 0 2 50n 1n 1n 630n 1u
vin vin 0 dc=2
li vin vswi 5u
meg vswi vge 0 0 nmos0 w=100m l=250n
```

(continued)

```
k1 li lo 1
lo 0 vswo 20u
ddo vswo vo fdiode1
co vo 0 5u
ro vo 0 10
.lib lib.txt
.tran 700u
.end
```

Tip: Plot v(vo), i(Li), i(Lo), v(vswi), and v(vswo) and view across 700 μs and from 695 to 700 μs.

C. **Conduction Modes**

In continuous conduction, the coupled inductor pair conducts a combined output current i_{XO} about an average $i_{XO(AVG)}$ that keeps i_{XO} above zero. When transferring little power, $i_{XO(AVG)}$ can be so low that i_{XO} reaches zero before t_{SW} lapses, like the thicker traces in Fig. 3.34 demonstrate. Since D_{DO} cannot conduct reverse current,

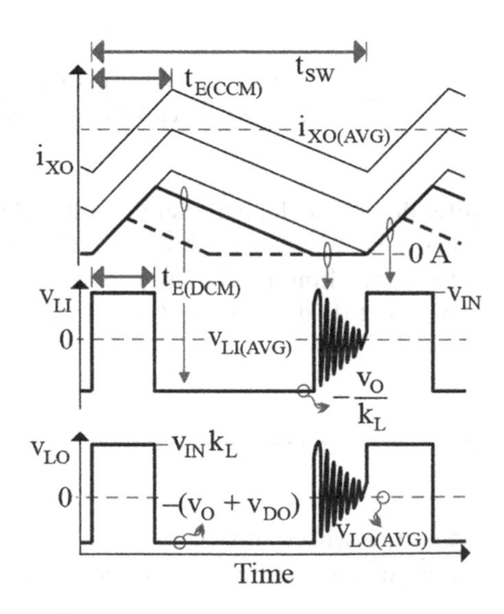

Fig. 3.34 Discontinuous-conduction waveforms in the flyback

i_{XO} reaches and remains near zero until the next t_{SW}. L_I:L_O is in discontinuous conduction when this happens.

Just after L_I:L_O drains, $-v_{LO}$ is v_{DO} over v_O and $-v_{LI}$ is a reverse k_L translation of v_{LO}. So when drained and disconnected, v_{SWI}'s parasitic capacitance impresses a voltage across L_I that draws power from v_{IN} and C_{SWI} and energizes L_I and later C_{SWI}. i_{LI} begins to drain L_I into C_{SWI} after v_{SWI} rises over v_{IN}.

When L_I depletes, v_{SWI} and v_{IN} impress a voltage across L_I that drains C_{SWI} into L_I and v_{IN}. L_I then drains into v_{IN} when v_{SWI} falls below v_{IN}. v_{SWI} is again below v_{IN} when L_I depletes, so the entire sequence repeats. v_{LI} oscillates with v_{SWI} this way about zero until resistances burn the energy or t_{SW} lapses.

v_O and D_{DO}'s v_{SWO} similarly impress a voltage across L_O when the core depletes that drains v_{SWO}'s C_{SWO} into L_O. L_O then drains into C_{SWO} to charge C_{SWO} in the negative direction, C_{SWO} drains back into L_O, L_O depletes into C_{SWO}, and the sequence repeats. v_{LO} oscillates with v_{SWO} about zero until resistances burn the energy or t_{SW} lapses.

Explore with SPICE:
See Appendix A for notes on SPICE simulations.

```
* Asynchronous Flyback in DCM
vge vge 0 dc=0 pulse 0 2 50n 1n 1n 100n 1u
vin vin 0 dc=2
li vin vli 5u
rli vli vswi 25
meg vswi vge 0 0 nmos0 w=100m l=250n
k1 li lo 1
lo 0 vlo 20u
rlo vlo vswo 25
ddo vswo vo fdiode1
vo vo 0 dc=6
.lib lib.txt
.tran 2u
.end
```

Tip: Plot i(Li), i(Lo), v(vswi), and v(vswo).

3.6.3 Synchronous Flyback

A. Power Stage

The difference between asynchronous and synchronous power supplies is what drains the core: diodes in one and transistors in the other. So the ground input FET M_{EI} that energizes the asynchronous flyback and implements S_{EI} in the ideal flyback

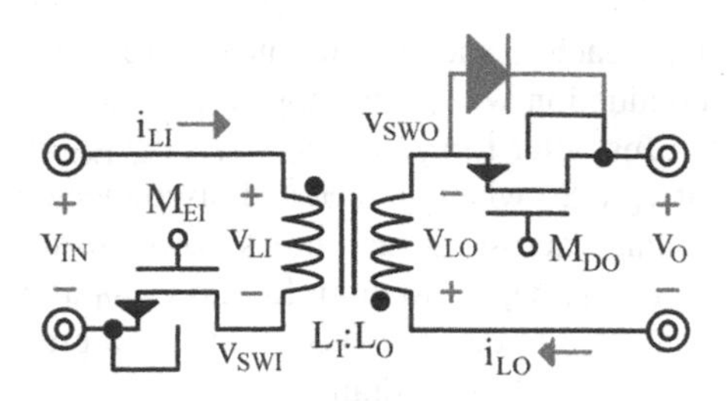

Fig. 3.35 Synchronous flyback

also energizes the synchronous counterpart in Fig. 3.35. The difference here is that M_{DO} drains L_O, which is the function of S_{DO} and D_{DO} in the ideal and asynchronous flybacks.

When closed, M_{DO}'s source and drain voltages v_S and v_D are close to v_O. M_{DO}'s gate voltage v_G can swing reliably between v_O and ground. With these voltages, the maximum gate drive of an NFET would be v_{GST}, $v_G - v_S - v_{TN}$, $v_O - v_O - v_{TN}$, or just $-v_{TN}$, which is too low to close an NFET. The maximum gate drive of a PFET would be v_{SGT}, $v_S - v_G - |v_{TP}|$, $v_O - 0 - |v_{TP}|$, or $v_O - |v_{TP}|$, which is greater and more likely to close the switch. This is why M_{DO} is oftentimes a PFET, because a high v_S calls for a high-side switch.

Since PMOS source terminals receive current, L_O feeds M_{DO}'s source and M_{DO}'s source arrow points into the transistor (into the channel through which i_L flows). The P-type drain terminal connects to the N-type body to block reverse i_O when v_{SWO} falls to $-v_{IN}k_L$. This connection also allows the surviving body diode to steer L_O's dead-time current into v_O.

B. **Duty-Cycle Translation**

The difference between asynchronous and synchronous operation across t_D is that M_{DO} drops a lower voltage than D_{DO} does. So v_{SWO} is close to v_O when M_{DO} closes within t_D. But since M_{EI} and M_{DO} open across the dead-time portions of t_D and M_{DO}'s body diode offers a drain path for i_{XO}, v_{SWO} reaches $v_O + v_{DO}$ across the t_{DT}'s within t_D.

This means that L_O drains two t_{DT} fractions of t_C with v_O plus a diode voltage and the remainder of t_D's fraction with v_O. This higher v_D drains the core faster than in the ideal flyback, but not as fast as in the asynchronous flyback because v_D is higher only across t_{DT}'s. t_D and t_C are therefore shorter, but not as short as in the asynchronous flyback. t_E is similarly a longer fraction of t_C, but also not as long a fraction as the asynchronous d_E' is.

As a result, v_{LI} is v_{IN} a t_E fraction of the t_C that t_{SW} sets, $-(v_O + v_{DO})/k_L$ two t_{DT} fractions, $-v_O/k_L$ the fraction of t_D that remains, and zero on average:

$$v_{LI(AVG)} \approx v_{IN}d_E'' - \left(\frac{v_O + v_{DO}}{k_L}\right)\left(\frac{2t_{DT}}{t_{SW}}\right) - \frac{v_O}{k_L}\left(\frac{t_D - 2t_{DT}}{t_{SW}}\right) = 0. \quad (3.76)$$

v_{LO} is similarly $v_{IN}k_L$ a t_E fraction of t_{SW}, $-(v_O + v_{DO})$ two t_{DT} fractions, $-v_O$ the fraction of t_D that remains, and zero on average:

$$v_{LO(AVG)} \approx v_{IN}k_Ld_E'' - (v_O + v_{DO})\left(\frac{2t_{DT}}{t_{SW}}\right) - v_O\left(\frac{t_D - 2t_{DT}}{t_{SW}}\right) = 0. \quad (3.77)$$

Since d_D'' is $1 - d_E''$, the diode raises v_D or v_O in the numerator of d_E' by a larger fraction than in the denominator that v_E's $v_{IN}k_L$ and v_D's v_O set:

$$\begin{aligned} d_E'' &= \frac{v_D}{v_E + v_D} \\ &\approx \frac{v_O/k_L}{v_{IN} + (v_O/k_L)} + \left[\frac{v_{DO}/k_L}{v_{IN} + (v_O/k_L)}\right]\left(\frac{2t_{DT}}{t_{SW}}\right) \\ &= \frac{v_O}{v_{IN}k_L + v_O} + \left(\frac{v_{DO}}{v_{IN}k_L + v_O}\right)\left(\frac{2t_{DT}}{t_{SW}}\right) \\ &= d_E + \left(\frac{v_{DO}}{v_{IN}k_L + v_O}\right)\left(\frac{2t_{DT}}{t_{SW}}\right) > d_E. \end{aligned} \quad (3.78)$$

Example 18: Determine d_E'' for the transformer in Example 15 when v_{IN} is 2 V, v_O is 6 V, D_{DO} drops 400 mV, t_{DT} is 50 ns, and t_{SW} is 1 μs.

Solution:

$$k_L = 2 \text{ from Example 15}$$

$$\begin{aligned} d_E'' &= \frac{v_D}{v_E + v_D} \\ &\approx \frac{v_O}{v_{IN}k_L + v_O} + \left(\frac{v_{DO}}{v_{IN}k_L + v_O}\right)\left(\frac{2t_{DT}}{t_{SW}}\right) \\ &= \frac{6}{2(2) + 6} + \left[\frac{800\text{m}}{2(2) + 6}\right]\left[\frac{2(50\text{n})}{1\mu}\right] = 61\% \end{aligned}$$

Note: d_E'' here is higher than d_E in the ideal example because D_{DO} raises v_D. But since the diode does so only two t_{DT} fractions of t_{SW}, t_D shortens and t_E's fraction of t_C rises less than d_E' in the asynchronous example.

Explore with SPICE:
See Appendix A for notes on SPICE simulations.

```
* Synchronous Flyback in CCM
vge vge 0 dc=0 pulse 0 2 50n 1n 1n 610n 1u
vgdb vgdb 0 dc=0 pulse 0 6 0n 1n 1n 710n 1u
vin vin 0 dc=2
li vin vswi 5u
meg vswi vge 0 0 fnmos0 w=100m l=250n
cswi vswi 0 1f
k1 li lo 1
lo 0 vswo 20u
mdo vo vgdb vswo vo fpmos0 w=300m l=250n
cswo vswo 0 1f
co vo 0 5u
ro vo 0 10
.lib lib.txt
.tran 700u
.end
```

Tip: Plot v(vo), i(Li), i(Lo), v(vswi), and v(vswo) and view across 700 μs and from 695 to 700 μs.

C. **Conduction Modes**

If the controller opens and closes M_{DO} when the asynchronous diode D_{DO} would, the only difference between asynchronous and synchronous operation is the voltage dropped across the switch: millivolts with FETs and 600–800 mV with diodes. But since M_{DO} is off across dead-time periods, M_{DO}'s body diode conducts across t_{DT}'s like D_{DO} in the asynchronous flyback.

If the controller does not open M_{DO} when i_{XO} reaches zero before t_{SW} lapses, M_{DO} lets i_{XO} reverse direction. Since M_{DO}'s body diode cannot conduct this negative i_{XO} across the t_{DT} that follows, M_{EI}'s body diode conducts i_{XI}'s $k_L i_{XO}$ into v_{IN}. So v_{SWI} falls a diode voltage across this t_{DT} and returns to zero when M_{EI} closes. Returning energy to v_{IN} this way means L_X transfers and burns more power than necessary.

Note one t_{DT} is in t_D and another in t_E when i_{XO} reverses, whereas without negative conduction, t_D includes both t_{DT}'s. And since v_{SWI} falls and v_{SWO} rises by a similar diode voltage across similar t_{DT}'s, diode effects on $v_{L(AVG)}$'s tend to cancel. So with negative conduction, d_E is close to the ideal case (and lower than d_E'').

Explore with SPICE:
See Appendix A for notes on SPICE simulations.

```
* Synchronous Flyback with Negative Conduction
vge vge 0 dc=0 pulse 0 2 50n 1n 1n 550n 1u
vgdb vgdb 0 dc=0 pulse 0 6 0n 1n 1n 650n 1u
vin vin 0 dc=2
li vin vswi 5u
meg vswi vge 0 0 fnmos0 w=100m l=250n
k1 li lo 1
lo 0 vswo 20u
mdo vo vgdb vswo vo fpmos0 w=300m l=250n
vo vo 0 dc=6
.ic i(lo)=-60m
.lib lib.txt
.tran 2u
.end
```

Tip: Plot i(Li), i(Lo), v(vswi), and v(vswo).

D. **Diode Conduction**

If M_{DO}'s threshold voltage is lower than a diode voltage, M_{DO} closes across t_{DT}'s when v_{SWO} climbs $|v_{TP}|$ over M_{DO}'s v_O-supplied gate voltage. M_{DO}'s body diode does not inject noise current into the substrate when this happens because it does not conduct. A Schottky diode across M_{DO} similarly channels dead-time current away from the body diode and the substrate into which the parasitic BJT present injects current.

Explore with SPICE:
Use the previous SPICE code, set V_{TN0} to 400 mV and V_{TP0} to −400 mV (use "nmos1" and "pmos1" models), and re-run the simulation.

3.7 Summary

Inductors that share magnetic space can energize and drain that space with voltages of opposing polarities. This is how switched inductors and transformers transfer input power to output loads. Unfortunately, series resistances burn some of this energy. And nearby coils and fast-changing currents restrict the medium through which their currents can flow. So the power lost to resistance in the coil increases with more nearby coils and higher switching frequency.

But since this loss is a small fraction of the power drawn, many consumer products use this method to transfer power from ac–dc chargers and internal batteries to electronic systems that require stable dc power supplies. In these applications, the inputs and outputs of the switched inductors are static or quasi-static voltages. Switchers use these dc voltages to ramp their inductor currents up and down. Since inductor current rises as much as it falls in steady state, these same voltages set the duty-cycle fractions of the switching period that energize and drain the inductors.

CMOS solutions use MOSFETs to energize inductors and diodes or MOSFETs to drain them. Which type of MOSFET to select depends on the gate drive available. To supply the energy these switches ultimately burn, switchers must energize inductors across a longer duty-cycle fraction of the switching period. This way, they can overcome the losses and deliver the power their loads require.

Switched inductors can buck and boost input voltages to lower and higher output levels with four switches. Removing the two input or two output switches sets an average voltage that only a higher voltage at the opposite end can balance. This is why two switches can buck or boost, but not both.

Asynchronous circuits drain the inductor with diodes and synchronous circuits with MOSFETs. To avoid momentary shorts, synchronous solutions insert dead times between the conduction periods of adjacent switches. Body, MOS, or Schottky diodes conduct the inductor current across these times.

In flybacks, an input inductor magnetizes the space that an output inductor drains. This way, input sources and output loads need not share a common ground (for galvanic isolation). But since parts of the input inductor do not couple to the output, engineers use snubbers to burn leftover energy.

Switched inductors are vital in electronic systems. One reason for this is they can boost input supply voltages to higher output levels, which is not possible with linear (non-switched) power stages. Another reason is they burn less power than their linear counterparts when transferring moderate to high power levels. This is why switched inductors are so pervasive in power supplies.

Power Losses

4

Abbreviations

BJT	Bipolar-junction transistor
CCM	Continuous-conduction mode
CMOS	Complementary MOS
DCM	Discontinuous-conduction mode
ESD	Electrostatic-discharge protection
ESR	Equivalent series resistance
FET	Field-effect transistor
FM	Frequency modulation
LED	Light-emitting diode
MOS	Metal–oxide–semiconductor
MPP	Maximum-power point
MPPT	MPP tracker
NMOS	N-channel MOSFET
PDCM	Pseudo-DCM
PFM	Pulse FM
PM	Peak modulation
PMOS	P-channel MOSFET
RMS	Root-mean-square
SL	Switched inductor
ZCS	Zero-current switching
ZVS	Zero-voltage switching
C_{CH}	Channel capacitance
C_{DB}	Drain–body capacitance
C_G	Gate capacitance
C_{GD}	Gate–drain capacitance
C_{GS}	Gate–source capacitance
C_{OL}	Overlap capacitance

G. A. Rincón-Mora, *Switched Inductor Power IC Design*,
https://doi.org/10.1007/978-3-030-95899-2_4

C_{SB}	Source–body capacitance
C_{SWI}	Input switch-node capacitance
C_{SWO}	Output switch-node capacitance
d_D	Drain duty cycle
d_E	Energize duty cycle
d_{IN}	Input duty cycle
D_{DG}	Ground drain diode
D_{DO}	Output drain diode
D_{DT}	Dead-time diode
d_O	Output duty cycle
f_{SW}	Switching frequency
i_D	Driver current
i_{DS}	Drain–source current
i_G	Gate current
i_{LD}	Load current
i_{IN}	Input current
i_L	Inductor current
$i_{L(HI)}$	Inductor current peak in CCM
$i_{L(MIN)}$	Minimum inductor current
$i_{L(LO)}$	Inductor current valley in CCM
$i_{L(PK)}$	Peak inductor current in DCM
i_O	Output current
i_{OFF}	Off current
i_{RR}	Reverse-recovery current
i_{SUB}	Substrate current
Δi_L	CCM inductor ripple current
i_Δ	Triangular current
I_S	Reverse saturation current
K'	Transconductance parameter
L_{CH}	Channel length
L_{MIN}	Minimum allowable oxide length
L_X	Switched/transfer inductor
P_B	Battery power
P_D	Drive power
P_{DD}	Diode drain power
P_{DT}	Dead-time power
P_G	Gate-charge power
P_{GI}	Driver gate-charge power
P_{IN}	Input power
P_{IV}	i_{DS}–v_{DS} overlap power
P_L	Inductor power
P_{LD}	Load power
P_{LOSS}	Power losses
P_{MOS}	MOS power
P_O	Output power

P_{OFF}	Cut-off power
P_R	Ohmic power
P_{RR}	Reverse-recovery power
P_{SWI}	Input switch-node power
P_{SWO}	Output switch-node power
q_{DIF}	Diffusion charge
q_{RR}	Reverse-recovery charge
R_{CH}	Channel resistance
R_D	Drain resistance
R_{DG}	Ground drain resistance
R_{DO}	Output drain resistance
R_E	Energize resistance
R_{EG}	Ground energize resistance
R_{EI}	Input energize resistance
$R_{L(AC)}$	Inductor's ac resistance
$R_{L(DC)}$	Inductor's dc resistance
R_N	Pull-down N-type resistance
R_{OFF}	Off resistance
R_P	Pull-up P-type resistance
R_S	Source resistance
S_{DG}	Ground drain switch
S_{DO}	Output drain switch
S_{EG}	Ground energize switch
S_{EI}	Input energize switch
t_C	Conduction time
t_D	Drain time
t_{DT}	Dead time
t_E	Energize time
t_{SW}	Switching period
T_J	Junction temperature
τ_F	Forward transit time
v_B	Battery voltage/battery
v_D	Drain voltage
v_{DD}	Power supply
v_{DS}	Drain–source voltage
$v_{DS(SAT)}$	Saturation voltage
v_E	Energize voltage
v_{GS}	Gate–source voltage
v_{IN}	Input voltage/input
v_O	Output voltage/output
v_S	Source voltage/source
v_{SWI}	Input switching node/voltage
v_{SWO}	Output switching node/voltage
v_T	MOS threshold voltage
v_{TH}	Gate–source threshold

V_{TO}	Zero-bias threshold
W_{CH}	Channel width
W_{CH}'	Optimal channel width
λ	Channel-length modulation parameter
η_C	Power-conversion efficiency
σ_{LOSS}	Fractional loss

Switched-inductor (SL) power supplies are pervasive in electronic systems because they output a large fraction of the power they draw from their inputs. The main reason for this is the voltages that switches drop are a very small fraction of the input and output voltages. So the inductor current draws and delivers a lot more input power into the output than switches consume.

Still, the heat that burning power generates can compromise electronic performance and mechanical integrity. And losing battery energy or ambient power to the switched inductor reduces the charge life or functionality of an electronic system. So understanding the nature, makeup, and sensitivity of these losses is important.

The most fundamental of these is *conduction power*. This is the power that components consume when they conduct inductor current. Series resistances, diodes, and transistors are to blame for this. Another loss is the power that gate drivers need to transition switches between states. Stray capacitances and large switches also leak power.

The operating mechanics of the switched inductor dictate how these components ultimately consume power. Quantifying losses, however, is not enough. Their significance ultimately rests on the applications they serve and the functionality they provide.

4.1 Power Conversion

Power-conversion efficiency η_C is the fraction of *input power* P_{IN} that the *input* v_{IN} delivers to the *output* v_O in Fig. 4.1:

$$\eta_C \equiv \frac{P_O}{P_{IN}} = \frac{P_O}{P_O + P_{LOSS}} = \frac{P_{IN} - P_{LOSS}}{P_{IN}} = 1 - \frac{P_{LOSS}}{P_{IN}} = 1 - \sigma_{LOSS}. \tag{4.1}$$

In addition to this *output power* P_O, P_{IN} also supplies the *power* P_{LOSS} *lost* to components in the circuit. So P_O outputs what is left: the difference $P_{IN} - P_{LOSS}$,

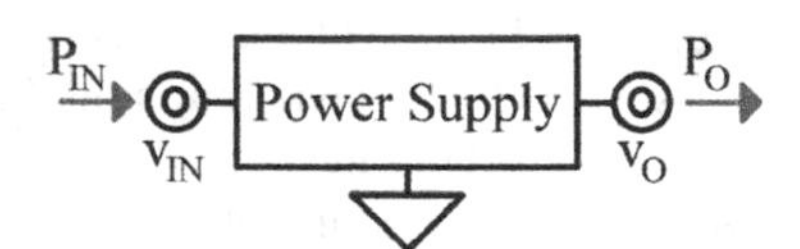

Fig. 4.1 Power supply

fractional loss σ_{LOSS} is the fraction of P_{IN} lost in P_{LOSS}, and η_C is below 100% by the amount σ_{LOSS} dictates. η_C and σ_{LOSS} are complementary measures of efficiency.

P_{IN} is critical in η_C and σ_{LOSS} because P_{LOSS} is a larger fraction of P_{IN} when the i_O that sets P_O and P_{IN} is lower. This is because i_{IN}'s average is an *input duty-cycle* d_{IN} fraction of i_L's average, which is a reverse *output duty-cycle* d_O translation of i_O's average. So when the load is light (i.e., i_O is low), P_{LOSS} is a higher fraction of P_{IN}, and η_C is, in consequence, lower by the higher σ_{LOSS} that P_{LOSS} and P_{IN} set:

$$\sigma_{LOSS} = \frac{P_{LOSS}}{P_{IN}} = \frac{P_{LOSS}}{i_{IN(AVG)} v_{IN}} = \frac{P_{LOSS}}{i_{L(AVG)} d_{IN} v_{IN}} = \frac{P_{LOSS}}{\left(i_{O(AVG)}/d_O\right) d_{IN} v_{IN}}. \tag{4.2}$$

But since boosts connect v_{IN} directly to the *transfer inductor* L_X (without input switches), d_{IN} is one, not a fraction. Similarly, d_O is one in bucks because they connect L_X directly to v_O (without output switches). In other words, i_O's translation to i_{IN} hinges on L_X's connectivity to v_{IN} and v_O.

4.1.1 Voltage Regulators and LED Drivers

Voltage regulators incorporate feedback loops that keep v_O near a prescribed target. This way, v_O hardly varies with the *load current* i_{LD} that v_O in Fig. 4.2 supplies. The v_O that *light-emitting diode* (LED) *drivers* establish is also steady because these drivers similarly keep the *output current* i_O near a target. Since v_O is steady either way and i_O and i_{LD} are independent variables, engineers calculate and show how η_C varies across i_O or the P_O that v_O outputs with i_O:

$$\eta_{C(R)} = \frac{P_O}{P_{IN}} = \frac{P_O}{P_O + P_{LOSS}} = \frac{i_{O(AVG)} v_O}{i_{O(AVG)} v_O + P_{LOSS}} = \frac{i_{LD} v_O}{i_{LD} v_O + P_{LOSS}}, \tag{4.3}$$

where v_O's static dc component V_O is typically much greater than v_O's dynamic ac variation Δv_O. v_{IN} is usually a good voltage source (with low *source resistance* R_S), so v_{IN} can supply all the P_{IN} that i_O with v_O and P_{LOSS} require.

4.1.2 Battery Chargers

Battery chargers normally incorporate feedback loops that keep i_O in Fig. 4.3 steady. This i_O charges a *battery* v_B across its operating range. Since i_O is steady and v_B

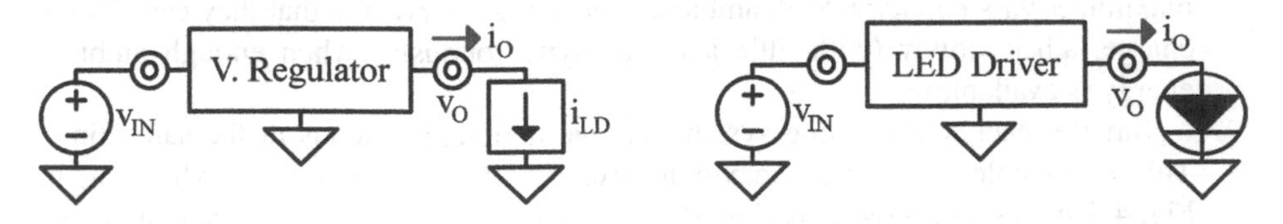

Fig. 4.2 Voltage regulator and LED driver

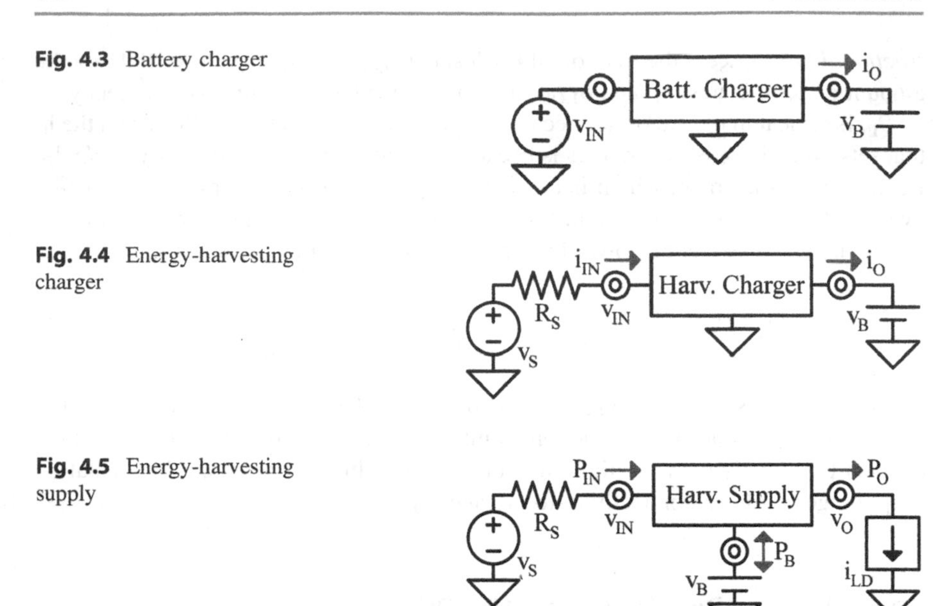

Fig. 4.3 Battery charger

Fig. 4.4 Energy-harvesting charger

Fig. 4.5 Energy-harvesting supply

climbs across a prescribed range, showing how η_C varies across v_B is often more revealing than across the P_O that i_O with v_B set:

$$\eta_{C(C)} = \frac{P_O}{P_{IN}} = \frac{P_O}{P_O + P_{LOSS}} = \frac{i_{O(AVG)} v_O}{i_{O(AVG)} v_O + P_{LOSS}} = \frac{i_{O(AVG)} v_B}{i_{O(AVG)} v_B + P_{LOSS}}, \quad (4.4)$$

where i_O's static dc component I_O is typically much greater than i_O's dynamic ac variation Δi_O.

v_{IN} is typically a low-resistance source that can supply all the P_{IN} that i_O with v_B and P_{LOSS} require. v_B's resistance is also low for a good battery. So $\eta_{C(C)}$ is ultimately the P_{IN} fraction that $i_{O(AVG)}$ and v_B's static component determine.

4.1.3 Energy Harvesters

The fundamental difference between an ambient-derived *source* v_S and a typical input is that v_S is deficient. In other words, part or all of P_O's range overloads v_S. This is why many *energy harvesters* are chargers that supply what v_S avails. So they cannot always output the i_O (in Fig. 4.4) that charges v_B quickly or the i_O that maximizes v_B's *capacity*. Still, ambient energy is so pervasive that they can always charge, albeit slowly (with little i_O) and asynchronously (when enough ambient energy is available).

Smarter energy-harvesting systems charge and supply loads at the same time. This is possible when P_{IN}'s maximum exceeds P_O's minimum. So when P_{IN} in Fig. 4.5 surpasses P_O by more than P_{LOSS}, the harvester supplies P_O and charges v_B

with excess P_{IN}. Otherwise, the harvester draws assistance from v_B, in which case P_{IN} and *battery power* P_B supply P_O and P_{LOSS}.

v_S here is the effective source that *transducers* establish when converting ambient energy into electrical power. R_S models the imperfections that current-limit v_S. v_{IN} therefore peaks to v_S when *input current* i_{IN} is zero and i_{IN} maxes to v_S/R_S when the harvester grounds v_{IN}.

R_S also limits P_{IN}. This R_S is also present in conventional regulators, LED drivers, and chargers. In these, however, $P_{IN(MAX)}$ is greater than $P_{O(MAX)}$, so P_O cannot overload P_{IN}. Ambient sources, on the other hand, do not always avail the same P_{IN}, so P_O in harvesters can and will at times overload P_{IN}.

A. **Maximum-Power Point**

The significance of power-conversion efficiency is that reducing losses conserves energy. η_C is less consequential in a harvester because unused ambient energy transforms into forms that the transducer cannot tap. So harvesters should convert and deliver all the power possible.

Good harvesters draw the P_{IN} that supplies the highest P_O. The feedback loops that keep them at the *maximum-power point* (MPP) are *MPP trackers* (MPPTs). $P_{O(MAX)}$ is therefore a good metric for harvesting efficacy. Since $P_{O(MAX)}$ reflects how much energy is available, $P_{O(MAX)}$ changes with ambient conditions.

The highest possible P_O results when η_C peaks at the MPP. At this point, the harvester draws the most P_{IN} and loses the least P_{LOSS}. When ambient conditions change, the P_{IN} that corresponds to the new MPP changes. So η_C shifts from its peak and $P_{O(MPP)}$ is no longer $P_{O(MAX)}$.

MPPTs usually keep P_O near $P_{O(MPP)}$ by adjusting P_{IN}. Engineers try to max η_C at the most probable P_{IN} so $P_{O(MPP)}$ matches $P_{O(MAX)}$ more frequently. The general aim, however, is to keep η_C high across P_{IN}'s range.

4.2 Operating Mechanics

The purpose of a switched inductor is to transfer v_{IN} energy to v_O. For this, *input* and *ground energize switches* S_{EI} and S_{EG} in Fig. 4.6 energize L_X from v_{IN} and *ground* and *output drain switches* S_{DG} and S_{DO} drain L_X into v_O in alternating phases. This way, v_{IN} produces an *inductor current* i_L that draws P_{IN} from v_{IN} and outputs P_O to v_O.

v_{IN} establishes an *energize voltage* v_E that raises i_L across *energize time* t_E in Fig. 4.7. v_O similarly sets an opposing *drain voltage* v_D that reduces i_L across *drain*

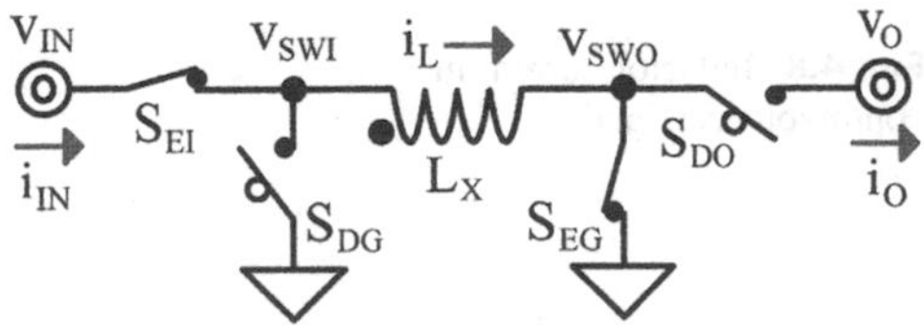

Fig. 4.6 Switched inductor

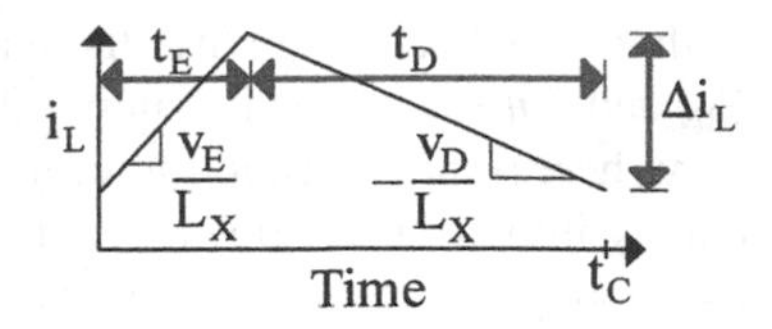

Fig. 4.7 Inductor current

time t_D. i_L rises and falls this way across t_C to produce a *ripple current* Δi_L that repeats across cycles:

$$\Delta i_L = \left(\frac{v_E}{L_X}\right)t_E = \left(\frac{v_D}{L_X}\right)t_D. \tag{4.5}$$

Here, *energize* and *drain duty cycles* d_E and d_D refer to corresponding t_E and t_D fractions of t_C: t_E/t_C and t_D/t_C.

So t_E-to-t_D's ratio follows d_E to d_D's and matches v_D to v_E's:

$$\frac{t_E}{t_D} = \frac{d_E}{d_D} = \frac{v_D}{v_E} = \frac{d_E}{1 - d_E}. \tag{4.6}$$

Since t_D is $t_C - t_E$ and d_D is $1 - d_E$, d_E is v_D's fraction of v_E and v_D:

$$d_E = \frac{v_D}{v_E + v_D}. \tag{4.7}$$

d_E is therefore a function of the v_E and v_D that v_{IN} and v_O set.

4.2.1 Continuous Conduction

In *continuous-conduction mode* (CCM), L_X conducts continuously across the entire *switching period* t_{SW}. This way, t_E and t_D in Fig. 4.8 establish a *conduction time* t_C that extends across t_{SW}. i_L's *CCM valley* $i_{L(LO)}$ is zero or higher, so $i_{L(AVG)}$ is half i_L's ripple Δi_L or higher.

4.2.2 Discontinuous Conduction

L_X conducts a fraction of t_{SW} in *discontinuous-conduction mode* (DCM). So t_C is less than t_{SW}, and i_L in Fig. 4.9 reaches zero at t_C and remains zero until t_{SW} lapses. $i_{L(AVG)}$ across t_{SW} is therefore a fraction of i_L's average $i_{C(AVG)}$ across t_C, which is half i_L's *DCM peak* $i_{L(PK)}$. $i_{L(PK)}$ is ultimately a reflection of the i_O that $i_{L(AVG)}$ across t_{SW} feeds.

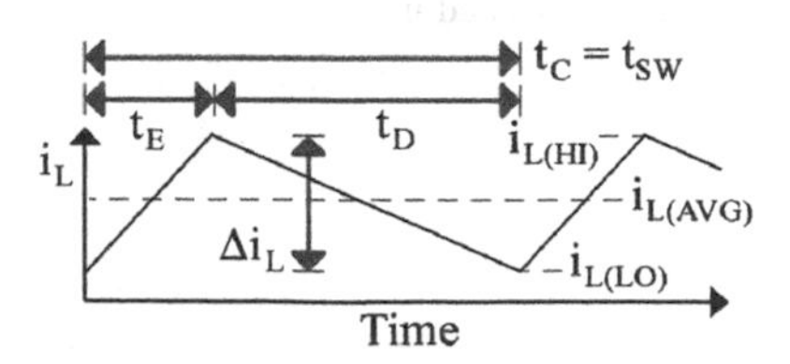

Fig. 4.8 Inductor current in continuous conduction

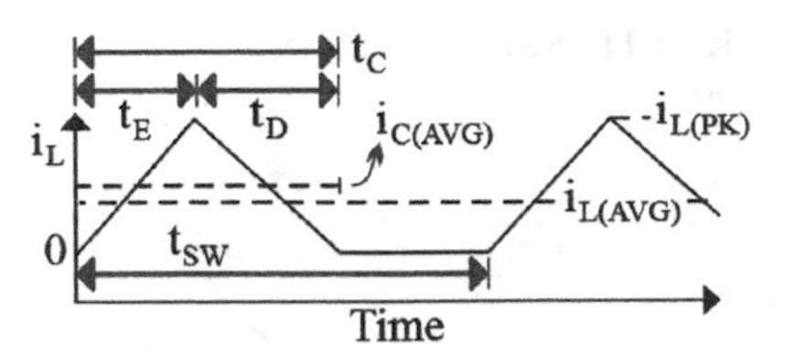

Fig. 4.9 Inductor current in discontinuous conduction

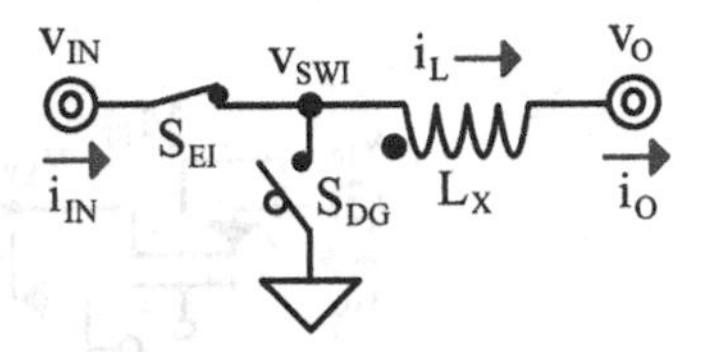

Fig. 4.10 Switched-inductor buck

Since t_E is a d_E fraction of t_C and v_E across L_X raises i_L to $i_{L(PK)}$ across t_E, t_C also scales with $i_{L(PK)}$:

$$t_C = \frac{t_E}{d_E} = \left(\frac{i_{L(PK)}}{d_E}\right)\left(\frac{L_X}{v_E}\right). \tag{4.8}$$

i_L averages half of $i_{L(PK)}$ across t_C and a t_C/t_{SW} fraction of that half across t_{SW}:

$$i_{L(AVG)} = i_{C(AVG)}\left(\frac{t_C}{t_{SW}}\right) = \left(\frac{i_{L(PK)}}{2}\right)\left(\frac{t_C}{t_{SW}}\right) = \frac{{i_{L(PK)}}^2 L_X}{2 d_E v_E t_{SW}}. \tag{4.9}$$

$i_{L(PK)}$ is therefore a square-root translation of $i_{L(AVG)}$:

$$i_{L(PK)} = \sqrt{2 d_E \left(\frac{v_E}{L_X}\right) t_{SW} i_{L(AVG)}} = \sqrt{2 d_E t_{SW}\left(\frac{v_E}{L_X}\right)\left(\frac{i_{O(AVG)}}{d_O}\right)}, \tag{4.10}$$

which is in turn a reverse *output duty-cycle* d_O translation of i_O. In short, t_C and $i_{L(PK)}$ scale with $\sqrt{i_{O(AVG)}}$.

Although not often the case, i_L can also fall to and remain at an $i_{L(MIN)}$ that is not zero. In these cases, i_L rises and falls with v_E and v_D across L_X and flattens with zero volts. This is *pseudo DCM*, which engineers often abbreviate to PDCM.

4.2.3 Circuit Variants

L_X in Fig. 4.6 can "step" v_{IN} down or up to a lower or higher v_O. If v_O is less than v_{IN}, $v_{IN} - v_O$ can establish the positive v_E needed to energize L_X. So removing S_{EG} and S_{DO} and connecting L_X to v_O transform the *buck–boost* in Fig. 4.6 into the *buck* in Fig. 4.10. In this case, d_O is the fraction of t_C that t_E and t_D together set, which is one. So $i_{L(AVG)}$ matches i_O, and since the *input duty-cycle* d_{IN} is a t_E fraction of t_C, i_{IN} is similarly a d_{IN} fraction of $i_{L(AVG)}$.

Fig. 4.11 Switched-inductor boost

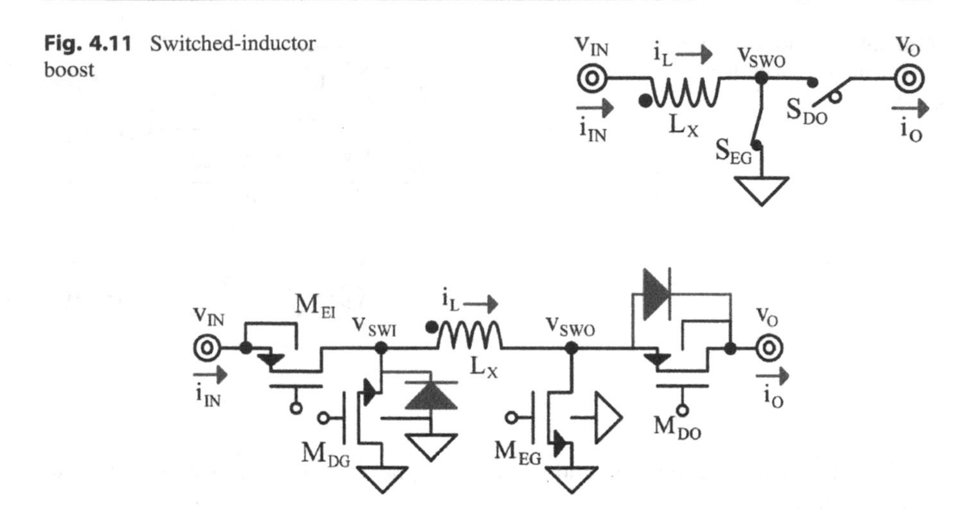

Fig. 4.12 CMOS switched inductor

When v_{IN} is less than v_O, $v_O - v_{IN}$ can set the v_D needed to drain L_X. So removing S_{EI} and S_{DG} and connecting v_{IN} to L_X transform the buck–boost into the *boost* in Fig. 4.11. Here, i_{IN} matches $i_{L(AVG)}$ and i_O is a d_O fraction of $i_{L(AVG)}$. Converting a buck–boost into a boost or a buck when possible is good because fewer switches occupy less space and require less power.

4.2.4 CMOS Implementation

N-channel metal–oxide–semiconductor (MOS) *field-effect transistors* (FETs) M_{DG} and M_{EG} in Fig. 4.12 implement the ground switches in Fig. 4.6 because *P-channel* MOS switches would require negative gate voltages to close. Similarly, P-channel transistors M_{EI} and M_{DO} normally realize the input and output switches because NMOS transistors would require above -v_{IN} and -v_O gate voltages to close. Paralleling an NMOS with M_{EI} or M_{DO} reduces resistance when v_O or v_{IN} is much higher than v_{IN} or v_O, in which case gate voltages can rise with v_O or v_{IN} well above v_{IN} or v_O. Note that all source-terminal arrows point in the direction they steer i_L.

A. Dead-Time Conduction

Dead time t_{DT} between the conduction periods of adjacent switches keep M_{EI}–M_{DG} and M_{EG}–M_{DO} from momentarily grounding v_{IN} and v_O, which could pull too much power from v_{IN} and v_O. Since v_{IN} directs i_L into v_O, M_{DG}'s and M_{DO}'s body diodes conduct i_L into v_O across these t_{DT}'s. The body connections shown ensure only these body diodes can conduct i_L.

When *MOS threshold voltages* v_T's are less than 500 mV or so, i_L discharges and charges capacitances at the *input* and *output switching nodes* v_{SWI} and v_{SWO} below and above the v_T's needed to engage M_{DG} and M_{DO}. So M_{DG} and M_{DO} conduct all or part of i_L across t_{DT}'s. Still, the effect is similar because M_{DG} and M_{DO} behave like diodes in this mode.

B. Duty Cycle

This diode action ultimately reduces the v_D that drains L_X. But since this happens across two small fractions of t_C and diode voltages are usually small fractions of v_E and v_D, the effect on d_E is normally low. The resulting d_E" is nevertheless greater than the ideal d_E by these fractions. In CCM, t_C extends across t_{SW}, so d_E is

$$\begin{aligned} d_{E(CCM)}'' &\approx \frac{v_D}{v_E + v_D} + \left(\frac{v_{DO} + v_{DG}}{v_E + v_D}\right)\left(\frac{2t_{DT}}{t_C}\right) \\ &= d_E + \left(\frac{v_{DO} + v_{DG}}{v_E + v_D}\right)\left(\frac{2t_{DT}}{t_{SW}}\right) > d_E. \end{aligned} \quad (4.11)$$

The effect is lower in DCM because the controller opens the drain switches when i_L is zero. So these diodes only conduct considerable i_L across the other t_{DT}:

$$d_{E(DCM)}'' \approx \frac{v_D}{v_E + v_D} + \left(\frac{v_{DO} + v_{DG}}{v_E + v_D}\right)\left(\frac{t_{DT}}{t_C}\right) = d_E + \left(\frac{v_{DO} + v_{DG}}{v_E + v_D}\right)\left(\frac{t_{DT}}{t_C}\right) < d_E''. \quad (4.12)$$

This t_{DT}, however, is a larger fraction of t_C because t_C ends before t_{SW} lapses.

C. Switching Voltages

When energize switches open, i_L pulls v_{SWI} low and v_{SWO} high until *ground* and *output drain diodes* D_{DG} and D_{DO} conduct i_L. So v_{SWI} falls from v_{IN} to $-v_{DG}$ and v_{SWO} rises from zero to v_{DO} over v_O when t_E ends in Fig. 4.13. v_{SWI} rises to zero and v_{SWO} falls to v_O a t_{DT} into t_D when S_{DG} and S_{DO} close.

Drain switches S_{DG} and S_{DO} open a t_{DT} before t_D ends, so D_{DG} and D_{DO} pull v_{SWI} a v_{DG} below ground and v_{SWO} a v_{DO} over v_O. And v_{SWI} climbs to v_{IN} and v_{SWO} falls to zero when energize switches S_{EI} and S_{EG} close at the beginning of t_E. This sequence repeats every t_{SW}.

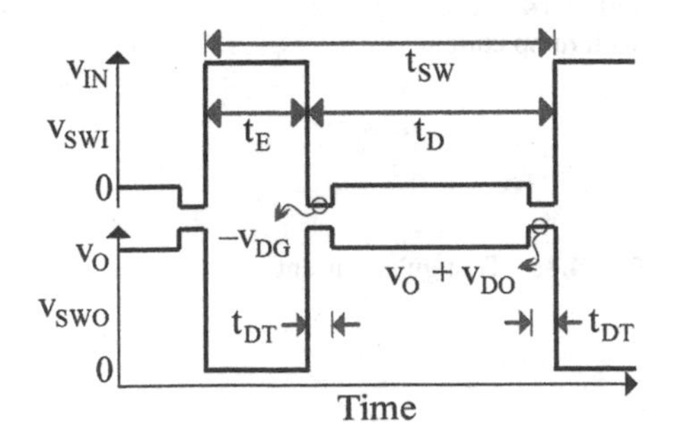

Fig. 4.13 Switching voltages

4.3 Ohmic Loss

Ohmic power P_R in this chapter refers to power that resistances in the switched inductor consume when conducting i_L. This is a loss because they burn v_{IN} power that v_O does not receive. The average power a device that conducts i_A and drops v_A consumes across time t_X is

$$P_{AX} \equiv P_{A(AVG)}\Big|_0^{t_X} = \frac{1}{t_X}\int_0^{t_X} P_A dt = \frac{1}{t_X}\int_0^{t_X} i_A v_A dt. \quad (4.13)$$

In truth, all power losses are fundamentally Ohmic in nature and all of them are ultimately lost in the form of heat. Here, Ohmic refers to resistance because Ohms is the unit that characterizes resistance. So any component that behaves like a resistor burns, from the perspective of this chapter, Ohmic power.

4.3.1 Ohmic Power

Since the voltage v_R across a resistor R_X that conducts i_X is $i_X R_X$, R_X power P_R is $i_X v_R$ and $i_X{}^2 R_X$. When i_X ramps linearly across time t_X like Fig. 4.14 shows, P_R climbs quadratically with i_X and P_R's average P_{RX} across t_X rises quadratically with i_X's *root-mean-square* (RMS) $i_{X(RMS)}$:

$$P_{RX} = \frac{1}{t_X}\int_0^{t_X} i_X v_R dt = \left(\frac{1}{t_X}\int_0^{t_X} i_X{}^2 dt\right) R_X = i_{X(RMS)}{}^2 R_X. \quad (4.14)$$

This means that P_{RX}'s rise across time accelerates with i_X.

A. **Triangular Current**

The *triangular current* i_Δ in Fig. 4.15 ramps across t_X to Δi_Δ. Squaring this i_Δ and averaging $i_\Delta{}^2$ across t_X reduces $i_{\Delta(RMS)}$ to

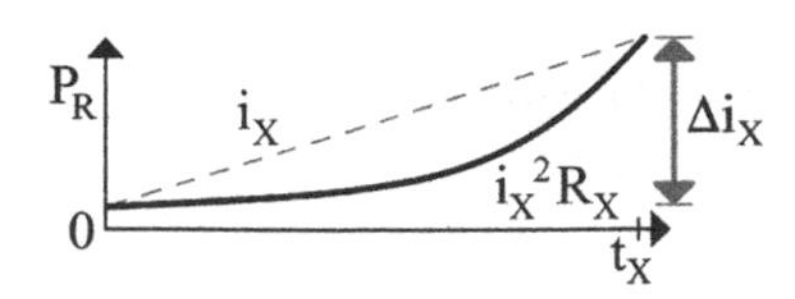

Fig. 4.14 Resistor power with ramp current

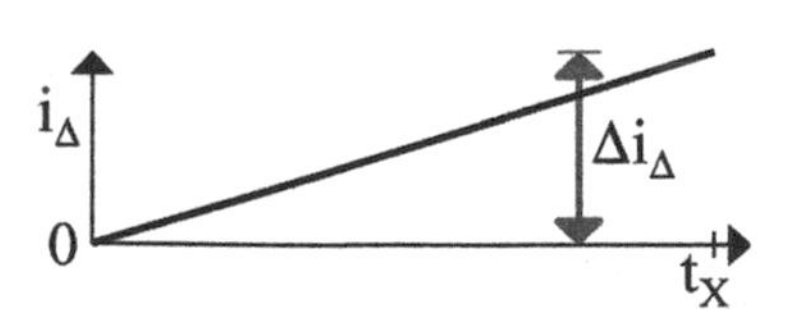

Fig. 4.15 Triangular current

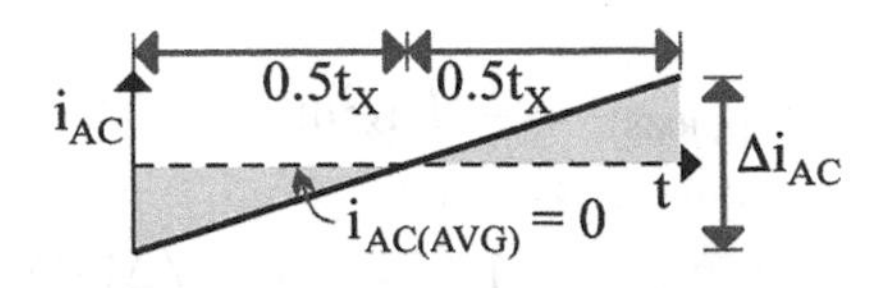

Fig. 4.16 Alternating triangular current

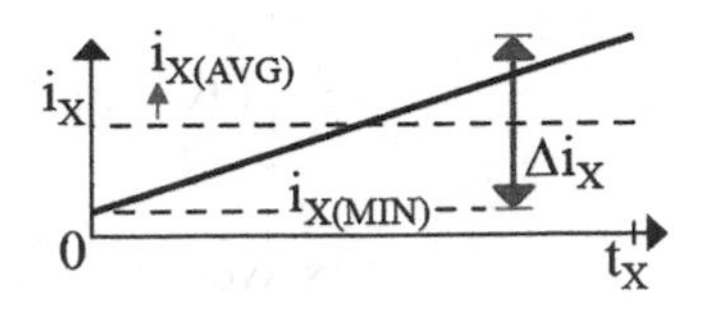

Fig. 4.17 Non-zero crossing ramp current

$$i_{\Delta(RMS)} = \sqrt{\frac{1}{t_X}\int_0^{t_X}\left(\frac{\Delta i_\Delta t}{t_X}\right)^2 dt} = \sqrt{\left(\frac{\Delta i_\Delta{}^2}{t_X{}^3}\right)\left(\frac{t_X{}^3}{3}\right)} = \frac{\Delta i_\Delta}{\sqrt{3}}. \quad (4.15)$$

So the RMS of a triangular current is a root-three fraction of its peak Δi_Δ.

B. Alternating Current

Positive and negative currents through a resistor burn power in the same way. RMS accounts for this because its squaring function desensitizes RMS from polarity. So when an *alternating current* i_{AC} is symmetrical, i_{AC}'s negative half burns as much power as i_{AC}'s positive half.

i_{AC} across each half in Fig. 4.16 is triangular like i_Δ in Fig. 4.15. So $i_{AC(RMS)}$ across each $0.5t_X$ is the same across t_X and like i_Δ across $0.5t_X$:

$$i_{AC(RMS)} \equiv i_{AC(RMS)}\big|_{t_X} = i_{AC(RMS)}\big|_{0.5t_X} = i_{\Delta(RMS)}\big|_{0.5t_X} = \frac{0.5\Delta i_{AC}}{\sqrt{3}}. \quad (4.16)$$

And since each half triangle traverses $0.5\Delta i_{AC}$, $i_{AC(RMS)}$ is a root-three fraction of half i_{AC}'s ripple Δi_{AC}.

C. Power Theorem

i_X in Fig. 4.17 ramps about an $i_{X(AVG)}$ that is greater than zero. Since i_X rises from $i_{X(MIN)}$ across Δi_X in t_X and $i_{X(MIN)}$ is half a ripple below i_X's average, i_X across time t is

$$i_X = i_{X(MIN)} + \left(\frac{\Delta i_X}{t_X}\right)t = i_{X(AVG)} - \frac{\Delta i_X}{2} + \left(\frac{\Delta i_X}{t_X}\right)t. \quad (4.17)$$

Squaring i_X and averaging $i_X{}^2$ across t_X reveals that $i_{X(RMS)}{}^2$ decomposes into average and alternating components $i_{X(AVG)}{}^2$ and $i_{AC(RMS)}{}^2$:

$$\begin{aligned} {i_{X(RMS)}}^2 &= \frac{1}{t_X}\int_0^{t_X} {i_X}^2 dt \\ &= \frac{1}{t_X}\left[{i_{X(MIN)}}^2 t + \left(\frac{{\Delta i_X}^2}{{t_X}^2}\right)\left(\frac{t^3}{3}\right) + 2 i_{X(MIN)}\left(\frac{\Delta i_X}{t_X}\right)\left(\frac{t^2}{2}\right)\right]\Bigg|_0^{t_X} \\ &= \frac{1}{t_X}\left[\left(i_{X(AVG)} - \frac{\Delta i_X}{2}\right)^2 t_X + \frac{{\Delta i_X}^2 t_X}{3} + \left(i_{X(AVG)} - \frac{\Delta i_X}{2}\right)\Delta i_X t_X\right] \\ &= {i_{X(AVG)}}^2 + \left(\frac{\Delta i_X}{2}\right)^2\left(1 + \frac{4}{3} - 2\right) = {i_{X(AVG)}}^2 + \left(\frac{0.5\Delta i_X}{\sqrt{3}}\right)^2 \\ &= {i_{X(AVG)}}^2 + {i_{AC(RMS)}}^2. \end{aligned} \tag{4.18}$$

When R_X conducts i_X across a t_X fraction of the switching period, R_X consumes a similar t_{SW} fraction of P_{RX}. So overall, P_R reduces to

$$\begin{aligned} P_R = P_{RX}\left(\frac{t_X}{t_{SW}}\right) &= {i_{X(RMS)}}^2 R_X\left(\frac{t_X}{t_{SW}}\right) \\ &= \left({i_{X(AVG)}}^2 + {i_{AC(RMS)}}^2\right) R_X\left(\frac{t_X}{t_{SW}}\right). \end{aligned} \tag{4.19}$$

This expression is very useful because it extrapolates the power that resistances burn when conducting across irregular fractions of t_{SW}.

4.3.2 Continuous Conduction

A. Switched Inductor

i_L in CCM ripples about an $i_{L(AVG)}$ that keeps $i_{L(LO)}$ at or above zero. L_X's *equivalent series resistance* (ESR) R_L conducts this i_L across all of t_{SW}. But since *skin* and *proximity effects* keep dynamic current near the edges of the coil, R_L is higher for the ripple than for the static component of i_L. So $i_{L(AVG)}$ burns power with the *inductor's dc resistance* $R_{L(DC)}$ and $i_{AC(RMS)}$ with the *inductor's* higher *ac resistance* $R_{L(AC)}$:

$$\begin{aligned} P_{RL} &\approx {i_{L(AVG)}}^2 R_{L(DC)} + {i_{AC(RMS)}}^2 R_{L(AC)} \\ &= \left(\frac{i_{O(AVG)}}{d_O}\right)^2 R_{L(DC)} + \left(\frac{0.5\Delta i_L}{\sqrt{3}}\right)^2 R_{L(AC)}, \end{aligned} \tag{4.20}$$

where $i_{L(AVG)}$ is a reverse d_O translation of $i_{O(AVG)}$ and $i_{AC(RMS)}$ is a root-three fraction of half Δi_L.

Example 1: Determine P_{RL} and σ_{RL} for L_X in an ideal buck–boost in CCM when v_{IN} is 2 V, v_O is 4 V, $i_{O(AVG)}$ is 250 mA, L_X is 10 μH, t_{SW} is 1 μs, and R_L is 200 mΩ.

Solution:

$$d_{IN} = d_E \approx \frac{v_O}{v_{IN} + v_O} = \frac{4}{2+4} = 67\% \quad \therefore \quad d_O = d_D = 1 - d_E \approx 1 - 67\% = 33\%$$

$$\Delta i_L = \left(\frac{v_E}{L_X}\right) t_E = \left(\frac{v_{IN}}{L_X}\right) d_E t_{SW} \approx \left(\frac{2}{10\mu}\right)(67\%)(1\mu) = 130 \text{ mA}$$

$$P_{RL} \approx \left[\left(\frac{i_{O(AVG)}}{d_O}\right)^2 + \left(\frac{0.5\Delta i_L}{\sqrt{3}}\right)^2\right] R_L = \left\{\left(\frac{250m}{33\%}\right)^2 + \left[\frac{0.5(130m)}{\sqrt{3}}\right]^2\right\}(200m)$$

$$= 120 \text{ mW}$$

$$\sigma_{RL} = \frac{P_{RL}}{(i_{O(AVG)}/d_O) d_{IN} v_{IN}} \approx \frac{120m}{(250m/33\%)(67\%)(2)} = 12\%$$

Note: R_L consumes roughly 12% of P_{IN}.

Explore with SPICE:
See Appendix A for notes on SPICE simulations.

```
* Ideal Buck-Boost with RL in CCM
vde de 0 dc=0 pulse 0 1 0 1n 1n 692n 1u
vin vin 0 dc=2
sei vin vswi de 0 sw1v
ddg 0 vswi idiode
lx vswi vl 10u
rl vl vswo 200m
seg vswo 0 de 0 sw1v
ddo vswo vo idiode
vo vo 0 dc=4
.ic i(lx)=700m
.lib lib.txt
```

(continued)

```
.tran 1u
.end
```

Tip: Plot i(Rl), v(vswi), and v(vswo); extract i(Rl)'s RMS; and use it to calculate $i_{R(RMS)}{}^2R_L$.

B. **Energize and Drain Resistances**

Energize switches conduct i_L across t_E only. So their current i_{RE} is i_L across t_E and zero across t_D, like Fig. 4.18 shows. *Energize resistance* R_E therefore consumes a t_E/t_{SW} fraction of the power R_E burns across t_E. Since i_{RE}'s average and ripple across t_E match i_L's $i_{L(AVG)}$ and Δi_L, *input* and *ground energize resistances* R_{EI} and R_{EG} in R_E dissipate

$$P_{RE} = i_{RE(RMS)}{}^2R_E = \left(i_{E(AVG)}{}^2 + i_{AC(RMS)}{}^2\right)R_E\left(\frac{t_E}{t_{SW}}\right)$$
$$= \left[\left(\frac{i_{O(AVG)}}{d_O}\right)^2 + \left(\frac{0.5\Delta i_L}{\sqrt{3}}\right)^2\right](R_{EI} + R_{EG})d_E, \tag{4.21}$$

where i_{RE}'s $i_{E(AVG)}$ and $i_{AC(RMS)}$ across t_E match i_L's across t_{SW}.

Drain components similarly conduct i_L across t_D only. Dead times, however, shorten the times that drain switches close. So drain-switch current i_{RD} is i_L across the t_D that excludes two t_{DT}'s: t_D' or $t_D - 2t_{DT}$ and zero across t_E. *Drain resistance* R_D therefore consumes a t_D'/t_{SW} fraction of the power R_D burns across t_D'. Since i_{RD}'s average across t_D' matches i_L's $i_{L(AVG)}$ when t_{DT}'s are symmetrical and i_{RD}'s ripple matches i_L's Δi_L when t_{DT}'s are much shorter than t_{SW}, *ground* and *output drain resistances* R_{DG} and R_{DO} in R_D burn:

$$\begin{aligned} P_{RD} &= i_{RD(RMS)}{}^2R_D \\ &= \left(i_{D(AVG)}{}^2 + i_{AC(RMS)}{}^2\right)R_D\left(\frac{t_D - 2t_{DT}}{t_{SW}}\right) \\ &\approx \left[\left(\frac{i_{O(AVG)}}{d_O}\right)^2 + \left(\frac{0.5\Delta i_L}{\sqrt{3}}\right)^2\right](R_{DG} + R_{DO})\left(d_D - \frac{2t_{DT}}{t_{SW}}\right), \end{aligned} \tag{4.22}$$

where i_{RD}'s $i_{D(AVG)}$ and $i_{AC(RMS)}$ across t_D roughly match i_L's across t_{SW}.

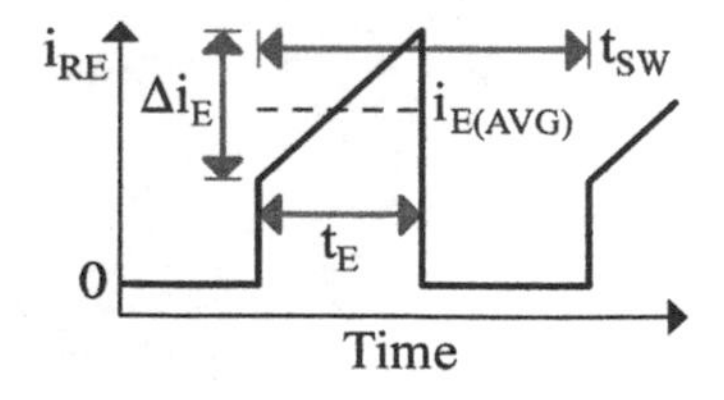

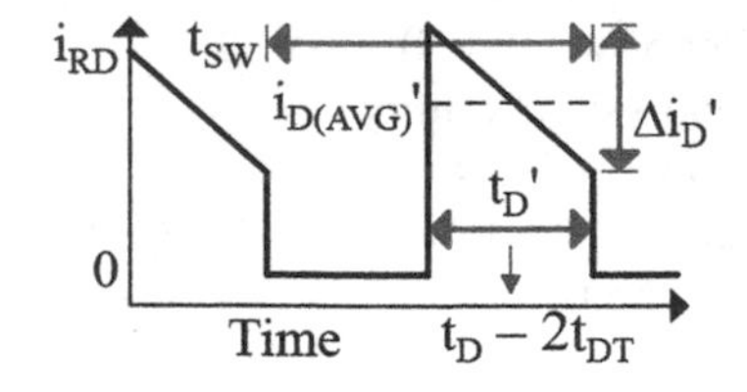

Fig. 4.18 Energize and drain resistor currents

Example 2: Determine P_{RE} and σ_{RE} for M_{EG} in the ideal buck–boost from Example 1 in CCM when R_{EG} is 200 mΩ.

Solution:

$$d_{IN} = d_E \approx 67\%, d_O = d_D \approx 32\% \text{ and } \Delta i_L \approx 130 \text{ mA from Example 1}$$

$$P_{RE} = \left[\left(\frac{i_{O(AVG)}}{d_O}\right)^2 + \left(\frac{0.5\Delta i_L}{\sqrt{3}}\right)^2\right] R_{EG} d_E$$
$$= \left\{\left(\frac{250m}{33\%}\right)^2 + \left[\frac{0.5(130m)}{\sqrt{3}}\right]^2\right\}(200m)(67\%) = 77 \text{ mW}$$

$$\sigma_{RE} = \frac{P_{RE}}{(i_{O(AVG)}/d_O)d_{IN}v_{IN}} = \frac{77m}{(250m/33\%)(67\%)(2)} = 7.6\%$$

Note: M_{EG} dissipates less of P_{IN} than R_L because M_{EG} conducts i_L a t_E fraction of t_{SW}. But since M_{DO} conducts the other t_D fraction of t_{SW}, M_{EG} and M_{DO} together can consume as much as R_L.

Explore with SPICE:
See Appendix A for notes on SPICE simulations.

```
* Ideal Buck-Boost with REG in CCM
vde de 0 dc=0 pulse 0 1 0 1n 1n 684n 1u
vin vin 0 dc=2
sei vin vswi de 0 sw1v
ddg 0 vswi idiode
lx vswi vswo 10u
seg vswo 0 de 0 sw1v200m
ddo vswo vo idiode
vo vo 0 dc=4
.ic i(lx)=700m
.lib lib.txt
```

(continued)

```
.tran 1u
.end
```

Tip: Plot i(Seg), v(vswi), and v(vswo); extract i(Seg)'s RMS; and use it to calculate $i_{R(RMS)}{}^2R_{EG}$.

C. **Output Capacitor**

The ultimate aim of voltage regulators and LED drivers is to supply i_O. Supplying this i_O, however, is impossible when the output switch is open. The purpose of C_O in Fig. 4.19 is to supply i_O when S_{DO} opens.

DC current into C_O must be zero for v_O and i_O to remain steady. So on average, S_{DO} outputs the i_O that feeds the load. But since S_{DO} opens a t_E fraction of t_{SW}, i_{DO}'s $i_{D(AVG)}$, which matches $i_{L(AVG)}$, supplies a correspondingly higher reverse d_O translation of i_O: i_O/d_O. i_{DO} is zero otherwise like Fig. 4.20 shows. So C_O supplies i_O when S_{DO} opens and receives i_O/d_O minus i_O otherwise. C_O's ESR R_C therefore burns power with i_O across t_E and with i_{DO} across t_O:

$$\begin{aligned} P_{RC(DO)} &= P_{RC}\big|_{t_E} + P_{RC}\big|_{t_O} \\ &= i_{O(AVG)}{}^2R_C\left(\frac{t_E}{t_{SW}}\right) + \left[\left(i_{D(AVG)} - i_{O(AVG)}\right)^2 + i_{AC(RMS)}{}^2\right]R_C\left(\frac{t_O}{t_{SW}}\right) \\ &= i_{O(AVG)}{}^2R_Cd_E + \left[\left(\frac{i_{O(AVG)}}{d_O} - i_{O(AVG)}\right)^2 + \left(\frac{0.5\Delta i_L}{\sqrt{3}}\right)^2\right]R_Cd_O, \end{aligned} \tag{4.23}$$

where $i_{O(AVG)}$ is equivalent to i_O in Fig. 4.19, R_C's $i_{C(AVG)}$ across t_O is $i_{DO(AVG)}$'s i_O/d_O minus i_O, and $i_{AC(RMS)}$ is a root-three fraction of half i_{DO}'s ripple Δi_L.

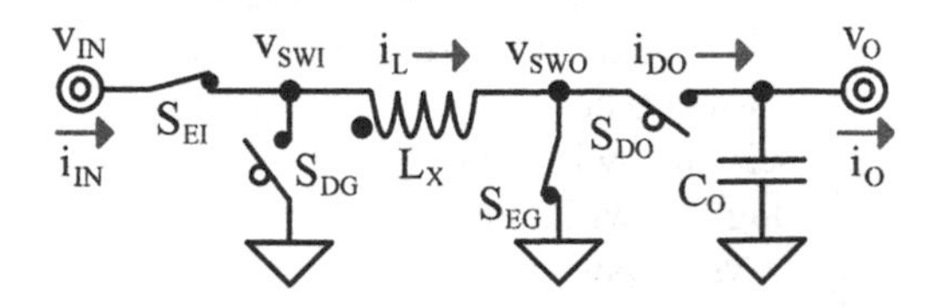

Fig. 4.19 Switched-inductor voltage regulator or LED driver

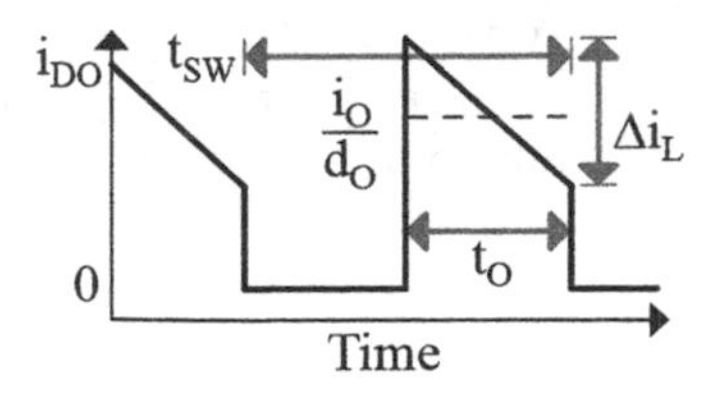

Fig. 4.20 Duty-cycled inductor drain current

Example 3: Determine P_{RC} and σ_{RC} for C_O in an ideal boost in CCM when v_{IN} is 2 V, v_O is 4 V, $i_{O(AVG)}$ is 250 mA, L_X is 10 μH, t_{SW} is 1 μs, and R_C is 200 mΩ.

Solution:

$$\text{Boost} \quad \therefore \quad d_{IN} = 1$$

$$d_E = \frac{v_D}{v_E + v_D} = \frac{v_O - v_{IN}}{v_O} = \frac{4-2}{4} = 50\% \quad \therefore \quad d_O = d_D = 1 - d_E = 1 - 50\% = 50\%$$

$$\Delta i_L = \left(\frac{v_E}{L_X}\right) t_E = \left(\frac{v_{IN}}{L_X}\right) d_E t_{SW} \approx \left(\frac{2}{10\mu}\right)(50\%)(1\mu) = 100 \text{ mA}$$

$$P_{RC} = i_{O(AVG)}{}^2 R_C d_E + \left[\left(\frac{i_{O(AVG)}}{d_O} - i_{O(AVG)}\right)^2 + \left(\frac{0.5\Delta i_L}{\sqrt{3}}\right)^2\right] R_C d_O$$

$$= (250\text{m})^2(200\text{m})(50\%) + \left\{\left(\frac{250\text{m}}{50\%} - 250\text{m}\right)^2 + \left[\frac{0.5(100\text{m})}{\sqrt{3}}\right]^2\right\}(200\text{m})(50\%)$$

$$= 13\,\text{mW}$$

$$\sigma_{RC} = \frac{P_{RC}}{\left(i_{O(AVG)}/d_O\right) d_{IN} v_{IN}} = \frac{13\text{m}}{(250\text{m}/50\%)(1)(2)} = 1.3\%$$

Note: R_C consumes 1.3% of P_{IN}.

Explore with SPICE:
See Appendix A for notes on SPICE simulations.

```
* Ideal Boost with RC in CCM
vde de 0 dc=0 pulse 0 1 0 1n 1n 504n 1u
vin vin 0 dc=2
lx vin vswo 10u
seg vswo 0 de 0 sw1v
ddo vswo vo idiode
```

(continued)

```
co vo vc 5u
rc vc 0 200m
io vo 0 dc=250m
.ic i(lx)=450m v(vo)=4
.lib lib.txt
.tran 1u
.end
```

Tip: Plot i(Rc) and v(vswo), extract i(Rc)'s RMS, and use it to calculate $i_{R(RMS)}{}^2R_C$.

In a buck, L_X connects to v_O. This means that i_L ripples about the average that supplies $i_{O(AVG)}$. In other words, i_O is $i_{L(AVG)}$, $i_{C(AVG)}$ is zero, and C_O conducts i_L's ripple. So the purpose of C_O in the buck is to supply and sink up to half i_L's ripple. Since half Δi_L is oftentimes much lower than the i_O that C_O in Fig. 4.19 supplies across t_E, P_{RC} in the buck in Fig. 4.21 is usually much lower than in the boost and buck–boost:

$$\begin{aligned} P_{RC(BK)} &= i_{C(RMS)}{}^2R_C = \left(i_{C(AVG)}{}^2 + i_{AC(RMS)}{}^2\right)R_C \\ &= \left[0^2 + \left(\frac{0.5\Delta i_L}{\sqrt{3}}\right)^2\right]R_C = \left(\frac{0.5\Delta i_L}{\sqrt{3}}\right)^2 R_C, \end{aligned} \tag{4.24}$$

where i_C's $i_{AC(RMS)}$ matches i_L's.

Example 4: Determine P_{RC} and σ_{RC} for C_O in an ideal buck in CCM when v_{IN} is 4 V, v_O is 2 V, $i_{O(AVG)}$ is 250 mA, L_X is 10 μH, t_{SW} is 1 μs, and R_C is 200 mΩ.

Solution:

$$\text{Buck} \quad \therefore \quad d_O = 1$$

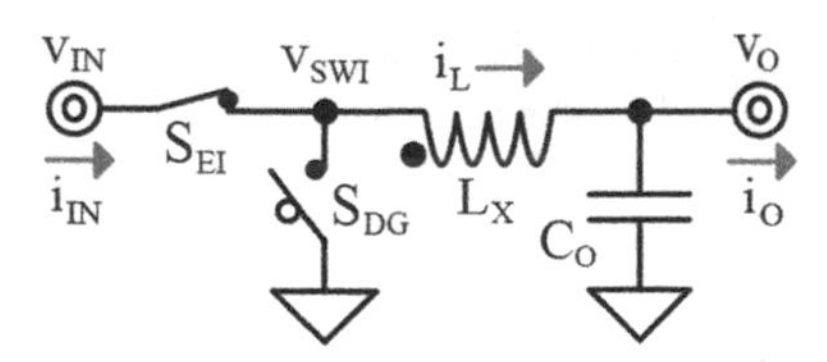

Fig. 4.21 Switched-inductor buck voltage regulator or LED driver

$$d_{IN} = d_E = \frac{v_D}{v_E + v_D} = \frac{v_O}{v_{IN}} = \frac{2}{4} = 50\% \quad \therefore \quad d_D = 1 - d_E = 1 - 50\% = 50\%$$

$$\Delta i_L = \left(\frac{v_E}{L_X}\right) t_E = \left(\frac{v_{IN} - v_O}{L_X}\right) d_E t_{SW} = \left(\frac{4-2}{10\mu}\right)(50\%)(1\mu) = 100\text{ mA}$$

$$P_{RC} = \left(\frac{0.5\Delta i_L}{\sqrt{3}}\right)^2 R_C = \left[\frac{0.5(100m)}{\sqrt{3}}\right]^2 (200m) = 170\ \mu W$$

$$\sigma_{RC} = \frac{P_{RC}}{\left(i_{O(AVG)}/d_O\right) d_{IN} v_{IN}} = \frac{170\mu}{(250m/1)(50\%)(4)} < 0.1\%$$

Note: R_C dissipates less of P_{IN} in the buck than in the boost because R_C conducts half of Δi_L only. In the boost, R_C conducts i_O-level currents.

Explore with SPICE:
See Appendix A for notes on SPICE simulations.

```
* Ideal Buck with RC in CCM
vde de 0 dc=0 pulse 0 1 0 1n 1n 500n 1u
vin vin 0 dc=4
sei vin vswi de 0 sw1v
ddg 0 vswi idiode
lx vswi vo 10u
co vo vc 5u
rc vc 0 200m
io vo 0 dc=250m
.ic i(lx)=200m v(vo)=2
.lib lib.txt
.tran 1u
.end
```

Tip: Plot i(Rc) and v(vswi), extract i(Rc)'s RMS, and use it to calculate $i_{R(RMS)}{}^2 R_C$.

4.3.3 Discontinuous Conduction

A. **Switched Inductor**

i_L in DCM rises to $i_{L(PK)}$ after t_E and falls to zero before t_{SW} ends. Since no part of i_L is steady across t_{SW}, all of i_L flows through the skin of L_X. This means that $R_{L(AC)}$ is the only part of L_X that consumes P_{RL}. This P_{RL} is a t_C/t_{SW} fraction of the RMS power burned across t_C:

$$\begin{aligned} P_{RL} &= i_{L(RMS)}{}^2 R_{L(AC)} \\ &= i_{C(RMS)}{}^2 R_{L(AC)} \left(\frac{t_C}{t_{SW}}\right) \\ &= \left(\frac{i_{L(PK)}}{\sqrt{3}}\right)^2 R_{L(AC)} \left(\frac{t_C}{t_{SW}}\right), \end{aligned} \tag{4.25}$$

where $i_{C(RMS)}$ is i_L's RMS across t_C, which is a root-three fraction of half i_L's DCM peak $i_{L(PK)}$.

Example 5: Determine P_{RL} and σ_{RL} for L_X in the ideal boost from Example 3 in DCM when $i_{O(AVG)}$ is 10 mA and $R_{L(AC)}$ is 200 mΩ.

Solution:

$$d_{IN} = 1, d_E = 50\%, \text{and } d_O = d_D = 50\% \text{ from Example 3}$$

$$i_{L(PK)} = \sqrt{2 d_E t_{SW} \left(\frac{v_E}{L_X}\right)\left(\frac{i_O}{d_O}\right)} = \sqrt{2(50\%)(1\mu)\left(\frac{2}{10\mu}\right)\left(\frac{10m}{50\%}\right)} = 63 \text{ mA}$$

$$t_C = \left(\frac{i_{L(PK)}}{d_E}\right)\left(\frac{L_X}{v_E}\right) = \left(\frac{63m}{50\%}\right)\left(\frac{10\mu}{2}\right) = 630 \text{ ns}$$

$$P_{RL} = \left(\frac{i_{L(PK)}}{\sqrt{3}}\right)^2 R_{L(AC)} \left(\frac{t_C}{t_{SW}}\right) = \left(\frac{63m}{\sqrt{3}}\right)^2 (200m)\left(\frac{630n}{1\mu}\right) = 170\ \mu\text{W}$$

$$\sigma_{RL} = \frac{P_{RL}}{\left(i_{O(AVG)}/d_O\right) d_{IN} v_{IN}} = \frac{170\mu}{(10m/50\%)(2)} = 0.4\%$$

Note: R_L dissipates 0.4% of P_{IN}.

Explore with SPICE:
See Appendix A for notes on SPICE simulations.

```
* Ideal Boost with RL in DCM
vde de 0 dc=0 pulse 0 1 0 1n 1n 315n 1u
vin vin 0 dc=2
lx vin vl 10u
rl vl vswo 200m
seg vswo 0 de 0 sw1v
ddo vswo vo idiode
co vo 0 5u
io vo 0 dc=10m
.ic i(lx)=0 v(vo)=4
.lib lib.txt
.tran 1u
.end
```

Tip: Plot i(Rl) and v(vswo), extract i(Rl)'s RMS, and use it to calculate $i_{R(RMS)}{}^2R_L$.

B. Energize and Drain Resistances

Energize and drain switches similarly consume t_E/t_{SW} and t_D/t_{SW} fractions of the RMS power R_E and R_D burn across t_E and t_D:

$$\begin{aligned} P_{RE} &= i_{RE(RMS)}{}^2R_E \\ &= i_{E(RMS)}{}^2R_E\left(\frac{t_E}{t_{SW}}\right) \\ &= \left(\frac{i_{L(PK)}}{\sqrt{3}}\right)^2(R_{EI}+R_{EG})d_E\left(\frac{t_C}{t_{SW}}\right) \end{aligned} \tag{4.26}$$

$$\begin{aligned} P_{RD} &= i_{RD(RMS)}{}^2R_D \\ &= i_{D(RMS)}{}^2R_D\left(\frac{t_D}{t_{SW}}\right) \\ &= \left(\frac{i_{L(PK)}}{\sqrt{3}}\right)^2(R_{DG}+R_{DO})d_D\left(\frac{t_C}{t_{SW}}\right). \end{aligned} \tag{4.27}$$

R_{EI} and R_{EG} in R_E burn P_{RE} and R_{DG} and R_{DO} in R_D burn P_{RD}. i_L's RMS across t_E, t_D, and t_C are all root-three fractions of $i_{L(PK)}$ because i_L is triangular and peaks to $i_{L(PK)}$ in all three cases.

C. **Output Capacitor**

In DCM, S_{DO} opens across t_E and the period that follows t_C before t_{SW} ends. This corresponds to the part of t_{SW} that excludes t_O. C_O supplies i_O across this time and the part of i_O that i_L does not when S_{DO} closes. So R_C consumes power with $i_{O(AVG)}$ across $t_{SW} - t_O$ and with $i_D - i_O$ across t_O:

$$P_{RC(DO)} = P_{RC}|_{t_{SW}-t_O} + P_{RC}|_{t_O}$$
$$= i_{O(AVG)}{}^2 R_C\left(\frac{t_{SW}-t_O}{t_{SW}}\right) + \left[\left(i_{D(AVG)} - i_{O(AVG)}\right)^2 + i_{AC(RMS)}{}^2\right] R_C\left(\frac{t_O}{t_{SW}}\right)$$
$$= i_{O(AVG)}{}^2 R_C\left[1 - d_O\left(\frac{t_C}{t_{SW}}\right)\right] + \left[\left(\frac{i_{L(PK)}}{2} - i_{O(AVG)}\right)^2 + \left(\frac{0.5 i_{L(PK)}}{\sqrt{3}}\right)^2\right] R_C d_O\left(\frac{t_C}{t_{SW}}\right). \quad (4.28)$$

Across t_O, i_D's average $i_{D(AVG)}$, which is half $i_{L(PK)}$, minus $i_{O(AVG)}$ burns steady power and i_D's ripple $i_{L(PK)}$ burns alternating power.

Example 6: Determine P_{RC} and σ_{RC} for C_O in the ideal boost from Examples 3 and 5 in DCM when R_C is 200 mΩ.

Solution:

$d_{IN} = 1, d_E = 50\%, d_O = d_D = 50\%, i_{L(PK)} = 63$ mA,
and $t_C = 630$ ns from Examples 3 and 5

$$P_{RC} = i_{O(AVG)}{}^2 R_C\left[1 - d_O\left(\frac{t_C}{t_{SW}}\right)\right] + \left[\left(\frac{i_{L(PK)}}{2} - i_{O(AVG)}\right)^2 + \left(\frac{0.5 i_{L(PK)}}{\sqrt{3}}\right)^2\right] R_C d_O\left(\frac{t_C}{t_{SW}}\right)$$
$$= (10\text{m})^2(200\text{m})\left[1 - 50\%\left(\frac{630\text{n}}{1\mu}\right)\right] + \left[\left(\frac{63\text{m}}{2} - 10\text{m}\right)^2 + \left(\frac{63\text{m}}{2\sqrt{3}}\right)^2\right](200\text{m})(50\%)\left(\frac{630\text{n}}{1\mu}\right)$$
$$= 64\ \mu\text{W}$$

$$\sigma_{RC} = \frac{P_{RC}}{\left(i_{O(AVG)}/d_O\right) d_{IN} v_{IN}} = \frac{64\mu}{(10\text{m}/50\%)(1)(2)} = 0.2\%$$

Note: R_C dissipates less of P_{IN} than R_L in Example 5 because R_C does not conduct the part of i_L that feeds i_O.

Explore with SPICE:
See Appendix A for notes on SPICE simulations.

```
* Ideal Boost with RC in DCM
vde de 0 dc=0 pulse 0 1 0 1n 1n 315n 1u
vin vin 0 dc=2
lx vin vswo 10u
seg vswo 0 de 0 sw1v
ddo vswo vo idiode
co vo vc 5u
rc vc 0 200m
io vo 0 dc=10m
.ic i(lx)=0 v(vo)=4
.lib lib.txt
.tran 1u
.end
```

Tip: Plot i(Rc) and v(vswo), extract i(Rc)'s RMS, and use it to calculate $i_{R(RMS)}{}^2R_C$.

In a buck, R_C conducts $i_L - i_O$ across t_{SW} because L_X connects to v_O. But since i_L is non-zero only across t_C, R_C consumes power with $i_L - i_O$ across t_C and with i_O across $t_{SW} - t_C$:

$$\begin{aligned}P_{RC(BK)} &= P_{RC}\big|_{tc} + P_{RC}\big|_{tsw-tc}\\ &= \left[\left(i_{C(AVG)} - i_{O(AVG)}\right)^2 + i_{AC(RMS)}{}^2\right]R_C\left(\frac{t_C}{t_{SW}}\right) + i_{O(AVG)}{}^2R_C\left(\frac{t_{SW}-t_C}{t_{SW}}\right)\\ &= \left[\left(\frac{i_{L(PK)}}{2} - i_{O(AVG)}\right)^2 + \left(\frac{0.5i_{L(PK)}}{\sqrt{3}}\right)^2\right]R_C\left(\frac{t_C}{t_{SW}}\right) + i_{O(AVG)}{}^2R_C\left(1 - \frac{t_C}{t_{SW}}\right)\end{aligned} \quad (4.29)$$

Across t_C, i_L's average $0.5i_{L(PK)}$ minus i_O burns steady power and i_L's ripple $i_{L(PK)}$ burns alternating power.

4.4 Diode Loss

4.4.1 Conduction Power

A static dc voltage v_X that conducts a ramping current like i_X in Fig. 4.17 burns the power that i_X's average across that time $i_{X(AVG)}$ into v_X sets:

$$P_{VX} = \frac{1}{t_X}\int_0^{t_X} i_X v_X dt = \left(\frac{1}{t_X}\int_0^{t_X} i_X dt\right) v_X = i_{X(AVG)} v_X. \quad (4.30)$$

This $i_{X(AVG)}$ is i_X's minimum $i_{X(MIN)}$ plus half i_X's variation Δi_X:

$$i_{X(AVG)} = \frac{1}{t_X}\int_0^{t_X} i_X dt = \frac{1}{t_X}\int_0^{t_X}\left[i_{X(MIN)} + \left(\frac{\Delta i_X}{t_X}\right)t\right]dt = i_{X(MIN)} + \frac{\Delta i_X}{2}. \quad (4.31)$$

So P_V's average P_{VX} climbs linearly with i_X's average $i_{X(AVG)}$.

4.4.2 Diode Drain Power

Asynchronous drain diodes conduct i_L across t_D. So on average, D_{DG} and D_{DO} burn power with v_{DG} and v_{DO} and i_L's average across a t_D fraction of t_{SW}. This t_D fraction is d_D in CCM and a t_C fraction of t_{SW} lower in DCM:

$$P_{DD} \approx i_{L(AVG)}(v_{DG} + v_{DO})\left(\frac{t_D}{t_{SW}}\right) = \left(\frac{i_{O(AVG)}}{d_O}\right)(v_{DG} + v_{DO})d_D\left(\frac{t_C}{t_{SW}}\right), \quad (4.32)$$

where $i_{L(AVG)}$ is a reverse d_O translation of $i_{O(AVG)}$.

This *diode drain power* P_{DD} is higher in CCM. This is because t_D scales with d_D and t_D is a larger fraction of t_{SW} in CCM than in DCM. The resulting fractional loss σ_{DD} can be significant when D_{DG} and D_{DO} drop 600–800 mV across more than 10% of t_{SW}.

Synchronous implementations are more efficient because M_{DG} and M_{DO} usually drop much lower voltages, on the order of millivolts. Dead-time diodes still conduct and burn power with v_{DG} and v_{DO} and i_L, but only across t_{DT}'s, which are typically shorter than t_D. These benefits fade, however, when t_{DT}'s become larger fractions of t_{SW}, which happens when f_{SW} is high.

4.4.3 Dead-Time Power

A. Continuous Conduction

Dead-time diodes conduct across t_{DT}'s the i_L that L_X holds before and after L_X energizes. Although i_L ramps over time, t_{DT}'s are usually so much shorter than t_{SW} that i_L is fairly steady across t_{DT}'s. So across the t_{DT} that follows t_E, when L_X begins to drain, i_L is fairly steady at $i_{L(HI)}$. i_L is similarly steady at $i_{L(LO)}$ across the t_{DT} that precedes t_E.

D_{DG} and D_{DO} consume power with $i_{L(HI)}$ and $i_{L(LO)}$ across these t_{DT}'s. Since i_L is nearly steady and diode voltages are largely insensitive to current variations, these diodes burn static power across t_{DT}'s and dead-time fractions across t_{SW}. So *dead-time power* P_{DT} is

$$\begin{aligned} P_{DT} &\approx i_{L(HI)}(v_{DG} + v_{DO})\left(\frac{t_{DT}}{t_{SW}}\right) + i_{L(LO)}(v_{DG} + v_{DO})\left(\frac{t_{DT}}{t_{SW}}\right) \\ &\approx 2i_{L(AVG)}(v_{DG} + v_{DO})\left(\frac{t_{DT}}{t_{SW}}\right) \\ &= \left(\frac{i_{O(AVG)}}{d_O}\right)(v_{DG} + v_{DO})\left(\frac{2t_{DT}}{t_{SW}}\right). \end{aligned} \tag{4.33}$$

Diode voltages can be so insensitive to current variations that D_{DG} and D_{DO} can drop similar v_{DG}'s and v_{DO}'s with $i_{L(HI)}$ and $i_{L(LO)}$. So $i_{L(HI)}$ and $i_{L(LO)}$ burn power with approximately equal t_{DT} fractions of similar v_{DG}'s and v_{DO}'s. And since $i_{L(HI)}$ and $i_{L(LO)}$ average $i_{L(AVG)}$, $i_{L(HI)}$ and $i_{L(LO)}$ add to $2i_{L(AVG)}$. This means that a reverse d_O translation of $i_{O(AVG)}$ burns P_{DT} with v_{DG} and v_{DO} across two t_{DT} fractions of t_{SW}.

Interestingly, P_{DT} is nearly a constant (current-independent) fraction of P_{IN}. This is because P_{DT} and P_{IN} both scale linearly with $i_{L(AVG)}$:

$$\sigma_{DT} = \frac{P_{DT}}{P_{IN}} = \frac{P_{DT}}{i_{IN(AVG)}v_{IN}} = \frac{P_{DT}}{i_{L(AVG)}d_{IN}v_{IN}} = \left(\frac{v_{DG} + v_{DO}}{d_{IN}v_{IN}}\right)\left(\frac{2t_{DT}}{t_{SW}}\right). \tag{4.34}$$

So P_{DT}'s fraction of P_{IN} ultimately hinges on the voltage and time fractions that diode voltages and t_{DT}'s establish with $d_{IN}v_{IN}$ and t_{SW}.

Example 7: Determine P_{DT} and σ_{DT} in an ideal boost in CCM when v_{IN} is 2 V, v_O is 4 V, $i_{O(AVG)}$ is 250 mA, L_X is 10 μH, t_{SW} is 1 μs, D_{DO} drops 800 mV, and t_{DT} is 50 ns.

Solution:

$$\text{Boost} \quad \therefore \quad d_{IN} = 1$$

$$d_E = \frac{v_D}{v_E + v_D} = \frac{v_O - v_{IN}}{v_O} + \left(\frac{v_{DO}}{v_O}\right)\left(\frac{2t_{DT}}{t_{SW}}\right) = \frac{4-2}{4} + \left(\frac{800m}{4}\right)\left[\frac{2(50n)}{1\mu}\right] = 52\%$$

$$\therefore \quad d_O = d_D = 1 - d_E = 1 - 52\% = 48\%$$

$$\Delta i_L = \left(\frac{v_E}{L_X}\right)t_E = \left(\frac{v_{IN}}{L_X}\right)d_E t_{SW} \approx \left(\frac{2}{10\mu}\right)(52\%)(1\mu) = 100 \text{ mA}$$

$$P_{DT} = \left(\frac{i_{O(AVG)}}{d_O}\right) v_{DO} \left(\frac{2t_{DT}}{t_{SW}}\right) = \left(\frac{250m}{48\%}\right)(800m)\left[\frac{2(50n)}{1\mu}\right] = 42 \text{ mW}$$

$$\sigma_{DT} = \frac{P_{DT}}{\left(i_{O(AVG)}/d_O\right) d_{IN} v_{IN}} \approx \frac{42m}{(250m/48\%)(1)(2)} = 4.0\%$$

Note: Dead-time diodes can consume considerable power.

Explore with SPICE:
See Appendix A for notes on SPICE simulations.

```
* Ideal Boost with Dead Time in CCM
vde de 0 dc=0 pulse 0 1 0 1n 1n 519n 1u
vdo do 0 dc=1 pulse 0 1 570n 1n 1n 381n 1u
vin vin 0 dc=2
lx vin vswo 10u
seg vswo 0 de 0 sw1v
sdo vswo vo do 0 sw1v
ddo vswo vo fdiode1
vo vo 0 dc=4
.ic i(lx)=470m
.lib lib.txt
.tran 1u
.end
```

Tip: Plot i(Ddo), v(vswo), and I(Ddo)*(v(vswo)-v(vo)) and extract the average of the product term.

B. **Discontinuous Conduction**

i_L in discontinuous conduction rises to $i_{L(PK)}$ after t_E and falls to zero across t_D before t_{SW} ends. So i_L is nearly $i_{L(PK)}$ across the first t_{DT} in t_D and nearly zero across the second. Dead-time diodes therefore consume noticeable power only across the first t_{DT}:

$$P_{DT} \approx i_{L(PK)}(v_{DG} + v_{DO})\left(\frac{t_{DT}}{t_{SW}}\right). \tag{4.35}$$

P_{DT}'s fraction of P_{IN} is the product of three ratios:

$$\sigma_{DT} = \frac{P_{DT}}{P_{IN}} = \frac{P_{DT}}{i_{IN}v_{IN}} = \frac{P_{DT}}{i_{L(AVG)}d_{IN}v_{IN}} = \left(\frac{i_{L(PK)}}{i_{L(AVG)}}\right)\left(\frac{v_{DG} + v_{DO}}{d_{IN}v_{IN}}\right)\left(\frac{t_{DT}}{t_{SW}}\right). \quad (4.36)$$

The current ratio is greater than one, the voltage ratio is lower than one, and the time ratio is usually a small fraction. So diode and t_{DT} fractions counter the effect of current gain. And since $i_{L(AVG)}$ scales faster with $i_{O(AVG)}$ than $i_{L(PK)}$ with $\sqrt{i_{O(AVG)}}$, P_{DT}'s fraction of P_{IN} falls with increasing $\sqrt{i_{O(AVG)}}$.

Example 8: Determine P_{DT} and σ_{DT} in an ideal buck in DCM when v_{IN} is 4 V, v_O is 2 V, $i_{O(AVG)}$ is 10 mA, L_X is 10 μH, t_{SW} is 1 μs, t_{DT} is 50 ns, and D_{DG} drops 700 mV.

Solution:

$$\text{Buck} \quad \therefore \quad d_O = 1$$

$$d_{IN} = d_E = \frac{v_D}{v_E + v_D} = \frac{v_O}{v_{IN}} + \left(\frac{v_{DG}}{v_{IN}}\right)\left(\frac{t_{DT}}{t_C}\right) \approx \frac{v_O}{v_{IN}} = \frac{2}{4} = 50\%$$

$$i_{L(PK)} = \sqrt{2d_E t_{SW}\left(\frac{v_E}{L_X}\right)i_O} \approx \sqrt{2(50\%)(1\mu)\left(\frac{4-2}{10\mu}\right)(10m)} = 45 \text{ mA}$$

$$P_{DT} = i_{L(PK)}v_{DG}\left(\frac{t_{DT}}{t_{SW}}\right) \approx (45m)(700m)\left(\frac{50n}{1\mu}\right) = 1.6 \text{ mW}$$

$$\sigma_{DT} = \frac{P_{DT}}{\left(i_{O(AVG)}/d_O\right)d_{IN}v_{IN}} \approx \frac{1.6m}{(10m/1)(50\%)(4)} = 8.0\%$$

Note: P_{DT} is a higher fraction of P_{IN} in DCM than in CCM because P_{IN} scales down faster with i_O than P_{DT} with $\sqrt{i_O}$.

Explore with SPICE:
See Appendix A for notes on SPICE simulations.

```
* Ideal Buck with Dead Time in DCM
vde de 0 dc=0 pwl 0 1 225n 1 225.1n 0
vdo do 0 dc=1 pwl 0 0 275n 0 275.1n 1 432n 1 432.1n 0
vin vin 0 dc=4
sei vin vswi de 0 sw1v
sdg vswi 0 do 0 sw1v
ddg 0 vswi fdiode1
cswi vswi 0 1p
lx vswi vo 10u
vo vo 0 dc=2
.ic i(lx)=0
.lib lib.txt
.tran 1u
.end
```

Tip: Plot i(Ddg), v(vswi), and i(Ddg)*v(vswi) and extract the average of the product term.

4.5 i_{DS}–v_{DS} Overlap Loss

i_{DS}–v_{DS} overlap power P_{IV} refers to the transitional power transistors consume when switching between on and off states. P_{IV} is essentially the power the *drain–source current* i_{DS} burns across the *drain–source voltage* v_{DS} the transistor drops when i_{DS} and v_{DS} transition. This loss hinges on the i_L that i_{DS} carries, the voltage v_{DS} collapses, and their transition times.

Since i_{DS} scales with *gate–source voltage* v_{GS}, the underlying goal of gate drivers is to transition gate voltages quickly. So they charge and discharge *gate–source* and *gate–drain oxide capacitances* C_{GS} and C_{GD} with low *pull-down* and *pull-up N-* and *P-type resistances* R_N and R_P. The resulting transitions should be (by design) shorter than dead times. This way, i_L is practically steady across these events.

4.5.1 Closing Switch

A. Power

When a switch is open, v_{GS} and i_{DS} are zero and v_{DS} is high at a level that other circuit components set. This v_{DS} is, in effect, the variation Δv_{SW} the switch collapses after it closes. So when closing, v_{GS} and i_{DS} climb and v_{DS} falls.

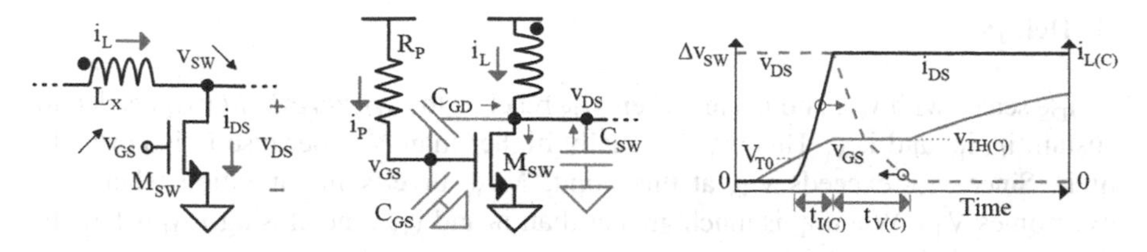

Fig. 4.22 Closing switch

In Fig. 4.22, M_{SW} closes as R_P charges C_{GS} and C_{GD}. i_{DS} climbs with v_{GS} after v_{GS} overcomes M_{SW}'s *zero-bias threshold* V_{T0}. v_{DS} falls later, when i_{DS} is high enough to sink the i_L that L_X carries and the i_{GD} and i_{SW} that C_{GD} and other switch-node capacitances C_{SW} need to discharge. When this happens, i_P slews C_{GD} at the v_{GS} (and i_{DS}) needed to sink i_L, i_P, and i_{SW}, so v_{GS} is fairly flat across this time. When i_L overwhelms i_P and i_{SW}, v_{DS} falls after i_{DS} reaches i_L.

Since v_{DS} remains at Δv_{SW} across the time t_I that i_{DS} needs to reach i_L, M_{SW} burns power P_I across t_I with Δv_{SW} and i_{DS}'s average:

$$\begin{aligned} P_I &= \frac{1}{t_I}\int_0^{t_I} i_{DS} v_{DS} dt \\ &= \left(\frac{1}{t_I}\int_0^{t_I} i_{DS} dt\right)\Delta v_{SW} \\ &\approx \left[\frac{1}{t_I}\int_0^{t_I}\left(\frac{i_L}{t_I{}^2}\right)t^2 dt\right]\Delta v_{SW} \approx \left(\frac{i_L}{3}\right)\Delta v_{SW}. \end{aligned} \tag{4.37}$$

i_{DS}'s rise is almost quadratic with time t because v_{GS} is close to linear when i_{DS} climbs to i_L and i_{DS} scales with $v_{GS}{}^2$ in MOSFET inversion. So across t_I, i_{DS}'s quadratic increase averages to about a third of i_L.

i_L also burns power P_V with v_{DS}'s average because i_{DS} is i_L across the time t_V that v_{DS} requires to collapse:

$$P_V = \frac{1}{t_V}\int_0^{t_V} i_{DS} v_{DS} dt \approx i_L\left(\frac{1}{t_V}\int_0^{t_V} v_{DS} dt\right) \approx i_L\left(\frac{\Delta v_{SW}}{2}\right). \tag{4.38}$$

v_{DS} averages half of Δv_{SW} because v_{DS} is largely linear. When combined, M_{SW} consumes 33% and 50% of $i_L\Delta v_{SW}$ across t_I and t_V, so P_{IV} across t_{SW} is:

$$P_{IV} = P_I\left(\frac{t_I}{t_{SW}}\right) + P_V\left(\frac{t_V}{t_{SW}}\right) \approx i_L\Delta v_{SW}\left(\frac{t_I}{3t_{SW}} + \frac{t_V}{2t_{SW}}\right). \tag{4.39}$$

B. **Delays**

i_{DS} scales with v_{GS} and maxes when v_{GS} reaches the *v_{GS} threshold* v_{TH} needed to sustain i_L, i_P, and i_{SW}. This v_{TH} is usually higher than V_{T0} because i_L is normally high. Since v_{DS} exceeds v_{GS} at this point, M_{SW} inverts in saturation when v_{GS} overcomes V_{T0}. When i_L is much greater than i_P and i_{SW}, the closing v_{TH} is largely the v_{GS} needed to sustain i_L:

$$v_{TH(C)} = v_{GS}\big|_{i_{L(C)}+i_P+i_{SW}} \approx V_{T0} + v_{DS(SAT)}\big|_{i_{L(C)}} \approx V_{T0} + \sqrt{\frac{2i_{L(C)}}{K'(W/L)}}, \tag{4.40}$$

where $v_{DS(SAT)}$ is the *saturation voltage* of M_{SW} in inversion.

i_{DS} does not rise much until M_{SW} inverts. So $t_{I(C)}$ is mostly the time v_{GS} requires to rise from V_{T0} to $v_{TH(C)}$. In other words, t_I is the fraction of the time t_{TH} needed to charge C_{GS} to $v_{TH(C)}$ that excludes the time t_{T0} needed to charge C_{GS} to V_{T0}. Since the gate driver's *power supply* v_{DD} and R_P require RC time t_X to charge C_{GS} and C_{GD} across Δv_X, t_X and $t_{I(C)}$ are

$$t_X = \tau_{RC} \ln\left(\frac{v_{DD}}{v_{DD} - \Delta v_X}\right) \tag{4.41}$$

and

$$t_{I(C)} \approx t_{TH} - t_{T0} = \tau_{RC(C)} \ln\left(\frac{v_{DD} - V_{T0}}{v_{DD} - v_{TH(C)}}\right), \tag{4.42}$$

where $\tau_{RC(C)}$ is the time constant of R_P and C_{GS} and C_{GD} in saturation:

$$\tau_{RC(C)} = R_P(C_{GS} + C_{GD}) \approx R_P\left[2C_{OL} + \left(\frac{2}{3}\right)C_{CH}\right], \tag{4.43}$$

and C_{OL} is *overlap capacitance*, C_{CH} is *channel capacitance*, and C_{GS} is C_{OL} plus two-thirds C_{CH} and C_{GD} is C_{OL} in saturated inversion.

v_{DS} falls when i_{DS} sinks i_L, C_{GD}'s i_P, and C_{SW}'s i_{SW}. Although i_{DS} is sensitive to v_{GS} and capable of sinking more current, R_P limits the current that feeds C_{GD}. So the time $t_{V(C)}$ that v_{DS} requires to collapse across Δv_{SW} is the time C_{GD}'s voltage v_C needs to traverse Δv_{SW}, which is a slew-rate translation of R_P's current i_P into C_{GD}:

$$\begin{aligned} t_{V(C)} &\approx \left(\frac{\Delta v_C}{i_P}\right)C_{GD} \\ &\approx \left(\frac{R_P}{v_{DD} - v_{GS}}\right)\left[C_{OL}\Delta v_{SW} + \left(\frac{0.5C_{CH}}{2}\right)v_{GS}\right] \\ &\approx \left(\frac{R_P}{v_{DD} - v_{TH(C)}}\right)\left[C_{OL}\Delta v_{SW} + \left(\frac{C_{CH}}{4}\right)v_{TH(C)}\right]. \end{aligned} \tag{4.44}$$

i_P slews C_{GD} at the v_{GS} needed to sustain i_L, i_P, and i_{SW}. So v_{GS} is steady at $v_{TH(C)}$ across t_V and $v_{DD} - v_{TH(C)}$ across R_P sets i_P. As v_{DS} collapses, M_{SW} transitions from saturation to triode, so C_{GD} receives half of C_{CH}.

The $0.25C_{CH}$ average of this $0.5C_{CH}$ therefore traverses the v_{DS} that transitions M_{SW} into triode, which starts when v_{DS} matches v_{GS} at $v_{TH(C)}$ when v_{GD} is zero. This is why v_{DS} collapses more quickly at the beginning of the transition: because C_{GD}'s saturated C_{OL} is easier to discharge than C_{GD}'s triode counterpart, which carries C_{OL} plus half of C_{CH}. In short, C_{GD} slows v_{DS}'s transition when v_{DS} falls below v_{GS}.

Example 9: Determine P_{IV} and σ_{IV} for M_{EG} in the ideal boost from Example 7 in CCM when M_{EG} closes, v_{DD} for M_{EG}'s gate driver is v_O, W_{EG} is 50 mm, L_{EG} is 250 nm, L_{OL} is 30 nm, V_{TN0} is 400 mV, K_N' is 200 μA/V^2, C_{OX}'' is 6.9 fF/μm^2, and R_P is 100 Ω.

Solution:

$$d_{IN} = 1, d_E = 52\%, d_O = d_D = 48\%, \text{and } \Delta i_L = 100 \text{ mA from Example 7}$$

$$M_{EG} \text{ closes when } i_L = i_{L(LO)}$$

$$i_{L(LO)} = \frac{i_O}{d_O} - \frac{\Delta i_L}{2} = \frac{250m}{48\%} - \frac{100m}{2} = 470 \text{ mA}$$

$$W_{CH} = W_{EG} = 50 \text{ mm}$$

$$L_{CH} = L_{EG} - 2L_{OL} = 250n - 2(30n) = 190 \text{ nm}$$

$$v_{DS(SAT)} \approx \sqrt{\frac{2i_{L(LO)}}{K_N'(W_{CH}/L_{CH})}} = \sqrt{\frac{2(470m)}{(200\mu)(50m/190n)}} = 130 \text{ mV}$$

$$v_{TH(C)} \approx V_{TN0} + v_{DS(SAT)} = 400m + 130m = 530 \text{ mV}$$

$$C_{OL} = C_{OX}''W_{CH}L_{OL} = (6.9m)(50m)(30n) = 10 \text{ pF}$$

$$C_{CH} = C_{OX}''W_{CH}L_{CH} = (6.9m)(50m)(190n) = 66 \text{ pF}$$

$$\tau_{RC(C)} = R_P\left[2C_{OL} + \left(\frac{2}{3}\right)C_{CH}\right] = (100)\left[2(10p) + \left(\frac{2}{3}\right)(66p)\right] = 6.4 \text{ ns}$$

$$t_{I(C)} \approx \tau_{RC(C)} \ln\left(\frac{v_{DD} - V_{T0}}{v_{DD} - v_{TH(C)}}\right) = (6.4n)\ln\left(\frac{4 - 400m}{4 - 530m}\right) = 240 \text{ ps}$$

$$v_{SWO} = v_O + v_{DO} = 4 + 800m = 4.8 \text{ V before } M_{EG} \text{ closes} \quad \therefore \quad \Delta v_{SWO} = 4.8 \text{ V}$$

$$t_{V(C)} \approx \left(\frac{R_P}{v_{DD} - v_{TH(C)}}\right)\left[C_{OL}\Delta v_{SWO} + \left(\frac{C_{CH}}{4}\right)v_{TH(C)}\right]$$
$$= \left(\frac{100}{4 - 530m}\right)\left[(10p)(4.8) + \left(\frac{66p}{4}\right)(530m)\right] = 1.6\text{ ns}$$

$$P_{IV} \approx i_{L(LO)}\Delta v_{SWO}\left(\frac{t_{I(C)}}{3t_{SW}} + \frac{t_{V(C)}}{2t_{SW}}\right) = (470m)(4.8)\left[\frac{240p}{(3)1\mu} + \frac{1.6n}{(2)1\mu}\right] = 2.0\text{ mW}$$

$$\sigma_{IV} = \frac{P_{IV}}{(i_{O(AVG)}/d_O)d_{IN}v_{IN}} = \frac{2.0m}{(250m/48\%)(1)(2)} = 0.2\%$$

Note: This P_{IV} excludes the power consumed when M_{EG} opens and other switches open and close.

Explore with SPICE:
See Appendix A for notes on SPICE simulations.

```
* I-V Overlap Loss when MEG Closes without Reverse Recovery
vdd vdd 0 dc=4
rp vdd vg 100
lx 0 vswo 10u
meg vswo vg 0 0 nmos1 w=50m l=250n
ddo vswo vo fdiode1
vo vo 0 dc=4
.ic i(lx)=470m v(vg)=0
.lib lib.txt
.tran 5n
.end
```

Tip: Plot id(Meg), v(vg), v(vswo), and id(Meg)*v(vswo), extract the average of the product term across 5 ns, and average across the 1-μs period (multiply by 5 ns of the 1-μs period or 0.5%) to determine P_{IV}. id(Meg) often includes C_{GD}'s current, so the extraction is not perfect. In this case, i_D is considerably greater than i_{GD}, so the approximation is fair.

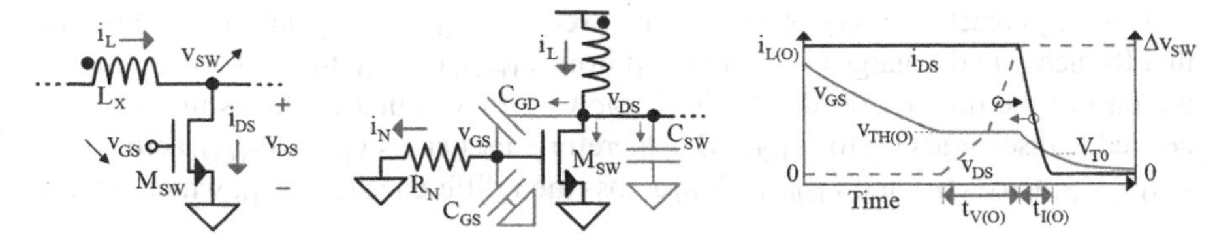

Fig. 4.23 Opening switch

4.5.2 Opening Switch

A. Power

After M_{SW} closes, v_{GS} is high, i_{DS} carries i_L, and v_{DS} is close to zero. To open M_{SW}, R_N in Fig. 4.23 collapses v_{GS}. i_{DS}, however, does not fall until v_{GS} drops below the v_{TH} needed to sink i_L minus C_{GD}'s i_N and C_{SW}'s i_{SW}. Even then, i_N slews C_{GD} at the v_{GS} (and i_{DS}) needed to sustain this i_{DS}. So v_{GS} is fairly flat when i_N and i_{SW} charge C_{GD} and C_{SW} across Δv_{SW}.

v_{DS} rises across the $t_{V(O)}$ that i_N needs to charge C_{GD} across the Δv_{SW} other circuit components set. P_V is the power v_{DS}'s average 50%Δv_{SW} burns across $t_{V(O)}$ with i_{DS} at i_L minus i_N and i_{SW}. When i_L overwhelms i_N and i_{SW}, v_{TH} is largely the v_{GS} needed to sustain i_L:

$$v_{TH(O)} = v_{GS}\Big|_{i_{L(O)}-i_N-i_{SW}} \approx V_{TO} + v_{DS(SAT)}\Big|_{i_{L(O)}} \approx V_{TO} + \sqrt{\frac{2 i_{L(O)}}{K'(W/L)}}. \tag{4.45}$$

Since v_{DS} stops rising after transitioning across Δv_{SW}, the effect of i_N on C_{GD} after $t_{V(O)}$ is to decrease v_{GS}. For this to happen, part of i_N must also discharge C_{GS}. The end result is that i_{DS} drops across the $t_{I(O)}$ that i_N needs to reduce v_{GS} from $v_{TH(O)}$ to V_{TO}. P_I is roughly the power i_{DS}'s quadratic average 33%i_L burns across this $t_{I(O)}$ with v_{DS} at Δv_{SW}.

B. Delays

Since v_{GS} is $v_{TH(O)}$ across $t_{V(O)}$, i_N is an Ohmic R_N translation of $v_{TH(O)}$. So $v_{TH(O)}/R_N$ slews C_{GD}'s C_{OL} across Δv_{SW} and C_{GD}'s channel average $0.25C_{CH}$ across the v_{DS} that transitions M_{SW} into triode, which starts when v_{DS} matches v_{GS}'s $v_{TH(O)}$. The resulting $t_{V(O)}$ is

$$\begin{aligned} t_{V(O)} &\approx \left(\frac{\Delta v_C}{i_N}\right) C_{GD} \approx \left(\frac{R_N}{v_{GS}}\right)\left[C_{OL}\Delta v_{SW} + \left(\frac{0.5C_{CH}}{2}\right) v_{GS}\right] \\ &\approx \left(\frac{R_N}{v_{TH(O)}}\right)\left[C_{OL}\Delta v_{SW} + \left(\frac{C_{CH}}{4}\right) v_{TH(O)}\right]. \end{aligned} \tag{4.46}$$

Once v_{DS} reaches Δv_{SW}, R_N's current i_N reduces v_{GS}. So i_{DS} falls across the $t_{I(O)}$ that R_N needs to discharge C_{GS} (and C_{GD}) from $v_{TH(O)}$ to V_{TO}. In other words, $t_{I(O)}$ is the part of the time t_{TO} needed to discharge C_{GS} to V_{TO} that excludes the time t_{TH} needed to discharge C_{GS} to $v_{TH(O)}$. C_{GS} therefore discharges $v_{DD} - v_{TH(O)}$ across t_{TH}, $v_{DD} - V_{TO}$ across t_{TO}, and $v_{TH(O)} - V_{TO}$ across their difference $t_{TO} - t_{TH}$, which is $t_{I(O)}$:

$$\begin{aligned} t_{I(O)} &\approx t_{TO} - t_{TH} \\ &= \tau_{RC(O)} \ln\left[\frac{v_{DD} - \left(v_{DD} - v_{TH(O)}\right)}{v_{DD} - \left(v_{DD} - V_{TO}\right)}\right] \\ &= \tau_{RC(O)} \ln\left(\frac{v_{TH(O)}}{V_{TO}}\right), \end{aligned} \tag{4.47}$$

where $\tau_{RC(O)}$ is the time constant of R_N and C_{GS} and C_{GD} in saturation:

$$\tau_{RC(O)} = R_N(C_{GS} + C_{GD}) \approx R_N\left[2C_{OL} + \left(\frac{2}{3}\right)C_{CH}\right]. \tag{4.48}$$

Note that R_P pre-charges C_{GS} to v_{DD} before R_N collapses C_{GS}'s v_{GS}.

Example 10: Determine P_{IV} and σ_{IV} for M_{EG} in the boost from Examples 7 and 9 in CCM when M_{EG} opens and R_N is 20 Ω.

Solution:

$$d_{IN} = 1, d_E = 52\%, d_O = d_D = 48\%, \text{and } \Delta i_L = 100 \text{ mA from Example 7}$$

$$L_{CH} = 190 \text{ nm}, C_{OL} = 10 \text{ pF}, \text{and } C_{CH} = 66 \text{ pF from Example 9}$$

$$M_{EG} \text{ opens when } i_L = i_{L(HI)}$$

$$i_{L(HI)} = \frac{i_O}{d_O} + \frac{\Delta i_L}{2} = \frac{250\text{m}}{48\%} + \frac{100\text{m}}{2} = 570 \text{ mA}$$

$$v_{DS(SAT)} \approx \sqrt{\frac{2i_{L(HI)}}{K_N'(W_{CH}/L_{CH})}} = \sqrt{\frac{2(570\text{m})}{(200\mu)(50\text{m}/190\text{n})}} = 150 \text{ mV}$$

$$v_{TH(O)} \approx V_{TN0} + v_{DS(SAT)} = 400\text{m} + 150\text{m} = 550 \text{ mV}$$

$$v_{SWO} = v_O + v_{DO} = 4 + 800\text{m} = 4.8 \text{ V after } M_{EG} \text{ opens} \quad \therefore \quad \Delta v_{SWO} = 4.8 \text{ V}$$

$$\begin{aligned} t_{V(O)} &\approx \left(\frac{R_N}{v_{TH(O)}}\right)\left[C_{OL}\Delta v_{SWO} + \left(\frac{C_{CH}}{4}\right)v_{TH(O)}\right] \\ &= \left(\frac{20}{550\text{m}}\right)\left[(10\text{p})(4.8) + \left(\frac{66\text{p}}{4}\right)(550\text{m})\right] = 2.1 \text{ ns} \end{aligned}$$

$$\tau_{RC(O)} = R_N\left[2C_{OL} + \left(\frac{2}{3}\right)C_{CH}\right] = (20)\left[2(10p) + \left(\frac{2}{3}\right)(66p)\right] = 1.3\text{ ns}$$

$$t_{I(O)} \approx \tau_{RC(O)} \ln\left(\frac{v_{TH(O)}}{V_{TN0}}\right) = (1.3n)\ln\left(\frac{550m}{400m}\right) = 410\text{ ps}$$

$$P_{IV} \approx i_{L(HI)}\Delta v_{SWO}\left(\frac{t_{I(O)}}{3t_{SW}} + \frac{t_{V(O)}}{2t_{SW}}\right) = (570m)(4.8)\left[\frac{410p}{(3)1\mu} + \frac{2.1n}{(2)1\mu}\right] = 3.2\text{ mW}$$

$$\sigma_{IV} = \frac{P_{IV}}{(i_{O(AVG)}/d_O)d_{IN}v_{IN}} = \frac{3.2m}{(250m/48\%)(1)(2)} = 0.3\%$$

Note: R_N is lower than R_P to balance response times (because $v_{TH(O)}$ is lower than $v_{DD} - v_{TH(C)}$ in t_V and v_{GS} reaches V_{TN0} near the end of R_N's exponential response, where the response is slower). M_{EG} still consumes more power opening than closing because i_{DS} is higher at $i_{L(HI)}$.

Explore with SPICE:
See Appendix A for notes on SPICE simulations.

```
* I-V Overlap Loss when MEG Opens
rn vg 0 20
lx 0 vswo 10u
meg vswo vg 0 0 nmos1 w=50m l=250n
ddo vswo vo diode1
vo vo 0 dc=4
.ic i(lx)=570m v(vg)=4
.lib lib.txt
.tran 9n
.end
```

Tip: Plot id(Meg), v(vg), v(vswo), and id(Meg)*v(vswo), extract the average of the product term across 9 ns, and multiply by 0.9% (9 ns of the 1-μs period). id(Meg) often includes C_{GD}'s current, so the extraction is not perfect. In this case, i_D is considerably greater than i_{GD}, so the approximation is fair.

4.5.3 Reverse Recovery

A. **Power**

Forward-biased in-transit charge across PN junctions reverses direction when diodes reverse-bias. In switched inductors, dead-time diodes carry this *reverse-recovery charge* q_{RR} when conducting i_L. q_{RR} is the charge in the junction that i_L feeds and *forward transit time* τ_F across the junction sets to $i_L\tau_F$.

The challenge with q_{RR} is that a switch must recover it when reverse-biasing a diode. In Fig. 4.24, for example, *dead-time diode* D_{DT} conducts i_L when M_{SW} is open. So for v_{DS} to fall when M_{SW} closes, i_{DS} must first rise to a peak $i_{DS(RR)}$ that sinks i_L, i_{GD}, i_{SW}, and q_{RR} held in D_{DT}. And a higher i_{DS} dissipates more P_{IV}.

As M_{SW} closes, i_{DS} climbs with v_{GS} mostly after v_{GS} overcomes V_{T0}. If i_L overwhelms i_P and i_{SW}, i_{DS} reaches i_L after $t_{I(C)}$ when v_{GS} reaches $v_{TH(C)}$. i_{DS} requires another t_{RR} to reach a level that can sink q_{RR}. Approximating i_{DS}'s rise past i_L to be linear with the slope $\partial i_{DS}/\partial t$ that i_{DS}'s quadratic climb $(i_L/t_{I(C)}{}^2)t^2$ reaches i_L at $t_{I(C)}$ yields a t_{RR} that is roughly a square-root translation of $t_{I(C)}$ and τ_F:

$$q_{RR} = \int_0^{t_{RR}} i_{DS}dt \approx \int_0^{t_{RR}} \left.\frac{\partial i_{DS}}{\partial t}\right|_{t_{I(C)}} tdt \approx \int_0^{t_{RR}} \left(\frac{2i_{L(C)}}{t_{I(C)}}\right) tdt = \left(\frac{i_{L(C)}}{t_{I(C)}}\right) t_{RR}{}^2 = i_{L(C)}\tau_F \quad (4.49)$$

$$t_{RR} \approx \sqrt{t_{I(C)}\tau_F}. \quad (4.50)$$

$i_{DS(RR)}$ is therefore a corresponding t_{RR} extension of i_L:

$$i_{DS(RR)} \approx i_{L(C)} + \left(\frac{2i_{L(C)}}{t_{I(C)}}\right) t_{RR} = i_{L(C)}\left(1 + 2\sqrt{\frac{\tau_F}{t_{I(C)}}}\right). \quad (4.51)$$

The v_{GS} that M_{SW} requires to sink this $i_{DS(RR)}$ is $v_{TH(RR)}$:

$$v_{TH(RR)} = V_{T0} + v_{DS(SAT)}\Big|_{i_{DS(RR)}} \approx V_{T0} + \sqrt{\frac{2i_{DS(RR)}}{K'(W/L)}}. \quad (4.52)$$

Since v_{DS} is steady at Δv_{SW} across $t_{I(C)}$ and t_{RR}, P_I' is the power $i_{DS(RR)}$'s quadratic average $33\% i_{DS(RR)}$ burns with Δv_{SW} across $t_{I(C)}$ and t_{RR}.

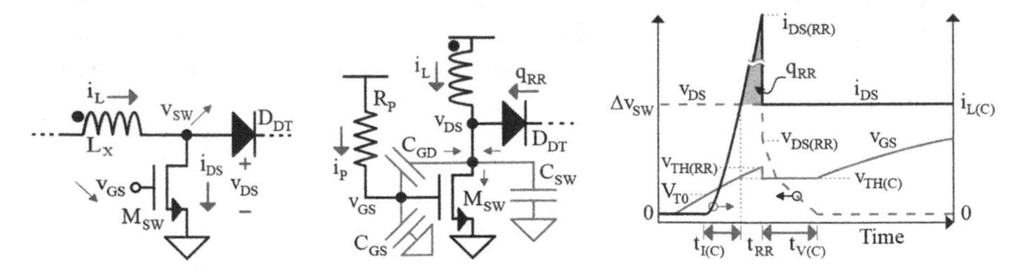

Fig. 4.24 Closing switch with reverse recovery

After i_{DS} recovers q_{RR}, $i_{DS(RR)}$ sinks more than i_L supplies, so the excess discharges C_{GD} and C_{SW}. The i_{GD} that $i_{DS(RR)}$, i_L, and i_{SW} avail is so much greater than i_P that the part of i_{GD} that excludes i_P discharges C_{GS}. C_{GD}, C_{SW}, and C_{GS} discharge this way until i_{DS} falls to the level that carries i_L, i_P, and i_{SW}, which happens when v_{GS} reaches $v_{TH(C)}$. Since i_P is much lower than i_{GD}, C_{GS} supplies most of the charge Δq_{GS} that discharges C_{GD} across Δv_{DG}, which amounts to Δv_{DS} minus Δv_{GS}:

$$\begin{aligned}\Delta q_{GD} &= C_{GD}\Delta v_{DG} = C_{GD}(\Delta v_{DS} - \Delta v_{GS}) \\ &= C_{GD}\left(\Delta v_{SW} - v_{DS(RR)} - \Delta v_{GS}\right) \\ &\approx \Delta q_{GS} = C_{GS}\Delta v_{GS} = C_{GS}\left(v_{TH(RR)} - v_{TH(C)}\right) \\ &= C_{GS}\Delta v_{TH},\end{aligned} \tag{4.53}$$

where C_{GS} discharges across Δv_{GS} or Δv_{TH} from $v_{TH(RR)}$ to $v_{TH(C)}$ and v_{DS} falls across Δv_{DS} from Δv_{SW} to $v_{DS(RR)}$. Solving this reveals v_{DS} falls to approximately

$$v_{DS(RR)} \approx \Delta v_{SW} - \left(\frac{C_{GS}}{C_{GD}} + 1\right)\Delta v_{TH}. \tag{4.54}$$

This transition to $v_{DS(RR)}$ is quick because i_{GD} is substantial.

After i_{DS} falls to a level that carries i_L, i_P, and i_{SW}, i_P discharges C_{GD} across the $t_{V(C)}'$ that collapses $v_{DS(RR)}$. $t_{V(C)}'$ is shorter than $t_{V(C)}$ because v_{DS} collapses a $v_{DS(RR)}$, that is lower than Δv_{SW}:

$$t_{V(C)}' \approx \left(\frac{R_P}{v_{DD} - v_{TH(C)}}\right)\left[C_{OL} v_{DS(RR)} + \left(\frac{C_{CH}}{4}\right) v_{TH(C)}\right]. \tag{4.55}$$

Since i_{DS} is steady near i_L across $t_{V(C)}'$ when i_L overwhelms i_P and i_{SW}, P_V' is largely the power i_L burns with $v_{DS(RR)}$'s average $50\% v_{DS(RR)}$. And P_{IV}' is a t_{SW} fraction of P_I' and P_V':

$$\begin{aligned}P_{IV}' &= P_I'\left(\frac{t_{I(C)} + t_{RR}}{t_{SW}}\right) + P_V'\left(\frac{t_{V(C)}'}{t_{SW}}\right) \\ &\approx \left(\frac{i_{DS(RR)}}{3}\right)\Delta v_{SW}\left(\frac{t_{I(C)} + t_{RR}}{t_{SW}}\right) + i_L\left(\frac{v_{DS(RR)}}{2}\right)\left(\frac{t_{V(C)}'}{t_{SW}}\right).\end{aligned} \tag{4.56}$$

Interestingly, q_{RR} not only raises the i_{DS} that consumes P_I' (to $i_{DS(RR)}$) and extends the time the switch burns P_I' (by t_{RR}) but also reduces the v_{DS} (to $v_{DS(RR)}$) that burns P_V'. The rise in i_{DS}, however, normally raises P_I' more than the fall in $v_{DS(RR)}$ reduces P_V'. So reducing q_{RR} usually saves power.

Example 11: Determine P_{IV} and σ_{IV} for M_{EG} in the ideal boost from Examples 7 and 9 in CCM when M_{EG} closes and D_{DO}'s τ_F is 5 ns.

Solution:

$$d_{IN} = 1, d_E = 52\%, d_O = d_D = 48\%, \text{and } \Delta i_L = 100 \text{ mA from Example 7}$$

$$i_{L(LO)} = 470 \text{ mA}, v_{TH(C)} = 530 \text{ mV}, \Delta v_{SWO} = 4.8 \text{ V}, L_{CH} = 190 \text{ nm},$$
$$C_{OL} = 10 \text{ pF}, C_{CH} = 66 \text{ pF}, R_P = 100\ \Omega, \tau_{RC} = 6.4 \text{ ns, and}$$

$$t_{I(C)} \approx 240 \text{ ps from Example 9}$$

$$t_{RR} \approx \sqrt{t_{I(C)}\tau_F} = \sqrt{(240p)(5n)} = 1.1 \text{ ns}$$

$$i_P \approx \frac{v_{DD} - v_{TH(C)}}{R_P} = \frac{4 - 530m}{100} = 35 \text{ mA}$$

$$i_{DS(RR)} \approx i_{L(LO)} + \left(\frac{2i_{L(LO)}}{t_{I(C)}}\right)t_{RR} = (470m)\left[1 + 2\left(\frac{1.1n}{240p}\right)\right] = 4.8 \text{ A}$$

$$v_{TH(RR)} \approx V_{TN0} + \sqrt{\frac{2i_{DS(RR)}}{K_N'(W_{CH}/L_{CH})}} = 400m + \sqrt{\frac{2(4.8)}{(200\mu)(50m/190n)}} = 830 \text{ mV}$$

$$v_{DS(RR)} = \Delta v_{SWO} - \left(\frac{C_{OL} + (2/3)C_{CH}}{C_{OL}} + 1\right)\left(v_{TH(RR)} - v_{TH(C)}\right)$$
$$= 4.8 - \left[\frac{10p + (2/3)(66p)}{10p} + 1\right](830m - 530m) = 2.9 \text{ V}$$

$$t_{V(C)}' \approx \left(\frac{1}{i_P}\right)\left[C_{OL}v_{DS(RR)} + \left(\frac{C_{CH}}{4}\right)v_{TH(C)}\right]$$
$$= \left(\frac{1}{35m}\right)\left[(10p)(2.9) + \left(\frac{66p}{4}\right)(530m)\right] = 1.1 \text{ ns}$$

$$P_{IV} \approx \left(\frac{i_{DS(RR)}}{3}\right)v_{SWO}\left(\frac{t_{I(C)} + t_{RR}}{t_{SW}}\right) + i_{L(LO)}\left(\frac{v_{DS(RR)}}{2}\right)\left(\frac{t_{V(C)}'}{t_{SW}}\right)$$
$$= \left(\frac{4.8}{3}\right)(4.8)\left(\frac{240p + 1.1n}{1\mu}\right) + (470m)\left(\frac{2.9}{2}\right)\left(\frac{1.1n}{1\mu}\right) = 11 \text{ mW}$$

$$\sigma_{IV} = \frac{P_{IV}}{\left(i_{O(AVG)}/d_O\right)d_{IN}v_{IN}} = \frac{11m}{(250m/48\%)(1)(2)} = 1.1\%$$

Note: This P_{IV} is higher than in Example 9 because $i_{DS(RR)}$ is higher than $i_{L(LO)}$ and $t_{I(C)} + t_{RR}$ is longer than $t_{I(C)}$ to a greater extent than $v_{DS(RR)}$ is lower than Δv_{SWO}.

Explore with SPICE:
Use the SPICE code from Example 9 and use "diode2" for D_{DO}'s model.

B. Approximation

This reverse-recovery analysis approximates the peak that i_{DS} reaches when recovering q_{RR}. Although not bad, this calculation requires several steps. Another approach is to approximate the behavior of i_{DS} in a way that, to some extent, preserves P_{IV}.

Like before, though, v_{DS} is close to Δv_{SW} when M_{SW} first closes and i_{DS} climbs to $i_{L(C)}$ across $t_{I(C)}$. But instead of i_{DS} climbing over $i_{L(C)}$ after $t_{I(C)}$ quadratically, assuming i_{DS} reaches and maxes to twice $i_{L(C)}$ quickly (like Fig. 4.25 shows) simplifies the analysis. This way, the *reverse-recovery current* i_{RR} that sinks q_{RR} is $i_{L(C)}$.

But since a quadratic rise in $i_{L(C)}$ recovers q_{RR} more quickly than the constant $i_{L(C)}$ assumed, t_{RR} is shorter than τ_F. Approximating t_{RR}' to half τ_F reduces *reverse-recovery power* P_{RR} in P_{IV} to $i_{L(C)}\Delta v_{SW}(\tau_F/t_{SW})$. Although not very accurate, this P_{RR} is not a bad approximation that is easy to calculate:

$$P_{RR} \approx i_{DS(RR)}'\Delta v_{SW}\left(\frac{t_{RR}'}{t_{SW}}\right) \approx 2i_{L(C)}\Delta v_{SW}\left(\frac{0.5\tau_F}{t_{SW}}\right) = i_{L(C)}\Delta v_{SW}\left(\frac{\tau_F}{t_{SW}}\right). \quad (4.57)$$

And instead of v_{DS} collapsing across a noticeable $t_{V(C)}$, assuming $i_{DS(RR)}$ is so high that it collapses v_{DS} across a negligible $t_{V(C)}$ is not unreasonable, especially when considering the t_{RR} that τ_F establishes with i_L. With this approximation, the part of $i_{DS(RR)}$ that excludes i_L discharges C_{GD} and C_{GS} to the extent v_{DS} collapses and v_{GS} falls to $v_{TH(C)}$ very quickly. P_{IV} therefore excludes the $P_{V(C)}$ that a negligibly short $t_{V(C)}$ induces:

$$\begin{aligned} P_{IV} &= P_{IV(C)} + P_{IV(O)} \\ &\approx P_{I(C)} + P_{RR} + P_{I(O)} + P_{V(O)} \\ &\approx i_{L(C)}\Delta v_{SW}\left(\frac{t_{I(C)}}{3t_{SW}} + \frac{\tau_F}{t_{SW}}\right) + i_{L(O)}\Delta v_{SW}\left(\frac{t_{I(O)}}{3t_{SW}} + \frac{t_{V(O)}}{2t_{SW}}\right). \end{aligned} \quad (4.58)$$

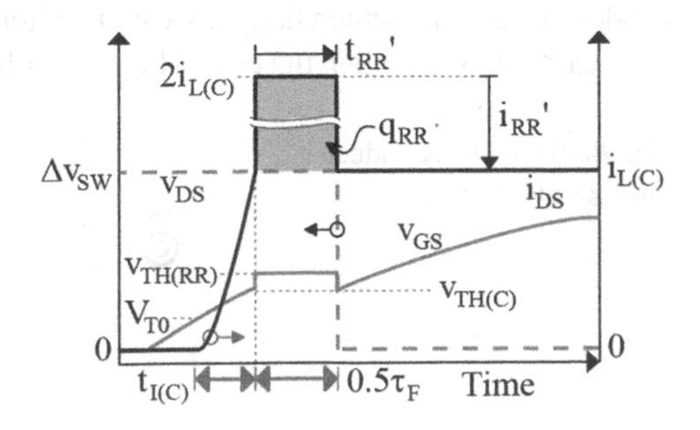

Fig. 4.25 Closing switch with approximated reverse recovery

All other close- and open-switch non-reverse-recovery components of P_{IV} remain the same: $P_{I(C)}$ across $t_{I(C)}$, $P_{I(O)}$ across $t_{I(O)}$, and $P_{V(O)}$ across $t_{V(O)}$.

Although behaviorally imprecise, this estimate is not bad. This is because q_{RR} balances i_{RR} and t_{RR} in P_{RR}: the shorter t_{RR} that a higher i_{DS} needs to recover q_{RR} consumes similar P_{RR} with Δv_{SW} as the longer t_{RR} that a lower i_{DS} requires. Halving τ_F approximates the t_{RR} that a quadratically increasing i_{DS} needs to recover q_{RR}.

Example 12: Approximate P_{IV} and σ_{IV} for M_{EG} in the ideal boost in CCM from Examples 7, 9, and 11 when M_{EG} closes and i_{EG} maxes to i_L.

Solution:

$$i_{L(LO)} = 470\text{ mA}, \Delta v_{SWO} = 4.8\text{ V}, t_{I(C)} = 240\text{ ps},$$
$$\text{and } t_{SW} = 1\ \mu\text{s from Examples 7, 9, and 11}$$

$$P_{IV} \approx i_{L(LO)}\Delta v_{SWO}\left(\frac{t_{I(C)}}{3t_{SW}} + \frac{\tau_F}{t_{SW}}\right) \approx (470\text{m})(4.8)\left[\frac{240\text{p}}{3(1\mu)} + \frac{5\text{n}}{1\mu}\right] = 12\text{ mW}$$

$$\sigma_{IV} = \frac{P_{IV}}{\left(i_{O(AVG)}/d_O\right)d_{IN}v_{IN}} = \frac{12\text{m}}{(250\text{m}/48\%)(1)(2)} = 1.2\%$$

Note: This σ_{IV} is only 0.1% higher than the more accurate σ_{IV} from Example 11, which is 0.9% higher than the σ_{IV} that excludes reverse recovery from Example 9.

C. **Implicit MOS Diodes**

The *diffusion charge* q_{DIF} that sets reverse-recovery charge is greatest when body diodes conduct i_L, which happens when switches are off. M_{DG} and M_{DO} in Fig. 4.26, for example, open when the controller grounds M_{DG}'s gate and connects M_{DO}'s gate

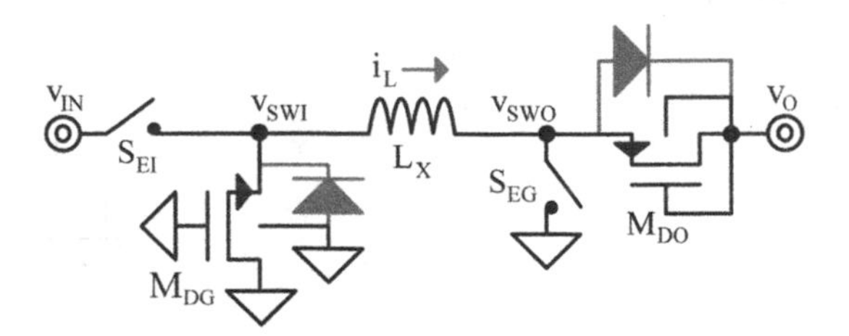

Fig. 4.26 Switched inductor during dead time

to v_O. This way, dead-time i_L discharges v_{SWI}'s C_{SWI} and charges v_{SWO}'s C_{SWO} until M_{DG}'s and M_{DO}'s body diodes conduct i_L.

When these diodes conduct, v_{SWI} falls a diode voltage below ground and v_{SWO} rises a diode voltage over v_O. Interestingly, M_{DG}'s v_{GS} and M_{DO}'s v_{SG} are positive under these conditions. Although sub-threshold current is not always negligible, M_{DG} and M_{DO} remain largely off when their threshold voltages match or surpass a PN diode voltage.

M_{DG} and M_{DO} start conducting some of i_L when threshold voltages are lower. So the currents that feed q_{RR}'s drop as V_{T0}'s fall below 600–700 mV. This diminishes the effect of q_{RR} on P_{IV}. The effect is minimal when the v_{GS} and v_{SG} that M_{DG} and M_{DO} need to sustain i_L are less than 600–700 mV, which can happen when V_{T0}'s are 300–500 mV.

M_{DG} and M_{DO}, however, also need q_{RR} to charge their C_{GS}'s across the v_{GS}'s needed to sustain i_L. Even when v_{GS} is below V_{T0}, C_{GS} still draws charge. So in effect, q_{RR} is the combined charge C_{GS} and the diode need to conduct i_L together.

q_{DIF} in q_{RR} falls with lower V_{T0}'s. Very low V_{T0}'s are uncommon because large low-V_{T0} switches leak too much current with zero v_{GS}. Although technologies and applications vary, 400–600-mV V_{T0}'s are more common. With these V_{T0}'s, the effect of q_{RR} on P_{IV} is still present, but not as severe.

Letting body diodes conduct poses another challenge: *substrate noise*. This is because MOSFETs on the substrate conduct body-diode current through the substrate. And MOSFETs in wells over the substrate activate vertical *bipolar-junction transistors* (BJTs) that inject current into the substrate. Switching events therefore produce, inject, and propagate noise energy across the substrate, coupling and affecting other circuits along the way.

So the implicit diode action of low-V_{T0} MOSFETs keeps body diodes from generating *substrate current* i_{SUB}. They also reduce the dead-time voltages that produce P_{DT}. And as already mentioned, they reduce the q_{RR} that burns P_{IV}.

Explore with SPICE:
See Appendix A for notes on SPICE simulations.

```
* I-V Overlap Loss when MEG Closes & VTP0 is -400 mV
vdd vdd 0 dc=4
rp vdd vg 100
lx 0 vswo 10u
meg vswo vg 0 0 nmos1 w=50m l=250n
ddo vswo vo diode2
mdo vo vo vswo vo pmos1 w=150m l=250n
vo vo 0 dc=4
.ic i(lx)=470m v(vg)=0
.lib lib.txt
.tran 5n
.end
```

(continued)

Tip: Plot id(Meg), v(vg), v(vswo), and id(Meg)*v(vswo), extract the average of the product term across 5 ns, and multiply by 0.5% (5 ns of the 1-μs period) to determine M_{EG}'s P_{IV}. id(Meg) often includes C_{GD}'s current, so the extraction is not perfect. In this case, i_D is considerably greater than i_{GD}, so the approximation is fair.

D. **Schottky Diodes**

Connecting *Schottky diodes* in parallel with M_{DG}'s and M_{DO}'s body diodes like Fig. 4.27 offers similar benefits. This is because they drop lower voltages than typical PN diodes without a junction that traps in-transit charge. So they steer current away from body diodes without trapping q_{DIF} or generating i_{SUB}. The only component that holds charge in Schottkys is depletion capacitance.

Good Schottky diodes, however, are not always available on-chip. And off-chip diodes require board space and money. Still, the advantages of lower noise and lower power can outweigh the added volume and cost of additional components or manufacturing steps.

4.5.4 Soft Switching

P_{IV} hinges on the currents and voltages that power switches carry and collapse. *Soft switching* refers to transitions that carry low currents or collapse low voltages. Although *zero-current switching* (ZCS) and *zero-voltage switching* (ZVS) are extreme cases, engineers often use ZVS and ZCS to refer to "soft" events. Either way, the net result is low P_{IV}.

A. **Zero-Voltage Switching**

Collapsing v_{IN} or v_O is not ZVS. Energize switches fall into this category. M_{EI}, for example, collapses v_{IN} plus v_{DG} because M_{EI} connects v_{IN} to v_{SWI} and D_{DG} conducts i_L before and after M_{EI} closes. M_{EG} similarly collapses v_O and v_{DO} because D_{DO} conducts i_L into v_O before and after M_{EG} closes. These are *hard-switching* examples.

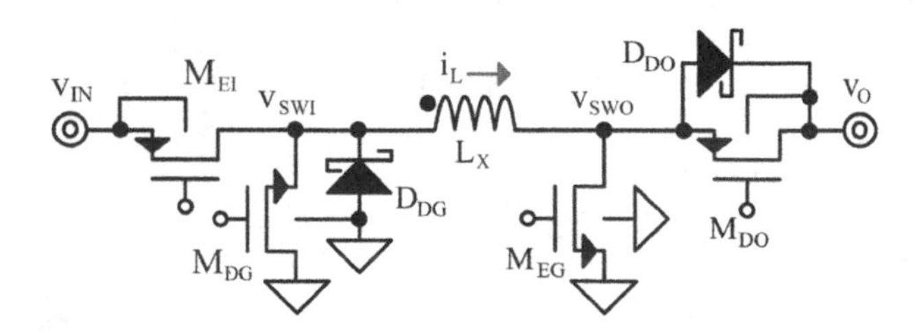

Fig. 4.27 Switched inductor with Schottky diodes

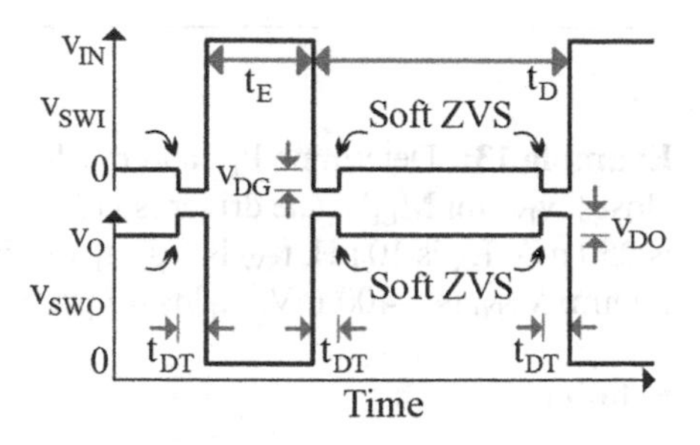

Fig. 4.28 Soft zero-voltage switching events

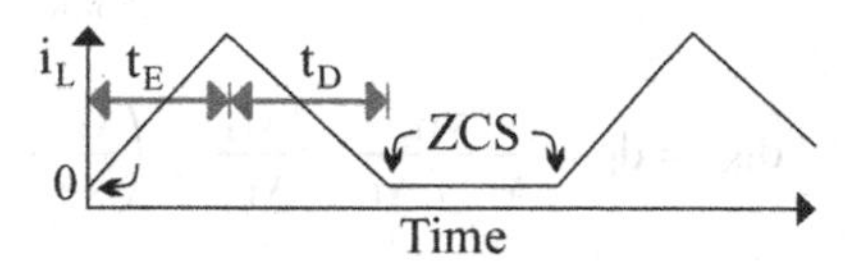

Fig. 4.29 Zero-current switching events

Collapsing diode voltages like drain switches do is "soft." M_{DG}, for one, collapses D_{DG}'s v_{DG} when M_{DG} grounds v_{SWI} a t_{DT} before and after t_E in Fig. 4.28. And M_{DO} collapses D_{DO}'s v_{DO} when M_{DO} connects v_{SWO} to v_O at those same times. These transitions soften further into ZVS when v_{DG} and v_{DO} are smaller fractions of v_{IN} and v_O.

B. **Zero-Current Switching**

ZCS happens in DCM when i_L reaches zero. Drain switches, for example, open with ZCS because i_L reaches zero when t_D ends in Fig. 4.29. Energize switches similarly close with ZCS because i_L is zero when t_E starts. So P_{IV} when t_D ends and t_E starts is usually negligibly low.

4.5.5 CMOS Expressions

NFETs are *active high*, which means they close when their gate voltages are high. Discussions, graphs, and expressions to this point reflect this context. In this light, v_{GS}, v_{DS}, and V_{T0} are positive values.

PFETs behave like NFETs, except they are *active low*, which is to say they close with low gate voltages. So pull-down resistors close them and pull-up resistors open them. CMOS stands for *complementary MOS* transistors for this reason: because NFETs and PFETs are complementary and available for use.

Not coincidentally, NFET expressions also apply to PFETs, except v_{GS}, v_{DS}, and V_{T0} for PFETs are negative, which is not always intuitive. Replacing v_{GS}, v_{DS}, and V_{T0} with v_{SG}, v_{SD}, and $|V_{T0}|$ or $|V_{TP0}|$ is more insightful because PFETs collapse v_{SD} when v_{SG} overcomes $|V_{T0}|$. But when generalizing, expressions with v_{GS}, v_{DS}, and V_{T0} apply to both: NFETs and PFETs.

Example 13: Determine P_{IV} and σ_{IV} for M_{EI} in the ideal buck in CCM when M_{EI} closes, v_{DD} for M_{EI}'s gate driver is v_{IN}, v_{IN} is 4 V, v_O is 2 V, v_{DG} is 800 mV, $i_{O(AVG)}$ is 250 mA, L_X is 10 μH, t_{SW} is 1 μs, t_{DT} is 50 ns, W_{EI} is 50 mm, L_{EI} is 250 nm, L_{OL} is 30 nm, V_{TP0} is −400 mV, K_P' is 40 μA/V^2, C_{OX}'' is 6.9 fF/μm^2, and R_N is 50 Ω.

Solution:

$$\text{Buck} \quad \therefore \quad d_O = 1$$

$$d_{IN} = d_E = \frac{v_D}{v_E + v_D} = \frac{v_O}{v_{IN}} + \left(\frac{v_{DG}}{v_{IN}}\right)\left(\frac{2t_{DT}}{t_{SW}}\right) = \frac{2}{4} + \left(\frac{800m}{4}\right)\left[\frac{2(50n)}{1\mu}\right] = 52\%$$

$$\therefore \quad d_D = 1 - d_E = 1 - 52\% = 48\%$$

$$\Delta i_L = \left(\frac{v_E}{L_X}\right)t_E = \left(\frac{v_{IN} - v_O}{L_X}\right)d_E t_{SW} \approx \left(\frac{4-2}{10\mu}\right)(52\%)(1\mu) = 100 \text{ mA}$$

$$i_{L(LO)} = \frac{i_{O(AVG)}}{d_O} - \frac{\Delta i_L}{2} = \frac{250m}{100\%} - \frac{100m}{2} = 200 \text{ mA}$$

$$W_{EI}, L_{EI}, L_{OL}, \text{and } C_{OX}'' \text{ match } M_{EG}\text{'s from Example 9}$$

$$\therefore \quad W_{CH} = 50 \text{ mm}, L_{CH} = 190 \text{ nm}, C_{OL} = 10 \text{ pF}, C_{CH} = 66 \text{ pF from Example 9}$$

$$M_{EI} \text{ closes when } i_L = i_{L(LO)}$$

$$v_{SD(SAT)} \approx \sqrt{\frac{2i_{L(LO)}}{K_P'(W_{CH}/L_{CH})}} = \sqrt{\frac{2(200m)}{(40\mu)(50m/190n)}} = 200 \text{ mV}$$

$$v_{TH(C)} \approx |V_{TP0}| + v_{SD(SAT)} = 400m + 200m = 600 \text{ mV}$$

$$\tau_{RC(C)} = R_N\left[2C_{OL} + \left(\frac{2}{3}\right)C_{CH}\right] = (50)\left[2(10p) + \left(\frac{2}{3}\right)(66p)\right] = 3.2 \text{ ns}$$

$$t_{I(C)} \approx \tau_{RC(C)} \ln\left(\frac{v_{DD} - |V_{TP0}|}{v_{DD} - v_{TH(C)}}\right) = (3.2n)\ln\left(\frac{4 - 400m}{4 - 600m}\right) = 180 \text{ ps}$$

$$v_{SWI} = -v_{DG} = -800 \text{ mV before } M_{EI} \text{ closes}$$

$$v_{SWI} \approx v_{IN} = 4 \text{ V after } M_{EI} \text{ closes}$$

$$\therefore \quad v_{SD} \text{ swings } \Delta v_{SWI} = v_{IN} - (-v_{DG}) = v_{IN} + v_{DG} = 4 + 800m = 4.8 \text{ V}$$

$$t_{V(C)} \approx \left(\frac{R_N}{v_{DD} - v_{TH(C)}}\right)\left[C_{OL}\Delta v_{SWI} + \left(\frac{C_{CH}}{4}\right)v_{TH(C)}\right]$$
$$= \left(\frac{50}{4 - 600m}\right)\left[(10p)(4.8) + \left(\frac{66p}{4}\right)(600m)\right] = 850 \text{ ps}$$

$$P_{IV} \approx i_{L(LO)}\Delta v_{SWI}\left(\frac{t_{I(C)}}{3t_{SW}} + \frac{t_{V(C)}}{2t_{SW}}\right) = (200m)(4.8)\left[\frac{180p}{(3)1\mu} + \frac{850p}{(2)1\mu}\right] = 470\ \mu W$$

$$\sigma_{IV} = \frac{P_{IV}}{(i_{O(AVG)}/d_O)d_{IN}v_{IN}} = \frac{470\mu}{(250m/1)(52\%)(4)} = 0.1\%$$

Note: This P_{IV} excludes the power consumed when M_{EI} opens and other switches open and close.

Explore with SPICE:

See Appendix A for notes on SPICE simulations.

```
* I-V Overlap Loss when MEI Closes without Reverse Recovery
vin vin 0 dc=4
rn vg 0 50
mei vswi vg vin vin pmos1 w=50m l=250n
ddg 0 vswi fdiode1
lx vswi vo 10u
vo vo 0 dc=2
.ic i(lx)=200m v(vg)=4
.lib lib.txt
.tran 5n
.end
```

Tip: Plot id(Mei), v(vg), v(vswi), and id(Mei)*(v(vin)-v(vswi)), extract the average of the product term across 5 ns, and multiply by 0.5% (5 ns of the 1-μs period). id(Mei) often includes C_{GD}'s current, so the extraction is not perfect. In this case, i_D is considerably greater than i_{GD}, so the approximation is fair.

4.6 Gate-Driver Loss

4.6.1 Gate Driver

Typical gate drivers use pull-up PFETs to charge gates and pull-down NFETs to discharge gates. Since PFETs are active low and NFETs are active high, a high input v_I in Fig. 4.30 shuts M_P and activates M_N, which grounds the *output* v_O. A low v_I does the opposite: shuts M_N and activates M_P, which pulls v_O to the power supply v_{DD}.

As v_I climbs from zero, M_N starts to conduct when v_I surpasses V_{TN0}. v_O transitions low when M_N pulls as much current as M_P can supply. The *shoot-through current* i_{ST} that M_N and M_P conduct maxes at this point. The *trip point* V_{TP} of the driver is the v_I that balances M_N's and M_P's strengths this way. So i_{ST} maxes when v_I reaches V_{TP} and v_O halves v_{DD}:

$$\begin{aligned} i_{ST(MAX)} &= i_N\Big|_{v_{DS}=\frac{v_{DD}}{2}}^{v_{GS}=V_{TP}} = \left(\frac{W_N}{L_N}\right)\left(\frac{K_N{}'}{2}\right)(V_{TP} - V_{TN0})^2\left[1 + \left(\frac{v_{DD}}{2}\right)\lambda_N\right] \\ &= i_P\Big|_{v_{SD}=\frac{v_{DD}}{2}}^{v_{SG}=V_{DD}-V_{TP}} = \left(\frac{W_P}{L_P}\right)\left(\frac{K_P{}'}{2}\right)(V_{DD} - V_{TP} - |V_{TP0}|)^2\left[1 + \left(\frac{v_{DD}}{2}\right)\lambda_P\right], \end{aligned} \tag{4.59}$$

where v_I at V_{TP} and v_O at half v_{DD} invert and saturate M_N and M_P.

A few points are worth noting. V_{TP} is half v_{DD} when V_{T0}'s and *channel-length modulation parameters* λ's match and M_P's W/L is greater than M_N's by the same amount that M_N's *transconductance parameter* K' is greater than M_P's. v_O can rail to zero and v_{DD} because, one transistor is on when the other one is off and *vice versa*. Static power when v_I is within a V_{T0} of the supplies is nearly zero because i_{ST} is close to zero. This is why this circuit is so attractive as a *digital inverter*, because power is zero when v_I is high or low.

The *gate capacitances* C_G that load the driver need M_P's i_P to charge and M_N's i_N to discharge. This means that i_N wastes power when i_P charges and i_P wastes power when i_N discharges. But if M_N's v_{DS} is low when C_G charges and M_P's v_{SD} is low when C_G discharges, their triode currents waste less power. This happens when v_O transitions more slowly than v_I, which results when M_P and M_N charge and discharge C_G slowly.

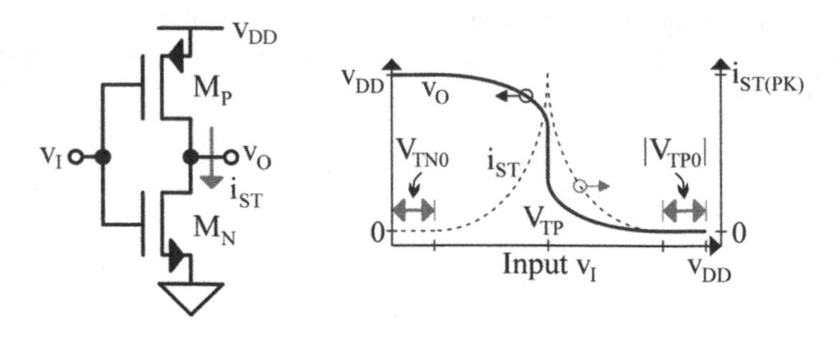

Fig. 4.30 Inverting gate driver

Across the t_I and t_V that switches carry i_L, the driver's v_O is V_{TO} to v_{TH} above ground or below v_{DD}. Since M_N's v_{GS} is v_{DD} and $v_{DS(SAT)}$ is $v_{DD} - V_{TN0}$, M_N's v_{DS} is lower than $v_{DS(SAT)}$ across t_I and t_V. M_P's v_{SD} is similarly lower than $v_{SD(SAT)}$'s $v_{DD} - |V_{TP0}|$. So M_N and M_P are in triode across t_I and t_V, when v_{DS} is largely v_{TH} or $v_{DD} - v_{TH}$, which means their triode resistances set the R_N and R_P that open and close power transistors:

$$R_{N/P} = \frac{v_{DS}}{i_{TRI}} = \frac{1}{(W_{CH}/L_{CH})K'(v_{GS} - V_{T0} - 0.5v_{DS})} = \left(\frac{L_{CH}}{W_{CH}}\right)\left[\frac{1}{K'(v_{DD} - V_{T0} - 0.5v_{DS})}\right]. \tag{4.60}$$

Example 14: Determine the W's for the gate driver that closes and opens M_{EG} in the ideal boost from Examples 7, 9, and 10 when v_{DD} is v_O, L's are 250 nm, L_{OL} is 30 nm, V_{TN0} is 400 mV, V_{TP0} is −400 mV, K_N' is 200 μA/V^2, and K_P' is 40 μA/V^2.

Solution:

$$v_{DD} = v_O = 4\ \text{V}, R_P = 100\ \Omega, v_{TH(C)} = 530\ \text{mV}, R_N = 20\ \Omega,$$

and $v_{TH(O)} = 550$ mV from Examples 9 and 10

$$L_{CH} = L - 2L_{OL} = 250\text{n} - 2(30\text{n}) = 190\ \text{nm}$$

$$M_P\text{'s } v_{SD} = v_{DD} - v_{TH(C)}$$

$$W_P = \frac{L_{CH}}{R_P K_P'\left[v_{DD} - |V_{TP0}| - 0.5\left(v_{DD} - v_{TH(C)}\right)\right]} = \frac{190\text{n}}{(100)(40\mu)[4 - 400\text{m} - 0.5(4 - 530\text{m})]} = 26\ \mu\text{m}$$

$$M_N\text{'s } v_{DS} = v_{TH(O)}$$

$$W_N = \frac{L_{CH}}{R_N K_N'\left(v_{DD} - V_{TN0} - 0.5v_{TH(O)}\right)} = \frac{190\text{n}}{(20)(200\mu)[4 - 400\text{m} - 0.5(550\text{m})]} = 14\ \mu\text{m}$$

Explore with SPICE:
See Appendix A for notes on SPICE simulations.

```
* Ideal Boost with Dead Time, MEG, & MEG's Driver in CCM
vdeb deb 0 dc=4 pulse 4 0 0 1n 1n 515n 1u
vdo do 0 dc=1 pulse 0 1 570n 1n 1n 385n 1u
vdd vdd 0 dc=4
mp vg deb vdd vdd pmos1 w=26u l=250n
mn vg deb 0 0 nmos1 w=14u l=250n
vin vin 0 dc=2
lx vin vswo 10u
meg vswo vg 0 0 nmos1 w=50m l=250n
sdo vswo vo do 0 sw1v
ddo vswo vo fdiode1
vo vo 0 dc=4
.ic i(lx)=470m
.lib lib.txt
.tran 1u
.end
```

Tip: Plot id(Meg), v(vg), and v(vswo) and view across 1 μs, first 20 ns, and between 520 and 540 ns.

4.6.2 Closing Switch

When closing M_{SW} in Fig. 4.31, M_P supplies the *gate current* i_G that C_{GB}, C_{GS}, and C_{GD} need and the i_{ST} that M_N leaks. i_{ST} is low and short-lived (by design) because M_{SW}'s low v_{GS} suppresses M_N's v_{DS} as v_I's fall collapses M_N's v_{GS}. i_G is much greater because M_P's v_{SG} and v_{SD} are high across the time M_N leaks i_{ST}. So the *driver current* i_D that v_{DD} supplies is mostly the charge q_G that C_{GB}, C_{GS}, and C_{GD} need to close M_{SW}.

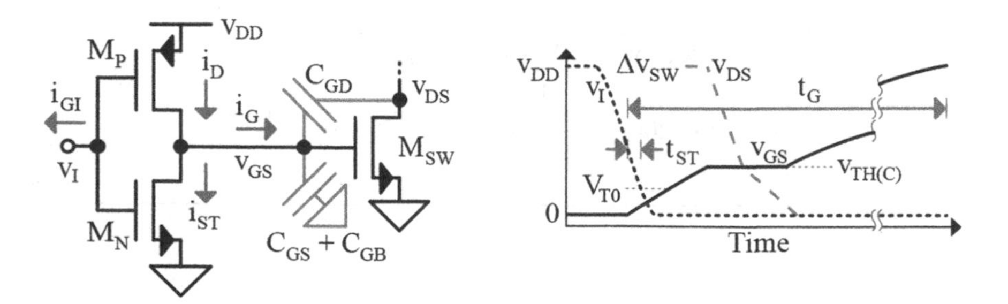

Fig. 4.31 Gate driver closing switch

For v_I to fall in the first place, the pre-driver must sink the charge q_{GI} that M_N and M_P's gate capacitances C_{GI} store when v_I is at v_{DD}. This means that the pre-driver drains and burns C_{GI}'s energy. The *drive power* P_D that v_{DD} supplies is therefore the power M_{SW} needs to close:

$$P_{D(C)} = P_G + P_{ST(C)} \approx P_G, \tag{4.61}$$

where *gate-charge power* P_G is the power v_{DD} supplies with q_G:

$$P_G = v_{DD} i_{G(AVG)} = v_{DD}\left(\frac{q_G}{t_{SW}}\right), \tag{4.62}$$

and $P_{ST(C)}$ is the power M_N leaks with the i_{LK} that v_{DD} supplies across t_{ST}:

$$P_{ST} = v_{DD} i_{LK(AVG)} = v_{DD}\left(\frac{1}{t_{SW}}\int_0^{t_{LK}} i_{LK}dt\right). \tag{4.63}$$

Components in C_{GB} and C_{GS} change as v_{GS} traverses v_{DD}. The reason for this is v_{GS}'s rise inverts M_{SW} into saturation (because v_{DS} is high), and afterwards, v_{DS}'s fall collapses M_{SW} into triode. So C_{GB}'s C_{CH} charges to V_{TO}, C_{GS}'s C_{OL} charges to V_{TO} (in sub-threshold), C_{GS}'s C_{OL} and two-thirds C_{CH} charge from V_{TO} to $v_{TH(C)}$ (in saturation), and C_{GS}'s C_{OL} and half C_{CH} charge from $v_{TH(C)}$ to v_{DD} in triode. The charge q_{GBS} that C_{GB} and C_{GS} draw is

$$q_{GBS} \approx C_{CH}V_{TO} + C_{OL}v_{DD} + \left(\frac{2}{3}\right)C_{CH}\left(v_{TH(C)} - V_{TO}\right) + \left(\frac{1}{2}\right)C_{CH}\left(v_{DD} - v_{TH(C)}\right). \tag{4.64}$$

i_G not only raises C_{GD}'s v_{GS} across v_{DD} but also collapses v_{DS} across v_{SW}, which means C_{GD} charges across v_{DD} and v_{SW}. C_{GD}'s C_{OL} charges with v_{GS} to v_{DD} and with v_{DS} across v_{SW}. Since C_{GD} begins to receive half C_{CH} when v_{DS} falls below v_{GS}'s $v_{TH(C)}$, half C_{CH}'s average $0.25C_{CH}$ charges across $v_{TH(C)}$. After v_{DS} falls, C_{GD}'s half of C_{CH} charges with v_{GS} from $v_{TH(C)}$ to v_{DD} (in triode):

$$q_{GD} \approx C_{OL}\left(v_{DD} + \Delta v_{SW}\right) + \left(\frac{1}{4}\right)C_{CH}v_{TH(C)} + \left(\frac{1}{2}\right)C_{CH}\left(v_{DD} - v_{TH(C)}\right). \tag{4.65}$$

i_G delivers both components: q_{GBS} and q_{GD}. Since v_{TH} is the v_{GS} needed to sustain i_L, which is $v_{DS(SAT)}$ over V_{TO}, q_G reduces to

$$
\begin{aligned}
q_G &= q_{GBS} + q_{GD} \\
&= C_{OL}(2v_{DD} + \Delta v_{SW}) + C_{CH}\left(v_{DD} + \frac{V_{T0}}{3} - \frac{v_{DS(SAT)}}{12}\right) \\
&\approx C_{OL}(2v_{DD} + \Delta v_{SW}) + C_{CH}\left(v_{DD} + \frac{V_{T0}}{3}\right) \\
&\approx (2C_{OL} + C_{CH})v_{DD} + C_{OL}\Delta v_{SW}.
\end{aligned}
\tag{4.66}
$$

Since $v_{DS(SAT)}$ is usually lower than V_{T0}, a 12th is negligibly lower.

Since M_{SW} is mostly in saturated inversion when v_{DS} collapses, C_{OL} in C_{GD} is, for the most part, the only component that charges across Δv_{SW}. Whereas all capacitances connected to v_G: C_{OL} and C_{CH} in C_{GS} and C_{GB} and C_{OL} in C_{GD}, charge across v_{DD} when v_G transitions. So reducing q_G to the charge these capacitances and swings require is not a bad approximation, especially when v_{DD} is considerably greater than a fourth of V_{T0}.

4.6.3 Opening Switch

When opening M_{SW} in Fig. 4.32, M_N sinks the i_G that drains C_{GB}, C_{GS}, and C_{GD} and the i_{ST} that M_P leaks. i_{ST} is low and short-lived (by design) because M_{SW}'s high v_{GS} suppresses M_P's v_{SD} as v_I's climb collapses M_P's v_{SG}. i_G is much greater because M_N's v_{GS} and v_{DS} are high across the time M_P leaks i_{ST}. So the P_{ST} that v_{DD} supplies with i_{ST} across t_{ST} is low.

For v_I to rise in the first place, the pre-driver must supply the q_{GI} needed to charge M_N and M_P's gate capacitances C_{GN} and C_{GP} to v_{DD}. Since M_{SW}'s v_{GS} (which sets M_N's v_{DS}) is close to v_{DD} across this transition, v_I's climb inverts M_N into saturation. C_{CHN} in C_{GB} therefore charges across V_{TN0}, C_{OLN} in C_{GSN} and C_{GDN} across v_{DD}, and (2/3)C_{CHN} in C_{GSN} across $v_{DD} - V_{TN0}$. Since M_N's v_{GS} matches v_{DS} at v_{DD} after v_I rises, v_{DS}'s fall charges C_{OLN} and half C_{CHN}'s average $0.25C_{CHN}$ in C_{GDN} across v_{DD}. So in all, C_{GN} requires

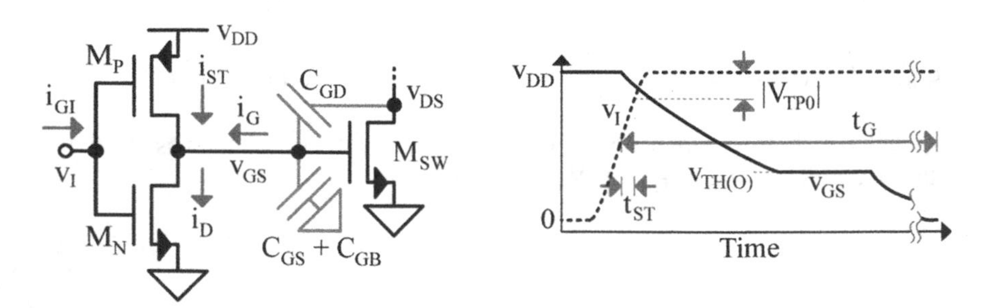

Fig. 4.32 Gate driver opening switch

$$q_{GN} \approx C_{CHN}V_{TN0} + C_{OLN}(3v_{DD}) + \left(\frac{2}{3}\right)C_{CHN}(v_{DD} - V_{TN0}) + \frac{C_{CHN}v_{DD}}{4} \approx C_{OLN}(3v_{DD}) + C_{CHN}\left(v_{DD} + \frac{V_{TN0}}{3}\right). \quad (4.67)$$

M_P starts in triode and ends in cut off when v_I climbs because M_{SW}'s v_{GS}, which is high across v_I's transition, suppresses M_P's v_{SD}. C_{OLP}'s in C_{GSP} and C_{GDP} therefore charge across v_{DD}, half C_{CHP}'s in C_{GSP} and C_{GDP} across $v_{DD} - |V_{TP0}|$, and C_{CH} in C_{GB} across $|V_{TP0}|$. After v_I rises, as M_{SW}'s v_{GS} falls and M_P's v_{SD} climbs in cut off, C_{OLP} in C_{GDP} charges across another v_{DD}. So C_{GP} requires

$$q_{GP} \approx C_{OLP}(3v_{DD}) + \left(\frac{C_{CHP}}{2} + \frac{C_{CHP}}{2}\right)(v_{DD} - |V_{TP0}|) + C_{CHP}|V_{TP0}| = (3C_{OLP} + C_{CHP})v_{DD}. \quad (4.68)$$

q_{GN} and q_{GP} are the q_{GI} that v_{DD} supplies C_{GI}. And P_{GI} is the *driver gate-charge power* that v_{DD} supplies with q_{GI}:

$$P_{GI} = v_{DD}i_{GI(AVG)} = v_{DD}\left(\frac{q_{GI}}{t_{SW}}\right) = v_{DD}\left(\frac{q_{GN} + q_{GP}}{t_{SW}}\right). \quad (4.69)$$

Together, C_{GI}'s q_{GI} and M_P's i_{ST} draw $P_{D(O)}$ from v_{DD} when M_{SW} opens:

$$P_{D(O)} = P_{GI} + P_{ST(O)}. \quad (4.70)$$

4.6.4 Driver Power

M_{SW}'s C_G receives charge q_G or C_Gv_{DD} to charge to v_{DD}. With this q_G, v_{DD} supplies q_Gv_{DD} or $C_Gv_{DD}^2$ and C_G stores $0.5C_Gv_{DD}^2$. This means that M_P burns half the energy v_{DD} supplies. M_N later consumes the other half when M_N drains C_G. So v_{DD} ultimately loses the P_G that C_G draws.

v_{DD} similarly loses the P_{GI} that C_{GI} requires. v_{DD} also leaks P_{ST} when closing and opening M_{SW}. But P_{ST} is so much lower than P_G (by design) that P_D reduces to the q_G and q_{GI} that C_G and C_{GI} need:

$$P_D = P_{D(C)} + P_{D(O)} \approx P_G + P_{GI} \approx v_{DD}\left(\frac{q_G + q_{GI}}{t_{SW}}\right) = v_{DD}\left(\frac{q_G + q_{GN} + q_{GP}}{t_{SW}}\right). \quad (4.71)$$

Example 15: Determine P_D and σ_D for M_{EG}'s driver in the ideal boost from Examples 7, 9, and 13.

Solution:

$$d_{IN} = 1, d_O = 48\%, v_{DD} = v_O = 4\text{ V}, V_{TN0} = 400\text{ mV},$$
$$\Delta v_{SWO} = v_O + v_{DO} = 4.8\text{ V}, C_{OX}'' = 6.9\text{ fF}/\mu\text{m}^2, L_{OL} = 30\text{ nm},$$
$$C_{OL} = 10\text{ pF, and } C_{CH} = 66\text{ pF from Examples 7 and 9}$$

$$L_{CH} = 190\text{ nm}, W_N = 14\ \mu\text{m, and } W_P = 26\ \mu\text{m from Example 13}$$

$$q_G \approx C_{OL}\Delta v_{SWO} + (2C_{OL} + C_{CH})v_{DD} = (10p)(4.8) + [2(10p) + 66p](4) = 390\text{ pC}$$

$$C_{OLN} = C_{OX}''W_N L_{OL} = (6.9m)(14\mu)(30n) = 2.9\text{ fF}$$

$$C_{CHN} = C_{OX}''W_N L_{CH} = (6.9m)(14\mu)(190n) = 18\text{ fF}$$

$$q_{GN} \approx C_{OLN}(3v_{DD}) + C_{CHN}\left(v_{DD} + \frac{V_{TN0}}{3}\right) = (2.9f)(3)(4) + (18f)\left(4 + \frac{400m}{3}\right) = 110\text{ fC}$$

$$C_{OLP} = C_{OX}''W_P L_{OL} = (6.9m)(26\mu)(30n) = 5.4\text{ fF}$$

$$C_{CHP} = C_{OX}''W_P L_{CH} = (6.9m)(26\mu)(190n) = 34\text{ fF}$$

$$q_{GP} \approx (3C_{OLP} + C_{CHP})v_{DD} = [3(5.4f) + 24f](4) = 160\text{ fC}$$

$$P_D \approx v_{DD}\left(\frac{q_G + q_{GN} + q_{GP}}{t_{SW}}\right) = 4\left(\frac{390p + 110f + 160f}{1\mu}\right) = 1.6\text{ mW}$$

$$\sigma_D = \frac{P_D}{(i_{O(AVG)}/d_O)d_{IN}v_{IN}} = \frac{1.6m}{(250m/48\%)(1)(2)} = 0.2\%$$

Note: q_G is usually much greater than q_{GI}. This P_D, which is 0.2% of P_{IN}, excludes the power M_{DO}'s driver consumes.

Explore with SPICE:
Use the SPICE code from Example 13.

Tip: Plot id(Meg), v(vg), v(vswo), and i(Vdd)*v(vdd) and extract the average of the product term.

4.7 Leaks

MOS transistors also incorporate *source–* and *drain–body junction capacitances* C_{SB} and C_{DB}. The terminals that do not connect to the switching nodes connect to ground, v_{IN}, or v_O, which are nearly fixed. So C_{SB} and C_{DB} add *input* and *output switch-node capacitances* C_{SWI} and C_{SWO} that charge and discharge with v_{SWI} and v_{SWO} in Fig. 4.33. *Electrostatic-discharge protection* (ESD), pads, pins, and board connections also add capacitance to C_{SWI} and C_{SWO}.

Charging C_{SWI} and C_{SWO} draws v_{IN} and L_X energy that v_O does not fully recover. Power switches burn this difference. These same switches also leak current when they are off, especially when they are large and hot. Although these losses are usually low, they become increasingly larger fractions of P_{IN} when P_O is lower, especially in low-power applications when their systems idle.

4.7.1 Input Switch-Node Capacitance

v_{SWI} rises and falls between $-v_{DG}$ and v_{IN} and between $-v_{DG}$ and ground. When M_{EI} closes to start t_E, v_{IN} supplies the charge needed to raise v_{SWI} from $-v_{DG}$ to v_{IN}: $v_{IN}q_{SWI}$ or $v_{IN}C_{SWI}(v_{IN} + v_{DG})$. During this transition, C_{SWI} loses the energy it held with $-v_{DG}$ and receives the energy it stores with v_{IN}: $-0.5C_{SWI}v_{DG}^2 + 0.5C_{SWI}v_{IN}^2$.

When M_{EI} opens to end t_E, as i_L collapses v_{SWI} to ground, L_X helps v_O recover the charge C_{SWI} held with v_{IN}. But as v_{SWI} falls below ground to $-v_{DG}$, L_X loses the energy C_{SWI} needs to charge to $-v_{DG}$. This is L_X energy v_O loses. So across this transaction, v_O recovers $0.5C_{SWI}v_{IN}^2$ and L_X loses $0.5C_{SWI}v_{DG}^2$.

After the first t_{DT} in t_D, M_{DG} burns the energy C_{SWI} holds with $-v_{DG}$. When M_{DG} opens to start the second t_{DT}, L_X loses the energy C_{SWI} needs to charge to $-v_{DG}$: $0.5C_{SWI}v_{DG}^2$. So across every cycle, v_{IN} loses $v_{IN}C_{SWI}(v_{IN} + v_{DG})$, v_O recovers

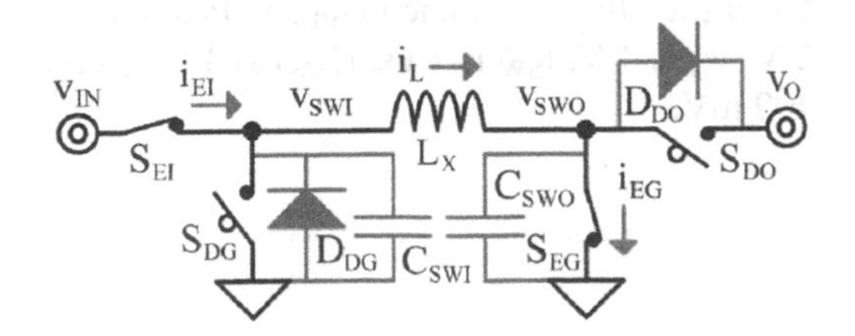

Fig. 4.33 Switched inductor with parasitic diodes and capacitances

$0.5C_{SWI}v_{IN}^2$, and L_X loses $0.5C_{SWI}v_{DG}^2$ two times. This means that the average *input switched-node power* P_{SWI} leaked across t_{SW} is

$$\begin{aligned} P_{SWI} &= E_{SWI}f_{SW} \\ &= \left(\frac{C_{SWI}}{t_{SW}}\right)\left[v_{IN}(v_{IN}+v_{DG}) - 0.5v_{IN}^2 + 2\left(0.5v_{DG}^2\right)\right] \\ &= \left(\frac{C_{SWI}}{t_{SW}}\right)\left(0.5v_{IN}^2 + v_{IN}v_{DG} + v_{DG}^2\right). \end{aligned} \tag{4.72}$$

4.7.2 Output Switch-Node Capacitance

v_{SWO} rises and falls between ground and v_{DO} over v_O and between v_{DO} over v_O and v_O. When M_{EG} opens to start t_D, L_X supplies the energy C_{SWO} needs to charge to v_{DO} over v_O: $0.5C_{SWO}(v_O + v_{DO})^2$. A t_{DT} after that, when M_{DO} closes, v_O receives the charge C_{SWO} outputs when v_{SWO} falls from v_{DO} over v_O to v_O. So of the energy L_X loses after raising v_{SWO} to v_{DO} over v_O, v_O recovers v_Oq_{SWO} or $v_OC_{SWO}v_{DO}$.

Later when M_{DO} opens, L_X supplies the energy C_{SWO} needs to raise v_{SWO} from v_O to v_{DO} over v_O. Here, L_X loses the energy C_{SWO} holds with v_{DO} over v_O that excludes the energy C_{SWO} stores with v_O. L_X therefore loses the difference: $0.5C_{SWO}(v_O + v_{DO})^2$ minus $0.5C_{SWO}v_O^2$.

M_{EG} burns the energy C_{SWO} holds with v_{DO} over v_O when M_{EG} collapses v_{SWO} to ground. So across every cycle, L_X loses $0.5C_{SWO}(v_O + v_{DO})^2$, v_O recovers $v_OC_{SWO}v_{DO}$, and L_X loses $0.5C_{SWO}(v_O + v_{DO})^2$ minus $0.5C_{SWO}v_O^2$. This means that the average *output switched-node power* P_{SWO} leaked across t_{SW} is

$$\begin{aligned} P_{SWO} &= E_{SWO}f_{SW} \\ &= \left(\frac{C_{SWO}}{t_{SW}}\right)\left\{0.5(v_O+v_{DO})^2 - v_Ov_{DO} + \left[0.5(v_O+v_{DO})^2 - 0.5v_O^2\right]\right\} \\ &= \left(\frac{C_{SWI}}{t_{SW}}\right)\left(0.5v_O^2 + v_{DO}^2 + v_Ov_{DO}\right). \end{aligned} \tag{4.73}$$

Example 16: Determine P_{SWI} and P_{SWO} in a synchronous buck–boost when v_{IN} is 2 V, v_O is 4 V, t_{SW} is 1 μs, C_{SWI} and C_{SWO} are 5 pF each, and D_{DG} and D_{DO} drop 400 mV.

Solution:

$$P_{SWI} = \left(\frac{C_{SWI}}{t_{SW}}\right)\left(0.5{v_{IN}}^2 + v_{IN}v_{DG} + {v_{DG}}^2\right)$$
$$= \left(\frac{5p}{1\mu}\right)\left[0.5(2)^2 + (2)(400m) + 400m^2\right] = 15\ \mu W$$

$$P_{SWO} = \left(\frac{C_{SWO}}{t_{SW}}\right)\left(0.5{v_O}^2 + {v_{DO}}^2 + v_O v_{DO}\right)$$
$$= \left(\frac{5p}{1\mu}\right)\left[0.5(4^2) + 400m^2 + (4)(400m)\right] = 49\ \mu W$$

Note: C_{SWO} leaks more than C_{SWI} because L_X helps v_O recover more when v_{SWI} collapses v_{IN} to ground than v_O recovers when v_{SWO} drops v_{DO}.

4.7.3 Cut-off Power

MOS current is zero when v_{GS} and v_{DS} are zero. Raising v_{DS} when v_{GS} is zero establishes an electric field that induces some i_{DS}. Body diodes also conduct zero current when their voltages are zero and close to *reverse saturation current* I_S when they reverse. These *off currents* i_{OFF} climb with *channel width* W_{CH} and *junction temperature* T_J. So the *off resistance* R_{OFF} that they exhibit falls with increasing W_{CH} and T_J.

Power-supply switches and body diodes usually leak noticeable i_{OFF} because they are large. Plus, the power they burn when they conduct heats them to an extent that keeps them hot when they open. A 30-mm-wide, 180-nm-long MOSFET, for example, can leak 80 nA with 1.8 V at 25°C and 800 nA at 125°C. This is 0.38–3.8 TΩ for each length-to-width (L/W) square.

In a switched inductor, energize switches are off across t_D. S_{EI} drops v_{IN} and S_{EG} drops v_O across the t_D that excludes dead times. When dead-time diodes conduct, S_{EI} drops v_{IN} and v_{DG} and S_{EG} drops v_O and v_{DO}. So the average cut-off power they consume across t_{SW} is

$$P_{OFF(E)} = \left(\frac{v_{IN}^2}{R_{EI(OFF)}} + \frac{v_O^2}{R_{EG(OFF)}}\right)\left(\frac{t_D - 2t_{DT}}{t_{SW}}\right) + \left[\frac{(v_{IN} + v_{DG})^2}{R_{EI(OFF)}} + \frac{(v_O + v_{DO})^2}{R_{EG(OFF)}}\right]\left(\frac{2t_{DT}}{t_{SW}}\right)$$
$$\approx \left(\frac{v_{IN}^2}{R_{EI(OFF)}} + \frac{v_O^2}{R_{EG(OFF)}}\right)\left(\frac{t_D}{t_{SW}}\right). \quad (4.74)$$

But since t_{DT}'s are small fractions of t_{SW} and diode voltages are usually lower than v_{IN} and v_O, the effects of v_{DG} and v_{DO} are minimal.

Drain switches are off across t_E and across dead times within t_D. S_{DG} drops v_{IN} and S_{DO} drops v_O across t_E and S_{DG} drops v_{DG} and S_{DO} drops v_{DO} across t_{DT}'s. So the average cut-off power they consume across t_{SW} is

$$P_{OFF(D)} = \left(\frac{v_{IN}^2}{R_{DG(OFF)}} + \frac{v_O^2}{R_{DO(OFF)}}\right)\left(\frac{t_E}{t_{SW}}\right) + \left[\frac{v_{DG}^2}{R_{DG(OFF)}} + \frac{v_{DO}^2}{R_{DO(OFF)}}\right]\left(\frac{2t_{DT}}{t_{SW}}\right)$$
$$\approx \left(\frac{v_{IN}^2}{R_{DG(OFF)}} + \frac{v_O^2}{R_{DO(OFF)}}\right)\left(\frac{t_E}{t_{SW}}\right). \quad (4.75)$$

The effects of v_{DG} and v_{DO} are minimal because D_{DG} and D_{DO} conduct short t_{DT} fractions of t_{SW} and drop lower voltages than v_{IN} and v_O.

Excluding dead times, S_{EI} and S_{DG} alternate conduction to v_{SWI} and S_{EG} and S_{DO} alternate conduction to v_{SWO}. So one switch is always off at v_{SWI} and one at v_{SWO}. The one at v_{SWI} drops v_{IN} and the one at v_{SWO} drops v_O. Since off resistances for similarly large switches are comparable, *cut-off power* P_{OFF} is largely consistent and similar across time:

$$P_{OFF} \approx \left(\frac{v_{IN}^2}{R_{EI(OFF)}} + \frac{v_O^2}{R_{EG(OFF)}}\right)\left(\frac{t_D}{t_{SW}}\right) + \left(\frac{v_{IN}^2}{R_{DG(OFF)}} + \frac{v_O^2}{R_{DO(OFF)}}\right)\left(\frac{t_E}{t_{SW}}\right)$$
$$\approx \frac{v_{IN}^2}{R_{I(OFF)}} + \frac{v_O^2}{R_{O(OFF)}}. \quad (4.76)$$

Example 17: Determine P_{OFF} in a buck–boost when v_{IN} is 2 V, v_O is 4 V, W's are 50 mm, L's are 250 nm, L_{OL} is 30 nm, T_J is 125°C, and $R_{OFF/SQ}$ at this T_J is 380 GΩ per L/W square.

Solution:

$$L_{CH} = L - 2L_{OL} = 250n - 2(30n) = 190 \text{ nm}$$

$$R_{OFF} = R_{OFF/SQ}\left(\frac{L_{CH}}{W_{CH}}\right) = (380G)\left(\frac{190n}{50m}\right) = 1.4\ M\Omega$$

$$P_{OFF} \approx \frac{v_{IN}^2}{R_{OFF}} + \frac{v_O^2}{R_{OFF}} = \frac{2^2}{1.4M} + \frac{4^2}{1.4M} = 14\ \mu W$$

4.8 Design

4.8.1 Optimal Power Setting

Power losses normally consume the lowest fraction of input power at a particular load level. With sufficient flexibility, engineers can define and set this *optimal output power* P_O'. In practice, however, applications and technologies impose operating conditions and parametric limits that constrain P_O'. But even then, design choices can still influence P_O'.

Setting P_O' to $P_{O(MAX)}$ saves the most power, but not the most energy, especially when $P_{O(MAX)}$ is an improbable extreme. The system saves the most energy when P_O' is at the most probable setting. If this setting is unknown, halfway between $P_{O(MIN)}$ and $P_{O(MAX)}$ is often a good alternative. Although more involved and less practical, P_O' can also be the P_O that produces the highest peak efficiency or the highest average efficiency.

4.8.2 Power Switch

MOSFETs require power in four ways: Ohmic power when they conduct i_L, gate-drive power when they close, i_{DS}–v_{DS} overlap power when they transition, and off power when they are open. Of these, P_{OFF} is usually a negligible part of P_O'. Although P_{IV} may not be as insignificant, gate drivers can reduce the impact of P_{IV} on P_O'. The only design variables that can suppress P_R and P_G are MOSFET dimensions.

Longer channels raise *channel resistance* R_{CH} and gate capacitance and, as a result, gate charge. So the R_{CH} and q_G that set P_R and P_G increase with *channel length* L_{CH}. This means that the *MOS power* P_{MOS} that P_R and P_G require is minimal when L is the *minimum allowable oxide length* L_{MIN} that can sustain v_{SW}'s swing without breakdown effects:

$$P_{MOS} \equiv P_R + P_G. \tag{4.77}$$

Wider channels decrease R_{CH} and increase C_G. So the R_{CH} that sets P_R in Fig. 4.34 falls with increasing W_{CH}'s as the q_G that sets P_G climbs:

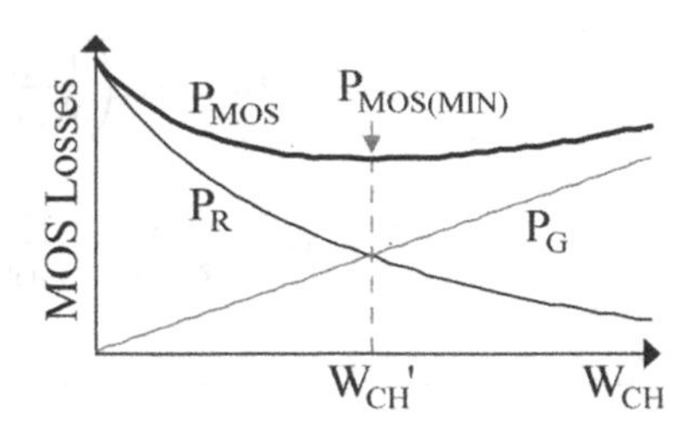

Fig. 4.34 Ohmic and gate-charge MOS power

$$P_R = i_{R(RMS)}{}^2 R_{CH} = \frac{k_R}{W_{CH}} \tag{4.78}$$

$$P_G = E_G f_{SW} = v_{DD} q_G f_{SW} = k_G W_{CH}, \tag{4.79}$$

where k_R and k_G are W_{CH}-independent coefficients. These opposing trends tend to desensitize P_{MOS} from W_{CH}.

Still, P_{MOS} falls with P_R when W_{CH} is narrow and rises with P_G when W_{CH} is wide. P_{MOS} is minimal when additional charge losses in P_G cancel P_R savings. This happens when P_{MOS}'s slope $\partial P_{MOS}/\partial W_{CH}$ is zero:

$$\left.\frac{\partial P_{MOS}}{\partial W_{CH}}\right|_{W_{CH}'} = \left.\frac{\partial P_R}{\partial W_{CH}}\right|_{W_{CH}'} + \left.\frac{\partial P_G}{\partial W_{CH}}\right|_{W_{CH}'} = -\frac{k_R}{W_{CH}'^2} + k_G = 0, \tag{4.80}$$

which results at the *optimal channel width* W_{CH}':

$$W_{CH}' = \sqrt{\frac{k_R}{k_G}}, \tag{4.81}$$

when P_R and P_G match:

$$P_R|_{W_{CH}'} = P_G|_{W_{CH}'} = \sqrt{k_R k_G}, \tag{4.82}$$

and $P_{MOS(MIN)}$ is twice the P_R or P_G that W_{CH}' sets:

$$P_{MOS(MIN)} = P_{MOS}|_{W_{CH}'} = P_R|_{W_{CH}'} + P_G|_{W_{CH}'} = 2P_R|_{W_{CH}'} = 2\sqrt{k_R k_G}. \tag{4.83}$$

In practice, on-chip, bond-wire, and board contacts and traces add resistance to switches. So Ohmic power is greater than the P_R that R_{CH} burns. Still, switches require the least power with W_{CH}'.

k_R and k_G are functions of parameters and variables that applications and manufacturing technologies frequently define. v_{IN} and v_O, for example, set duty cycle. Accuracy and noise sensitivity often specify or constrain current ripple and switching frequency. And device parameters dictate the shortest dead time that keeps adjacent switches from cross-conducting. In other words, k_R and k_G are often pre-set.

Example 18: Determine M_{DG}'s optimal W_{DG}, L_{DG}, P_{MOS}, and σ_{MOS} in a buck–boost in CCM when v_{DD} for M_{DG}'s gate driver is v_{IN}, v_{IN} is 2 V, v_O is 4 V, D_{DG} and D_{DO} drop 400 mV, $i_{O(AVG)}$ is 250 mA, L_X is 10 μH, t_{SW} is 1 μs, t_{DT} is 50 ns, L_{MIN} is 250 nm, L_{OL} is 30 nm, V_{TN0} is 400 mV, K_N' is 200 μA/V^2, and C_{OX}'' is 6.9 fF/μm^2.

Solution:

$$d_{IN} = d_E \approx \frac{v_O}{v_{IN} + v_O} + \left(\frac{v_{DO} + v_{DG}}{v_{IN} + v_O}\right)\left(\frac{2t_{DT}}{t_{SW}}\right)$$

$$= \frac{4}{2+4} + \left(\frac{400m + 400m}{2+4}\right)\left[\frac{2(50n)}{1\mu}\right] = 68\%$$

$$\therefore \quad d_O = d_D = 1 - d_E \approx 1 - 68\% = 32\%$$

$$\Delta i_L = \left(\frac{v_E}{L_X}\right)t_E = \left(\frac{v_{IN}}{L_X}\right)d_E t_{SW} \approx \left(\frac{2}{10\mu}\right)(68\%)(1\mu) = 140\text{ mA}$$

$$L_{DG} \equiv L_{MIN} = 250\text{ nm} \quad \therefore \quad L_{CH} = L_{DG} - 2L_{OL} = 250n - 2(30n) = 190\text{ nm}$$

$$R_{DG} \approx \left(\frac{L_{CH}}{W_{CH}}\right)\left[\frac{1}{K_N'(v_{DD} - V_{TN0})}\right] = \left(\frac{190n}{W_{CH}}\right)\left[\frac{1}{(200\mu)(2 - 400m)}\right] = \frac{590\mu}{W_{CH}}$$

$$P_R = \left[\left(\frac{i_O}{d_O}\right)^2 + \left(\frac{0.5\Delta i_L}{\sqrt{3}}\right)^2\right] R_{DG}\left(d_D - \frac{2t_{DT}}{t_{SW}}\right)$$

$$\approx \left\{\left(\frac{250m}{32\%}\right)^2 + \left[\frac{0.5(140m)}{\sqrt{3}}\right]^2\right\}\left(\frac{590\mu}{W_{CH}}\right)\left[32\% - \frac{2(50n)}{1\mu}\right] = \frac{79\mu}{W_{CH}}$$

$$C_{OL} = C_{OX}''W_{CH}L_{OL} = (6.9m)W_{CH}(30n) = (210p)W_{CH}$$

$$C_{CH} = C_{OX}''W_{CH}L_{CH} = (6.9m)W_{CH}(190n) = (1.3n)W_{CH}$$

v_{SWI} is $-v_{DG}$ before M_{EI} closes and v_{IN} after M_{EI} closes

$$\therefore \quad \Delta v_{SWI} = v_{IN} + v_{DG} = 2 + 400m = 2.4\text{ V}$$

$$q_G \approx C_{OL}\Delta v_{SWI} + (2C_{OL} + C_{CH})v_{DD}$$

$$= \{(210p)(2) + [2(210p) + 1.3n](2)\}W_{CH}$$

$$= (3.9n)W_{CH}$$

$$P_G = v_{DD}\left(\frac{q_G}{t_{SW}}\right) \approx (2)\left[\frac{(3.9n)W_{CH}}{1\mu}\right] = (7.8m)W_{CH}$$

$$W_{DG} \equiv W_{CH}' = \sqrt{\frac{k_R}{k_G}} = \sqrt{\frac{79\mu}{7.8m}} = 100\ \text{mm}$$

$$R_{DG} \approx \frac{590\mu}{W_{DG}} = 5.9\ \text{m}\Omega$$

$$P_{MOS} \approx 2\sqrt{k_R k_G} = 2\sqrt{(79\mu)(7.8m)} = 1.6\ \text{mW}$$

$$\sigma_{MOS} = \frac{P_{MOS}}{(i_{O(AVG)}/d_O)d_{IN}v_{IN}} = \frac{1.6m}{(250m/32\%)(68\%)(2)} = 0.2\%$$

Note: P_{MOS}, which includes the P_R that M_{DG} burns and the P_G the driver supplies, is much lower than P_R in Example 2 because R_{CH} is much lower. This P_{MOS} excludes the dead-time power M_{DG} consumes across t_{DT}'s, when M_{DG} conducts i_L like a 400-mV diode.

Explore with SPICE:
See Appendix A for notes on SPICE simulations.

```
* Ideal Buck-Boost with Dead Time & MDG in CCM
vde de 0 dc=0 pulse 0 1 0 1n 1n 673n 1u
vido ido 0 dc=2 pulse 0 2 730n 1n 1n 227n 1u
vdd vdd 0 dc=2
sh vdd do ido 0 sw2v
sl do 0 vdd ido sw2v
vin vin 0 dc=2
sei vin vswi de 0 sw1v
mdg 0 do vswi 0 nmos1 w=100m l=250n
lx vswi vswo 10u
seg vswo 0 de 0 sw1v
ddo vswo vo idiode
vo vo 0 dc=4
.ic i(lx)=710m
.lib lib.txt
.tran 1u
.end
```

(continued)

Tip: Plot id(Mdg), v(vswi), and id(Mdg)*v(vswi) and view between 731 and 951 ns, extract the average of the product term, and multiply by 22% (220 ns of the 1-μs period) to determine M_{DG}'s P_R. Plot v(do) and i(Vdd)*v(vdd), and extract the average of the product term to determine M_{DG}'s P_G.

4.8.3 Gate Driver

Similar transition times balance propagation delays and distribute switching losses across the switching period. This way, dead times and response time are consistent and peak transient power balances across switching events. Pull-up and pull-down resistances in the gate driver (which transistor W/L's, K_N' and K_P', v_{GS} and v_{SG}, and V_{TN0} and V_{TP0} set) match opposing v_{DS} transition times $t_{V(C)}$ and $t_{V(O)}$ when R_P-to-R_N's ratio is

$$\begin{aligned}\frac{R_P}{R_N} &= \left(\frac{W_N}{W_P}\right)\left(\frac{L_P}{L_N}\right)\left(\frac{K_N'}{K_P'}\right)\left[\frac{v_{DD} - V_{TN0} - 0.5v_{TH(O)}}{v_{DD} - |V_{TP0}| - 0.5\left(v_{DD} - v_{TH(C)}\right)}\right] \\ &\equiv \left(\frac{v_{DD} - v_{TH(C)}}{v_{TH(O)}}\right)\left(\frac{C_{OL}v_{SW} + 0.25C_{CH}v_{TH(O)}}{C_{OL}v_{SW} + 0.25C_{CH}v_{TH(C)}}\right).\end{aligned} \tag{4.84}$$

Although not perfectly matched, these resistances produce similar i_{DS} transitions $t_{I(C)}$ and $t_{I(O)}$. And since dead-time diodes set v_{DS} before and after switches close, v_{DS} swings the same Δv_{SW} when closing and opening. So excluding the reverse-recovery power that body diodes induce (i.e., assuming the switches are low-V_{TO} MOSFETs or low-voltage Schottkys parallel the body diodes), the overlap power that switches consume when closing and opening reduces to

$$\begin{aligned}P_{IV} &= P_{I(C)}\left(\frac{t_{I(C)}}{t_{SW}}\right) + P_{V(C)}\left(\frac{t_{V(C)}}{t_{SW}}\right) + P_{I(O)}\left(\frac{t_{I(O)}}{t_{SW}}\right) + P_{V(O)}\left(\frac{t_{V(O)}}{t_{SW}}\right) \\ &\approx \left(i_{L(LO)} + i_{L(HI)}\right)\Delta v_{SW}\left(\frac{t_I}{3t_{SW}} + \frac{t_V}{2t_{SW}}\right) \\ &= 2\left(\frac{i_{O(AVG)}}{d_{DO}}\right)\Delta v_{SW}\left(\frac{t_I}{3t_{SW}} + \frac{t_V}{2t_{SW}}\right) = \frac{k_{IV}}{W_N}.\end{aligned} \tag{4.85}$$

Switches either energize (raise $i_{L(LO)}$ to $i_{L(HI)}$) or drain L_X (reduce $i_{L(HI)}$ to $i_{L(LO)}$), so they steer $i_{L(LO)}$ in one transition and $i_{L(HI)}$ in the other. Since $i_{L(LO)}$ and $i_{L(HI)}$ average $i_{L(AVG)}$, their sum is twice the $i_{L(AVG)}$ that $i_{O(AVG)}$ and d_O set. The only design variable left is the R_N that sets R_P, t_I, and t_V.

Since W_N relates to R_N and W_P to the R_P that R_N sets, R_N determines the total gate capacitance of the driver. Increasing W_N not only reduces the R_N that shortens t_I and

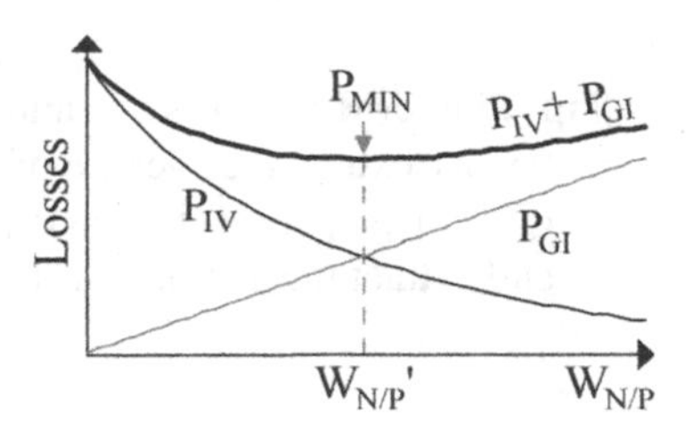

Fig. 4.35 Overlap and driver gate-charge power

t_V and reduces P_{IV} but also increases the C_{GI} that v_{DD} charges. So P_{GI} in the driver climbs with W_N as P_{IV} falls:

$$P_{GI} = v_{DD}\left(\frac{q_{GI}}{t_{SW}}\right) = v_{DD}\left(\frac{q_{GN} + q_{GP}}{t_{SW}}\right) = k_{GI}W_N, \tag{4.86}$$

where k_{IV} and k_{GI} are W_N-independent coefficients for P_{IV} and P_{GI}.

Longer channels increase channel resistance and gate capacitance, and as a result, gate charge. So the R_P, R_N, and q_{GI} that set P_{IV} and P_{GI} climb with L_{CH}. This means that P_{IV} and P_{GI}'s sum is lowest when L is the L_{MIN} that can sustain v_{DD} without breakdown effects.

P_{GI}'s rise with W_N in Fig. 4.35 opposes P_{IV}'s fall like P_G and P_R in P_{MOS}. So P_{IV} and P_{GI}'s sum is minimal when additional charge losses cancel P_{IV} savings. This happens when the slope of P_{IV} and P_{GI}'s sum is zero, which results when W_N is at the optimal width W_N' and P_{IV} and P_{GI} match:

$$\left.\frac{\partial P_{IV}}{\partial W_N}\right|_{W_N'} + \left.\frac{\partial P_{GI}}{\partial W_N}\right|_{W_N'} = -\frac{k_{IV}}{W_N'^2} + k_{GI} = 0, \tag{4.87}$$

$$W_N' = \sqrt{\frac{k_{IV}}{k_{GI}}}, \tag{4.88}$$

$$P_{IV}|_{W_N'} = P_{GI}|_{W_N'} = \sqrt{k_{IV}k_{GI}}. \tag{4.89}$$

The W_P and W_N that raise P_{GI} to the point P_{GI} matches P_{IV} can be so high that i_P and i_N can reach i_L levels. This amplifies the effects of i_P and i_N on the closing and opening thresholds that v_{GS} reaches when v_{DS} transitions. Although incorporating this shift in v_{TH} is not impossible, the $v_{TH(C)}$ and $v_{TH(O)}$ that set P_{IV} in Section 5 are good first-order approximations. Computer models and simulations can account for the rest, including sub-threshold and weak-inversion effects.

Example 19: Determine the optimal W's and L's, P_{IV}, and P_{GI} for the gate driver that switches M_{EG} in the boost from Examples 7, 9, 10, and 13 when L_{MIN} is 250 nm and $i_{O(AVG)}$ is 250 mA.

Solution:

$$d_E = 52\%, d_O = d_D = 48\%, \text{and } \Delta i_L = 100 \text{ mA from Example 7}$$

$$v_{DD} = v_O = 4 \text{ V}, V_{TN0} = 400 \text{ mV}, K_N' = 200 \text{ µA/V}^2, L_{OL} = 30 \text{ nm},$$
$$C_{OX}'' = 6.9 \text{ fF/µm}^2, C_{OL} = 10 \text{ pF}, C_{CH} = 66 \text{ pF}, \Delta v_{SWO} = 4.8 \text{ V},$$
$$\text{and } v_{TH(C)} = 530 \text{ mV from Example 9}$$

$$v_{TH(O)} = 550 \text{ mV from Example 10}$$

$$V_{TP0} = -400 \text{ mV and } K_P' = 40 \text{ µA/V}^2 \text{ from Example 13}$$

$$L \equiv L_{MIN} = 250 \text{ nm} \quad \therefore \quad L_{CH} = L - 2L_{OL} = 250\text{n} - 2(30\text{n}) = 190 \text{ nm}$$

$$R_P \approx \frac{L_{CH}}{W_P K_P'\left[v_{DD} - |V_{TP0}| - 0.5\left(v_{DD} - v_{TH(C)}\right)\right]} = \frac{190\text{n}}{W_P(40\text{µ})[4 - 400\text{m} - 0.5(4 - 530\text{m})]} = \frac{2.6\text{m}}{W_P}$$

$$R_N \approx \frac{L_{CH}}{W_N K_N'\left(v_{DD} - V_{TN0} - 0.5 v_{TH(O)}\right)} = \frac{190\text{n}}{W_N(200\text{µ})[4 - 400\text{m} - 0.5(550\text{m})]} = \frac{290\text{µ}}{W_N}$$

$$\frac{R_P}{R_N} \equiv \left(\frac{v_{DD} - v_{TH(C)}}{v_{TH(O)}}\right)\left(\frac{C_{OL}\Delta v_{SWO} + 0.25 C_{CH} v_{TH(O)}}{C_{OL}\Delta v_{SWO} + 0.25 C_{CH} v_{TH(C)}}\right) = \left(\frac{4 - 530\text{m}}{550\text{m}}\right)\left[\frac{10\text{p}(4.8) + 0.25(66\text{p})(550\text{m})}{10\text{p}(4.8) + 0.25(66\text{p})(530\text{m})}\right] = 6.3$$

$$\rightarrow \quad \frac{R_P}{R_N} = \left(\frac{2.6\text{m}}{W_P}\right)\left(\frac{W_N}{290\text{µ}}\right) = 6.3 \quad \therefore \quad W_P = 1.4 W_N$$

$$t_V \approx \left(\frac{R_N}{v_{TH(O)}}\right)\left[C_{OL}\Delta v_{SWO} + \left(\frac{C_{CH}}{4}\right)v_{TH(O)}\right]$$
$$= \left(\frac{290\mu/W_N}{550m}\right)\left[(10p)(4.8) + \left(\frac{66p}{4}\right)(550m)\right] = \frac{30f}{W_N}$$

$$t_{I(O)} \approx R_N\left[2C_{OL} + \left(\frac{2}{3}\right)C_{CH}\right]\ln\left(\frac{v_{TH(O)}}{V_{TN0}}\right)$$
$$= \left(\frac{290\mu}{W_N}\right)\left[2(10p) + \left(\frac{2}{3}\right)(66p)\right]\ln\left(\frac{550m}{400m}\right) = \frac{5.9f}{W_N}$$

$$P_{IV} \approx 2\left(\frac{i_{O(AVG)}}{d_{DO}}\right)\Delta v_{SWO}\left(\frac{t_I}{3t_{SW}} + \frac{t_V}{2t_{SW}}\right)$$
$$= 2\left(\frac{250m}{48\%}\right)(4.8)\left[\frac{5.9f}{3(1\mu)} + \frac{30f}{2(1\mu)}\right]\left(\frac{1}{W_N}\right) = \frac{85n}{W_N}$$

$$q_{GP} = (3C_{OLP} + C_{CHP})v_{DD}$$
$$= C_{OX}''W_P(3L_{OL} + L_{CH})v_{DD}$$
$$= (6.9m)(1.4W_N)[3(30n) + 190n](4) = (11n)W_N$$

$$q_{GN} = C_{OLN}(3v_{DD}) + C_{CHN}\left(v_{DD} + \frac{V_{TN0}}{3}\right)$$
$$= C_{OX}''W_N\left[L_{OL}(3v_{DD}) + L_{CH}\left(v_{DD} + \frac{V_{TN0}}{3}\right)\right]$$
$$= (6.9m)W_N\left[(30n)(3)(4) + (190n)\left(4 + \frac{400m}{3}\right)\right]$$
$$= (7.9n)W_N$$

$$P_{GI} = v_{DD}\left(\frac{q_{GP} + q_{GN}}{t_{SW}}\right) = (4)\left(\frac{11n + 7.9n}{1\mu}\right)W_N = (76m)W_N$$

$$W_N \equiv W_N' = \sqrt{\frac{k_{IV}}{k_{GI}}} = \sqrt{\frac{85n}{76m}} = 1.1\text{ mm}$$

$$W_P = 1.4W_N = 1.4(1.1m) = 1.5\text{ mm} \quad \therefore \quad P_{IV} \approx 80\ \mu W \quad \text{and} \quad P_{GI} \approx 84\ \mu W$$

Note: W_N and W_P are much wider than in Example 13 and P_{IV} and P_{GI} are much lower than P_{IV}'s in Examples 9 and 10.

Explore with SPICE:
See Appendix A for notes on SPICE simulations.

```
* Ideal Boost with Dead Time, MEG, & MEG's Optimal Driver in CCM
vde de 0 dc=0 pulse 0 4 0 1n 1n 519n 1u
vdo do 0 dc=1 pulse 0 1 570n 1n 1n 381n 1u
vdd1 vdd1 0 dc=4
sh deb vdd1 vdd2 de sw4v
sl deb 0 de 0 sw4v
vdd2 vdd2 0 dc=4
mp vg deb vdd2 vdd2 pmos1 w=1.5m l=250n
mn vg deb 0 0 nmos1 w=1.1m l=250n
vin vin 0 dc=2
lx vin vswo 10u
meg vswo vg 0 0 nmos1 w=50m l=250n
sdo vswo vo do 0 sw1v
ddo vswo vo fdiode1
vo vo 0 dc=4
.ic i(lx)=470m
.lib lib.txt
.tran 1u
.end
```

Tip: Plot id(Meg), v(vg), and v(vswo) and zoom into transitions to explore. Plot i(Sh), v(deb), and i(Vdd1)*v(vdd1), and extract the average of the product term to determine P_{GI}. Isolating P_{IV} is challenging because id(Meg) often includes C_{GD}'s current, which in this case is a significant fraction of id(Meg).

4.8.4 Operation

A. Switch Configuration

Direct v_{IN}–L_X–v_O connections in bucks and boosts deliver v_{IN} power that L_X does not transfer. This is because v_{IN} supplies v_O as L_X energizes in bucks and as L_X drains in boosts. This way, bucks and boosts supply more power than L_X transfers.

In bucks, the *inductor power* P_L that v_O receives with $i_{L(AVG)}$ across a t_D fraction of t_{SW} is less than the d_E fraction $i_{L(AVG)}$ draws from v_{IN}. L_X in boosts similarly delivers less P_L with the $i_{L(AVG)}$ that v_{IN} supplies across a t_E fraction of t_{SW} than $i_{L(AVG)}$ draws from v_{IN}. In both cases, P_{IN} supplies more than P_L:

$$P_{IN(BK)} = i_{IN(AVG)}v_{IN} = i_{L(AVG)}d_E v_{IN} = i_{O(AVG)}d_E v_{IN}, \tag{4.90}$$

$$\begin{aligned} P_{L(BK)} &= i_{L(AVG)}v_D\left(\frac{t_D}{t_{SW}}\right) = i_{O(AVG)}v_O d_D \\ &= i_{O(AVG)}v_E d_E = i_{O(AVG)}(v_{IN} - v_O)d_E, \end{aligned} \tag{4.91}$$

$$P_{IN(BST)} = i_{IN(AVG)}v_{IN} = i_{L(AVG)}v_{IN}, \tag{4.92}$$

$$P_{L(BST)} = i_{L(AVG)}v_E\left(\frac{t_E}{t_{SW}}\right) = i_{IN(AVG)}v_{IN}d_E. \tag{4.93}$$

The buck–boost, on the other hand, delivers across t_D the energy v_{IN} feeds L_X across t_E. So P_L is the P_{IN} that v_{IN} supplies. This means that, for the same P_O, L_X in bucks and boosts transfers less energy than in buck–boosts. Since lower inductor energy translates to lower i_L, bucks and boosts burn less Ohmic, dead-time, and overlap power. These *direct* v_{IN}–L_X–v_O *transfers* are one reason why bucks and boosts are more efficient than buck–boosts.

The other reason is fewer switches. Buck–boosts need two more switches than bucks and boosts, which require additional Ohmic, gate-drive, dead-time, overlap, and switch-node power. This is why engineers normally resort to buck–boosts only when absolutely necessary, when v_{IN}'s and v_O's operating ranges overlap.

When a buck–boost is unavoidable, behaving like a buck when bucking and like a boost when boosting saves energy. This way, by switching two of the four switches while keeping a third closed and the fourth open, gate-drive power is lower. And with the lower i_L that results, Ohmic loss in R_L is also lower.

So when bucking, the controller should switch S_{EI} and S_{DG}, open S_{EG}, and close S_{DO}. S_{EI} should similarly close, S_{DG} open, and S_{EG} and S_{DO} switch when boosting. The controller should switch all transistors only when v_{IN} and v_O are close.

B. Discontinuous Conduction

In discontinuous conduction, i_L rises to $i_{L(PK)}$ and falls to zero across t_C before t_{SW} ends. The average Ohmic power $P_{R(C)}$ that a switch consumes across t_C is a squared RMS translation of the current R_E or R_D carry across t_E or t_D, which is a root-three reflection of $i_{L(PK)}$:

$$P_{R(C)} \approx i_{E/D(RMS)}{}^2 R_{E/D}\left(\frac{t_{E/D}}{t_C}\right) = \left(\frac{i_{L(PK)}}{\sqrt{3}}\right)^2 R_{E/D}\left(\frac{t_{E/D}}{t_C}\right) = \frac{k_{RC}}{W_{CH}}. \tag{4.94}$$

The average gate-charge power $P_{G(C)}$ needed across t_C is the power v_{DD} supplies when feeding gate charge q_G into the gate:

$$P_{G(C)} = v_{DD}i_{G(AVG)} = v_{DD}\left(\frac{q_G}{t_C}\right) = k_{GC}W_{CH}. \tag{4.95}$$

Since R_{CH} and $P_{R(C)}$ fall and q_G and $P_{G(C)}$ rise with wider W_{CH}'s, $P_{R(C)}$ and $P_{G(C)}$'s sum $P_{MOS(C)}$ is minimal with the width that flattens the slope of $P_{MOS(C)}$ to zero. This optimal W_{CH}' is the square-root ratio of the Ohmic and gate-drive coefficients k_{RC} and k_{GC}:

$$\left.\frac{\partial P_{MOS(C)}}{\partial W_{CH}}\right|_{W_{CH}'} = \left.\frac{\partial P_{R(C)}}{\partial W_{CH}}\right|_{W_{CH}'} + \left.\frac{\partial P_{G(C)}}{\partial W_{CH}}\right|_{W_{CH}'} = -\frac{k_{RC}}{W_{CH}'^2} + k_{GC} = 0 \quad (4.96)$$

$$W_{E/D} \equiv W_{CH}' = \sqrt{\frac{k_{RC}}{k_{GC}}}. \quad (4.97)$$

With this W_{CH}, $P_{R(C)}$ and $P_{G(C)}$ match, reducing $P_{MOS(C)}$ to

$$P_{MOS(C)} = P_{R(C)} + P_{G(C)} = 2P_{R(C)} = 2P_{G(C)} = 2\sqrt{k_{RC}k_{GC}}. \quad (4.98)$$

So across t_{SW}, the optimal switch consumes $P_{MOS(C)}$ a t_C fraction of t_{SW}:

$$P_{MOS} = P_{MOS(C)}\left(\frac{t_C}{t_{SW}}\right) = 2\left(\sqrt{k_{RC}k_{GC}}\right)t_C f_{SW}. \quad (4.99)$$

Across each t_{SW}, P_O outputs the energy L_X collects with $i_{L(PK)}$ and the additional power $P_{E/D}$ that v_{IN} in bucks and boosts supplies. This $P_{E/D}$ is what i_L's average $0.5i_{L(PK)}$ supplies v_O across t_E in bucks or draws from v_{IN} across t_D in boosts. Either way, the resulting P_O climbs with *switching frequency* f_{SW}:

$$P_O \approx \frac{E_L}{t_{SW}} + P_{E/D}\left(\frac{t_{E/D}}{t_{SW}}\right) \approx \left[\left(\frac{1}{2}\right)L_X i_{L(PK)}^2 + \left(\frac{i_{L(PK)}}{2}\right)v_{O/IN}t_{E/D}\right]f_{SW}. \quad (4.100)$$

This is fortunate because P_{MOS}, P_{DT}, P_{IV}, P_{GI}, and P_{SW} also scale with f_{SW}.

So if L_X's energy E_L is constant (with fixed t_E, $i_{L(PK)}$, t_D, and t_C), power-conversion efficiency would be similarly independent of f_{SW}, and as a result, of P_O, because P_O and losses scale with f_{SW}:

$$\eta_C = \frac{P_O}{P_{IN}} \approx \frac{P_O}{P_O + P_{MOS} + P_{DT} + P_{IV} + P_{GI} + P_{SW}} \propto \frac{f_{SW}}{f_{SW}} \neq f(P_O). \quad (4.101)$$

And with W_{CH}', this η_C would also be optimally high.

For this, the controller should adjust t_{SW} (not d_E), which is a form of *frequency modulation* (FM). *Constant on-time control* and *pulse-FM* (PFM), for example, fix t_E and vary the frequency that L_X delivers energy packets. *Burst mode* is a variation that adjusts either the number of consecutive energy packets delivered between conduction gaps or the conduction gap between consecutive energy packets.

Figure 4.36 shows the measured efficiency of a photovoltaic battery-charging voltage regulator that adjusts the frequency of energy packets in discontinuous conduction. η_C is nearly constant at 95% when L_X is $3 \times 3 \times 1.5$ mm^3. η_C is lower but still constant when L_X is $1.6 \times 0.8 \times 0.8$ mm^3 because a smaller L_X is more resistive and therefore lossier. η_C drops when the *load power* P_{LD} that sets P_O nears zero because the controller consumes quiescent power that does not scale with f_{SW} (or P_O).

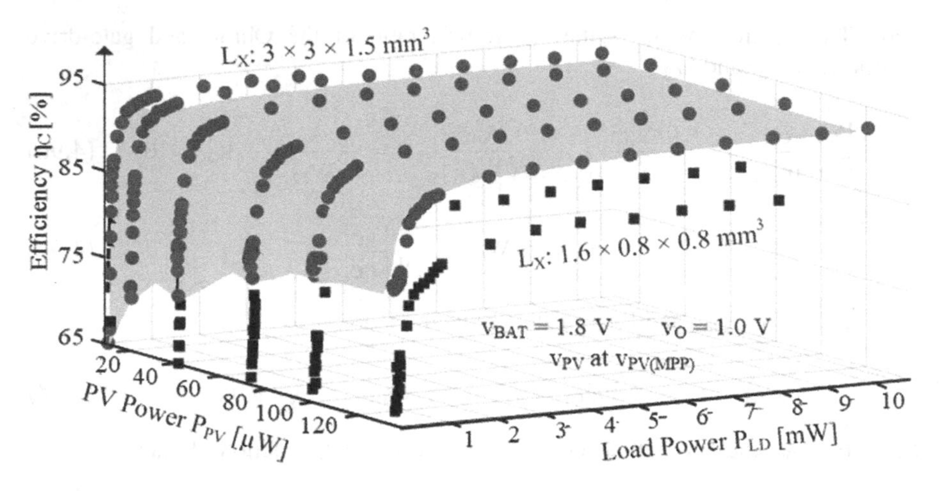

Fig. 4.36 Measured efficiency of frequency-modulated DCM system

4.8.5 Power-Conversion Efficiency

Power-conversion efficiency refers to the fraction of P_{IN} that P_O outputs. This η_C is ultimately a reflection of fractional losses. So increasing η_C amounts to reducing σ_{LOSS}.

A. **Discontinuous Conduction**

In DCM, P_R scales with $i_{L(PK)}{}^2$, t_C, and f_{SW}; P_{DT} and P_{IV} scale with $i_{L(PK)}$ and f_{SW}; and P_G, P_{GI}, and P_{SW} scale with f_{SW}. P_{CNTRL} and P_{OFF} are largely independent of $i_{L(PK)}$ and f_{SW}. And P_{IN} scales with the i_O that sets P_O.

Frequency Modulation With FM, $i_{L(PK)}$ and t_C are constant and f_{SW} scales with i_O. This way, P_{CNTRL} and P_{OFF} scale with $i_O{}^0$ and P_R, P_{DT}, P_{IV}, P_G, P_{GI}, P_{SW}, and P_{IN} with $i_O{}^1$. So their fractional losses σ_{DCM0} and σ_{DCM1} scale with $i_O{}^{-1}$ and $i_O{}^0$. This means σ_{DCM0} falls with i_O and σ_{DCM1} is close to constant:

$$\sigma_{DCM0} \approx \frac{P_{CNTRL} + P_{OFF}}{P_{IN}} \propto \frac{i_O{}^0}{i_O{}^1} \propto \frac{1}{i_O} \tag{4.102}$$

$$\sigma_{DCM1} \approx \frac{P_R + P_{DT} + P_{IV} + P_G + P_{GI} + P_{SW}}{P_{IN}} \propto \frac{i_O{}^1}{i_O{}^1} \neq f(i_O). \tag{4.103}$$

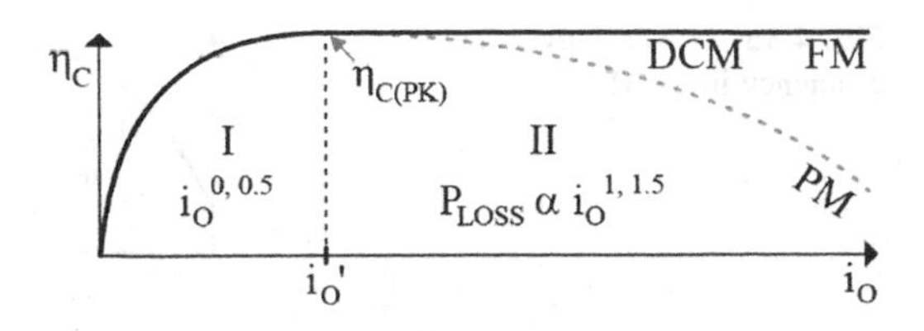

Fig. 4.37 Power-conversion efficiency in DCM

When lightly loaded, i_O-dependent losses are so low that P_{CNTRL} and P_{OFF} dominate. In this region in Fig. 4.37, η_C rises because σ_{DCM0} falls with i_O. η_C eventually peaks and flattens in Region II when P_R, P_{DT}, P_{IV}, P_G, P_{GI}, and P_{SW} dominate because σ_{DCM1} is constant across that range.

Peak Modulation *Peak modulation* (PM) is another way of controlling P_O. In this scheme, f_{SW} is constant. And the controller adjusts $i_{L(PK)}$ so $i_{L(AVG)}$ delivers the $i_{O(AVG)}$ the load demands.

This way, $i_{L(PK)}$ and t_C scale with $\sqrt{i_O}$. And P_G, P_{GI}, P_{SW}, P_{CNTRL}, and P_{OFF} scale with $i_O{}^0$, P_{DT} and P_{IV} with $i_O{}^{0.5}$, and P_R with $i_O{}^{1.5}$. So σ_{DCM0}, $\sigma_{DCM0.5}$, and $\sigma_{DCM1.5}$ scale with $i_O{}^{-1}$, $i_O{}^{-0.5}$, and $i_O{}^{0.5}$. In other words, σ_{DCM0} and $\sigma_{DCM0.5}$ fall and $\sigma_{DCM1.5}$ rises with i_O:

$$\sigma_{DCM0} \approx \frac{P_G + P_{GI} + P_{SW} + P_{CNTRL} + P_{OFF}}{P_{IN}} = k_{D0}\left(\frac{i_O{}^0}{i_O{}^1}\right) = \frac{k_{D0}}{i_O}, \tag{4.104}$$

$$\sigma_{DCM0.5} \approx \frac{P_{DT} + P_{IV}}{P_{IN}} = k_{D0.5}\left(\frac{i_O{}^{0.5}}{i_O{}^1}\right) = \frac{k_{D0.5}}{\sqrt{i_O}}, \tag{4.105}$$

$$\sigma_{DCM1.5} \approx \frac{P_R}{P_{IN}} = k_{D1.5}\left(\frac{i_O{}^{1.5}}{i_O{}^1}\right) = k_{D1.5}\sqrt{i_O}. \tag{4.106}$$

When lightly loaded, i_O reduces P_R to such an extent that P_{DT}, P_{IV}, P_G, P_{GI}, P_{SW}, P_{CNTRL}, and P_{OFF} dominate. η_C rises in Region I in Fig. 4.37 because $\sigma_{DCM0.5}$ and σ_{DCM0} fall with i_O. η_C falls in Region II when P_R dominates because $\sigma_{DCM1.5}$ climbs with i_O.

Their combined σ_{LOSS} reaches its minimum when its slope flattens with respect to i_O. This happens when $\sigma_{CCM1.5}$'s rise cancels σ_{DCM0} and $\sigma_{DCM0.5}$'s fall. At this optimal i_O', η_C maxes (with one minus this σ_{LOSS} at i_O'):

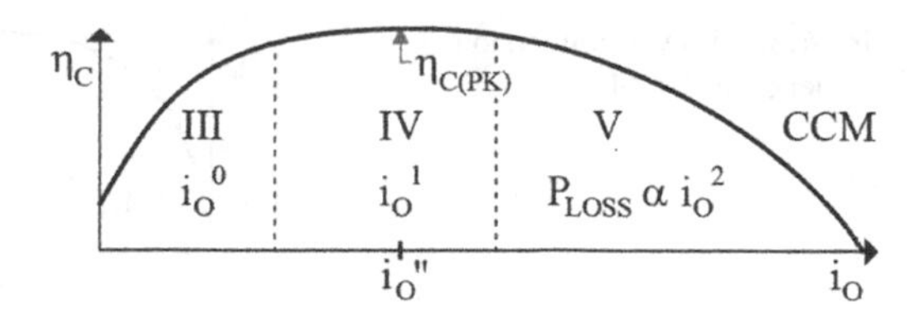

Fig. 4.38 Power-conversion efficiency in CCM

$$\left.\frac{\partial\sigma_{DCM0}}{\partial i_O}+\frac{\partial\sigma_{DCM0.5}}{\partial i_O}+\frac{\partial\sigma_{DCM1.5}}{\partial i_O}\right|_{i_O'}=-\frac{k_{D0}}{{i_O'}^2}-\frac{k_{D0.5}}{2\sqrt{{i_O'}^3}}+\frac{k_{D1.5}}{2\sqrt{i_O'}}=0. \quad (4.107)$$

So η_C peaks when P_R balances P_{DT}, P_{IV}, P_G, P_{GI}, P_{SW}, P_{CNTRL}, and P_{OFF}.

B. Continuous Conduction

In CCM, $P_{R(AC)}$, P_G, P_{GI}, P_{SW}, P_{CNTRL}, and P_{OFF} scale with i_O^0; P_{DT}, P_{IV}, and P_{IN} with i_O^1; and $P_{R(DC)}$ with i_O^2. So their corresponding fractional losses σ_{CCM0}, σ_{DCM1}, and σ_{DCM2} scale with i_O^{-1}, i_O^0, and i_O^1. In other words, σ_{DCM0} falls and σ_{DCM2} rises with i_O and σ_{DCM1} is close to constant:

$$\sigma_{CCM0}\approx\frac{P_{R(AC)}+P_G+P_{GI}+P_{SW}+P_{CNTRL}+P_{OFF}}{P_{IN}}=\frac{k_{C0}}{i_O}, \quad (4.108)$$

$$\sigma_{CCM1}\approx\frac{P_{DT}+P_{IV}}{P_{IN}}=k_{C1}\left(\frac{i_O}{i_O}\right)=k_{C1}, \quad (4.109)$$

$$\sigma_{CCM2}\approx\frac{P_{R(DC)}}{P_{IN}}=k_{C2}\left(\frac{i_O^2}{i_O}\right)=k_{C2}i_O. \quad (4.110)$$

When i_O is low, $P_{R(AC)}$, P_G, P_{GI}, P_{SW}, P_{CNTRL}, and P_{OFF} often outweigh i_O-dependent losses. η_C rises in Region III in Fig. 4.38 because σ_{DCM0} falls with i_O. η_C flattens in Region IV when P_{DT} and P_{IV} dominate because σ_{DCM1} is largely insensitive to i_O. η_C falls in Region V when $P_{R(DC)}$ dominates because σ_{DCM2} climbs with i_O.

Their combined σ_{LOSS} reaches its minimum when its slope flattens with respect to i_O. This happens when σ_{CCM2}'s rise cancels σ_{DCM0}'s fall. At this optimal i_O'', η_C maxes (with one minus this σ_{LOSS} at i_O''):

$$\left.\frac{\partial\sigma_{CCM0}}{\partial i_O}+\frac{\partial\sigma_{CCM1}}{\partial i_O}+\frac{\partial\sigma_{CCM2}}{\partial i_O}\right|_{i_O''}=-\frac{k_{C0}}{{i_O''}^2}+0+k_{C2}=0, \quad (4.111)$$

$$i_O''=\sqrt{\frac{k_{C0}}{k_{C2}}}, \quad (4.112)$$

$$\sigma_{CCM0}\big|_{i_O''} = \sigma_{CCM2}\big|_{i_O''} = \sqrt{k_{C0}k_{C2}}, \quad (4.113)$$

$$\eta_{C(PK)} = 1 - \sigma_{CCM0}\big|_{i_O''} - \sigma_{CCM1}\big|_{i_O''} - \sigma_{CCM2}\big|_{i_O''}$$
$$= 1 - \sigma_{CCM1} - 2\sqrt{k_{C0}k_{C2}}. \quad (4.114)$$

So η_C peaks when $P_{R(DC)}$ balances $P_{R(AC)}$, P_G, P_{GI}, P_{SW}, P_{CNTRL}, and P_{OFF}.

C. **Response**

The operating range of a load can sometimes exclude η_C's outer regions. Wireless microsensors, for example, can exclude Regions IV–V because output power is never high. Less mobile higher-power applications, on the other hand, can exclude Regions I–II because their loads are never low.

Although not often the case, η_C can peak twice when f_{SW} is constant: once in DCM and another time in CCM. η_C can also shrink or skip regions. η_C can, for example, reduce or jump Regions II–III when P_{DT} and P_{IV} are heavy or Regions III–IV when P_{DT} and P_{IV} are light and P_R is heavy.

Maxing η_C is ultimately more important than peaking it. In other words, reducing σ_{LOSS} across loads saves more energy than peaking η_C at a particular load. This is because the highest η_C for a particular i_O is not necessarily where η_C peaks.

4.9 Summary

Switched-inductor power supplies are popular in electronics because they condition and transfer most of the power they receive. Power-conversion efficiency is high because losses are low. Still, resistances, diodes, and transistors consume power that the output does not receive.

Resistances in switches, inductors, and capacitors burn Ohmic power when they conduct. The current that sets this conduction power decomposes into static and alternating components. The dc portion is ultimately a reflection of the load. The ac portion accounts for ripples, including those that connecting and disconnecting the load creates.

Diodes that conduct dead-time current when switches are off consume power. These diodes conduct twice every switching cycle in continuous conduction. In discontinuous conduction, they only conduct once per cycle because inductor current is zero after L_X drains.

Transistors burn i_{DS}–v_{DS} overlap power when they switch. Input, output, and diode voltages set the voltage they swing and inductor current sets the current they conduct. Recovering in-transit charge held in diodes increases the current and time they conduct. Although gate drivers ensure these transitions are short, overlap power is not always negligibly low.

Luckily, diode voltages are usually so much lower than input and output voltages that transitioning across these voltages burns a small fraction of drawn input power.

Overlap power is similarly low when inductor current reaches zero in discontinuous conduction. Soft-switching events like these are desirable in power supplies.

Gate drivers and pre-drivers draw and burn the supply power needed to switch transistors: half when charging gates and the other half when draining gates. They also leak shoot-through power. This leakage is low, however, when input and output voltage transitions do not overlap.

Switch-node capacitances leak the energy they need to charge. Although the inductor helps the output recover some of this energy, switches ultimately burn most of it. Large power transistors also leak current in cut off, especially when they are hot.

Although not always known, minimizing losses at the most probable load level saves the most energy. Ohmic and gate-charge power in MOSFETs can balance at this load setting with a particular channel width. Gate drivers can similarly balance driver gate-charge and i_{DS}–v_{DS} overlap power with specific N- and P-channel widths.

Since bucks steer input power into the output when they energize and boosts when they drain, bucks and boosts deliver more power than their inductor transfer. So bucks and boosts need less current to draw and deliver power than buck–boosts. Bucks and boosts also need fewer switches. And with lower current and fewer switches, Ohmic losses are lower.

Efficiency is usually low when loads are light because quiescent power in the controller is a large fraction of the input power drawn. Efficiency is also low when loads are heavy because dc Ohmic losses scale faster with output power than input power does. Frequency modulation in discontinuous conduction flattens efficiency when MOS Ohmic and gate-charge losses balance. Efficiency peaks in continuous conduction when ac ripple and charge losses similarly balance dc Ohmic losses.

High efficiency is important in voltage regulators, LED drivers, and chargers because it saves energy. Although not to the same extent, it is also important in energy harvesters. Output power is ultimately more important in harvesters because ambient energy is not always available. Still, output power is higher when efficiency peaks at the maximum-power point, which is not always possible, especially when the ambient source and the corresponding maximum-power point drift over time.

Frequency Response

5

Abbreviations

CCM	Continuous-conduction mode
CMOS	Complementary metal–oxide–semiconductor
DCM	Discontinuous-conduction mode
LED	Light-emitting diode
SL	Switched inductor
RSS	Root sum of squares
A_0	Zero-/low-frequency gain
A_{GI}	Input transconductance
A_{GO}	Output transconductance
A_{HF}	High-frequency gain
A_{II}	Input current gain
A_{IO}	Output current gain
A_N	Norton gain
A_T	Thévenin gain
A_{VI}	Input voltage gain
A_{VO}	Output voltage gain
A_{ZI}	Input transimpedance
A_{ZO}	Output transimpedance
C_B	Bypass capacitor
C_C	Couple capacitor
C_O	Output capacitor
C_S	Shunt capacitor
d_D	Drain duty cycle
d_O	Output duty cycle
d_E	Energize duty cycle
d_E'	Energize command
Δi_L	CCM current ripple

G. A. Rincón-Mora, *Switched Inductor Power IC Design*,
https://doi.org/10.1007/978-3-030-95899-2_5

f_{LC}	Transitional LC (resonant) frequency
f_O	Operating frequency
f_{SW}	Switching frequency
C_G	Gate capacitance
i	Imaginary unit
i_{DO}	Duty-cycled current
i_{IN}	Input current
i_L	Inductor current
i_N	Norton current
i_O	Output current
i_s	Small-signal current source
L_C	Couple inductor
L_{DO}	Duty-cycled inductance
L_S	Shunt inductor
L_X	Switched (transfer) inductor
p_C	Capacitor pole
p_L	Inductor pole
p_{LC}	LC double pole
p_{SW}	Switching pole
p_X	Reversal pole
Q_{LC}	LC peak quality
R_C	Couple resistor/capacitor resistance
R_{DO}	Duty-cycled resistance
R_I	Current-limit resistor
R_{IN}	Input resistance
R_L	Inductor resistance
R_{LD}	Load resistance
R_{LD}'	Equivalent load resistance
R_N	Norton resistance
R_O	Output resistance
R_P	Parallel resistance
R_S	Series resistance
R_T	Thévenin resistance
R_V	Voltage-limit resistor
s_A	Analog signal
s_C	Control signal
s_I	Input signal
s_O	Output signal
t_C	Conduction time
t_D	Drain time
t_E	Energize time
t_O	Output conduction time
t_{SW}	Switching period
v_D	Drain voltage
v_E	Energize voltage

v_{IN}	Input voltage/input
v_L	Inductor voltage
v_O	Output voltage/output
v_s	Small-signal voltage source
v_T	Thévenin voltage
ω_O	Angular frequency
z_C	Capacitor zero
z_{DO}	Duty-cycled zero
Z_{DO}	Duty-cycled impedance
z_L	Inductor zero
z_X	Reversal zero

The principal aim of power supplies is to transfer input power to the output. Conditioning this power into a form that the load can receive and use is a crucial component of this directive. This is what turns *voltage regulators* into voltage sources and *battery chargers* and *light-emitting diode* (LED) *drivers* into current sources.

To suppress deviations, power supplies incorporate feedback loops that monitor and oppose variations in the output. This opposition is ultimately a reaction to a disturbance. So understanding how *switched inductors* (SL) respond to the adjustments that disturbances prompt is critical.

Since fluctuations decompose into frequency components, *frequency response* describes how systems react to dynamic variations. This response is well-understood in linear systems. Nonlinear systems like the switched inductor are not so straightforward. In these cases, isolating and modeling the dynamic elements peel away the complexities of nonlinearity.

5.1 Two-Port Models

Two-port models are four-component networks that predict the reverse and forward response of a circuit when stimulated and loaded. The input and output of the model are interdependent resistive voltage or current sources. The input loads the circuit that drives the network and models *feedback* effects and the output drives the load of the network and models *forward translations.*

The fundamental advantage of these two-port models is simplicity, because four components can emulate the effects of complex circuits. This is possible because the components model orthogonal effects. In other words, each component models what the others do not.

Each interdependent source incorporates two components: resistance and gain. Resistance models loading in the absence of gain and gain models amplification in the absence of a load. Extracting one parameter therefore requires test conditions that nullify the other.

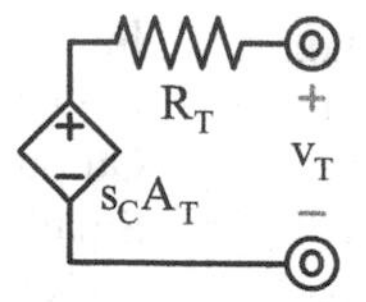

Fig. 5.1 Thévenin voltage source

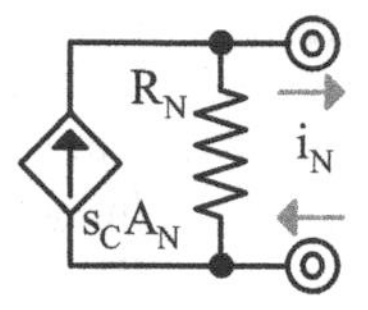

Fig. 5.2 Norton current source

5.1.1 Primitives

A. **Voltage Source**

The *Thévenin model* is a dependent voltage source in series with a resistor. The *Thévenin gain* A_T is the unloaded gain translation to the *Thévenin voltage* v_T in Fig. 5.1 and the *Thévenin resistance* R_T is the resistance into the circuit. When loaded, v_T manifests the effects of A_T and R_T.

To nullify the effects of R_T on v_T, R_T should drop zero volts. This happens when R_T's current is zero, which results when the load is absent. So removing the load eliminates the effects of R_T when deriving A_T. Zeroing A_T's *control signal* s_C similarly extinguishes the effects of A_T on v_T when extracting R_T.

B. **Current Source**

The *Norton model* is a dependent current source paralleled by a resistor. The *Norton gain* A_N is the zero-volt gain translation to the *Norton current* i_N in Fig. 5.2 and the *Norton resistance* R_N is the resistance into the circuit. When loaded, i_N manifests the effects of A_N and R_N.

To nullify the effects of R_N on i_N, R_N should not conduct current. This happens when R_N's voltage is zero, which results when the output terminals short. So shorting the output removes the effects of R_N when deriving A_N. Zeroing A_N's s_C similarly eliminates the effects of A_N on i_N when extracting R_N.

5.1.2 Bidirectional Models

A. **Impedance: Voltage**

The *impedance voltage model* in Fig. 5.3 uses voltage sources to model the input and output. *Input* and *output transimpedances* A_{ZI} and A_{ZO} model feedback and forward

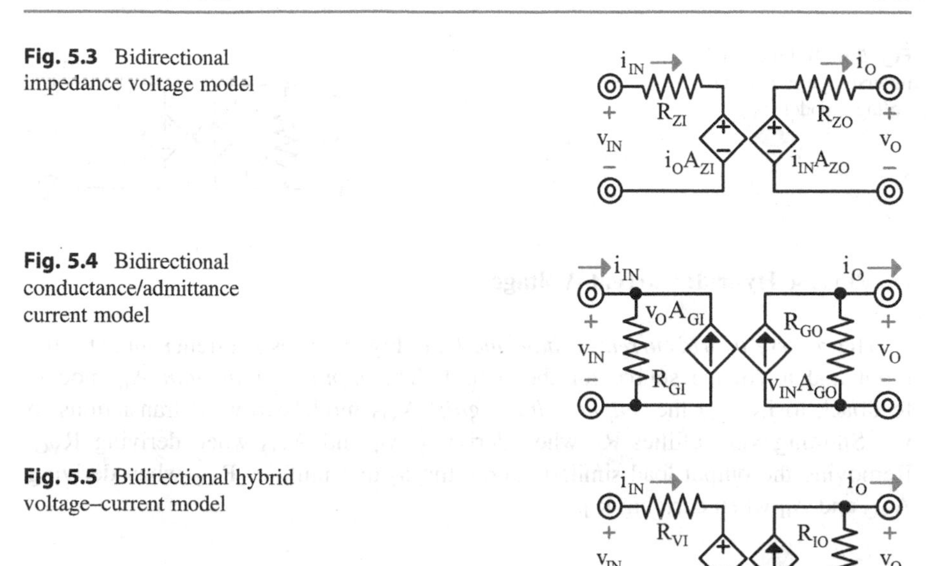

Fig. 5.3 Bidirectional impedance voltage model

Fig. 5.4 Bidirectional conductance/admittance current model

Fig. 5.5 Bidirectional hybrid voltage–current model

translations to the *input voltage* v_{IN} and *output voltage* v_O. And resistances R_{ZI} and R_{ZO} model loading effects.

The elegance of this model is that all extractions require open-circuit conditions. Removing the input source, for example, eliminates the *input current* i_{IN} that nullifies R_{ZI} when deriving A_{ZI} and A_{ZO} when deriving R_{ZO}. Similarly, removing the output load zeros the *output current* i_O that nullifies R_{ZO} when deriving A_{ZO} and A_{ZI} when deriving R_{ZI}.

B. Conductance/Admittance: Current

The *conductance* or *admittance current model* in Fig. 5.4 uses current sources to model the input and output. In this case, *input* and *output transconductances* A_{GI} and A_{GO} model feedback and forward translations to i_{IN} and i_O and all extractions require short-circuit conditions. Shorting v_{IN} nullifies R_{GI} when deriving A_{GI} and A_{GO} when deriving R_{GO}. Shorting v_O similarly nullifies R_{GO} when deriving A_{GO} and A_{GI} when deriving R_{GI}.

C. Hybrid: Voltage–Current

The *hybrid voltage–current model* in Fig. 5.5 uses a voltage source for the input and a current source for the output. The *input voltage gain* A_{VI} models feedback to v_{IN} and the *output current gain* A_{IO} models forward translations to i_O. Here, removing the input source eliminates the i_{IN} that nullifies R_{VI} when deriving A_{VI} and A_{IO} when deriving R_{IO}. And shorting v_O nullifies R_{IO} when deriving A_{IO} and A_{VI} when deriving R_{VI}.

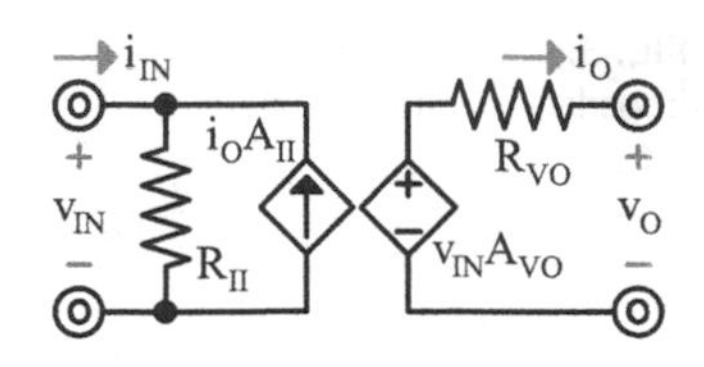

Fig. 5.6 Bidirectional reverse-hybrid current–voltage model

D. **Reverse Hybrid: Current–Voltage**

The *reverse-hybrid current–voltage model* in Fig. 5.6 uses a current source for the input and a voltage source for the output. The *input current gain* A_{II} models feedback to i_{IN} and the *output voltage gain* A_{VO} models forward translations to v_O. Shorting v_{IN} nullifies R_{II} when deriving A_{II} and A_{VO} when deriving R_{VO}. Removing the output load similarly zeros the i_O that nullifies R_{VO} when deriving A_{VO} and A_{II} when deriving R_{II}.

Example 1: Extract hybrid voltage–current parameters for the impedance voltage model.

Solution:

$$R_{VI} \equiv \left.\frac{v_{IN}}{i_{IN}}\right|_{v_O=0} = \frac{i_{IN}R_{ZI} + i_OA_{ZI}}{i_{IN}} = R_{ZI} + \frac{(i_{IN}A_{ZO}/R_{ZO})A_{ZI}}{i_{IN}} = R_{ZI} + \left(\frac{A_{ZO}}{R_{ZO}}\right)A_{ZI}$$

$$A_{VI} \equiv \left.\frac{v_{IN}}{v_O}\right|_{i_{IN}=0} = \left.\frac{i_OA_{ZI}}{v_O}\right|_{i_{IN}=0} = \left.\left(\frac{i_{IN}A_{ZO} - v_O}{R_{ZO}}\right)\left(\frac{A_{ZI}}{v_O}\right)\right|_{i_{IN}=0}$$

$$= -\left(\frac{v_O}{R_{ZO}}\right)\left(\frac{A_{ZI}}{v_O}\right) = -\frac{A_{ZI}}{R_{ZO}}$$

$$A_{IO} \equiv \left.\frac{i_O}{i_{IN}}\right|_{v_O=0} = \frac{i_{IN}A_{ZO}/R_{ZO}}{i_{IN}} = \frac{A_{ZO}}{R_{ZO}}$$

$$R_{IO} \equiv \left.\frac{v_O}{-i_O}\right|_{i_{IN}=0} = \left.\frac{i_{IN}A_{ZO} - i_OR_{ZO}}{-i_O}\right|_{i_{IN}=0} = R_{ZO}$$

Note: Two-port models are transformable.

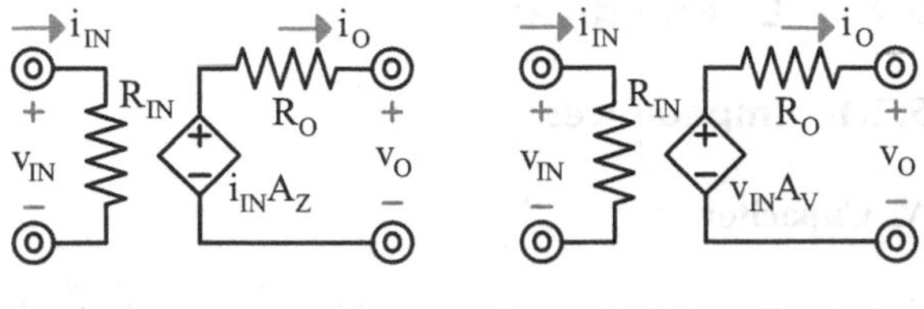

Fig. 5.7 Forward voltage-sourced models

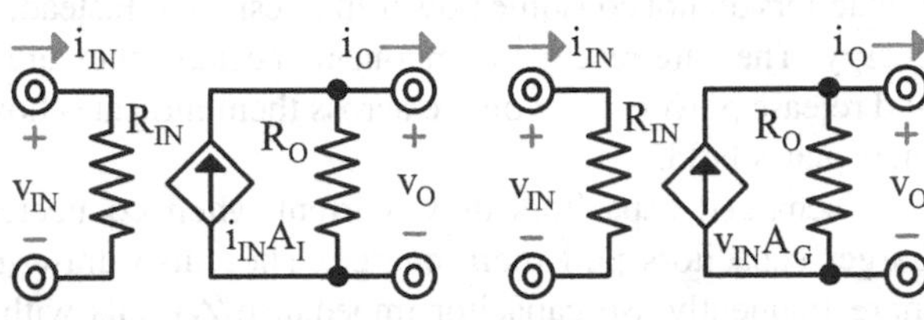

Fig. 5.8 Forward current-sourced models

5.1.3 Forward Models

Many circuits incorporate little to no feedback. In such cases, modeling the negligible feedback component is an unnecessary complication. This is why forward-only models are so popular: because they are simple.

Without feedback, the current or voltage source that models the input reduces to the *input resistance* R_{IN} shown in Figs. 5.7 and 5.8. Since no other component models the input, test conditions do not apply to R_{IN}. This means that R_{IN} is the same in all forward models.

Since v_{IN} and i_{IN} are Ohmic R_{IN} translations of one another, zeroing one eliminates the other. So nulling forward translations ultimately produces the same effect on the output. This means that the *output resistance* R_O is also the same in all forward models.

The only variation in forward models is the forward translation. This forward translation is a series voltage when using a voltage source and a parallel current when using a current source. And it responds to v_{IN} or an Ohmic R_{IN} translation of v_{IN}, which is i_{IN}.

Example 2: Extract voltage-driven current-source parameters for the impedance voltage model when feedback effects are negligible.

Solution:

Negligible feedback $\therefore$

$$A_{ZI} \approx 0$$

$$R_{IN} \approx R_{ZI}$$

$$R_O \approx R_{ZO}$$

$$A_G \equiv \left.\frac{i_O}{v_{IN}}\right|_{v_O=0} \approx \frac{i_{IN}A_{ZO}/R_{ZO}}{i_{IN}R_{ZI} + i_O A_{ZI}} \approx \frac{A_{ZO}}{R_{ZI}R_{ZO}}$$

5.2 LC Primitives

5.2.1 Impedances

A. Capacitor

Capacitors do not consume power like resistors. Instead, they draw, hold, and supply energy. They are reactive components because they are reacting when they receive and release power. The voltage across them indicates how much electrostatic energy their plates hold.

Discharged capacitors draw current when connected across a voltage source. Larger capacitors pull more charge. They also draw more current when charged more frequently. So capacitor impedance Z_C falls with increasing capacitance C_X and *operating frequency* f_O:

$$Z_C = \frac{1}{sC_X} = \frac{1}{i\omega_O C_X} = \frac{1}{i(2\pi f_O)C_X} \propto \frac{1}{f_O}, \tag{5.1}$$

where "s" in the *Laplace domain* is $i\omega_O$ or $i(2\pi f_O)$, "i" is the *imaginary unit* whose square i^2 is -1, ω_O is *angular frequency* in radians per second, f_O is in cycles per second, and each cycle is 2π radians long (i.e., 2π radians per cycle).

Capacitors are like switches that close and short with f_O. They open at low f_O, close and shunt resistances as f_O climbs, and short at high f_O. Parallel resistors fade when capacitors shunt them and capacitors effectively short when series resistances overwhelm them.

B. Inductor

Inductors are also reactive because they also draw, store, and deliver energy. In their case, current indicates how much magnetic energy their windings and core hold. They draw this current when connected across a voltage source.

This current scales with voltage, is lower with more windings and larger loops, and rises with time. So inductor impedance Z_L climbs with the inductance L_X that windings set and the shorter energize periods that higher f_O avails. In short, Z_L is

$$Z_L = sL_X = i\omega_O L_X = i(2\pi f_O)L_X \propto f_O. \tag{5.2}$$

Inductors are like switches that open with f_O. They close at low f_O, open and overcome resistances as f_O climbs, and altogether open at high f_O. Series resistors effectively short when inductors overcome them and inductors open and fade when parallel resistors limit their voltage.

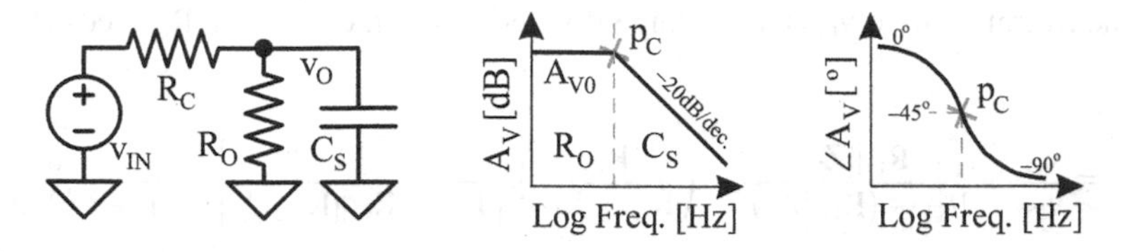

Fig. 5.9 Shunt capacitor

5.2.2 Shunt Capacitor

A. Response

Nodes in a circuit incorporate the parasitic capacitance that components and traces on a board add. These are *shunt capacitors* because they "shunt" current and energy away from their intended recipients. C_S in Fig. 5.9 is one such example because C_S steers current away from R_O.

Gain The effect of C_S on v_O is low at low f_O because C_S opens at low f_O. So the *zero-* or *low-frequency gain* A_{V0} to v_O is the voltage-divided fraction of v_{IN} that *couple resistor* R_C feeds R_O:

$$A_{V0} \approx \frac{R_O}{R_C + R_O}. \tag{5.3}$$

But as Z_S falls with f_O, C_S sinks more and more current away from R_O. And with less current available, R_O drops a lower voltage.

C_S begins to dominate the response at the f_O that C_S's impedance Z_s or $1/sC_S$ shunts the parallel resistance that R_O and R_C into v_{IN} establish across C_S:

$$Z_S = \frac{1}{sC_S}\bigg|_{f_O \geq \frac{1}{2\pi(R_O||R_C)C_S} \equiv p_C} \leq R_O||R_C. \tag{5.4}$$

So R_O sets the gain below this *capacitor pole* p_C and C_S above it. Since Z_S shunts R_O and R_C surpasses Z_S past this p_C, the gain to v_O past p_C reduces to Z_S/R_C, which scales with $1/f_O$:

$$A_V = \frac{R_O||Z_S}{R_C + (R_O||Z_S)}\bigg|_{f_O > p_C} \approx \frac{Z_S}{R_C} = \frac{1}{sR_C C_S} \propto \frac{1}{f_O}. \tag{5.5}$$

This gain A_V drops 10× or 20 dB when f_O climbs 10× or a decade. In short, A_V falls when C_S shunts the resistance at v_O.

The overall gain to v_O is the voltage-divided fraction of v_{IN} that R_C feeds R_O and C_S:

$$A_V \equiv \frac{v_O}{v_{IN}} = \frac{R_O||Z_S}{R_C + (R_O||Z_S)} = \left(\frac{R_O}{R_C + R_O}\right)\left[\frac{1}{1 + s(R_O||R_C)C_S}\right] = \frac{A_{V0}}{1 + s/2\pi p_C}. \tag{5.6}$$

The gain drops with f_O in "s" when $s(R_O \parallel R_C)C_S$ exceeds 1, $2\pi f_O$ in "s" overcomes $2\pi p_C$, or more simply, f_O surpasses p_C, as already stated. A_V's magnitude $|A_V|$ is the ratio of the *root sum of squares* (RSS) of the real and imaginary components in A_{V0} and $1 + s(R_O \parallel R_C)C_S$ or $1 + s/2\pi p_C$:

$$|A_V| = \frac{\sqrt{{A_{V0}}^2 + 0^2}}{\sqrt{1^2 + (f_O/p_C)^2}} = \frac{A_{V0}}{\sqrt{1^2 + (f_O/p_C)^2}}. \tag{5.7}$$

So $|A_V|$ is nearly A_{V0} a decade below p_C, $\sqrt{2}$ or 3 dB lower at p_C, and almost 10× or 20 dB lower a decade past p_C.

Phase Since C_S requires time to charge (and raise v_O), C_S delays v_{IN}-to-v_O translations. The delay between v_{IN} and v_O sinusoids when $1/sC_S$ swamps $R_O \parallel R_C$ effects is 90° of the 360° cycle. This lagging (negative) delay halves when $1/sC_S$ matches $R_O \parallel R_C$ and fades when C_S opens. So A_V's phase $\angle A_V$ is nearly 0° a decade below p_C, −45° at p_C, and almost −90° a decade past p_C:

$$\angle A_V = -\tan^{-1}\left(\frac{f_O}{p_C}\right). \tag{5.8}$$

Transitional frequencies that prompt gain and phase to drop 20 dB per decade and up to 90° this way are *poles*.

B. Current-Limit Resistor

The shunting effects of capacitors fade when they short. And they effectively short when resistors limit their current. Consider C_S and *current-limit resistor* R_I in Fig. 5.10. C_S shunts resistances past p_C and shorts past an f_O that R_I sets.

Gain R_I adds to the resistance R_O and R_C into v_{IN} present. So C_S and R_I shunt R_O with R_C into v_{IN} when Z_S falls below the parallel resistance that R_I and R_O with R_C set. And A_V falls past the p_C that their RC frequency determines:

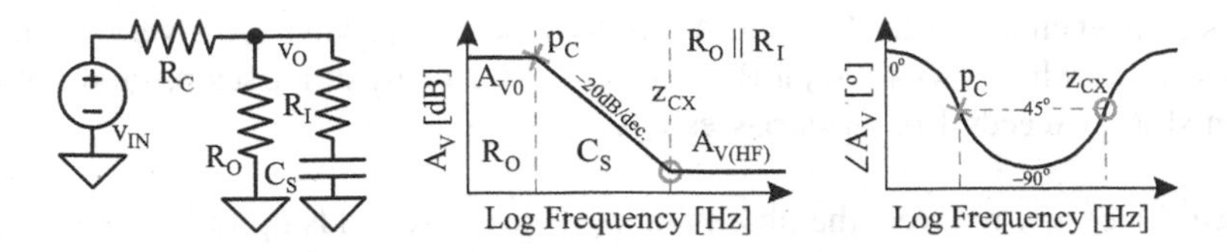

Fig. 5.10 Current-limited shunt capacitor

$$Z_S = \frac{1}{sC_S}\bigg|_{f_O \geq \frac{1}{2\pi[R_I + (R_O \| R_C)]C_S} \equiv p_C} \leq R_I + (R_O \| R_C). \tag{5.9}$$

R_I lowers the p_C that C_S induces because R_I adds series resistance to C_S.

The gain to v_O drops past p_C as Z_S falls with $1/f_O$. When C_S shorts, R_I parallels R_O, so the *high-frequency* voltage-divided *gain* $A_{V(HF)}$ flattens to

$$A_{V(HF)} \approx \frac{R_O \| R_I}{R_C + (R_O \| R_I)}. \tag{5.10}$$

This happens because R_I limits the current that C_S can pull.

So the effects of p_C fade when C_S shorts with respect to R_I:

$$Z_S = \frac{1}{sC_S}\bigg|_{f_O \geq \frac{1}{2\pi R_I C_S} \equiv z_{CX}} \leq R_I. \tag{5.11}$$

In eliminating the effects of p_C, R_I is effectively raising gain 20 dB per decade and recovering up to 90° of phase. Transitional frequencies that prompt gain and phase to climb this way are *zeros*. This particular zero is a *reversal zero* because resistors that current-limit shunt capacitors reverse the effects of capacitor poles.

The overall voltage gain to v_O is the voltage-divided fraction of v_{IN} that R_C feeds R_O and C_S with R_I:

$$\begin{aligned} A_V \equiv \frac{v_O}{v_{IN}} &= \frac{R_O \| (Z_S + R_I)}{R_C + [R_O \| (Z_S + R_I)]} \\ &= \left(\frac{R_O}{R_C + R_O}\right)\left(\frac{1 + sR_IC_S}{1 + s[R_I + (R_O \| R_C)]C_S}\right) \\ &= A_{V0}\left(\frac{1 + s/2\pi z_{CX}}{1 + s/2\pi p_C}\right). \end{aligned} \tag{5.12}$$

A_V is A_{VO} when f_O is very low and $A_{VO}p_C/z_{CX}$ or $A_{V(HF)}$ when f_O is very high. A_V falls with f_O when $s[R_I + (R_O \parallel R_C)]C_S$ exceeds 1 or f_O surpasses p_C and flattens when sR_IC_S exceeds 1 or f_O surpasses z_{CX}.

Phase Since z_{CX} recovers the phase that p_C loses, $\angle A_V$ adds up to 90° of phase at higher f_O. So when p_C is much lower than z_{CX}, $\angle A_V$ is close to 0° a decade below p_C, −45° at p_C, and almost −90° a decade past p_C:

$$\angle A_V = -\tan^{-1}\left(\frac{f_O}{p_C}\right) + \tan^{-1}\left(\frac{f_O}{z_{CX}}\right). \tag{5.13}$$

$\angle A_V$ does not recover much phase a decade below z_{CX}, recovers 45° at z_{CX}, and recovers almost all 90° a decade past z_{CX}.

Example 3: Determine A_{VO}, $A_{V(HF)}$, and related poles and zeros when R_C is 50 Ω, R_O is 100 Ω, C_S is 5 µF, and R_I is 10 mΩ.

Solution:

$$A_{VO} \approx \frac{R_O}{R_C + R_O} = \frac{100}{50 + 100} = 670\ \text{mV/V} = -3.5\ \text{dB}$$

$$p_C = \frac{1}{2\pi[R_I + (R_O \| R_C)]C_S} = \frac{1}{2\pi[10\text{m} + (100\|50)](5\mu)} = 950\ \text{Hz}$$

$$z_{CX} = \frac{1}{2\pi R_I C_S} = \frac{1}{2\pi(10\text{m})(5\mu)} = 3.2\ \text{MHz}$$

$$A_{V(HF)} \approx \frac{R_O \| R_I}{R_C + (R_O \| R_I)} = \frac{100\|10\text{m}}{50 + (100\|10\text{m})} = 200\ \mu\text{V/V} = -74\ \text{dB}$$

Note: In practice, R_I and R_O can be the parasitic resistance that C_S incorporates and the load resistance that a circuit presents.

Explore with SPICE:
See Appendix A for notes on SPICE simulations.

```
* Shunt Capacitor
vin vin 0 dc=0 ac=1
rc vin vo 50
ro vo 0 100
ri vo vc 10m
cs vc 0 5u
.ac dec 1000 10 100e6
.end
```

Tip: Plot v(vo) in dB to inspect A_V.

5.2.3 Couple Capacitor

A. **Response**

Capacitors can also short terminals together. In this sense, they are like switches that shunt and short the infinite resistance between disconnected nodes. Consider the *couple capacitor* C_C in Fig. 5.11. C_C feeds current i_C as it connects and eventually shorts v_{IN} to v_O.

Gain Z_C is so high at very low f_O that C_C drops almost all of v_{IN}. Since R_O drops a very small fraction of v_{IN}, the low-f_O gain to v_O is very low. In fact, Z_C is so much greater than R_O that Z_C overwhelms R_O. This means, the voltage gain to v_O reduces to R_O/Z_C and, as a result, scales with f_O:

$$A_V = \left.\frac{R_O}{Z_C + R_O}\right|_{f_O < p_{CX}} \approx \frac{R_O}{Z_C} = sR_OC_C \propto f_O. \tag{5.14}$$

So A_V climbs 20 dB per decade as C_C shunts. This very low-f_O rise is like the effect of a *capacitor zero* z_{CO} at the *origin* (lower left-hand corner) of the graph.

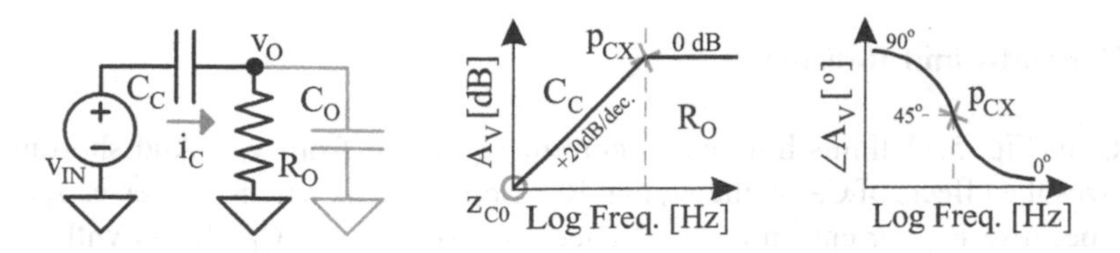

Fig. 5.11 Couple capacitor

As Z_C falls with increasing f_O, C_C drops a diminishing fraction of v_{IN}. At very high f_O, C_C drops so little of v_{IN} that the high-f_O gain $A_{V(HF)}$ to v_O nears 1 or 0 dB. So C_C drops more of v_{IN} at low f_O and R_O at high f_O.

C_C loses dominance to R_O when $1/sC_C$ shunts (falls below) R_O:

$$Z_C = \frac{1}{sC_C}\bigg|_{f_O \ge \frac{1}{2\pi R_O C_C} \equiv p_{CX}} \le R_O. \tag{5.15}$$

The f_O where this happens is a *reversal pole* because removing the effects of z_{CO} is like decreasing gain 20 dB per decade. But more to the point, C_C effectively shorts v_{IN} to v_O when C_C shorts with respect to the resistance at v_O, which is how p_{CX} reverses z_{CO}.

The voltage-divided gain to v_O is the v_{IN} fraction that C_C feeds R_O:

$$\begin{aligned} A_V &\equiv \frac{v_O}{v_{IN}} = \frac{R_O}{Z_C + R_O} = \frac{R_O}{1/sC_C + R_O} = \frac{sR_OC_C}{1 + sR_OC_C} \\ &= A_{V(HF)}\left(\frac{s/2\pi p_{CX}}{1 + s/2\pi p_{CX}}\right). \end{aligned} \tag{5.16}$$

The gain is low when f_O is low and climbs with f_O until sR_OC_C exceeds 1 or f_O surpasses p_{CX}. $|A_V|$ is the ratio of the root sum of squares of real and imaginary components in sR_OC_C and $1 + sR_OC_C$:

$$|A_V| = A_{V(HF)}\left[\frac{\sqrt{0^2 + (f_O/p_{CX})^2}}{\sqrt{1^2 + (f_O/p_{CX})^2}}\right] = A_{V(HF)}\left[\frac{f_O/p_{CX}}{\sqrt{1^2 + (f_O/p_{CX})^2}}\right]. \tag{5.17}$$

So $|A_V|$ is near $A_{V(HF)}$'s 1 a decade above p_{CX}, $\sqrt{2}$ or 3 dB lower at p_{CX}, and almost $10\times$ or 20 dB lower a decade below p_{CX}.

Phase Since the i_C that C_C feeds climbs with f_O, i_C replenishes the i_O that a shunt capacitor C_O at v_O pulls. This is like recovering the phase that C_O loses. But since p_{CX} reverses this recovery, z_{CO} starts $\angle A_V$ at 90° and p_{CX} reduces $\angle A_V$ to +45° at p_{CX} and to nearly 0° a decade past p_{CX}:

$$\angle A_V = 90° - \tan^{-1}\left(\frac{f_O}{p_{CX}}\right). \tag{5.18}$$

B. Current-Limit Resistor

R_O in Fig. 5.11 limits how much current C_C draws from v_{IN}. And shorting C_C removes the effects of C_C on the v_O that R_O drops. This is why R_O reverses z_{CO} with p_{CX}: because R_O current-limits C_C. Stated in another way, C_C shunts with f_O and shorts past p_{CX}.

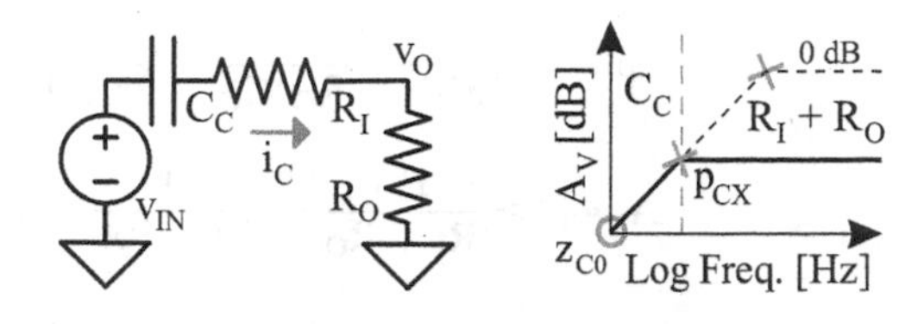

Fig. 5.12 Current-limited couple capacitor

Adding another resistor further current-limits C_C. Consider R_I in Fig. 5.12. Z_C is so high at low f_O that C_C drops most of v_{IN}. As f_O climbs, C_C drops a diminishing fraction of v_{IN}, so the voltage gain to v_O climbs with f_O. This is the effect of the z_{C0} that C_C induces as C_C shunts.

But when C_C shorts with respect to R_I and R_O:

$$Z_C = \left.\frac{1}{sC_C}\right|_{f_O \geq \frac{1}{2\pi(R_I+R_O)C_C} \equiv p_{CX}} \leq R_I + R_O, \tag{5.19}$$

A_V flattens to the voltage-divided fraction of v_{IN} that R_I feeds R_O:

$$A_{V(HF)} \approx \frac{R_O}{R_I + R_O}. \tag{5.20}$$

R_I and R_O remove z_{C0} because R_I and R_O limit the current C_C feeds. And p_{CX} and $A_{V(HF)}$ are lower with R_I because C_C shorts with respect to a higher resistance, and once shorted, R_I drops a part of v_{IN} that R_O does not drop.

Overall, A_V is the voltage-divided v_{IN} fraction that C_C and R_I feed R_O:

$$\begin{aligned} A_V \equiv \frac{v_O}{v_{IN}} &= \frac{R_O}{Z_C + R_I + R_O} \\ &= \frac{sR_OC_C}{1 + s(R_I + R_O)C_C} = A_{V(HF)}\left(\frac{s/2\pi p_{CX}}{1 + s/2\pi p_{CX}}\right). \end{aligned} \tag{5.21}$$

A_V is low when f_O is low and climbs with f_O until $s(R_I + R_O)C_C$ exceeds 1 or f_O surpasses p_{CX}, past which A_V flattens to $A_{V(HF)}$. z_{C0} starts $\angle A_V$ at 90° and p_{CX} reduces $\angle A_V$ to +45° at p_{CX} and to nearly 0° a decade past p_{CX}.

Example 4: Determine $A_{V(HF)}$ and related poles and zeros when C_C is 5 μF, R_I is 50 Ω, and R_O is 100 Ω.

Solution:

z_{C0} at the origin

$$p_{CX} = \frac{1}{2\pi(R_I + R_O)C_C} = \frac{1}{2\pi(50 + 100)(5\mu)} = 210 \text{ Hz}$$

$$A_{V(HF)} \approx \frac{R_O}{R_I + R_O} = \frac{100}{50 + 100} = 670 \text{ mV/V} = -3.5 \text{ dB}$$

Explore with SPICE:
See Appendix A for notes on SPICE simulations.

```
* Couple Capacitor
vin vin 0 dc=0 ac=1
cc vin vc 5u
ri vc vo 50
ro vo 0 100
.ac dec 1000 1 100k
.end
```

Tip: Plot v(vo) in dB to inspect A_V.

5.2.4 Couple Inductor

A. **Response**

Gain Interconnecting inductors impede and limit current flow as they open with f_O. Consider the *couple inductor* L_C in Fig. 5.13. Z_C is so low at low f_O that L_C drops a very small fraction of v_{IN}. Since R_O drops almost all of v_{IN}, the low-f_O gain A_{V0} to v_O is nearly 1 or 0 dB.

As Z_C climbs with f_O, L_C drops an increasing fraction of v_{IN}. At very high f_O, L_C drops so much of v_{IN} that the gain to v_O is very low. So R_O drops more of v_{IN} at low f_O and L_C at high f_O. L_C begins to dominate when sL_C overcomes R_O:

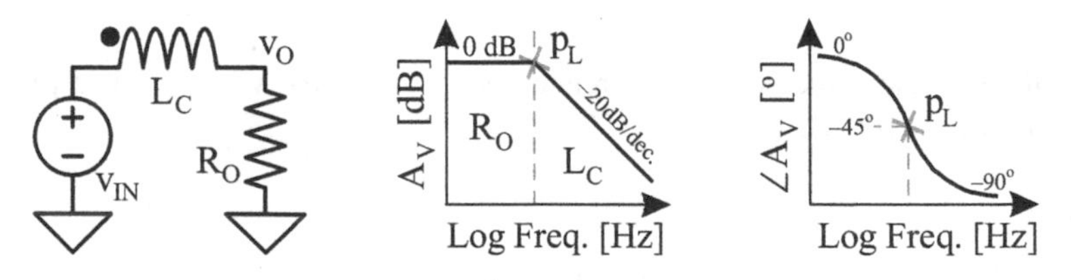

Fig. 5.13 Couple inductor

$$Z_C = sL_C\big|_{f_O \geq \frac{R_O}{2\pi L_C} \equiv p_L} \geq R_O. \tag{5.22}$$

Z_C is so much greater than R_O at high f_O that Z_C overwhelms R_O and the voltage gain to v_O reduces to R_O/sL_C, which scales with $1/f_O$:

$$A_V = \left.\frac{R_O}{Z_C + R_O}\right|_{f_O > p_L} \approx \frac{R_O}{Z_C} = \frac{R_O}{sL_C} \propto \frac{1}{f_O}. \tag{5.23}$$

Past this *inductor pole* p_L, A_V drops 20 dB when f_O increases a decade. A_V falls because L_C dominates when Z_C overcomes the resistance at v_O.

The gain to v_O is the voltage-divided v_{IN} fraction that L_C feeds R_O:

$$A_V \equiv \frac{v_O}{v_{IN}} = \frac{R_O}{Z_C + R_O} = \frac{R_O}{sL_C + R_O} = \frac{1}{1 + sL_C/R_O} = \frac{A_{V0}}{1 + s/2\pi p_L}. \tag{5.24}$$

The gain drops with f_O when sL_C/R_O exceeds 1 or f_O surpasses p_L. $|A_V|$ is the ratio of the root sum of squares of the real and imaginary components in A_{V0}'s 1 and 1 + sL_C/R_O or 1 + $s/2\pi p_L$:

$$|A_V| = \frac{\sqrt{A_{V0}{}^2 + 0^2}}{\sqrt{1^2 + (f_O/p_L)^2}} = \frac{A_{V0}}{\sqrt{1^2 + (f_O/p_L)^2}}. \tag{5.25}$$

So $|A_V|$ is nearly A_{V0}'s 1 upto a decade below p_L, $\sqrt{2}$ or 3 dB lower at p_L, and almost 10× or 20 dB lower a decade past p_L.

Phase Since L_C requires time to grow its current i_C, L_C delays v_{IN}-to-i_C translations. The delay between v_{IN} and i_C sinusoids after sL_C swamps R_O is 90° of the 360° cycle. This lagging (negative) delay halves when sL_C matches R_O and fades when L_C shorts. So A_V's phase $\angle A_V$ is close to 0° a decade below p_L, −45° at p_L, and almost −90° a decade past p_L:

$$\angle A_V = -\tan^{-1}\left(\frac{f_O}{p_L}\right). \tag{5.26}$$

B. Voltage-Limit Resistor

The effects of inductors overcoming resistances disappear when inductors open. They open when resistors limit the voltage they drop. Consider L_C and *voltage-limit resistor* R_V in Fig. 5.14. L_C overcomes series resistances past p_L and opens past an f_O that R_V sets.

Gain L_C drops a negligible fraction of v_{IN} at low f_O, so A_{V0} is nearly one. A_V falls when Z_C begins to drop a noticeable part of v_{IN}. This happens when sL_C overcomes the resistance that R_O and R_V into v_{IN} establish across L_C:

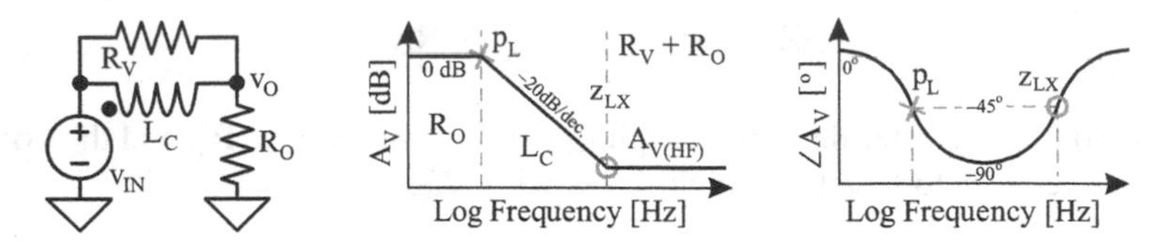

Fig. 5.14 Voltage-limited couple inductor

$$Z_C = sL_C\big|_{f_O \geq \frac{R_O||R_V}{2\pi L_C} \equiv p_L} \geq R_O||R_V. \tag{5.27}$$

Notice that adding R_V to the circuit reduces the p_L that L_C induces.

L_C eventually opens when R_V limits the voltage L_C drops, which happens when sL_C surpasses R_V, which is the resistance that parallels L_C:

$$Z_C = sL_C\big|_{f_O \geq \frac{R_V}{2\pi L_C} \equiv z_{LX}} \geq R_V. \tag{5.28}$$

Past this reversal zero, R_V and R_O alone drop v_{IN}, so the high-f_O A_V flattens to

$$A_{V(HF)} \approx \frac{R_O}{R_V + R_O}. \tag{5.29}$$

In eliminating the effects of p_L, R_V is effectively raising gain 20 dB per decade and recovering up to 90° of phase. So resistors that limit the voltage that couple inductors drop reverse inductor poles.

Overall, A_V is the voltage-divided v_{IN} fraction that L_C and R_V feed R_O:

$$A_V = \frac{R_O}{(Z_C||R_V) + R_O} = \frac{1 + sL_C/R_V}{1 + sL_C/(R_O||R_V)} = A_{V0}\left(\frac{1 + s/2\pi z_{LX}}{1 + s/2\pi p_L}\right). \tag{5.30}$$

A_V is A_{V0}'s 1 or 0 dB when f_O is very low and $A_{V0}p_L/z_{LX}$ or $A_{V(HF)}$ when f_O is very high. A_V falls with f_O when $sL_C/(R_O \parallel R_V)$ exceeds 1 or f_O surpasses p_L and flattens when sL_C/R_V exceeds 1 or f_O surpasses z_{LX}.

$|A_V|$ is the ratio of the root sum of squares of real and imaginary components in $1 + sL_C/R_V$ and $1 + sL_C/(R_O \parallel R_V)$:

$$|A_V| = A_{V0}\left[\frac{\sqrt{1^2 + (f_O/z_{LX})^2}}{\sqrt{1^2 + (f_O/p_L)^2}}\right]. \tag{5.31}$$

When z_{LX} is much higher than p_L, $|A_V|$ is nearly A_{V0} a decade below p_L, $\sqrt{2}$ or 3 dB lower at p_L, and almost 10× or 20 dB lower a decade past p_L. $|A_V|$ is $\sqrt{2}$ or 3 dB above $A_{V(HF)}$ at z_{LX} and nearly $A_{V(HF)}$ a decade past z_{LX}.

Phase Since z_{LX} recovers the phase that p_L loses, $\angle A_V$ adds up to 90° of phase at high f_O. So when p_L is much lower than z_{LX}, $\angle A_V$ is close to 0° a decade below p_L, −45° at p_L, and almost −90° a decade past p_L:

$$\angle A_V = -\tan^{-1}\left(\frac{f_O}{p_L}\right) + \tan^{-1}\left(\frac{f_O}{z_{LX}}\right). \quad (5.32)$$

$\angle A_V$ recovers 45° at z_{LX} and almost all 90° a decade past z_{LX}.

Example 5: Determine A_{V0}, $A_{V(HF)}$, and related poles and zeros when L_C is 10 μH, R_V is 1 kΩ, and R_O is 100 Ω.

Solution:

$$A_{V0} \approx 1\ V/V = 0\ dB$$

$$p_L = \frac{R_O \| R_V}{2\pi L_C} = \frac{100\|1k}{2\pi(10\mu)} = 1.4\ MHz$$

$$z_{LX} = \frac{R_V}{2\pi L_C} = \frac{1k}{2\pi(10\mu)} = 16\ MHz$$

$$A_{V(HF)} \approx \frac{R_O}{R_V + R_O} = \frac{100}{1k + 100} = 91\ mV/V = -21\ dB$$

Explore with SPICE:
See Appendix A for notes on SPICE simulations.

```
* Couple Inductor
vin vin 0 dc=0 ac=1
lc vin vo 10u
rv vin vo 1k
ro vo 0 100
.ac dec 1000 10k 1g
.end
```

Tip: Plot v(vo) in dB to inspect A_V.

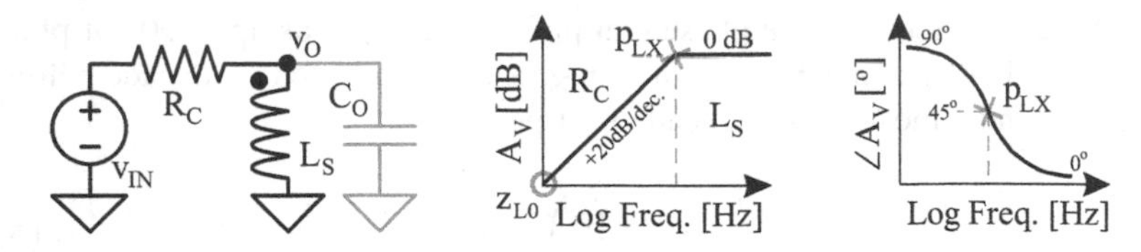

Fig. 5.15 Shunt inductor

5.2.5 Shunt Inductor

A. **Response**

Shunt inductors block and avail current as they impede current flow and open with f_O. So inductors that shunt v_O unload with f_O. Consider the *shunt inductor* L_S in Fig. 5.15.

Gain Z_S is so low at very low f_O that L_S drops a small fraction of v_{IN}. Since R_C drops almost all of v_{IN}, the low-f_O gain to v_O is very low. This gain scales with Z_S as f_O climbs. Z_S is so low that R_C overwhelms Z_S, so the voltage gain to v_O reduces to Z_S/R_C, which scales with f_O:

$$A_V = \left.\frac{Z_S}{R_C + Z_S}\right|_{f_O < p_{LX}} \approx \frac{Z_S}{R_C} = \frac{sL_S}{R_C} \propto f_O. \tag{5.33}$$

So as L_C opens, A_V rises 20 dB when f_O increases a decade. This very low-f_O rise is like the effect of an *inductor zero* z_{LO} at the origin of the graph.

As Z_S climbs with f_O, L_S drops an increasing fraction of v_{IN}. At very high f_O, L_S drops so much of v_{IN} that the high-f_O gain $A_{V(HF)}$ to v_O nears 1 or 0 dB. So R_C drops more of v_{IN} at low f_O and L_S at high f_O. L_S begins to dominate when sL_S overcomes R_C:

$$Z_S = sL_S\Big|_{f_O \geq \frac{R_C}{2\pi L_S} \equiv p_{LX}} \geq R_C. \tag{5.34}$$

The f_O where this happens is a reversal pole because removing the effects of z_{LO} is like decreasing gain 20 dB per decade. But more to the point, L_S essentially opens when Z_S overcomes the series resistance R_C sets.

The voltage-divided gain to v_O is the v_{IN} fraction that R_C feeds L_S:

$$\begin{aligned} A_V \equiv \frac{v_O}{v_{IN}} &= \frac{Z_S}{R_C + Z_S} = \frac{sL_S}{R_C + sL_S} \\ &= \frac{sL_S/R_C}{1 + sL_S/R_C} = A_{V(HF)}\left(\frac{s/2\pi p_{LX}}{1 + s/2\pi p_{LX}}\right). \end{aligned} \tag{5.35}$$

The gain is low at low f_O and climbs with f_O until sL_S/R_O exceeds 1 or f_O surpasses p_{LX}. $|A_V|$ is the ratio of the root sum of squares of real and imaginary components in sL_S/R_O and $1 + sL_S/R_O$:

$$|A_V| = A_{V(HF)}\left[\frac{\sqrt{0^2 + (f_O/p_{LX})^2}}{\sqrt{1^2 + (f_O/p_{LX})^2}}\right] = A_{V(HF)}\left[\frac{f_O/p_{LX}}{\sqrt{1^2 + (f_O/p_{LX})^2}}\right]. \quad (5.36)$$

So $|A_V|$ is near $A_{V(HF)}$'s one a decade above p_{LX}, $\sqrt{2}$ or 3 dB lower at p_{LX}, and almost 10× or 20 dB lower a decade below p_{LX}.

Phase Since the current L_S pulls falls with f_O, L_S avails the i_O that a shunt capacitor C_O at v_O pulls. This is like recovering the phase C_O loses. But since p_{LX} reverses this recovery, z_{LO} starts $\angle A_V$ at 90° and p_{LX} reduces $\angle A_V$ to +45° at p_{LX} and to nearly 0° a decade past p_{LX}:

$$\angle A_V = 90° - \tan^{-1}\left(\frac{f_O}{p_{LX}}\right). \quad (5.37)$$

B. Voltage-Limit Resistor

Resistors that voltage-limit inductors remove their effects. R_C in Fig. 5.15, for example, limits the voltage L_S drops. And voltage-limiting L_S removes the effects of L_S on v_O. This is why R_C reverses z_{LO} with p_{LX}: because L_S disconnects when R_C limits the voltage L_S drops. In other words, L_S overcomes series resistances as f_O climbs and opens past an f_O that R_V sets.

Paralleling a resistor across L_S further limits L_S's voltage. Consider R_V in Fig. 5.16. Z_S is so low at low f_O that L_S drops a small fraction of v_{IN}. As f_O climbs, L_S drops an increasing fraction of v_{IN}, so the gain to v_O climbs with f_O. This is the effect of the z_{LO} that L_S induces as L_S opens.

But when L_S overcomes the resistance R_V and R_C into v_{IN} set:

$$Z_S = sL_S\big|_{f_O \ge \frac{R_C||R_V}{2\pi L_S} \equiv p_{CX}} \le R_C||R_V, \quad (5.38)$$

A_V flattens to the voltage-divided fraction of v_{IN} that R_C feeds R_V:

$$A_{V(HF)} \approx \frac{R_V}{R_C + R_V}. \quad (5.39)$$

R_C and R_V remove z_{LO} because R_C and R_V limit the voltage L_S drops. Here, R_V reduces the resistance that L_S overcomes, and with it, the p_{LX} and $A_{V(HF)}$ that R_C and R_V set.

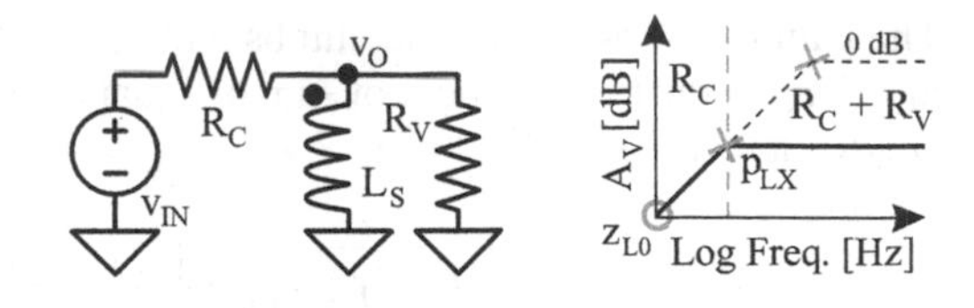

Fig. 5.16 Voltage-limited shunt inductor

Overall, A_V is the voltage-divided v_{IN} fraction that R_C feeds L_S and R_V:

$$\begin{aligned} A_V \equiv \frac{v_O}{v_{IN}} = \frac{Z_S || R_V}{R_C + (Z_S || R_V)} &= \left(\frac{R_V}{R_C + R_V}\right)\left[\frac{sL_S/(R_C || R_V)}{1 + sL_S/(R_C || R_V)}\right] \\ &= A_{V(HF)}\left(\frac{s/2\pi p_{LX}}{1 + s/2\pi p_{LX}}\right). \end{aligned} \tag{5.40}$$

A_V is low at low f_O and climbs with f_O until $sL_S/(R_C \parallel R_V)$ exceeds 1 or f_O surpasses p_{LX}, past which A_V flattens to $A_{V(HF)}$. z_{LO} starts $\angle A_V$ at 90° and p_{LX} reduces $\angle A_V$ to +45° at p_{LX} and to nearly 0° a decade past p_{LX}.

Example 6: Determine $A_{V(HF)}$ and related poles and zeros when R_C is 50 Ω, L_S is 10 μF, and R_V is 100 Ω.

Solution:

z_{LO} at the origin

$$p_{LX} = \frac{R_C || R_V}{2\pi L_S} = \frac{50 || 100}{2\pi(10\mu)} = 530\ \text{kHz}$$

$$A_{V(HF)} \approx \frac{R_V}{R_C + R_V} = \frac{100}{50 + 100} = 670\ \text{mV/V} = -3.5\ \text{dB}$$

Explore with SPICE:
See Appendix A for notes on SPICE simulations.

```
* Shunt Inductor
vin vin 0 dc=0 ac=1
rc vin vo 50
ls vo 0 10u
rv vo 0 100
.ac dec 1000 1k 100e6
.end
```

Tip: Plot v(vo) in dB to inspect A_V.

5.3 Bypass Capacitors

5.3.1 Bypassed Resistor

A. Response

Capacitors that bypass couple resistors add energy to their outputs. Since adding this energy increases gain, these capacitors induce zeros. Consider the couple resistor in Fig. 5.17 and the *bypass capacitor* C_B that bypasses R_C.

Gain Z_B is so high at low f_O that R_O drops the fraction of v_{IN} that R_C's current i_C sets. So the low-f_O gain to v_O is the v_{IN} fraction that i_C feeds R_O:

$$A_{V0} \approx \frac{i_C R_O}{v_{IN}} = \frac{R_O}{R_C + R_O}. \tag{5.41}$$

But since Z_B falls as f_O climbs, C_B feeds more and more bypass current i_B to v_O. v_O grows with i_B when the i_B overcomes the i_C that v_{IN} and v_O set:

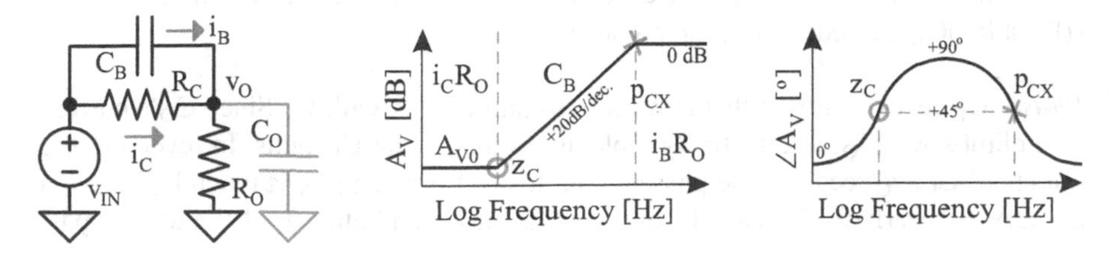

Fig. 5.17 Bypass capacitor

$$i_B = \frac{v_{IN} - v_O}{Z_B} = (v_{IN} - v_O)sC_B\bigg|_{f_O \geq \frac{1}{2\pi R_C C_B} \equiv z_C} \geq i_C = \frac{v_{IN} - v_O}{R_C}. \quad (5.42)$$

Since the gain scales with f_O in i_B's $(v_{IN} - v_O)sC_B$, A_V rises 20 dB when f_O increases a decade over this z_C.

More simply stated, the gain to v_O climbs when C_B shunts R_C:

$$Z_B = \frac{1}{sC_B}\bigg|_{f_O \geq \frac{1}{2\pi R_C C_B} \equiv z_C} \leq R_C. \quad (5.43)$$

C_B bypasses R_C when this happens. The rise in A_V that results is the effect of the z_C that C_B induces as C_B shunts.

Z_B falls to such an extent at very high f_O that C_B drops a negligible fraction of v_{IN}, letting R_O drop the rest. This means that i_B alone feeds R_O and A_V reaches and flattens near a high-f_O gain $A_{V(HF)}$ that approaches 1 or 0 dB. This happens when C_B shorts the parallel resistance R_O and R_C into v_{IN} set:

$$Z_B = \frac{1}{sC_B}\bigg|_{f_O \geq \frac{1}{2\pi (R_O \| R_C) C_B} \equiv p_{CX}} \leq R_O \| R_C. \quad (5.44)$$

Since shunting effects vanish when capacitors short, C_B shunts past a z_C that p_{CX} reverses when C_B shorts. In practice, the *output capacitor* C_O helps C_B short R_C and R_O. From the perspective of C_O, p_{CX} is the shunting pole that C_O produces with the help of C_B.

The overall gain to v_O is the voltage-divided fraction of v_{IN} that R_C's i_C and C_B's i_B feed R_O:

$$\begin{aligned} A_V \equiv \frac{v_O}{v_{IN}} &= \frac{(i_C + i_B)R_O}{v_{IN}} \\ &= \frac{R_O}{(Z_B \| R_C) + R_O} = \left(\frac{R_O}{R_C + R_O}\right)\left[\frac{1 + sR_C C_B s}{1 + s(R_C \| R_O)C_B}\right] \\ &= A_{V0}\left(\frac{1 + s/2\pi z_C}{1 + s/2\pi p_{CX}}\right). \end{aligned} \quad (5.45)$$

A_V is A_{V0} when f_O is very low and $A_{V0}p_{CX}/z_C$ or $A_{V(HF)}$'s one when f_O is very high. A_V climbs with f_O when $sR_C C_B$ exceeds 1 or f_O surpasses z_C and flattens when $s(R_C \| R_O)C_B$ exceeds 1 or f_O surpasses p_{CX}.

Phase C_B feeds a current that reinforces i_C and climbs with f_O. Since C_O's current also climbs with f_O, i_B effectively replenishes the i_C that C_O pulls. In reversing the effects of C_O, z_C recovers the phase a pole loses. So when p_{CX} is much higher than z_C, $\angle A_V$ is nearly 0° a decade below z_C, +45° at z_C, and almost +90° a decade past z_C:

$$\angle A_V = +\tan^{-1}\left(\frac{f_O}{z_C}\right) - \tan^{-1}\left(\frac{f_O}{p_{CX}}\right). \quad (5.46)$$

$\angle A_V$ then loses 45° at p_{CX} and almost all 90° a decade past p_{CX}.

B. Current-Limit Resistor

Resistors that current-limit capacitors remove their effects. R_C and R_O in Fig. 5.17, for example, limit how much current C_B conducts. And when C_B shorts, the effects of C_B on the v_O that R_C feeds R_O fade. This is why R_O and R_C reverse z_{CO} with p_{CX}: because R_C and R_O limit the current C_B feeds.

Adding another resistor further current-limits C_B. Consider R_I in Fig. 5.18. Z_B is so high at low f_O that A_{V0} is the voltage-divided fraction of v_{IN} that R_C's i_C feeds R_O. A_V, however, rises when C_B and R_I shunt R_C, when happens when Z_B falls below R_I and R_C. The f_O where this happens is the z_C that C_B induces as C_B shunts:

$$Z_B = \left.\frac{1}{sC_B}\right|_{f_O \geq \frac{1}{2\pi(R_I+R_C)C_B} \equiv z_C} \leq R_I + R_C. \quad (5.47)$$

But when C_B shorts the parallel resistance R_I and R_O with R_C into v_{IN} set, the effects of z_C reverse:

$$Z_B = \left.\frac{1}{sC_B}\right|_{f_O \geq \frac{1}{2\pi[R_I+(R_O\|R_C)]C_B} \equiv p_{CX}} \leq R_I + (R_O\|R_C). \quad (5.48)$$

And A_V flattens to the voltage-divided v_{IN} fraction that R_C and R_I feed R_O, which the resistive current i_R that R_C, R_I, and R_O establish ultimately determines:

$$A_{V(HF)} \approx \frac{i_R R_O}{v_{IN}} = \frac{R_O}{(R_C\|R_I) + R_O}. \quad (5.49)$$

R_I and R_O with R_C remove z_C because they limit the current C_B feeds. p_{CX} and $A_{V(HF)}$ are lower with R_I because C_B shunts a higher resistance, and after C_B shorts, R_I and R_C drop a part of v_{IN} that R_O does not drop. In short, z_C appears when C_B and R_I shunt R_C and p_{CX} reverses z_C when C_B shorts.

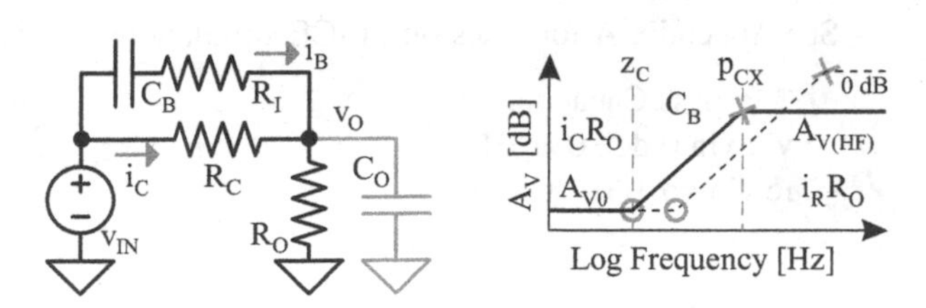

Fig. 5.18 Current-limited bypass capacitor

The overall gain to v_O is the v_{IN} fraction R_C and C_B with R_I feed R_O:

$$\begin{aligned} A_V \equiv \frac{v_O}{v_{IN}} &= \frac{R_O}{[(Z_B + R_I)||R_C] + R_O} \\ &= \left(\frac{R_O}{R_C + R_O}\right)\left\{\frac{1 + s(R_I + R_C)C_B}{1 + s[R_I + (R_C||R_O)]C_B}\right\} \\ &= A_{V0}\left(\frac{1 + s/2\pi z_C}{1 + s/2\pi p_{CX}}\right). \end{aligned} \tag{5.50}$$

A_V is A_{V0} when f_O is very low and $A_{V0}p_{CX}/z_C$ or $A_{V(HF)}$ when f_O is very high. A_V climbs with f_O when $s(R_I + R_C)C_B$ exceeds 1 or f_O surpasses z_C and flattens when $s[R_I + (R_C \parallel R_O)]C_B$ exceeds 1 or f_O surpasses p_{CX}.

Example 7: Determine A_{V0}, $A_{V(HF)}$, and related poles and zeros when R_C is 1 kΩ, C_B is 5 pF, R_I is 10 kΩ, and R_O is 1 kΩ.

Solution:

$$A_{V0} \approx \frac{R_O}{R_C + R_O} = \frac{1k}{100k + 1k} = 9.9\ mV/V = -40\ dB$$

$$z_C = \frac{1}{2\pi(R_I + R_C)C_B} = \frac{1}{2\pi(10k + 100k)(5p)} = 290\ kHz$$

$$p_{CX} = \frac{1}{2\pi[R_I + (R_C||R_O)]C_B} = \frac{1}{2\pi[10k + (100k||1k)](5p)} = 2.9\ MHz$$

$$A_{V(HF)} \approx \frac{R_O}{(R_C||R_I) + R_O} = \frac{1k}{(100k||10k) + 1k} = 99\ mV/V = -20\ dB$$

Explore with SPICE:
See Appendix A for notes on SPICE simulations.

```
* Bypass Capacitor
vin vin 0 dc=0 ac=1
rc vin vo 100k
```

(continued)

```
cb vin vc 5p
ri vc vo 10k
ro vo 0 1k
.ac dec 1000 10k 100e6
.end
```

Tip: Plot v(vo) in dB to inspect A_V.

5.3.2 Bypassed Amplifier

A. In-Phase Capacitor

Capacitors that bypass amplifiers also add energy to their outputs. Since adding this energy increases gain, these capacitors also induce zeros. Consider the non-inverting amplifier that the forward model in Fig. 5.19 emulates and the bypass capacitor C_B that bypasses it. The amplifier is non-inverting because a rise in v_{IN} increases the i_{GO} that feeds R_O, which raises v_O.

C_B's Z_B is so high at low f_O that A_{G0}'s i_{G0} into R_O sets v_O at low f_O. As Z_B falls with f_O, C_B feeds more and more i_B to v_O. v_O grows with i_B when i_B surpasses i_{G0}. So A_{G0} sets v_O at low f_O and C_B raises v_O at high f_O.

To determine the f_O when C_B raises v_O, consider the forward model that incorporates C_B into A_G in Fig. 5.20. Since A_G accounts for A_{G0} and C_B, i_G's i_{G0} and i_B are the currents that result when v_O shorts. This means i_B is an Ohmic $1/sC_B$ translation of C_B's voltage $v_{IN} - v_O$ when v_O is zero, i_{G0} is an A_{G0} translation of v_{IN}, and i_B surpasses i_{G0} when C_B's conductance sC_B overcomes A_{G0}:

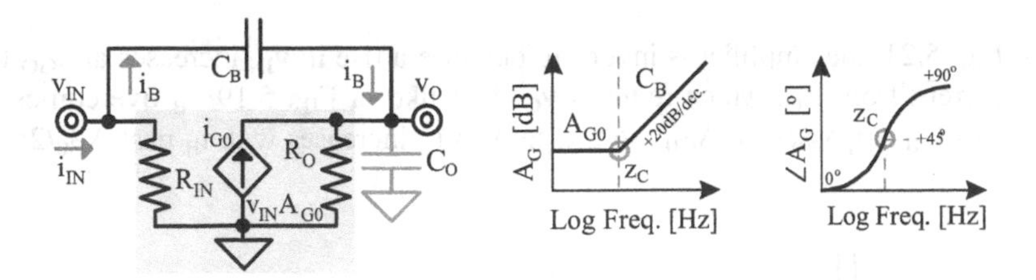

Fig. 5.19 In-phase bypass capacitor across amplifier

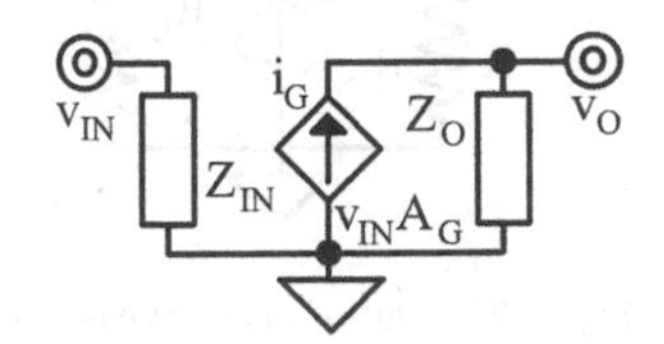

Fig. 5.20 Forward current-sourced model of bypassed amplifier

$$i_B = \frac{v_{IN} - v_O}{Z_B}\Big|_{v_O=0} = v_{IN}sC_B\Big|_{f_O \geq \frac{A_{G0}}{2\pi C_B} \equiv z_C} \geq i_{G0} = v_{IN}A_{G0}. \tag{5.51}$$

A_G rises 20 dB per decade past this z_C because i_B scales with f_O in $v_{IN}sC_B$.

Since i_B and i_{G0} rise and fall with v_{IN}, i_B reinforces i_{G0}. So like zeros in RLC networks, the i_B that sets z_C does not invert the polarity of v_O. These are *in-phase zeros* because low- and high-f_O gains are in-phase. z_C is in-phase in Fig. 5.19 because i_B is in-phase with the i_{G0} that C_B bypasses.

The overall two-port gain A_G to i_G's i_{G0} and i_B is the transconductance and Ohmic v_{IN} translations that A_{G0} and $1/sC_B$ set when v_O is zero:

$$\begin{aligned} A_G &\equiv \frac{i_G}{v_{IN}}\Big|_{v_O=0} = \frac{i_{G0} + i_B}{v_{IN}}\Big|_{v_O=0} \\ &= A_{G0} + \frac{1}{Z_B} = A_{G0}\left(1 + \frac{sC_B}{A_{G0}}\right) = A_{G0}\left(1 + \frac{s}{2\pi z_C}\right). \end{aligned} \tag{5.52}$$

A_G climbs past z_C, which is in-phase because z_C does not invert A_G at high f_O. In-phase zeros are *left-half-plane zeros* because the "s" that zeroes their gain factors ($1 + sC_B/A_{G0}$ in this case) is negative ($-C_B/A_G$ in this case). So "s" is left of the zero axis that splits the "s" plane.

C_B feeds a current that climbs with f_O. Since C_O's current also climbs with f_O, i_B effectively replenishes the i_C that C_O pulls. This means that z_C cancels the shunting effects of C_O. In other words, z_C recovers the phase that a pole loses. So $\angle A_G$ is nearly 0° a decade below z_C, +45° at z_C, and almost +90° a decade past z_C:

$$\angle A_G = +\tan^{-1}\left(\frac{f_O}{z_C}\right). \tag{5.53}$$

B. Out-of-Phase Capacitor

In Fig. 5.21, the amplifier is inverting because a rise in v_{IN} increases an i_{GO} that pulls current from R_O, which reduces v_O. But like in Fig. 5.19, i_B overcomes i_{G0} when $v_{IN}sC_B$ surpasses $v_{IN}A_{G0}$. So the gain to i_G increases with i_B past $A_{G0}/2\pi C_B$.

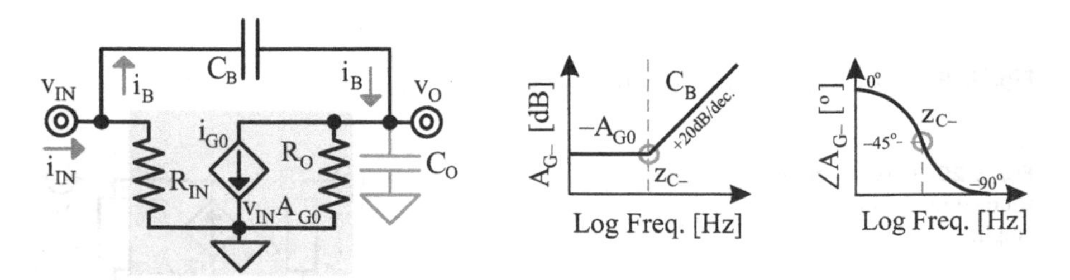

Fig. 5.21 Out-of-phase bypass capacitor across amplifier

The effects of i_B and i_{G0} on v_O in Fig. 5.21, however, oppose because C_B feeds i_B and A_{G0} pulls i_{G0}. This means that i_B and i_{G0} are out-of-phase, and the resulting zero inverts 180° while at the same time recovering the 90° that C_O loses. So the phase shift of the resulting gain A_{G-} is nearly 0° a decade below z_{C-}, −90° + 45° or −45° at z_{C-}, and almost −180° + 90° or −90° a decade above z_{C-}:

$$\angle A_{G-} = -\tan^{-1}\left(\frac{f_O}{z_{C-}}\right). \tag{5.54}$$

The overall gain to i_G's i_B and i_{G0} is the Ohmic and transconductance v_{IN} translations that $1/sC_B$ and A_{G0} set when v_O is zero:

$$\begin{aligned} A_{G-} &\equiv \left.\frac{i_G}{v_{IN}}\right|_{v_O=0} \equiv \left.\frac{i_B - i_{G0}}{v_{IN}}\right|_{v_O=0} \\ &= \frac{1}{Z_B} - A_{G0} = -A_{G0}\left(1 - \frac{sC_B}{A_{G0}}\right) = -A_{G0}\left(1 - \frac{s}{2\pi z_{C-}}\right). \end{aligned} \tag{5.55}$$

Although $|A_{G-}|$ still climbs with f_O after f_O surpasses z_{C-}, A_{G-} inverts at high f_O: from $-A_{G0}$ to $+A_{G0}f_O/z_{C-}$. z_{C-} is out-of-phase because z_{C-} inverts the polarity of A_{G-}. *Out-of-phase zeros* like this z_{C-} are *right-half-plane zeros* because the "s" that zeroes their gain factors ($1 - sC_B/A_{G0}$ in this case) is positive ($+C_B/A_{G0}$ in this case). So "s" is right of the zero axis that splits the "s" plane.

C. Other Effects

Since C_B also steers i_B away from R_{IN}, Z_{IN} in Fig. 5.20 incorporates a shunt capacitor that models the effects of C_B and A_G on v_{IN}. Z_O similarly includes a shunt capacitor that models the effects of C_B on v_O when disabling A_G (with a grounded v_{IN}). So in addition to z_C, C_B adds shunt capacitances that help establish input and output poles p_{IN} and p_O.

Current-limiting C_B with a series resistor removes the effects of C_B at higher f_O, when C_B shorts. So R_I in Fig. 5.22 induces a pole p_{CX} and two zeros z_{INX} and z_{OX} that reverse the effects of p_{IN} on v_{IN}, z_C on A_G, and p_O on v_O. Other resistances and capacitances that connect to v_{IN} and v_O can not only alter these but also produce additional poles and zeros.

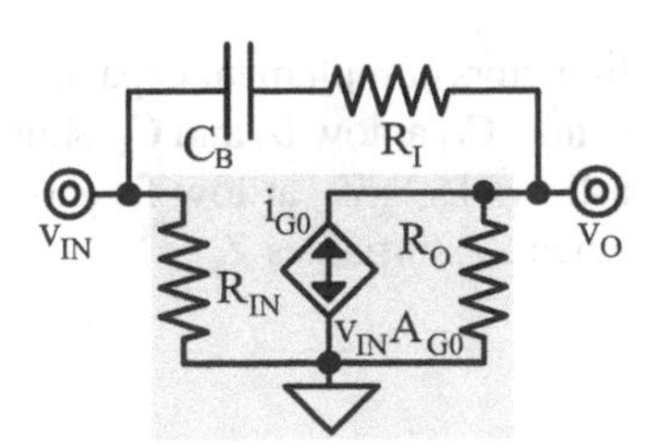

Fig. 5.22 Current-limited bypass capacitor across amplifier

Example 8: Determine A_{V0} and z_C across an inverting amplifier when A_{G0} is 250 μS, R_O is 100 Ω, and C_B is 10 pF.

Solution:

$$A_{V0} \approx -A_{G0}R_O = -(250\mu)(100) = -25\ \text{mV/V} = -32\ \text{dB}$$

$$z_{C-} = \frac{A_{G0}}{2\pi C_B} = \frac{250\mu}{2\pi(10p)} = 3.9\ \text{MHz}$$

Explore with SPICE:
See Appendix A for notes on SPICE simulations.

```
* Bypassed Amplifier
vin vin 0 dc=0 ac=1
cb vin vo 10p
g0 vo 0 vin 0 250u
ro vo 0 100
.ac dec 1000 100k 10g
.end
```

Tip: Plot v(vo) in dB to inspect A_V.

5.4 LC Circuits

5.4.1 Current-Sourced LC

A. Parallel Impedance

Inductors complement capacitors. Consider the parallel combination in Fig. 5.23. L_X shunts C_O at low f_O and C_O shunts L_X at high f_O. So their combined impedance $Z_{L||C}$ follows L_X's Z_L at low f_O and C_O's Z_C at high f_O. $Z_{L||C}$ transitions from Z_L to Z_C when Z_L surpasses Z_C:

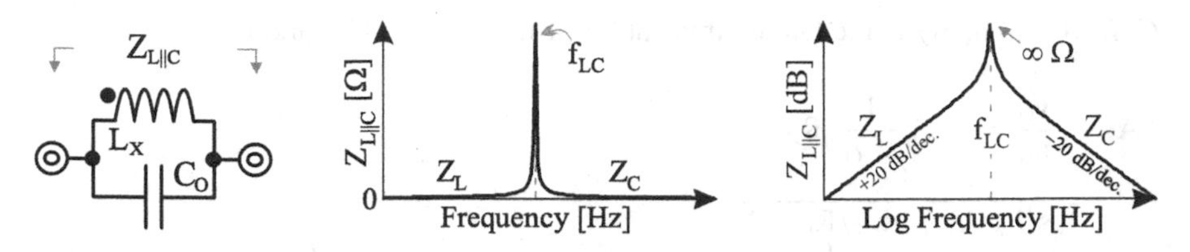

Fig. 5.23 Parallel LC impedance

Fig. 5.24 Current-sourced LC

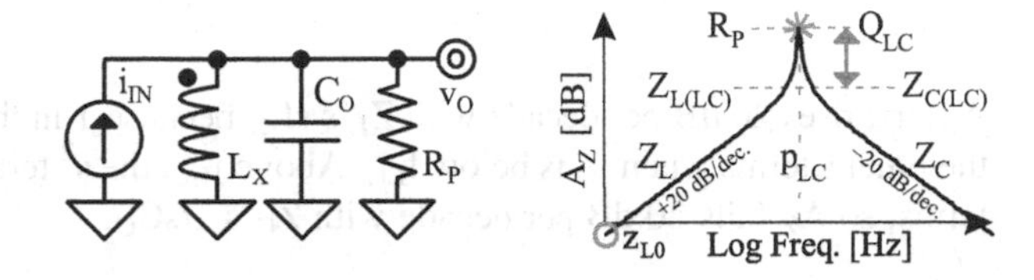

$$Z_L = sL_X\Big|_{f_O \geq \frac{1}{2\pi\sqrt{L_X C_O}} = f_{LC}} \geq Z_C = \frac{1}{sC_O}. \tag{5.56}$$

This happens past their *transitional LC (resonant) frequency* f_{LC}.

Since i^2 is -1, $Z_{L||C}$ skyrockets at this same f_{LC}:

$$Z_{L||C} = sL_X || \frac{1}{sC_O} = \frac{sL_X}{1 + s^2 L_X C_O}\Bigg|_{f_{LC}} = \frac{(i2\pi f_{LC})L_X}{1-1} \to \infty. \tag{5.57}$$

This is because Z_L and Z_C are equal complements at f_{LC}. So $Z_{L||C}$ climbs with Z_L 20 dB per decade below f_{LC}, opens (peaks towards infinity) at f_{LC}, and drops with Z_C 20 dB per decade above f_{LC}.

B. **Ohmic Response**

i_{IN} in Fig. 5.24 feeds an LC structure that includes a *parallel resistance* R_P. L_X shunts R_P at low f_O and C_O shunts R_P at high f_O. The roles reverse at f_{LC}, when $Z_{L||C}$ is so high (i.e., open circuit) that R_P shunts L_X and C_O. So the v_O that i_{IN} sets rises with Z_L below f_{LC}, peaks with R_P at f_{LC}, and falls with Z_C past f_{LC}:

$$v_{O(PK)} = i_{IN}(Z_L || Z_C || R_P)\Big|_{f_{LC}} = i_{IN} R_P. \tag{5.58}$$

The Ohmic gain A_Z to v_O follows the combined impedance. Z_L raises A_Z as L_X opens from z_{L0}, p_L reverses the effects of z_{L0} when L_X disconnects, and Z_C reduces A_Z past p_C as C_O shunts. Since L_X and C_O interact at f_{LC}, reversing z_{L0} and reducing A_Z coincide at f_{LC}. So p_L and p_C appear at f_{LC} in the form of an *LC double pole* p_{LC}.

Gain Overall, A_Z is the Ohmic translation of i_{IN} into L_X, C_O, and R_P:

$$\begin{aligned} A_Z &\equiv \frac{v_O}{i_{IN}} = sL_X \| \frac{1}{sC_O} \| R_P \\ &= \frac{sL_X}{s^2 L_X C_O + sL_X/R_P + 1} \\ &= \frac{sL_X}{\left(\frac{s}{2\pi f_{LC}}\right)^2 + \left(\frac{s}{2\pi f_{LC}}\right)\left(\frac{f_{LC}}{f_{LP}}\right) + 1} \equiv \frac{Z_L}{\left(\frac{s}{2\pi f_{LC}}\right)^2 + \left(\frac{s}{2\pi f_{LC}}\right)\left(\frac{1}{Q_{LC}}\right) + 1}. \end{aligned} \tag{5.59}$$

A_Z first rises 20 dB per decade with Z_L's sL_X because 1 in the denominator swamps the other s terms when f_O is below f_{LC}. Above f_{LC}, the s^2 term overwhelms the other terms, so A_Z falls 20 dB per decade with Z_C's $1/sC_O$.

Peak A_Z reaches R_P at f_{LC} because L_X and C_O open at f_{LC} (i. e., $s^2L_XC_O$ is i^2 or -1):

$$A_{Z(LC)} = \left. \frac{sL_X}{-1 + sL_X/R_P + 1} \right|_{f_{LC}} = Z_{L(LC)} Q_{LC} = R_P. \tag{5.60}$$

This $A_{Z(LC)}$ is Q_{LC} above Z_L's projection $Z_{L(LC)}$ to f_{LC} when A_Z rises and falls with Z_L and Z_C. Stated in another way, A_Z peaks when L_X and C_O interact at f_{LC}.

For this, R_P should not voltage-limit L_X (past f_{LP}) below f_{LC}. This is like saying C_O shunts R_P (past f_{CP}) below f_{LC}. This is because R_P is greater than Z_L (and Z_C) at f_{LC} when Z_L cannot overcome R_P. This way, when L_X overcomes R_P above f_{LC}, A_Z follows Z_L to f_{LC}.

A_Z peaks over $Z_{L(LC)}$ when f_{CP} precedes f_{LC} and f_{LP} follows f_{LC}. In this light, f_{LP}/f_{LC} and f_{LC}/f_{CP} in Q_{LC} indicate the *quality* and magnitude of the *peak*. A_Z peaks over Z_L's projection when R_P raises this Q_{LC} over one:

$$Q_{LC} \equiv \frac{f_{LP}}{f_{LC}} = \frac{f_{LC}}{f_{CP}} = R_P \sqrt{\frac{C_O}{L_X}}. \tag{5.61}$$

But if Z_L overcomes R_P below f_{LC}, f_{LP} sets a p_L that precedes f_{LC}. Since R_P is less than Z_L (and Z_C) at f_{LC} this way, C_O shunts R_P past a p_C that f_{CP} sets above f_{LC}. p_L and p_C split away from f_{LC} this way because R_P keeps L_X and C_O from interacting at f_{LC}.

C. Resistive Effects

In practice, electrical components incorporate series resistance. L_X and C_O in Fig. 5.24, for example, include the *inductor* and *capacitor resistances* R_L and R_C shown in Fig. 5.25. Circuits that connect to v_O also add *load resistance* R_{LD}.

Peaked R_L alters A_Z's low-f_O gain. Since L_X shorts and C_O opens at low f_O, the low-f_O gain to v_O is an Ohmic R_L and R_{LD} translation of i_{IN}:

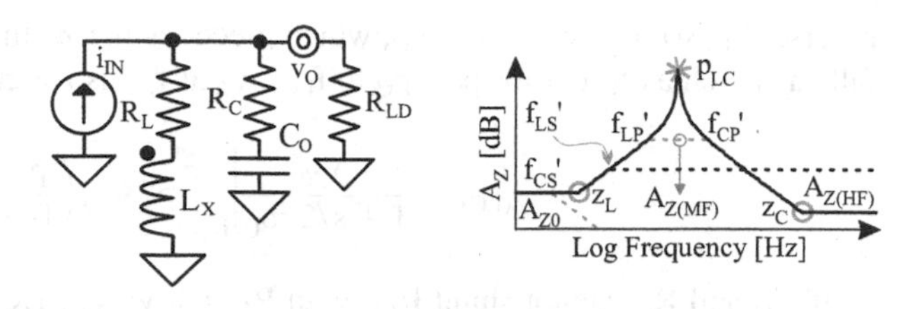

Fig. 5.25 Loaded current-sourced LC with parasitic resistances

$$A_{Z0} \approx R_L || R_{LD}. \tag{5.62}$$

A_Z climbs past the z_L that f_L sets when sL_X overcomes R_L:

$$Z_L = sL_X \Big|_{f_O \geq \frac{R_L}{2\pi L_X} \equiv f_L} \geq R_L. \tag{5.63}$$

In power supplies, this f_L is usually lower than f_{LC} because R_L is very low.

R_C alters A_Z's high-f_O response. A_Z falls with Z_C past f_{LC} until C_O shorts, which happens when R_C limits the current C_O can sink:

$$Z_C = \frac{1}{sC_O}\Bigg|_{f_O \geq \frac{1}{2\pi R_C C_O} \equiv z_C} \leq R_C. \tag{5.64}$$

This z_C in power supplies is typically above f_{LC} because R_C is also very low. After C_O shorts, $A_{Z(HF)}$ becomes an Ohmic R_C and R_{LD} translation of i_{IN}:

$$A_{Z(HF)} \approx R_C || R_{LD}. \tag{5.65}$$

A_Z peaks when L_X and C_O interact at f_{LC}. For this, L_X and C_O should eclipse the effects of R_L, R_C, and R_{LD}. So L_X should overcome R_L below f_{LC}, C_O and R_C should shunt R_{LD} below f_{LC}, and C_O should short above f_{LC}. This way, R_L and R_{LD} fade below f_{LC} and the effects of R_C surface above f_{LC}, so Q_{LC} exceeds one.

Unpeaked Scenarios If L_X cannot overcome R_L below f_{LC}, C_O and R_C shunt R_{LD} with R_L first:

$$Z_C = \frac{1}{sC_O}\Bigg|_{f_O \geq \frac{1}{2\pi[R_C + (R_{LD}||R_L)]C_O} \equiv f_{CS}} \leq R_C + (R_{LD}||R_L). \tag{5.66}$$

This f_{CS} sets a p_C that is below f_{LC} because R_L is higher than Z_L at f_{LC} (i.e., L_X is still a short below f_{LC}). In this scenario, C_O keeps L_X from inducing z_L and the p_L that reverses z_L, so p_{LC} reduces to p_C, which precedes the z_C that reverses p_C. Since A_Z falls as much as f_O climbs past p_C to f_{LC}, A_Z at f_{LC} is roughly $A_{Z0}(p_C/f_{LC})$:

$$A_{Z(LC)} = \left.\frac{A_{Z0}}{1 + s/2\pi p_C}\right|_{f_{LC}} \approx A_{Z0}\left(\frac{p_C}{f_{LC}}\right). \tag{5.67}$$

If C_O and R_C cannot shunt R_{LD} with R_L below f_{LC}, L_X overcomes R_L and R_{LD} past f_{LP}:

$$Z_L = sL_X|_{f_O \geq \frac{R_L + R_{LD}}{2\pi L_X} \equiv f_{LP}} \leq R_L + R_{LD}. \tag{5.68}$$

This f_{LP} sets a p_L that is below f_{LC} because R_{LD} is less than Z_C at f_{LC} (i.e., C_O is still open below f_{LC}). Since L_X opens and R_{LD} is less than Z_L at f_{LC}, C_O and R_C shunt R_{LD} without R_L above f_{LC}:

$$Z_C = \left.\frac{1}{sC_O}\right|_{f_O \geq \frac{1}{2\pi(R_C + R_{LD})C_O} \equiv f_{CP}} \leq R_C + R_{LD}. \tag{5.69}$$

So f_{LP} sets a p_L that reverses z_L and precedes f_{LC} and f_{CP} a p_C that follows f_{LC} and precedes the z_C that reverses p_C. Since A_Z rises as much as f_O climbs past z_L and f_{LC} is above p_L, $A_{Z(LC)}$ and the mid-f_O gain $A_{Z(MF)}$ are $A_{Z0}(p_L/z_L)$:

$$A_{Z(MF)} = A_{Z(LC)} = A_{Z0}\left.\left(\frac{1 + s/2\pi z_L}{1 + s/2\pi p_L}\right)\right|_{f_{LC}} \approx A_{Z0}\left(\frac{p_L}{z_L}\right). \tag{5.70}$$

If C_O shunts and shorts (with respect to R_C) below f_{LC}, L_X overcomes R_L and R_{LD} with R_C past the p_L that f_{LS} sets:

$$Z_L = sL_X\Big|_{f_O \geq \frac{R_L + (R_{LD}||R_C)}{2\pi L_X} \equiv f_{LS}} \leq R_L + (R_{LD}||R_C). \tag{5.71}$$

In this case, R_C keeps C_O from inducing p_C and the z_C that reverses p_C, so p_{LC} reduces to p_L, which follows the z_L that p_L reverses. When f_{LC} is above p_L, $A_{Z(LC)}$ matches the $A_{Z(HF)}$ that R_C and R_{LD} set when C_O shorts. If f_{LC} is below p_L, $A_{Z(LC)}$ climbs f_{LC}/z_L over A_{Z0} when f_O rises from z_L to f_{LC}.

Example 9: Determine A_{Z0}, A_Z at f_{LC}, $A_{Z(HF)}$, and related poles and zeros when L_X is 10 μH, R_L is 50 mΩ, C_O is 5 μF, R_C is 10 mΩ, and R_{LD} is 100 Ω.

Solution:

$$A_{Z0} \approx R_L \parallel R_{LD} = 50m \parallel 100 = 50\ mV/A = -26\ dB$$

$$z_L = f_L = \frac{R_L}{2\pi L_X} = \frac{50m}{2\pi(10\mu)} = 800\ Hz$$

$$f_{CP} = \frac{1}{2\pi(R_C + R_{LD})C_O} = \frac{1}{2\pi(10m + 100)(5\mu)} = 320\ Hz$$

$$f_{LC} = \frac{1}{2\pi\sqrt{L_X C_O}} = \frac{1}{2\pi\sqrt{(10\mu)(5\mu)}} = 22\ kHz$$

$$z_C = \frac{1}{2\pi R_C C_O} = \frac{1}{2\pi(10m)(5\mu)} = 3.2\ MHz$$

$$f_{CP} < z_L < f_{LC} < z_C \quad \therefore \quad z_L < p_L = p_C = p_{LC} = f_{LC} < z_C$$

$$A_{Z(LC)} = R_P \approx R_{LD} = 100\ V/A = 40\ dB$$

$$A_{Z(HF)} \approx R_C \parallel R_{LD} = 10m \parallel 100 = 10\ mV/A = -40\ dB$$

Note: $A_{Z(LC)}$ is lower than 100 V/A because R_L and R_C reduce the R_P that R_{LD} establishes at f_{LC}.

Example 10: Determine A_{Z0}, A_Z at f_{LC}, $A_{Z(HF)}$, and related poles and zeros for Example 9 when R_L is 10 Ω.

Solution:

$$f_{CP} = 320\ Hz, f_{LC} = 22\ kHz, z_C = 3.2\ MHz,$$
$$\text{and } A_{Z(HF)} \approx 10\ m\Omega \text{ from Example 9}$$

$$A_{Z0} = R_L \parallel R_{LD} = 10 \parallel 100 = 9.1\ V/A = 19\ dB$$

$$z_L = \frac{R_L}{2\pi L_X} = \frac{10}{2\pi(10\mu)} = 160 \text{ kHz}$$

$$f_{LC} < z_L < z_C \quad \therefore \quad p_C \approx f_C < f_{LC} < z_C$$

$$p_C \approx f_{CS} = \frac{1}{2\pi[R_C + (R_{LD}||R_L)]C_O} = \frac{1}{2\pi[10m + (100||10)](5\mu)} = 3.5 \text{ KHz}$$

$$A_{Z(LC)} \approx A_{Z0}\left(\frac{p_C}{f_{LC}}\right) = (9.1)\left(\frac{3.5k}{22k}\right) = 1.4 \text{ V/A} = 2.9 \text{ dB}$$

Note: A_V does not peak at f_{LC} because R_L is so high that L_X cannot overcome R_L and R_{LD} before Z_L can surpass Z_C.

Example 11: Determine the R_{LD} needed to suppress A_V's peak at f_{LC} in Example 9.

Solution:

$$z_L = 800 \text{ Hz}, f_{LC} = 22 \text{ kHz}, \text{and } z_C = 3.2 \text{ MHz from Example 9}$$

$$\text{Peak fades when} \quad p_L \approx f_{LP} \le f_{LC} \quad \text{or} \quad f_{LC} \le p_C \approx f_{CP}$$

$$Q_{LC} = \frac{f_{LP}}{f_{LC}} \approx \frac{(R_L + R_{LD})/2\pi L_X}{f_{LC}} = \frac{(50m + R_{LD})/2\pi(10\mu)}{22k} \le 1$$

$$\therefore \quad R_{LD} \le 1.3\ \Omega$$

$$z_L < p_L \approx f_{LP} \le f_{LC} \le p_C \approx f_{CP} < z_C$$

Note: This R_{LD} is fairly low for a 10-μH, 5-μF power supply.

Example 12: Determine A_{Z0}, A_Z at f_{LC}, $A_{Z(HF)}$, and related poles and zeros for Example 9 when R_C is 10 Ω.

Solution:

$$A_{Z0} \approx 50\ \text{mV/A}, z_L = 800\ \text{Hz, and } f_{LC} = 22\ \text{kHz from Example 9}$$

$$f_{CP} = \frac{1}{2\pi(R_C + R_{LD})C_O} = \frac{1}{2\pi(10 + 100)(5\mu)} = 290\ \text{Hz}$$

$$z_C = \frac{1}{2\pi R_C C_O} = \frac{1}{2\pi(10)(5\mu)} = 3.2\ \text{kHz}$$

$$z_L < f_{CP} < z_C < f_{LC} \quad \therefore \quad z_L < f_{LC} < p_L \approx f_{LS}$$

$$p_L \approx f_{LS} = \frac{R_L + (R_{LD} \| R_C)}{2\pi L_X} = \frac{50\text{m} + (100\|10)}{2\pi(10\mu)} = 150\ \text{kHz}$$

$$A_{Z(LC)} \approx A_{Z0}\left(\frac{f_{LC}}{z_L}\right) = (50\text{m})\left(\frac{22\text{k}}{800}\right) = 1.4\ \text{V/A} = 2.9\ \text{dB}$$

$$A_{Z(HF)} \approx R_C \parallel R_{LD} = 10 \parallel 100 = 9.1\ \text{V/A} = 19\ \text{dB}$$

Explore with SPICE:
See Appendix A for notes on SPICE simulations.

```
* Current-Sourced LC
iin 0 vo dc=0 ac=1
rl vo vl 50m
*rl vo vl 10
lx vl 0 10u
rc vo vc 10m
*rc vo vc 10
co vc 0 5u
rld vo 0 100
*rld vo 0 1.3
.ac dec 1000 10 100e6
.end
```

Tip: Plot v(vo) in dB and comment/un-comment rl, rc, and rld to see their effects on A_Z.

5.4.2 Voltage-Sourced LC

A. **Series Impedance**

Series LC structures complement their parallel counterparts. Consider L_X and C_O in Fig. 5.26. L_X shorts when C_O opens at low f_O and L_X opens when C_O shorts at high f_O. So their combined impedance Z_{LC} is mostly C_O's Z_C at low f_O and L_X's Z_L at high f_O. Z_{LC} transitions from Z_C to Z_L when Z_L surpasses Z_C, which happens past their f_{LC}.

Since i^2 is -1, Z_{LC} reaches zero at this same f_{LC}:

$$Z_{LC} = sL_X + \frac{1}{sC_O} = \left.\frac{s^2L_XC_O + 1}{sC_O}\right|_{f_{LC}} = \frac{1-1}{(i2\pi f_{LC})C_O} = 0. \tag{5.72}$$

This is because Z_L and Z_C are equal complements at f_{LC}. So Z_{LC} falls with Z_C 20 dB per decade below f_{LC}, shorts at f_{LC}, and climbs with Z_L 20 dB per decade above f_{LC}, which is the opposite of what happens in $Z_{L||C}$.

B. **Ohmic Response**

v_{IN} in Fig. 5.27 feeds a series LC structure that includes *series resistance* R_S. Z_C swamps R_S at low f_O and Z_L swamps R_S at high f_O. The roles reverse at f_{LC}, because R_S swamps Z_{LC}'s zero impedance. So v_{IN} supplies an i_{IN} that climbs with $1/Z_C$ below f_{LC}, peaks with $1/R_S$ at f_{LC}, and falls with $1/Z_L$ past f_{LC}:

$$i_{IN(PK)} = \left.\frac{v_{IN}}{Z_L + R_S + Z_C}\right|_{f_{LC}} = \frac{v_{IN}}{R_S}. \tag{5.73}$$

The Ohmic gain translation to i_{IN} follows the combined conductance of the structure. sC_O in $1/Z_C$ raises A_G as C_O shunts (above z_{CO}), p_C reverses z_{CO} when C_O shorts, and $1/sL_X$ in $1/Z_L$ reduces A_G past p_L as L_X opens. Since L_X and C_O interact at f_{LC}, p_C reverses z_{CO} and p_L reduces A_G at f_{LC}. So p_C and p_L coincide at f_{LC} in the form of a double pole p_{LC}.

Gain Overall, A_G is an Ohmic L_X, R_S, and C_O translation of v_{IN}:

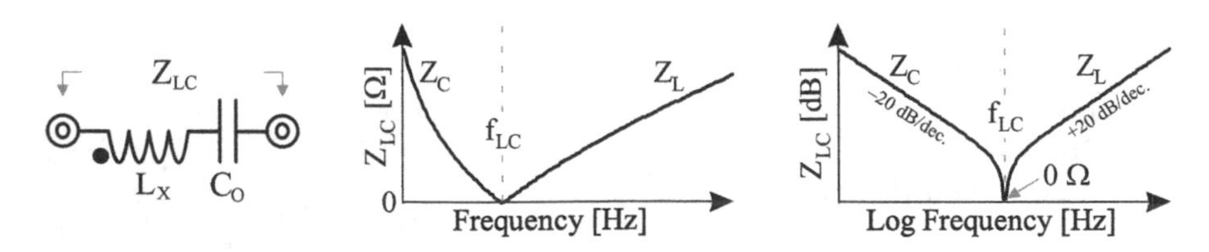

Fig. 5.26 Series LC impedance

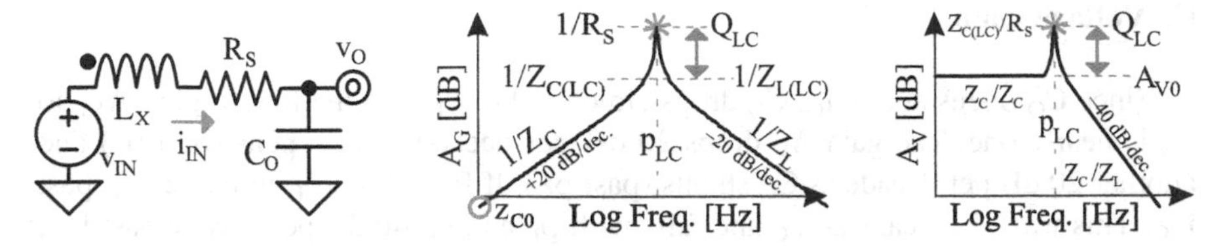

Fig. 5.27 Voltage-sourced LC

$$A_G \equiv \frac{i_{IN}}{v_{IN}} = \frac{1}{sL_X + R_S + 1/sC_O} = \frac{sC_O}{s^2L_XC_O + sR_SC_O + 1}$$
$$\equiv \frac{sC_O}{\left(\frac{s}{2\pi f_{LC}}\right)^2 + \left(\frac{s}{2\pi f_{LC}}\right)\left(\frac{f_{LC}}{f_{CS}}\right) + 1} \tag{5.74}$$
$$\equiv \frac{1/Z_C}{\left(\frac{s}{2\pi f_{LC}}\right)^2 + \left(\frac{s}{2\pi f_{LC}}\right)\left(\frac{1}{Q_{LC}}\right) + 1}.$$

A_G climbs 20 dB per decade with $1/Z_C$'s sC_O because 1 swamps other s terms when f_O is below f_{LC}. Above f_{LC}, the s^2 term overwhelms other terms in the denominator, so A_G falls 20 dB per decade with $1/Z_L$'s $1/sL_X$.

Peak A_G reaches $1/R_S$ at f_{LC} because L_X and C_O short at f_{LC} ($s^2L_XC_O$ is i^2 or -1):

$$A_{G(LC)} = \left.\frac{sC_O}{-1 + sR_SC_O + 1}\right|_{f_{LC}} = \frac{Q_{LC}}{Z_{C(LC)}} = \frac{1}{R_S}. \tag{5.75}$$

$A_{G(LC)}$ is Q_{LC} above $1/Z_C$'s projection $1/Z_{C(LC)}$ to f_{LC} when A_G rises and falls with Z_C and Z_L. In other words, A_G peaks when C_O and L_X interact at f_{LC}.

For this, R_S should not current-limit C_O (past f_{CS}) below f_{LC}. This is like saying L_X surpasses R_S (past f_{LS}) below f_{LC}. This is because R_S is less than Z_C (and z_L) at f_{LC} when C_O cannot short R_S. This way, A_G follows $1/Z_C$ to f_{LC}.

So A_G peaks above $1/Z_{C(LC)}$ when f_{LS} precedes f_{LC} and f_{CS} follows f_{LC}. In this light, f_{CS}/f_{LC} and f_{LC}/f_{LS} in Q_{LC} indicate the quality and magnitude of the peak. Peaking occurs when R_S raises this Q_{LC} over one:

$$Q_{LC} \equiv \frac{f_{CS}}{f_{LC}} = \frac{f_{LC}}{f_{LS}} = \frac{1}{R_S}\sqrt{\frac{L_X}{C_O}}. \tag{5.76}$$

But if R_S current-limits C_O before f_{LC}, f_{CS} sets a p_C that precedes f_{LC}. Since R_S is higher than Z_C (and Z_L) at f_{LC} this way, L_X overcomes R_S past a p_L that f_{LS} sets above f_{LC}. p_C and p_L split away from f_{LC} this way because R_S keeps C_O and L_X from interacting at f_{LC}.

C. Voltage Gain

Since C_O opens at low f_O, C_O drops almost all of v_{IN}. So the low-f_O gain A_{V0} to v_O is nearly one. The gain A_V drops 20 dB per decade as L_X opens (past p_L) and another 20 dB per decade as C_O shunts (past p_C). If R_S is low, Z_L surpasses Z_C past f_{LC}. This way, A_V peaks at f_{LC} and falls with p_L and p_C 40 dB per decade past f_{LC}.

A_V is ultimately the v_O that A_G's i_{IN} into Z_C's $1/sC_O$ establishes:

$$\begin{aligned} A_V &\equiv \frac{v_O}{v_{IN}} = \left(\frac{i_L}{v_{IN}}\right)\left(\frac{v_O}{i_L}\right) = A_G Z_C \\ &= \frac{1/sC_O}{sL_X + R_S + 1/sC_O} \\ &= \frac{1}{s^2 L_X C_O + sR_S C_O + 1} \equiv \frac{A_{V0}}{\left(\frac{s}{2\pi f_{LC}}\right)^2 + \frac{1}{Q_{LC}}\left(\frac{s}{2\pi f_{LC}}\right) + 1}. \end{aligned} \tag{5.77}$$

So all the elements in A_G appear in A_V. Except, the Z_C that turns i_{IN} into v_O cancels the z_{CO} that $1/Z_C$ induces in A_G below f_{LC}. This is why A_V is Z_C/Z_C or one below f_{LC}, peaks Q_{LC} over A_{V0}'s projection (one) to Z_C/R_S at f_{LC}, and falls with Z_C/Z_L past f_{LC}. In short, the basic difference between A_G and A_V is z_{CO}'s presence in A_G.

D. Resistive Effects

Peaked In practice, L_X and C_O incorporate R_L and R_C in Fig. 5.28 and circuits load v_O with R_{LD}. R_L and R_{LD} alter A_G's and A_V's low-f_O gains. Since L_X shorts and C_O opens at low f_O, A_{G0} is the i_{IN} that R_L and R_{LD} draw from v_{IN} and A_{V0} is the v_O that i_{IN} drops across R_{LD}:

$$A_{V0} = A_{G0} R_{LD} \approx \frac{R_{LD}}{R_L + R_{LD}}. \tag{5.78}$$

R_L is typically so low in power supplies that A_{G0} nears $1/R_{LD}$ and A_{V0} is close to unity.

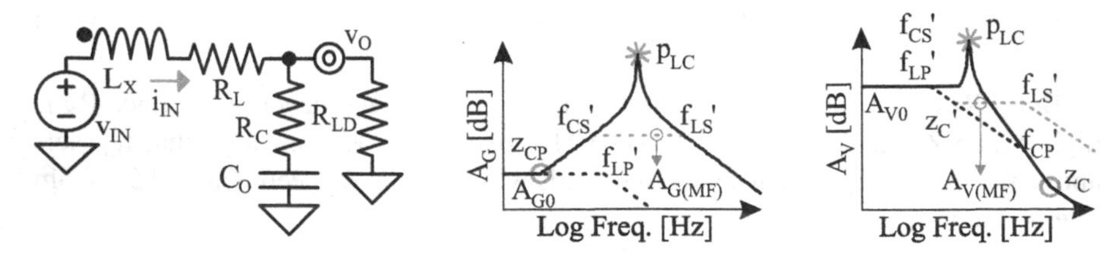

Fig. 5.28 Loaded voltage-sourced LC with parasitic resistances

A_G climbs over A_{G0} past the z_{CP} that f_{CP} sets when C_O and R_C shunt R_{LD}. This effect is absent in A_V because C_O also shunts the impedance A_G drives in A_V. Their effects cancel because Z_C falls as A_G rises.

In A_V, the effects of C_O fade past z_C when R_S limits the current C_O can pull. So A_V's drop eventually reduces to 20 dB per decade. This z_C is higher than f_{LC} when R_C is low.

A_G and A_V peak when L_X and C_O interact at f_{LC}, which is to say, when L_X is the dominant coupling impedance and C_O is the dominant shunting impedance at f_{LC}. For this, L_X and C_O should eclipse the effects of R_L, R_C, and R_{LD}. So L_X should overcome R_L below f_{LC}, C_O and R_C should shunt R_{LD} below f_{LC}, and C_O should short above f_{LC}. This way, R_L and R_{LD} fade below f_{LC} and the effects of R_C surface above f_{LC}, so Q_{LC} exceeds one.

Unpeaked Scenarios If L_X cannot surpass R_L (past f_L) below f_{LC}, C_O and R_C shunt R_{LD} with R_L into v_{IN}. This f_{CS} sets a p_C that is below f_{LC} because R_L is higher than Z_L (and Z_C) at f_{LC} (i.e., L_X is still a short). L_X overcomes this higher R_L when C_O shunts past a p_L that f_{LS} sets above f_{LC}. This way, p_C and p_L split away from f_{LC}.

If C_O and R_C cannot shunt R_{LD} below f_{LC}, L_X overcomes R_L and R_{LD}. This f_{LP} sets a p_L that is below f_{LC} because R_{LD} is less than Z_C (and Z_L) at f_{LC} (i.e., C_O is still open). With L_X open, C_O and R_C in A_V shunt this lower R_{LD} above f_{LC}. So f_{CP} sets a p_C in A_V that follows f_{LC} and precedes the z_C that reverses p_C.

If C_O shunts and shorts below f_{LC}, f_{CS} sets a p_C that A_V's z_C reverses below f_{LC}. R_C is higher than Z_C (and Z_L) at f_{LC} because C_O shorts below f_{LC}. So L_X overcomes R_L and R_{LD} with this higher R_C past an f_{LS} that sets a p_L above f_{LC}. When this R_C is high, C_O shunts and shorts similar resistances. p_C and z_C are close when this happens, so their effects tend to cancel.

When L_X overcomes R_L and R_{LD} before C_O and R_C shunt R_{LD}, A_G climbs over A_{G0} past p_C as much as f_O rises past z_{CP} to p_C, so $A_{G(MF)}$ is roughly $A_{G0}(p_C/z_{CP})$. A_V similarly falls below A_{V0} past p_C or p_L as much as f_O climbs past p_C or p_L to f_{LC}, so $A_{V(LC)}$ is $A_{V0}(p_C/f_{LC})$ or $A_{V0}(p_L/f_{LC})$. Since f_{LC} is above p_C and z_C when C_O shunts and shorts below f_{LC}, $A_{V(LC)}$ and $A_{V(MF)}$ are $A_{V0}(p_C/z_C)$.

Example 13: Determine A_{G0} and A_{V0}, A_G and A_V at f_{LC}, and related poles and zeros when L_X is 10 μH, R_L is 50 mΩ, C_O is 5 μF, R_C is 10 mΩ, and R_{LD} is 100 Ω.

Solution:

$$f_{CP} = 320\text{ Hz}, f_L = 800\text{ Hz}, f_{LC} = 22\text{ kHz, and } z_C = 3.2\text{ MHz from Example 9}$$

$$A_{G0} \approx \frac{1}{R_L + R_{LD}} = \frac{1}{50m + 100} = 10 \text{ mA/V} = -40 \text{ dB}$$

$$A_{V0} = A_{G0}R_{LD} \approx (10m)(100) = 1 \text{ V/V} = 0 \text{ dB}$$

$$f_{CP} < f_L < f_{LC} < z_C \quad \therefore \quad z_{CP} \approx f_{CP} < p_L = p_C = p_{LC} = f_{LC} < z_C$$

$$A_{G(LC)} \approx \frac{1}{R_L + R_C} = \frac{1}{50m + 10m} = 17 \text{ A/V} = 25 \text{ dB}$$

$$A_{V(LC)} = A_{G(LC)}\left(\frac{1}{2\pi f_{LC}C_O}\right) \approx \frac{17}{2\pi(22k)(5\mu)} = 25 \text{ V/V} = 28 \text{ dB}$$

Note: z_{CP} is absent in the A_GZ_C that sets A_V because Z_C's fall cancels A_G's rise. z_C is absent in A_G because A_G is independent of C_O past f_{LC}. $A_{G(LC)}$ and $A_{V(LC)}$ are lower than the predicted 17 A/V or 25 dB and 25 V/V or 28 dB because the R_S that R_L and R_C set excludes the current R_{LD} steers away from the LC network.

Example 14: Determine A_{G0} and A_{V0}, A_G and A_V at f_{LC}, and related poles and zeros for Example 13 when R_L is 10 Ω.

Solution:

$$f_{CP} = 320 \text{ Hz}, f_{LC} = 22 \text{ kHz, and } z_C = 3.2 \text{ MHz from Example 9}$$

$$A_{G0} \approx \frac{1}{R_L + R_{LD}} = \frac{1}{10 + 100} = 9.1 \text{ mA/V} = -41 \text{ dB}$$

$$A_{V0} = A_{G0}R_{LD} \approx (9.1m)(100) = 910 \text{ mV/V} = -0.82 \text{ dB}$$

$$f_L = \frac{R_L}{2\pi L_X} = \frac{10}{2\pi(10\mu)} = 160 \text{ kHz} \rightarrow f_{LC} < f_L < z_C$$

$$\therefore \quad f_{CS} = \frac{1}{2\pi[R_C + (R_{LD}||R_L)]C_O} = \frac{1}{2\pi[10m + (100||10)](5\mu)} = 3.5 \text{ kHz}$$

$$f_{LS} = \frac{R_L + (R_C||R_{LD})}{2\pi L_X} = \frac{10 + (10m||100)}{2\pi(10\mu)} = 160 \text{ kHz}$$

$$z_{CP} \approx f_{CP} < p_C \approx f_{CS} < f_{LC} < p_L \approx f_{LS} < z_C$$

$$A_{G(LC)} \approx A_{G0}\left(\frac{p_C}{z_{CP}}\right) = (9.1m)\left(\frac{3.5k}{320}\right) = 100\ mA/V = -20\ dB$$

$$A_{V(LC)} \approx A_{V0}\left(\frac{f_{CS}}{f_{LC}}\right) = (910m)\left(\frac{3.5k}{22k}\right) = 140\ mV/V = -17\ dB$$

Note: z_{CP} is absent in A_V and z_C is absent in A_G.

Example 15: Determine the R_{LD} needed to suppress A_G's and A_V's peaks at f_{LC} in Example 13.

Solution:

$$f_{LC} = 22\ kHz \text{ from Example 9}$$

$$Q_{LC} = \frac{f_{CS}}{f_{LC}} = \frac{1}{2\pi(R_C + R_{LD})C_O f_{LC}} = \frac{1}{2\pi(10m + R_{LD})(5\mu)(22k)} \leq 1$$

$$\therefore\quad R_{LD} \leq 1.4\Omega$$

Note: This R_{LD} is fairly low for a 10-μH, 5-μF power supply.

Example 16: Determine A_{G0} and A_{V0}, A_G and A_V at f_{LC}, and related poles and zeros for Example 13 when R_L is 50 Ω and R_C is 10 Ω.

Solution:

$$f_{LC} = 22\ kHz \text{ from Example 9}$$

$$A_{G0} \approx \frac{1}{R_L + R_{LD}} = \frac{1}{50 + 100} = 6.7\ \text{mA/V} = -44\ \text{dB}$$

$$A_{V0} = A_{G0}R_{LD} \approx (6.7\text{m})(100) = 670\ \text{mV/V} = -3.5\ \text{dB}$$

$$f_{CP} = \frac{1}{2\pi(R_C + R_{LD})C_O} = \frac{1}{2\pi(10 + 100)(5\mu)} = 290\ \text{Hz}$$

$$f_L = \frac{R_L}{2\pi L_X} = \frac{50}{2\pi(10\mu)} = 780\ \text{kHz} > f_{LC}$$

$$\therefore \quad f_{CS} = \frac{1}{2\pi[R_C + (R_{LD}||R_L)]C_O} = \frac{1}{2\pi[10 + (100||50)](5\mu)} = 740\ \text{Hz}$$

$$z_C = \frac{1}{2\pi R_C C_O} = \frac{1}{2\pi(10)(5\mu)} = 3.2\ \text{kHz}$$

$$f_{LS} = \frac{R_L + (R_C||R_{LD})}{2\pi L_X} = \frac{50 + (10||100)}{2\pi(10\mu)} = 940\ \text{kHz}$$

$$\rightarrow \qquad z_{CP} \approx f_{CP} < p_C \approx f_{CS} < z_C < f_{LC} < p_L \approx f_{LS}$$

$$A_{G(LC)} \approx A_{G0}\left(\frac{p_C}{z_{CP}}\right) = (6.7\text{m})\left(\frac{740}{290}\right) = 17\ \text{mA/V} = -35\ \text{dB}$$

$$A_{V(LC)} \approx A_{V0}\left(\frac{p_C}{z_C}\right) = (670\text{m})\left(\frac{740}{3.2\text{k}}\right) = 160\ \text{mV/V} = -16\ \text{dB}$$

Note: z_{CP} is absent in A_V and z_C is absent in A_G.

Explore with SPICE:
See Appendix A for notes on SPICE simulations.

```
* Voltage-Sourced LC
vin vin 0 dc=0 ac=1
lx vin vl 10u
rl vl vo 50m
*rl vl vo 10
*rl vl vo 50
rc vo vc 10m
```

(continued)

```
*rc vo vc 10
co vc 0 5u
rld vo 0 100
*rld vo 0 1.4
.ac dec 1000 10 100e6
.end
```

Tip: Plot i(Lx) and v(vo) in dB and comment/un-comment rl's, rc, and rld to see their effects on A_G and A_V.

5.4.3 LC Tank

Energy offers another perspective of the peaking that results when L_X and C_O interact at f_{LC}. L_X holds magnetic energy $0.5L_Xi_L^2$ with i_L and C_O stores electrostatic energy $0.5C_Ov_C^2$ with v_C. At f_{LC}, i_L charges and discharges C_O and v_C energizes and drains L_X. As this happens, i_L draws v_{IN} energy.

This way, L_X and C_O become an *LC tank* that collects and exchanges v_{IN} energy. LC energy grows until i_L is so high that resistances in the network burn the energy v_{IN} supplies. This is how resistances limit and dampen the peaking in i_L and v_C at the *LC resonant frequency* f_{LC}.

5.4.4 Phase Shift

The gain of parallel and series LC structures peaks when Z_L and Z_C swap dominance at f_{LC}. This happens when Z_L and Z_C eclipse resistive effects. So the phase shift across p_{LC} is the phase difference that Z_L and Z_C produce.

Since Z_L and Z_C set v_O in parallel networks, sL_X induces a z_{LO} that adds 90° and $1/sC_O$ produces a p_{CO} that sheds 90°. Series combinations are similar because the $1/Z_C$ and $1/Z_L$ that set i_L induce a z_{CO} and a p_{LO} that also add and shed 90°. When gain shifts between Z_L or $1/Z_C$ and Z_C or $1/Z_L$, signals lose the 90° that z_{LO} or z_{CO} adds and the 90° that p_{CO} or p_{LO} deducts. So LC structures lose 180° across p_{LC}.

When resistive effects are negligible, this shift is abrupt. As resistances dampen the interaction, phase shifts more gradually. When Q_{LC} is one, for example, the p_L that L_X establishes and the p_C that C_O sets coincide at f_{LC} without peaking the gain.

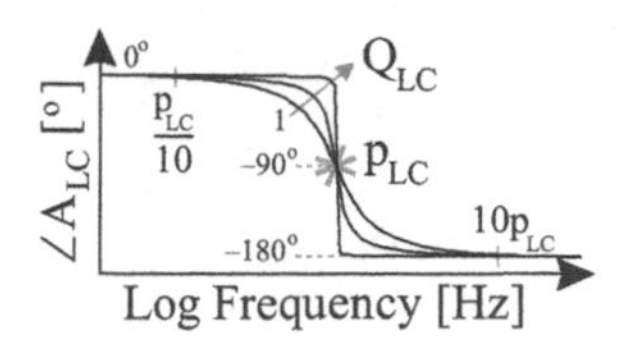

Fig. 5.29 LC phase shift

So p_L and p_C start and stop losing noticeable phase in Fig. 5.29 a decade below and above f_{LC}. As Q_{LC} increases, the transition band narrows and the edge sharpens.

The arc tangent is useful when expressing phase because it outputs up to ±90° with ± operands. In the case of LC phase $\angle A_{LC}$, subtracting 90° removes the offset $\tan^{-1}$ produces when f_O is less than f_{LC}. This way, the result matches the 0° to −180° range that L_X and C_O shift:

$$\angle A_{LC} = \tan^{-1}\left[Q_{LC}\left(\frac{f_{LC}}{f_O} - \frac{f_O}{f_{LC}}\right)\right] - 90^\circ. \tag{5.79}$$

f_{LC}/f_O minus f_O/f_{LC} inside the $\tan^{-1}$ term senses how f_O relates to f_{LC} (determining the polarity of the $\tan^{-1}$) and Q_{LC} magnifies the difference. So when Q_{LC} is high, small positive and negative differences output close to 90° – 90°, which is 0°, and −90° – 90°, which equals −180°. Large differences produce similar results when Q_{LC} is low.

5.5 Switched Inductor

5.5.1 Signal Translations

The underlying aim of power supplies is to regulate the output voltage v_O in voltage regulators and the output current i_O in battery chargers and LED drivers. For that, they incorporate a *controller* that compares v_O or i_O with a reference, amplifies the error, and uses the result to adjust and steady v_O or i_O. Together, the switched inductor and controller close a *feedback loop* that regulates v_O or i_O by sensing and responding to small variations in the output.

A. **Analog Response**

How v_O or i_O responds to changing operating conditions hinges on *small-signal translations*. Luckily, feedback loops suppress variations to such a degree that linear projections approximate them fairly well. The linear slope of the exponential-like

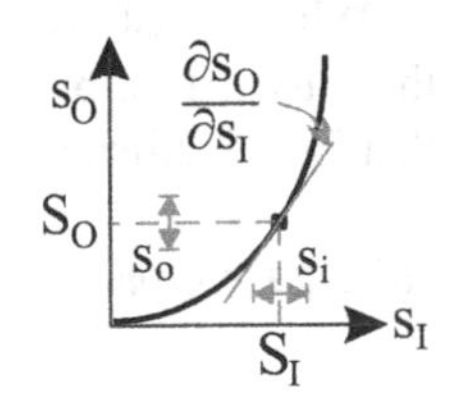

Fig. 5.30 Linear projection of small-signal variation

response in Fig. 5.30, for example, can project small variations in the *input signal* s_I to the *output signal* s_O that are very close to s_O's actual variations.

s_i represents small signals in s_I that the slope (partial derivative) of s_I at s_I's static point S_I projects to small signals s_o in s_O:

$$s_o = s_i A_X \approx s_i \left(\frac{\partial s_O}{\partial s_I}\right). \tag{5.80}$$

This linear projection is fairly close to the actual translation A_X because s_i and s_o are small variations about S_I and S_O. The approximation loses accuracy when variations grow to become larger fractions of S_I and S_O.

Nomenclature Lowercase variables with uppercase subscripts include small- and large-signal components. Uppercase variables with uppercase subscripts are the static steady-state components. And lowercase variables with lowercase subscripts are the dynamic small-signal components. When combined, S_A and s_a components complete the *analog signal* s_A:

$$s_A = S_A + \partial s_A = S_A + s_a, \tag{5.81}$$

where ∂s_A also refers to small variations in s_A.

B. **Switched Response**

Switched inductors energize and drain with large volt-level *energize* and *drain voltages* v_E and v_D in alternating phases of a switching cycle. The resulting *inductor voltage* v_L pulses volts and swings the *inductor current* i_L across amp-level ramps. These variations in v_L and i_L are large signal.

But when conditions settle, the power supply reaches a steady state that pulses v_L and ripples i_L to peaks and about averages that do not vary much over time. So cycles repeat and variations between cycles fade. This is the static periodic steady state of the switched inductor.

When operating conditions vary, the controller uses the error it senses in the output to adjust the switching cycle by a small amount. This small variation in amplitude, duty cycle, or frequency is the dynamic manifestation of the feedback command. So cycle-to-cycle variations in v_L and i_L reflect small-signal translations across the switched inductor.

5.5.2 Small-Signal Model

The switched inductor is a network of switches that connects the *transfer inductor* L_X in Fig. 5.31 to the *input* v_{IN}, *output* v_O, and ground. The digital input that adjusts

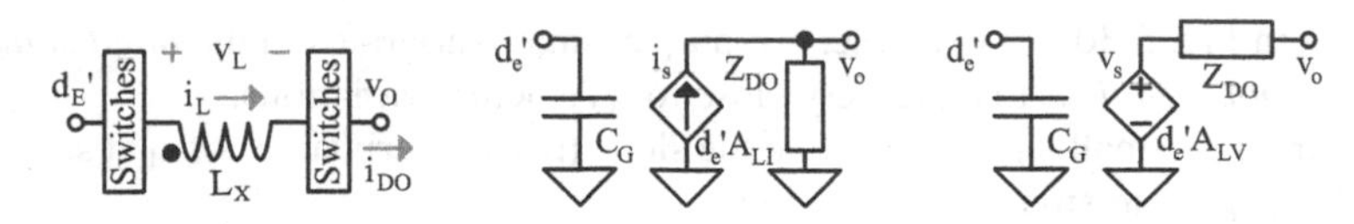

Fig. 5.31 Small-signal model of the switched inductor

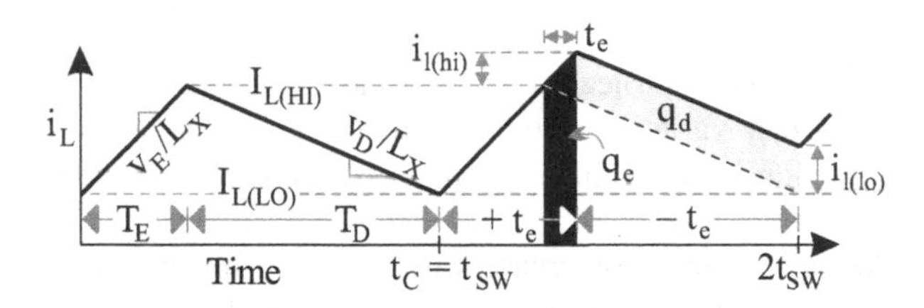

Fig. 5.32 Small-signal inductor-current variation in CCM

the connectivity of the network is the *energize command* d_E'. d_E' feeds logic that configures the switches so L_X energizes when d_E' is high.

d_E' is the fraction of the *switching period* t_{SW} that energizes L_X. This d_E' in modern power supplies connects to the gates of *complementary metal–oxide–semiconductor* (CMOS) transistors. These gates are capacitive and largely insensitive to the behavior of the network. So d_E' connects to *gate capacitance* C_G and the small-signal model that incorporates this C_G excludes a feedback translation.

Variations in d_E' ultimately alter the *duty-cycled current* i_{DO} that the network outputs. Z_{DO} is the *impedance* that d_O *duty-cycles* in the absence of d_E' variations d_e'. A_{LI} is the small-signal translation that projects small d_E' variations d_e' to small i_{DO} variations i_{do} without v_O variations v_o. A_{LV} is the unloaded voltage that A_{LI} produces across Z_{DO}, which is $A_{LI}Z_{DO}$.

A. Continuous Conduction

L_X in *continuous-conduction mode* (CCM) conducts continuously. This is because L_X energizes and drains continuously across alternating *energize* and *drain times* t_E and t_D of t_{SW} in Fig. 5.32. i_L climbs when v_E energizes L_X and falls when v_D drains L_X. d_E' in CCM is also the *energize duty cycle* d_E because L_X's *conduction time* t_C extends through t_{SW}, so t_E's fraction of t_C and t_{SW} is the same.

i_L is also i_{DO} in *bucks* because L_X energizes and drains directly into v_O. i_{DO} in *boosts* and *buck–boosts* is a *drain duty-cycle* d_D fraction of this i_L because L_X connects to v_O only when L_X drains. In other words, the *output duty cycle* d_O is one for bucks and d_D for boost-derived topologies, which duty-cycle L_X into v_O. So generally, i_{DO} is a d_O translation of i_L:

$$i_{DO} = i_L d_O, \tag{5.82}$$

where d_O in boost-based power supplies in CCM is d_D's t_D/t_{SW} and t_O/t_{SW} and t_O is the *output conduction time* that d_O steers i_L into v_O.

Duty-Cycled Impedance Z_{DO} is the output impedance of the switched inductor in the absence of d_E' variations. More specifically, Z_{DO} is the impedance into L_X that results when t_O and t_{SW} are static and d_O connects L_X to v_O. In this light, Z_{DO} is the *duty-cycled inductance* L_{DO} that loads v_o across T_{SW} with the energy L_X draws across T_O.

This is equivalent to saying the energy E_{LO} that L_{DO} draws from v_o across T_{SW} is the energy E_{LX} that L_X extracts across T_O. From this perspective, i_{LX} is the current L_X draws from v_o across T_O: a D_O fraction of T_{SW}, and E_{LX} is the energy L_X collects after i_L reaches i_{LX}:

$$E_{LX} = \left(\frac{1}{2}\right)L_X i_{LX}^2 = \left(\frac{1}{2}\right)L_X\left[\left(\frac{v_O}{L_X}\right)T_O\right]^2 = \left(\frac{v_O^2}{2}\right)\left(\frac{D_O^2}{L_X}\right)T_{SW}^2. \quad (5.83)$$

i_{LO}, on the other hand, is the current L_{DO} draws from v_o across T_{SW} and E_{LO} is the energy L_{DO} collects after i_L reaches i_{LO}:

$$E_{LO} = \left(\frac{1}{2}\right)L_{DO} i_{LO}^2 = \left(\frac{1}{2}\right)L_{DO}\left[\left(\frac{v_O}{L_{DO}}\right)T_{SW}\right]^2 = \left(\frac{v_O^2}{2}\right)\left(\frac{1}{L_{DO}}\right)T_{SW}^2. \quad (5.84)$$

But since E_{LO} is the energy E_{LX} loads (i. e., E_{LO} and E_{LX} equal), L_{DO} is a reverse quadratic D_O translation of L_X:

$$L_{DO} = \frac{L_X}{D_O^2}. \quad (5.85)$$

Although often negligible, R_L also loads v_o when L_X connects to v_O. Like with L_X, R_L loads v_o across T_O the power *duty-cycled resistance* R_{LO} loads across T_{SW}. Since R_L's P_{RL} is a T_O/T_{SW} fraction of v_o^2/R_L, R_{LO}'s P_{RO} is v_o^2/R_{LO}, and P_{RO} is the power R_{LO} loads, R_{LO} is a reverse D_O translation of R_L:

$$P_{RL} = \left(\frac{v_O^2}{R_L}\right)\left(\frac{T_O}{T_{SW}}\right) = \frac{v_O^2}{R_L/D_O}, \quad (5.86)$$

$$P_{RO} = \frac{v_O^2}{R_{LO}}, \quad (5.87)$$

$$R_{LO} = \frac{R_L}{D_O}. \quad (5.88)$$

This R_{LO} is greater than R_L when S_{DO} duty-cycles L_X into v_O because R_{LO} alternates between R_L (a D_O fraction of T_{SW}) and infinity (across what remains of T_{SW}).

Gain When d_E' rises in CCM, t_E lengthens by the same amount t_D shortens. v_L is therefore positive (with v_E) longer than v_L is negative (with v_D). This net rise in v_L across L_X increases i_L. This is the Ohmic translation that A_{LI} and A_{LV} model.

Generally, i_L is an Ohmic Z_L translation of v_L. v_L is in turn a d_E fraction of v_E and an inverting d_D fraction of v_D. But since d_D in CCM is the fraction of t_{SW} that excludes t_E, i_L is ultimately a d_E translation of v_E and v_D into Z_L:

$$i_L = \frac{v_L}{Z_L} = \frac{v_E d_E - v_D d_D}{Z_L} = \frac{v_E d_E - v_D(1 - d_E)}{Z_L} = \frac{(v_E + v_D)d_E - v_D}{Z_L}. \tag{5.89}$$

Under static conditions (when small variations fade), v_E and v_D fractions in v_L cancel because d_E is $v_D/(v_E + v_D)$. This is equivalent to saying v_L averages zero and L_X shorts at low f_O. Dynamic (nonzero f_O) adjustments to d_E alter this v_L. So d_e' produces a small v_L variation v_l that induces a corresponding small i_L variation i_l.

A_L is the part of A_{LI} that projects d_e' to i_l when small v_O variations v_o are absent. Since v_{IN} is an independent voltage source, small v_{IN} variations v_{in} are zero. So v_E and v_D in A_L's projection $\partial i_L/\partial d_E'$ are static and i_l's short-circuit component i_l' (when v_o is zero) is

$$i_l' \equiv i_l\Big|_{v_o=0} = d_e' A_L \approx d_e'\left(\frac{\partial i_L}{\partial d_E}\right)\Big|_{v_o=0} = d_e'\left(\frac{V_E + V_D}{Z_L}\right). \tag{5.90}$$

In boost-derived supplies, L_X only delivers charge across t_D. So of the q_l that i_l' garners, i_d delivers the part that t_D carries. When averaged across T_{SW} and duty-cycled across T_O, this i_d approximates to a D_O fraction of i_l'.

Extending t_E in boost-derived supplies also shortens t_D (and t_O) when t_{SW} is constant. So v_O loses charge q_e to t_E across t_E's small extension t_e in Fig. 5.32 when i_L reaches $i_{L(HI)}$. This q_e across t_{SW} is a t_e current i_e that i_{do} loses:

$$i_e = \frac{q_e}{T_{SW}} = \frac{t_e i_{L(HI)}}{T_{SW}} \approx \frac{t_e I_{L(HI)}}{T_{SW}} = d_e' I_{L(HI)}, \tag{5.91}$$

where $i_{L(HI)}$'s static component $I_{L(HI)}$ is much greater than $i_{L(HI)}$'s dynamic counterpart $i_{l(hi)}$. Bucks do not lose this i_e because L_X in bucks connects to v_O also across t_E.

A_{LI} translates d_e' to i_{do} when disabling Z_{DO} with zero v_o. This short-circuit i_{do} is a *small-signal current source* i_s. Since i_L and d_O in i_{DO} both vary with d_E, i_s carries two components:

$$\begin{aligned} i_s \equiv i_{do}\Big|_{v_o=0} &= d_e' A_{LI} = d_e'\left(\frac{\partial i_{DO}}{\partial d_E}\right)\Big|_{v_o=0} \\ &= d_e'\left[\left(\frac{\partial i_L}{\partial d_E}\right) D_O + \left(\frac{\partial d_{DO}}{\partial d_E}\right) I_{L(HI)}\right]\Big|_{v_o=0} \\ &= i_l' D_O - d_e' I_{L(HI)} = i_d - i_e. \end{aligned} \tag{5.92}$$

The first component is a D_O fraction of i_l', which is the t_d current i_d that q_d in Fig. 5.32 delivers. The second term is zero for bucks because d_O is one and an inverting d_e' fraction of i_L at $I_{L(HI)}$ for boosts because d_O is d_D's $1 - d_E$. This last component is the i_e that i_s loses to t_E, which is the q_e lost by $t_e I_{L(HI)}$ across T_{SW}.

Luckily, i_d is much higher than i_e at low f_O because Z_L in the i_l' that sets i_d is very low. i_l', however, falls with f_O 20 dB per decade as Z_L's sL_X climbs. Since i_e is constant, i_s drops with i_l' until i_d falls below i_e (past z_{DO}):

$$i_d = i_l' D_O \approx d_e' \left(\frac{V_E + V_D}{sL_X}\right) D_O \bigg|_{f_O \geq \left(\frac{V_E+V_D}{2\pi L_X}\right)\left(\frac{D_O}{I_{L(HI)}}\right) \equiv \frac{R_{LD}'}{2\pi L_X} \equiv z_{DO} \approx \left(\frac{V_E+V_D}{2\pi L_X}\right)\left(\frac{D_O^2}{I_O}\right)}$$
$$\leq i_e \approx d_e' I_{L(HI)}. \qquad (5.93)$$

This *duty-cycled zero* z_{DO} is a reversal out-of-phase zero because i_s stops falling with the p_L that sL_X sets and i_e inverts i_s.

When the *CCM current ripple* Δi_L is a small fraction of $i_{L(AVG)}$, $I_{L(HI)}$ and $I_{L(HI)} D_O$ near I_L and I_O's $I_L D_O$. In this light, z_{DO} surfaces when L_X overcomes the *equivalent load resistance* R_{LD}' that v_L's V_E and V_D, D_O, and i_O's average I_O set. Bucks do not exhibit this z_{DO} because they do not lose i_e.

A_{LI}'s translation to i_s is therefore a duty-cycled Z_L Ohmic translation of V_E and V_D that i_l' sets and z_{DO} inverts and reverses with increasing f_O:

$$A_{LI} \equiv \frac{i_s}{d_e'}\bigg|_{v_o=0} \approx \left(\frac{V_E + V_D}{Z_L}\right) D_O \left(1 - \frac{s}{2\pi z_{DO}}\right). \qquad (5.94)$$

A_{LV} is the unloaded Ohmic translation that A_{LI} feeds Z_{DO}:

$$A_{LV} \equiv \frac{v_s}{d_e'}\bigg|_{i_o=0} = A_{LI} Z_{DO} = A_{LI}\left(\frac{Z_L}{D_O^2}\right) \approx \left(\frac{V_E + V_D}{D_O}\right)\left(1 - \frac{s}{2\pi z_{DO}}\right). \qquad (5.95)$$

Part of this Z_{DO} in A_{LI} cancels the Z_L and D_O that translates d_e' to i_l' and i_l' to i_s. So A_{LV} is a reverse D_O translation of V_E and V_D that z_{DO} inverts and increases with f_O.

A_{LV}'s independence of Z_L in Z_{DO} indicates the switched inductor in CCM is more a voltage-sourced inductor than the current-sourced inductor A_{LI} models. From this view point, the role of the switcher is to set a *small-signal voltage source* v_s that drives and feeds an inductor L_{DO} into v_o. This v_s is largely a D_O translation of the V_E and V_D that the switcher applies to L_X.

The *signal-flow graph* in Fig. 5.33 is an insightful way of summarizing and visualizing these translations. This graph breaks and traces individual components to i_s. All translations are operation-based, from the behavior of v_L and i_L across t_{SW} cycles.

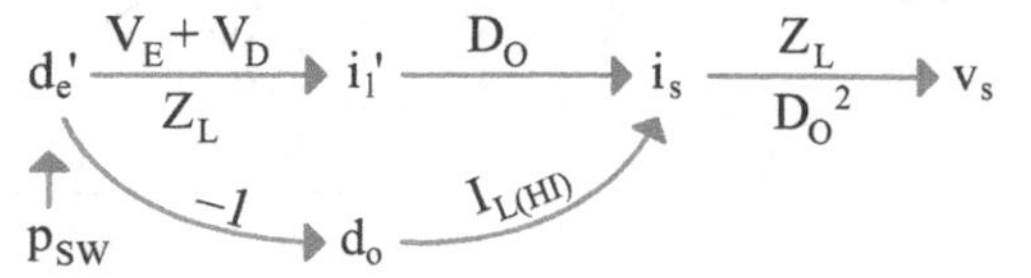

Fig. 5.33 Signal-flow graph of the switched inductor in CCM

This graph shows that d_E' variations produce i_L and d_O variations that propagate to i_{DO} (via i_s). i_{do}'s short-circuit component i_s is in part a D_O fraction of i_l', which is an Ohmic translation of d_e', and in part an inverting $I_{L(HI)}$ translation of d_e'. d_o inverts and keeps i_s from falling with i_l' when D_O's fraction of i_l' falls below d_o's inverting translation. And v_s is an Ohmic Z_{DO} translation of i_s. But as mentioned earlier, the inverting bypass path that d_o feeds i_s is absent in bucks because v_O receives all of i_l'.

B. Discontinuous Conduction

L_X in *discontinuous-conduction mode* (DCM) conducts a fraction of t_{SW}. This is because L_X energizes and drains across t_C before t_{SW} in Fig. 5.34 ends. So i_L climbs with v_E, falls with v_D, and flattens before and until another t_{SW} cycle begins.

Several observations are worth noting. First, d_E and d_D are t_E and t_D fractions of a t_C that is shorter than t_{SW}. So d_E is not the t_E/t_{SW} that the input d_E' commands.

Second, L_X energizes and depletes every cycle. So L_X delivers all the charge it collects. This means i_s does not lose the i_e that inverts and alters i_s in CCM. As a result, A_{LI} and A_{LV} in DCM exclude z_{DO}.

Third, extending t_E in DCM extends t_D when t_{SW} is constant. This is because adding t_e raises $i_{L(PK)}$, and with it, the $0.5L_Xi_{L(PK)}{}^2$ energy that L_X collects. So L_X requires more t_D to drain.

In fact, t_D and t_C scale proportionately with t_E within t_{SW} because the v_E and v_D that project i_L are static. So the fraction of v_E that d_e applies to L_X cancels the fraction of v_D that d_d applies. This means that the resulting small-signal voltage v_l across L_X is zero:

$$\frac{v_l}{Z_L} = \frac{V_E d_e - V_D d_d}{Z_L} = 0. \tag{5.96}$$

Losing sensitivity to f_O this way removes inductive effects from A_{LI}, A_{LV}, and Z_{DO}.

Since v_E and v_D projections are static, variations in t_E induce variations in t_C that track T_E and T_C and the D_E that T_E and T_C set:

$$d_E = \frac{T_E + t_e}{T_C + t_c} = \frac{T_E}{T_C} = \frac{t_e}{t_c} = D_E. \tag{5.97}$$

So d_E is static, which means d_e is zero. And t_e's fraction of T_E matches t_c's fraction of T_C:

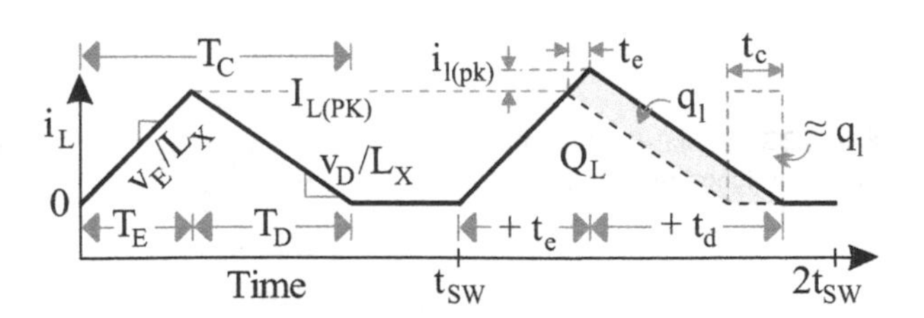

Fig. 5.34 Small-signal inductor-current variation in DCM

$$k_d \equiv \frac{t_e}{T_E} = \frac{t_c}{T_C} = \frac{i_{l(pk)}}{I_{L(PK)}}. \tag{5.98}$$

These fractions (defined as k_d here) also match $i_{l(pk)}$'s fraction of $I_{L(PK)}$ because v_E and v_D projections of i_L are steady.

Duty-Cycled Impedance Z_{DO} is the output impedance of the switched inductor in the absence of d_E' variations d_e'. So Z_{DO} is the impedance into L_X that results when t_C and t_{SW} are static and L_X connects to v_O a d_O fraction of t_C. Since Z_{DO} is insensitive to f_O (because L_X's v_I is zero), Z_{DO} is a *duty-cycled resistance* R_{DO} that loads v_o across T_{SW} with the energy L_X draws across D_O's fraction of T_C.

In this light, i_{LX} is the current L_X draws from v_o across D_O's fraction of T_C and E_{LX} is the energy L_X collects after i_L reaches i_{LX}:

$$E_{LX} = \left(\frac{1}{2}\right)L_X i_{LX}^2 = \left(\frac{1}{2}\right)L_X\left[\left(\frac{v_O}{L_X}\right)D_O T_C\right]^2 = \left(\frac{v_O^2}{2}\right)\left(\frac{D_O^2}{L_X}\right)T_C^2. \tag{5.99}$$

This E_{LX} is the power P_{RO} that R_{DO} burns across T_{SW}:

$$P_{RO} = \frac{v_O^2}{R_{DO}} = \frac{E_{LX}}{T_{SW}} = \left(\frac{v_O^2}{2}\right)\left(\frac{D_O^2}{L_X}\right)\left(\frac{T_C^2}{T_{SW}}\right). \tag{5.100}$$

So R_{DO} is a frequency-independent translation of L_{DO} in CCM, T_C, and T_{SW}. R_L's duty-cycled translation raises this R_{DO}, but usually not by much:

$$R_{DO} = 2\left(\frac{L_X}{D_O^2}\right)\left(\frac{T_{SW}}{T_C^2}\right) + \frac{R_L}{D_O} \approx 2L_{DO}\left(\frac{T_{SW}}{T_C^2}\right). \tag{5.101}$$

Gain A_{LI} translates d_e' to i_{do} when disabling Z_{DO} with zero v_o. While i_{DO} outputs all the charge q_L that L_X collects, i_{do} delivers the part of q_L that excludes the static component Q_L. Since i_L ramps in straight lines, q_L and Q_L are the areas (current into time) under the triangles that i_L and I_L outline.

q_L's t_C and $i_{L(PK)}$ are T_C and $I_{L(PK)}$ plus corresponding k_d fractions of T_C and $I_{L(PK)}$ because t_c and $i_{l(pk)}$ scale with T_C and $I_{L(PK)}$:

$$q_L = 0.5t_C i_{L(PK)} = 0.5T_C(1 + k_d)I_{L(PK)}(1 + k_d) \tag{5.102}$$

$$\begin{aligned} q_l = q_L - Q_L &= 0.5T_C I_{L(PK)}(1 + k_d)^2 - 0.5T_C I_{L(PK)} \\ &= 0.5T_C I_{L(PK)}\left(k_d^2 + 2k_d\right) \\ &\approx T_C I_{L(PK)} k_d = t_c I_{L(PK)}. \end{aligned} \tag{5.103}$$

Since k_d is a small fraction and t_e/t_c is D_E, $2k_d$ swamps k_d^2 and t_c is t_e/D_E. So q_l is roughly the rectangular area that t_c and $I_{L(PK)}$ outline and i_s is q_l/T_{SW}:

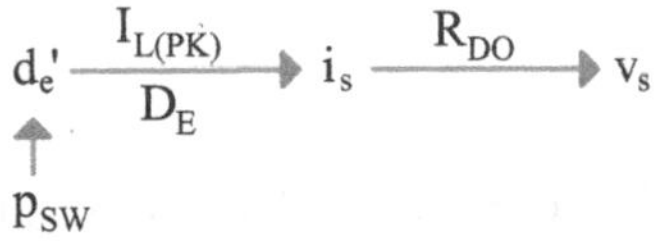

Fig. 5.35 Signal-flow graph of the switched inductor in DCM

$$i_s = i_{do}\big|_{v_o=0} = \frac{q_l}{T_{SW}} \approx \frac{t_c I_{L(PK)}}{T_{SW}} = \frac{t_e I_{L(PK)}}{D_E T_{SW}} = d_e'\left(\frac{I_{L(PK)}}{D_E}\right). \tag{5.104}$$

Like Fig. 5.35 shows, A_{LI} is ultimately the $I_{L(PK)}/D_E$ translation q_l sets:

$$A_{LI} \equiv \frac{i_s}{d_e'}\bigg|_{v_o=0} \approx \frac{I_{L(PK)}}{D_E}. \tag{5.105}$$

And A_{LV} is the unloaded Ohmic translation that A_{LI} feeds Z_{DO}'s R_{DO}:

$$A_{LV} \equiv \frac{v_s}{d_e'}\bigg|_{i_o=0} = A_{LI}Z_{DO} = A_{LI}R_{DO} \approx \left(\frac{I_{L(PK)}}{D_E}\right)R_{DO}. \tag{5.106}$$

Still, the independence of A_{LI} and R_{DO} indicates a current-sourced resistor is a better model. Notice A_{LI}, A_{LV}, and R_{DO} are all independent of f_O.

C. **Switching Pole**

The switcher requires up to one switching cycle after t_E ends to adjust the next t_E. This adjustment in t_E is the switcher's response to variations in d_E'. Even with multiple sub-cycle d_E' variations, the switcher's response is still one t_E adjustment. This is like saying the switcher delays and masks (and suppresses) d_E' variations that are faster than the *switching frequency* f_{SW} that t_{SW} sets.

Since poles also delay and suppress higher-f_O signals, a *switching pole* p_{SW} can model this behavior. Although the delay and suppression are not necessarily linear or consistent, p_{SW} is a useful, albeit imperfect, way of indicating the switcher filters higher-f_O signals. Note this p_{SW} affects the d_e' that feeds and propagates to i_l', d_o, i_s, and v_s.

5.5.3 Power Stage

Switched inductors deliver v_{IN} energy with i_{DO}. Engineers add C_O in Fig. 5.36 to voltage regulators and LED drivers to supply i_O when d_O disconnects L_X from v_O. In the case of chargers, C_O is the effective capacitance of the battery, which is usually very high. R_L and R_C are the parasitic series resistances in L_X and C_O. i_{LD} is the static part of the load and R_{LD} is the part that responds to small v_O variations v_o.

The small-signal gain to v_O is critical in voltage regulators because the feedback loop regulates this v_O. In LED drivers, i_O is more important because the brightness

Fig. 5.36 Power stage

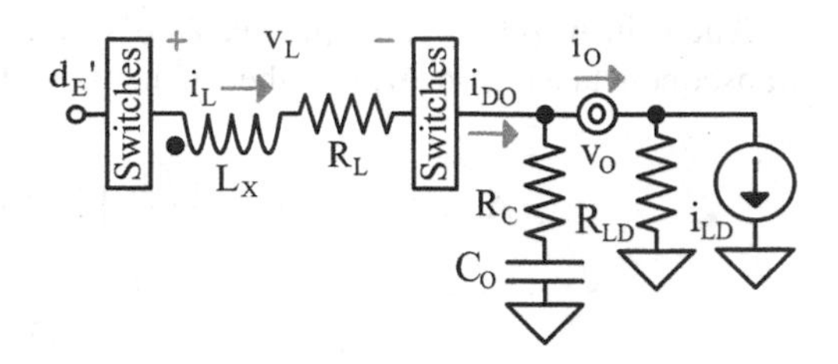

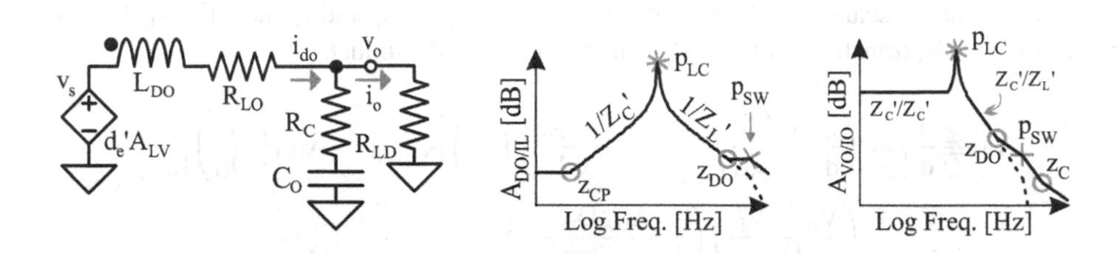

Fig. 5.37 Small-signal model of the power stage in CCM

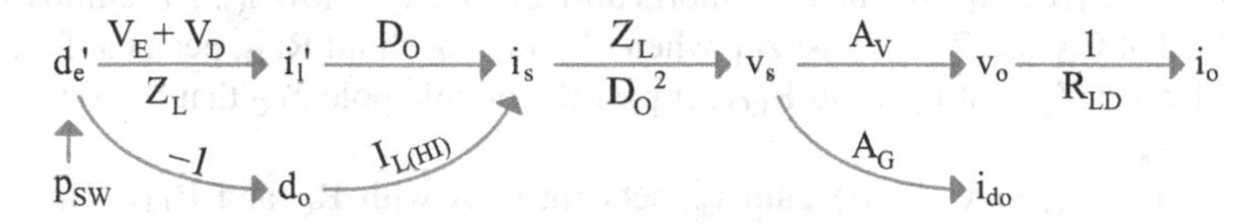

Fig. 5.38 Signal-flow graph of the power stage in CCM

and spectrum of the emitted light depend on the i_O the LEDs receive. i_{DO} is critical in chargers because i_{DO} is what feeds the battery. i_L is also important when systems control the i_L that L_X conducts. This is why small-signal d_E' translations to v_O, i_O, i_{DO}, and i_L are relevant.

A. Continuous Conduction

Frequency Response The switched inductor in continuous conduction is a voltage-sourced inductor. So the switcher in Fig. 5.36 sets a small-signal voltage v_s in Fig. 5.37 that drives L_{DO} and R_{LO} into R_{LD} and C_O with R_C. i_{LD} is absent because i_{LD} does not respond to small signals.

Small d_E' variations d_e' in Figs. 5.36, 5.37, and 5.38 feed and propagate to i_l', i_s, v_s, i_{do} and v_o, and i_o. i_l' and i_s are the short-circuit components of i_l and i_{do} that result when v_o is zero. d_o bypasses i_l' and inverts i_s past z_{DO} only in boost-derived topologies, when d_O is a fraction of t_{SW}.

The gain A_{DO} to i_{do} incorporates A_{LV}'s translation to v_s and the voltage-sourced transconductance gain A_G into the LC network to i_{do}:

$$\begin{aligned} A_{DO} &\equiv \frac{i_{do}}{d_e'} = \left(\frac{i_l'}{d_e'}\right)\left(\frac{i_s}{i_l'}\right)\left(\frac{v_s}{i_s}\right)\left(\frac{i_{do}}{v_s}\right) = A_{LV}A_G \\ &= \left(\frac{V_E + V_D}{D_O}\right)\left(\frac{1 - s/2\pi z_{DO}}{1 + s/2\pi p_{SW}}\right)A_G. \end{aligned} \quad (5.107)$$

Since i_l and i_{do} scale with their short-circuit counterparts i_l' and i_s, the gain A_{IL} to i_l is a reverse D_O translation of i_{do} without the z_{DO} that d_O induces:

$$\begin{aligned} A_{IL} &\equiv \frac{i_l}{d_e'} = \left(\frac{i_{do}}{d_e'}\right)\left(\frac{i_l}{i_{do}}\right)\Bigg|_{\text{No } z_{DO}} = \left(\frac{i_{do}}{d_e'}\right)\left(\frac{i_l'}{i_s}\right)\Bigg|_{\text{No } z_{DO}} = A_{DO}\left(\frac{1}{D_O}\right)\Bigg|_{\text{No } z_{DO}} \\ &= \left(\frac{V_E + V_D}{D_O}\right)\left(\frac{A_G/D_O}{1 + s/2\pi p_{SW}}\right). \end{aligned} \quad (5.108)$$

A_G is $1/[(R_L/D_O) + R_{LD})]$ when L_X shorts and C_O opens at low f_O. A_G climbs with the $1/Z_C'$ that C_O and R_C set past z_{CP} when C_O and R_C shunt R_{LD}, peaks at f_{LC}, and falls with the $1/Z_L'$ that L_{DO} and R_{LO} set past the double pole p_{LC} that f_{LC} sets when Z_L' surpasses Z_C'.

The gain A_{VO} to v_o is the gain i_{do} sets into C_O with R_C and R_{LD}. This gain incorporates A_{LV}'s translation to v_s and the voltage-sourced voltage gain A_V across the LC network to v_o:

$$\begin{aligned} A_{VO} &\equiv \frac{v_o}{d_e'} = \left(\frac{i_l}{d_e'}\right)\left(\frac{i_s}{i_l}\right)\left(\frac{v_s}{i_s}\right)\left(\frac{v_o}{v_s}\right) \\ &= A_{DO}[(Z_C + R_C)||R_{LD}] = A_{LV}A_G[(Z_C + R_C)||R_{LD}] \\ &= A_{LV}A_V = \left(\frac{V_E + V_D}{D_O}\right)\left(\frac{1 - s/2\pi z_{DO}}{1 + s/2\pi p_{SW}}\right)A_G\left\{\frac{(Z_C + R_C)||R_{LD}}{Z_{LO} + R_{LO} + [(Z_C + R_C)||R_{LD}]}\right\}. \end{aligned} \quad (5.109)$$

And the gain A_{IO} to i_o is an Ohmic R_{LD} translation of A_{VO}:

$$A_{IO} \equiv \frac{i_o}{d_e'} = \left(\frac{v_o}{d_e'}\right)\left(\frac{i_o}{v_o}\right) = A_{VO}\left(\frac{1}{R_{LD}}\right) = A_{LV}A_V\left(\frac{1}{R_{LD}}\right). \quad (5.110)$$

A_V is close to one when L_X shorts and C_O opens at low f_O and R_{LO} is much lower than R_{LD}. A_V peaks at f_{LC}, falls 40 dB per decade past the double pole p_{LC} that f_{LC} sets when L_{DO} opens and C_O shunts, and falls 20 dB per decade past z_C when C_O shorts.

A_G and A_V behave this way when L_X overcomes R_L and C_O and R_C shunt R_{LD} below f_{LC} and C_O shorts with respect to R_C above f_{LC}, which is often the case in power supplies. A_{LV} to v_s is largely a reverse D_O translation of V_E and V_D. In A_{DO}, A_{VO}, and A_{IO}, however, this A_{LV} inverts and rises past z_{DO} when d_o bypasses i_l. This out-of-phase zero z_{DO} is absent in bucks because L_X feeds v_O all of i_l. In all cases, the switcher delays and suppresses at- and above-f_{SW} signals.

Example 17: Determine A_{DO} and A_{VO} at low f_O and related poles and zeros in CCM when v_E is 2 V, v_D is 4 V, d_E is 67%, d_O is 33%, $i_{L(HI)}$ is 190 mA, and t_{SW} is 1 μs with the L_X, R_L, C_O, R_C, and R_{LD} specified in Example 9.

Solution:

$$A_{LVO} = \frac{V_E + V_D}{D_O} = \frac{2+4}{33\%} = 18 \text{ V/V} = 25 \text{ dB}$$

$$A_{GO} \approx \frac{1}{(R_L/D_O) + R_{LD}} = \frac{1}{(50m/33\%) + 100} = 10 \text{ mA/V} = -40 \text{ dB}$$

$$A_{VO} = A_{GO}R_{LD} \approx (10m)(100) = 1 \text{ V/V} = 0 \text{ dB}$$

$$A_{DOO} = A_{LVO}A_{GO} \approx (18)(10m) = 180 \text{ mA/V} = -15 \text{ dB}$$

$$A_{VOO} = A_{LVO}A_{VO} \approx (18)(1) = 18 \text{ V/V} = 25 \text{ dB}$$

$$f_{CP} = 320 \text{ Hz and } z_C = 3.2 \text{ MHz from Example 9}$$

$$L_{DO} = \frac{L_X}{D_O^2} = \frac{10\mu}{33\%^2} = 92 \text{ μH}$$

$$f_L = \frac{R_L/D_O}{2\pi L_{DO}} = \frac{50m/33\%}{2\pi(92\mu)} = 260 \text{ Hz}$$

$$f_{LC} = \frac{1}{2\pi\sqrt{L_{DO}C_O}} = \frac{1}{2\pi\sqrt{(92\mu)(5\mu)}} = 7.4 \text{ kHz}$$

$$z_{DO} = \left(\frac{V_E + V_D}{2\pi L_X}\right)\left(\frac{D_O}{I_{L(HI)}}\right) = \left[\frac{2+4}{2\pi(10\mu)}\right]\left(\frac{33\%}{190m}\right) = 170 \text{ kHz}$$

$$f_{SW} = \frac{1}{t_{SW}} = \frac{1}{1\mu} = 1 \text{ MHz}$$

$$f_{CP} < f_L < f_{LC} < z_{DO} < f_{SW} < z_C$$

$$\therefore \quad z_{CP} \approx f_{CP} < p_{LC} = f_{LC} < z_{DO} < p_{SW} \approx f_{SW} < z_C$$

Note: z_{CP} is absent in A_{VO} and z_C is absent in A_{DO} because A_V excludes z_{CP} and A_G excludes z_C. z_C is largely inconsequential because it appears above f_{SW}.

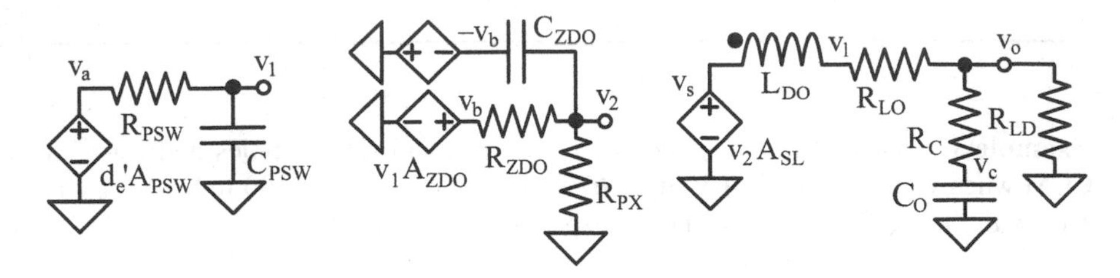

Fig. 5.39 Duty-cycled CCM frequency-response model of the switched inductor

Circuit Model The circuit in Fig. 5.39 models the duty-cycled CCM response of the switched inductor. A_{PSW} buffers d_e' (with a gain of one) so C_{PSW} can establish p_{SW} in v_1 when C_{PSW} shunts R_{PSW} into A_{PSW}. This way, A_{PSW} is 0 dB at low frequency and falls 20 dB per decade past p_{SW}:

$$\left.\frac{1}{sC_{PSW}}\right|_{f_O \geq \frac{1}{2\pi R_{PSW} C_{PSW}} \equiv p_{SW}} \leq R_{PSW}. \tag{5.111}$$

A_{ZDO} buffers v_1 into the voltage divider R_{ZDO} and R_{PX} implement so C_{ZDO} can inject z_{DO} into v_2. A_{ZDO} recovers the gain lost across R_{ZDO} and R_{PX} so the low-f_O gain to v_2 is one. This way, the overall low-f_O gain from d_e' to v_2 is also one:

$$A_{ZDO} \equiv \frac{R_{ZDO} + R_{PX}}{R_{PX}}. \tag{5.112}$$

C_{ZDO} feeds an inverting (out-of-phase) v_b signal that bypasses R_{ZDO} past z_{DO}. R_{PX} reverses this z_{DO} past p_{ZX} when C_{ZDO} shorts with respect to R_{PX} and R_{ZDO} into A_{ZDO}. Since this p_{ZX} is an artificial byproduct of the model, R_{PX} should be low enough to push p_{ZX} over p_{SW}, where its presence is largely inconsequential:

$$\left.\frac{1}{sC_{ZDO}}\right|_{f_O \geq \frac{1}{2\pi R_{ZDO} C_{ZDO}} \equiv z_{DO}} \leq R_{ZDO} \tag{5.113}$$

$$\left.\frac{1}{sC_{ZDO}}\right|_{f_O \geq \frac{1}{2\pi (R_{ZDO} || R_{PX}) C_{ZDO}} \equiv p_{ZX} > p_{SW}} \leq R_{ZDO} || R_{PX}. \tag{5.114}$$

Explore with SPICE:
See Appendix A for notes on SPICE simulations.

```
* Switched Inductor: Duty-Cycled CCM Frequency-Response Model
vde de 0 dc=0 ac=1
epsw va 0 de 0 1
```

(continued)

```
rpsw va v1 1
cpsw v1 0 159.2n
ezdo vb 0 v1 0 1001
enzdo nvb 0 vb 0 -1
rzdo vb v2 1
czdo nvb v2 940n
rpx v2 0 1m
esl vs 0 v2 0 18
ldo vs vl 92u
rlo vl vo 150m
rc vo vc 10m
co vc 0 5u
rld vo 0 100
.ac dec 1000 10 10e6
.end
```

Tip: Plot v(vs), i(Ldo), v(vo), and i(Rld) in dB to inspect A_{LV}, A_{DO}, A_{VO}, and A_{IO}.

Example 18: Determine A_{IL} and A_{IO} at low f_O and related poles and zeros in CCM when v_E and v_D are 2 V, d_E is 50%, d_O is 100%, and $i_{L(HI)}$ is 130 mA with the L_X, R_L, C_O, R_C, t_{SW}, and R_{LD} specified in Example 17.

Solution:

$$f_{CP} = 320\text{ Hz}, f_L = 800\text{ Hz, and } z_C = 3.2\text{ MHz from Example 9}$$

$$f_{SW} = 1\text{ MHz from Example 17}$$

$$L_{DO} = \frac{L_X}{{D_O}^2} = \frac{L_X}{1^2} = L_X = 10\ \mu\text{H} \quad \therefore \quad f_{LC} = 22\text{ kHz from Example 9}$$

$$A_{LV0} = V_E + V_D = 2 + 2 = 4\text{ V/V} = 12\text{ dB}$$

$$R_{LO} = \frac{R_L}{D_O} = \frac{R_L}{1} = R_L$$

$$A_{G0} = \frac{1}{R_L + R_{LD}} = \frac{1}{50m + 100} = 10\ mA/V = -40\ dB$$

$$A_{V0} = A_{G0}R_{LD} = (10m)(100) = 1\ V/V = 0\ dB$$

$$A_{IL0} = A_{LV0}A_{G0} = (4)(10m) = 40\ mA/V = -28\ dB$$

$$A_{IO0} = \frac{A_{LV0}A_{V0}}{R_{LD}} = \frac{(4)(1)}{100} = 40\ mA/V = -28\ dB$$

$$f_L < f_{CP} < f_{LC} < f_{SW} < z_C \quad \therefore \quad z_{CP} \approx f_{CP} < p_{LC} = f_{LC} < p_{SW} \approx f_{SW} < z_C$$

Note: z_{DO} is absent because d_o does not bypass i_l and A_{IL0} and A_{IO0} match because i_L flows to R_{LD} at low f_O.

Explore with SPICE:
Remove A_{ZDO}, C_{ZDO}, R_{ZDO}, and R_{PX} from the model and feed v_1 into A_{SL} to exclude z_{DO} from the response. See Appendix A for notes on SPICE simulations.

```
* Switched Inductor: CCM Frequency-Response Model of the Buck
vde de 0 dc=0 ac=1
epsw va 0 de 0 1
rpsw va v1 1
cpsw v1 0 159.2n
esl vs 0 v1 0 4
lx vs vl 10u
rl vl vo 50m
rc vo vc 10m
co vc 0 5u
rld vo 0 100
.ac dec 1000 10 10e6
.end
```

Tip: Plot v(vs), i(Lx), v(vo), and i(Rld) in dB to inspect A_{LV}, A_{DO}, A_{VO}, and A_{IO}.

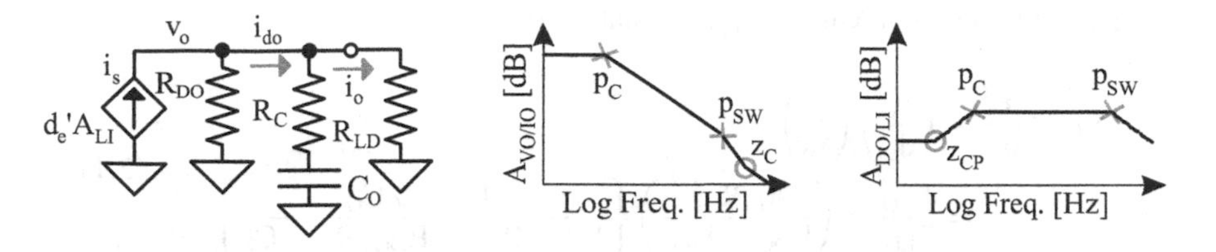

Fig. 5.40 Small-signal model of the power stage in DCM

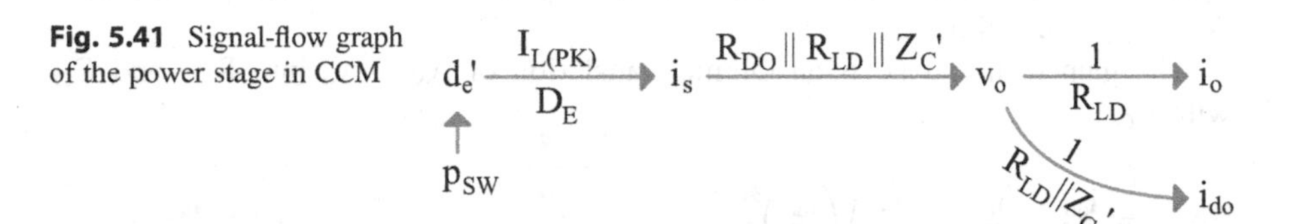

Fig. 5.41 Signal-flow graph of the power stage in CCM

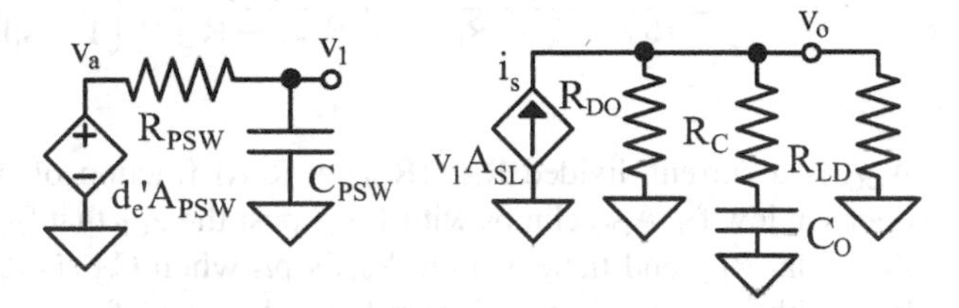

Fig. 5.42. DCM frequency-response model of the switched inductor

B. **Discontinuous Conduction**

Frequency Response The switched inductor in discontinuous conduction is a current-sourced resistor. So a short-circuit current i_s and R_{DO} in Fig. 5.40 output a small-signal current i_{do} into R_{LD} and C_O with R_C. i_{LD} is absent because i_{LD} is static.

d_E' variations d_e' in Figs. 5.36, 5.41, and 5.42 propagate to i_s, v_o, i_{do}, and i_o. The gain to v_o is A_{LI}'s i_s into R_{DO}, R_{LD}, and C_O with R_C:

$$\begin{aligned} A_{VO} \equiv \frac{v_o}{d_e{}'} &= \left(\frac{i_s}{d_e{}'}\right)\left(\frac{v_o}{i_s}\right) = A_{LI}[R_{DO}||R_{LD}||(Z_C + R_C)] \\ &= \left(\frac{I_{L(PK)}}{D_E}\right)\left(\frac{R_{DO}||R_{LD}}{1 + s/2\pi p_{SW}}\right)\left\{\frac{1 + sR_CC_O}{1 + s[R_C + (R_{DO}||R_{LD})]C_O}\right\}. \end{aligned} \tag{5.115}$$

A_{VO} is A_{LI}'s $I_{L(PK)}/D_E$ into R_{DO} and R_{LD} when C_O opens at low f_O. A_{VO} falls past p_C when C_O and R_C shunt R_{DO} and R_{LD}, z_C reverses p_C when C_O shorts with respect to R_C, and p_{SW} accelerates A_{VO}'s fall past f_{SW}. Z_C' in Fig. 5.41 is C_O's impedance with R_C.

The gain to i_o is an Ohmic R_{LD} translation of A_{VO}'s v_o:

$$A_{IO} \equiv \frac{i_o}{d_e{}'} = \left(\frac{v_o}{d_e{}'}\right)\left(\frac{i_o}{v_o}\right) = \frac{A_{VO}}{R_{LD}} = \left(\frac{A_{LI}R_{DO}}{R_{DO} + R_{LD}}\right)\left\{\frac{1 + sR_CC_O}{1 + s[R_C + (R_{DO}||R_{LD})]C_O}\right\}. \tag{5.116}$$

A_{IO} is a current-divided $R_{DO}/(R_{DO} + R_{LD})$ fraction of A_{LI}'s $I_{L(PK)}/D_E$ when C_O opens at low f_O. Since A_{IO} is just an Ohmic R_{LD} translation of A_{VO}, A_{IO} includes A_{VO}'s p_C, z_C, and p_{SW}.

The gain to i_l and i_{do} is an Ohmic translation of A_{VO}'s v_o into R_{LD} and C_O with R_C:

$$A_{DO} \equiv \frac{i_{do}}{d_e{}'} = \frac{i_l}{d_e{}'} = \left(\frac{v_o}{d_e{}'}\right)\left(\frac{i_{do}}{v_o}\right) = \frac{A_{VO}}{R_{LD}||(Z_C + R_C)} = \left(\frac{A_{LI}R_{DO}}{R_{DO} + R_{LD}}\right)\left\{\frac{1 + s(R_C + R_{LD})C_O}{1 + s[R_C + (R_{DO}||R_{LD})]C_O}\right\}. \tag{5.117}$$

A_{DO} is a current-divided $R_{DO}/(R_{DO} + R_{LD})$ fraction of A_{LI}'s $I_{L(PK)}/D_E$ when C_O opens at low f_O. A_{DO} climbs with $1/Z_C{}'$ past the z_{CP} that f_{CP} sets when C_O and R_C in $Z_C{}'$ shunt R_{LD} and flattens with A_{VO}'s p_C when C_O shorts with respect to R_C and R_{DO} with R_{LD}. A_{VO}'s p_{SW} then reduces A_{DO} past f_{SW}.

Example 19: Determine A_{VO} and A_{DO} at low f_O and related poles and zeros in DCM when d_E and d_O are 50%, t_C is 570 ns, t_{SW} is 1 μs, $i_{L(PK)}$ is 57 mA, and R_{LD} is 500 Ω with the L_X, R_L, C_O, and R_C specified in Example 9.

Solution:

$$R_{DO} = 2\left(\frac{L_X}{D_O{}^2}\right)\left(\frac{T_{SW}}{T_C{}^2}\right) + \frac{R_L}{D_O} = 2\left(\frac{10\mu}{50\%^2}\right)\left(\frac{1\mu}{570n^2}\right) + \frac{50m}{50\%} = 250\ \Omega$$

$$A_{LI} \approx \frac{I_{L(PK)}}{D_E} = \frac{57m}{50\%} = 110\ \text{mA/V} = -19\ \text{dB}$$

$$A_{VO0} = A_{LI}(R_{DO} \parallel R_{LD}) = (110m)(250 \parallel 500) = 18\ V/V = 25\ dB$$

$$A_{DO0} = \frac{A_{VO0}}{R_{LD}} = \frac{18}{500} = 36\ mA/V = -29\ dB$$

$$z_{CP} = \frac{1}{2\pi(R_C + R_{LD})C_O} = \frac{1}{2\pi(10m + 500)(5\mu)} = 64\ Hz$$

$$p_C = \frac{1}{2\pi[R_C + (R_{DO} \| R_{LD})]C_O} = \frac{1}{2\pi[10m + (250 \| 500)](5\mu)} = 190\ Hz$$

$$p_{SW} \approx f_{SW} = 1\ MHz \text{ from Example 17}$$

$$z_C = 3.2\ MHz \text{ from Example 9}$$

Note: A_{DO} excludes z_C. z_{CP} is in A_{DO} and absent in A_{VO} because A_{DO} is an Ohmic translation of A_{VO} that C_O and R_C set.

Circuit Model The circuit in Fig. 5.42 models the DCM response of the switched inductor. A_{PSW} buffers d_e' so C_{PSW} can establish p_{SW} when C_{PSW} shunts R_{PSW} into A_{PSW}. A_{PSW} is one so the low-f_O gain from d_e' to v_1 is one. This way, p_{SW} is the only effect of the buffer stage on the overall gain of the model.

Explore with SPICE:

See Appendix A for notes on SPICE simulations.

```
* Switched Inductor: DCM Frequency-Response Model
vde de 0 dc=0 ac=1
epsw va 0 de 0 1
rpsw va v1 1
cpsw v1 0 159.2n
gsl 0 vs v1 0 110m
rdo vs 0 250
vio vs vo dc=0
rc vo vc 10m
co vc 0 5u
rld vo 0 500
.ac dec 1000 10 10e6
.end
```

Tip: Plot i(Gis), i(Vio), v(vo), and i(Rld) in dB to inspect A_{LI}, A_{DO}, A_{VO}, and A_{IO}.

5.6 Summary

Two-port models are two- to four-component networks that can model the behavior of almost any circuit. Their inputs and outputs are interdependent sources with impedances. These networks work because sources model what impedances do not. In other words, sources are the voltages or currents that result when the effects of impedances are absent and impedances are the Ohmic translations that result when disabling sources. This way, input–output combinations can model feedback and forward translations.

Impedances can be resistive, capacitive, and inductive. Resistance is the part that does not scale with frequency. Capacitors and inductors are, in a way, like Ohmic switches. This is because capacitors close and inductors open as frequency increases. They also delay translations because capacitors and inductors require time to energize. So capacitor voltages and inductor currents lag the currents and voltages that drive them.

Couple capacitors feed current and shunt capacitors pull current. And couple inductors impede current and shunt inductors avail current. Poles are the transitional frequencies that result when gain and the phase shift that delay produces in sinusoids fall. Zeros oppose these effects: they raise gain and recover phase by the same amount poles lose them.

The effects that capacitors and inductors produce when they shunt and overcome resistances reverse when they short and open. So resistors that current-limit capacitors and voltage-limit inductors reverse the poles and zeros that capacitors and inductors produce. And bypass capacitors add zeros when they feed more current than the circuits they bypass. These zeros are out-of-phase when they invert translations.

Poles and zeros in LC circuits hinge on the frequency where inductor impedance overcomes capacitor impedance. At this f_{LC}, Z_L and Z_C are equal complements, so parallel LC networks open and series LC networks short (as much as their parallel and series resistors allow). This is how parallel and series combinations can peak their voltages and currents at f_{LC}.

But peaking results only when resistors do not keep inductors and capacitors from interacting at f_{LC}. So to peak, capacitors should shunt parallel resistances, inductors should overcome series resistances, and capacitors should not short below f_{LC}. If not, the peak fades and the double pole that produces the peak splits or reduces to one lower-frequency pole.

Switched inductors in power supplies behave this way because they feed output capacitors that help supply a load. Although inductor voltage and current are largely nonlinear, variations across cycles are small, and in consequence, almost linear. Switched inductors respond and propagate small signals like LC networks for this reason.

Switchers in continuous conduction duty-cycle energize and drain voltages across the inductor that feeds the output and its capacitor. The voltage-sourced LC networks that result normally peak at f_{LC} because series resistances are usually low and load resistance is a few Ohms or higher. A zero reverses the pole that the output capacitor induces when the capacitor shorts. And the switcher delays and suppresses sub-cycle variations that would otherwise generate signals above the switching frequency.

Boost-derived switched inductors in continuous conduction add an out-of-phase zero. This is because their outputs receive current only when the inductor drains. Since extending energize time shortens drain time, raising inductor energy reduces drain current. This sacrifice eventually overcomes the gain because inductors carry less energy at higher frequencies (i.e., inductor impedance climbs with frequency).

Switched inductors are less sensitive to frequency in discontinuous conduction. This is because energize and drain times scale together. So variations in one do not oppose the other (which is what causes the out-of-phase zero in the first place). And average inductor voltage is always the same (whose variation in continuous conduction is responsible for inductive effects). So in this mode, switched inductors are current-sourced resistors that output capacitors shunt and eventually short to produce a pole and a zero. Like in CCM, the switcher delays and suppresses signals that surpass the switching frequency.

Feedback Control

6

Abbreviations

ADC	Analog–digital converter
CCM	Continuous-conduction mode
DCM	Discontinuous-conduction mode
DSP	Digital-signal processor
GBW	Gain–bandwidth product
GM	Gain margin
LED	Light-emitting diode
LSB	Least-significant bit
OA	Operational amplifier/op amp
OTA	Operational transconductance amplifier
PM	Phase margin
PWM	Pulse-width modulator
SL	Switched inductor
A_0	Zero-/low-frequency gain
$A_β$	Feedback gain
A_{CL}	Closed-loop gain
A_{DIG}	Digital gain
A_E	Error amplifier/amp
A_F	Overall forward gain
A_{FW}	Forward gain
A_G	Transconductance gain
A_{LG}	Loop gain
A_{PRE}	Pre-amplifier/pre-amp gain
A_{PWM}	PWM gain
A_S	Stabilizer gain
A_{SL}	Switched-inductor gain
A_V	Amplifier voltage gain

G. A. Rincón-Mora, *Switched Inductor Power IC Design*,
https://doi.org/10.1007/978-3-030-95899-2_6

β_{FB}	Feedback translation/scaler
C_X	Parasitic capacitance
C_O	Output capacitor
d_O	Output duty cycle
d_E	Energize duty cycle
d_E'	Energize duty-cycled command
Δi_{LD}	Load dump
f_{0dB}	Unity-gain frequency
$f_{180°}$	Inversion frequency
f_{BW}	Bandwidth frequency
$f_{BW(CL)}$	Closed-loop bandwidth
f_{LC}	Transitional LC (resonant) frequency
f_O	Operating frequency
f_{SW}	Switching frequency
i_{FB}	Feedback current
i_I	Input current
i_L	Inductor current
i_{LD}	Load current
$i_{L(PK)}$	Peak inductor current in DCM
i_O	Output current
i_s	Small-signal current source
L_{DO}	Duty-cycled inductance in CCM
N_{CLK}	Number of clock cycles
N_{LSB}	Number of LSBs
p_A	Amplifier pole
p_{BW}	Bandwidth-setting pole
p_C	Capacitor pole
p_L	Inductor pole
p_O	Output pole
p_{PWM}	PWM pole
p_{SW}	Switching pole
Q_{LC}	LC quality factor
R_C	Capacitor resistance
R_{DO}	Duty-cycled resistance in DCM
R_{IN}	Input resistance
R_L	Inductor resistance
R_{LO}	Duty-cycled inductor resistance
R_{LD}	Load resistance
R_O	Output resistance
R_S	Series resistance
s_E	Error signal
s_I	Input signal
s_O	Output signal
s_{FB}	Feedback signal

While the underlying aim of power supplies is to transfer power, the more visible and distinguishable objective is conditioning it into a form that the load can use. In most cases, this amounts to setting and regulating the voltage or current with which the power supply delivers power. *Voltage regulators*, for example, set voltages and *battery chargers* and *light-emitting diode* (LED) *drivers* set currents.

This responsibility of setting the output rests on the *feedback controller*. Its basic function is to adjust the switching action of the power supply so the output nears its target. This way, the controller counters the deviations that external factors (like load variations) would otherwise produce in the output.

But when delayed, rather than correcting the error, adjustments can reinforce and amplify the error. Avoiding this is not so easy when considering the *switched inductor* (SL) already delays the response and circuits in the controller require time to react. So in addition to sensing and correcting fluctuations in the output, the feedback controller must also manage delays.

6.1 Negative Feedback

6.1.1 Model

Negative feedback sets and keeps the output from deviating. It does this by sensing and comparing the output against a reference input and using the error sensed to adjust the output. Sensing, comparing, and adjusting the output closes a feedback loop in Fig. 6.1 that couples the input to the output.

Amplifying the error reduces deviations in the output. And translating the output before the comparison extends how far the output can reach. In other words, the efficacy and flexibility of the loop center on gain and translation.

In all, feedback loops perform four basic functions: sense, translate, compare, and amplify. The *sampler* in Fig. 6.2 senses the *output signal* s_O and the *feedback scaler*

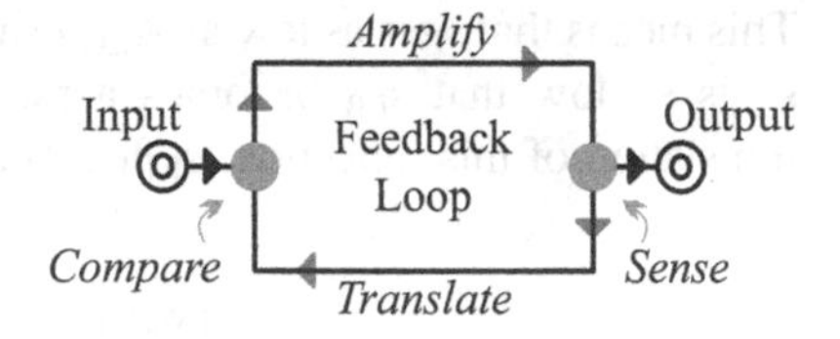

Fig. 6.1 Feedback actions

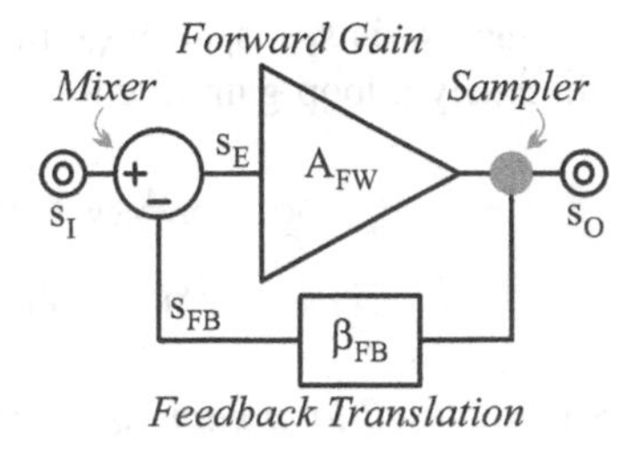

Fig. 6.2 Inverting feedback loop

β_{FB} translates s_O to the *feedback signal* s_{FB}. The *mixer* compares this s_{FB} to the *input signal* s_I and the *forward gain* A_{FW} amplifies the *error signal* s_E that results. So by definition, β_{FB} and A_{FW} are

$$\beta_{FB} \equiv \frac{s_{FB}}{s_O} \tag{6.1}$$

and

$$A_{FW} \equiv \frac{s_O}{s_E} = \frac{s_O}{s_I - s_{FB}}. \tag{6.2}$$

s_I, s_{FB}, and s_E carry the same dimensional units because the mixer can only compare and output signals of the same nature. The *loop gain* A_{LG} is the gain across the loop, which by definition, is also the gain from s_E to s_{FB}:

$$A_{LG} \equiv \frac{s_{FB}}{s_E} = \frac{s_{FB}}{s_I - s_{FB}} = A_{FW}\beta_{FB}. \tag{6.3}$$

The loop is inverting because s_E inverts s_{FB} fluctuations. This negative feedback action is what counters deviations in the output.

6.1.2 Translations

Since s_E is the difference between s_I and s_{FB} and A_{LG} translates s_E to s_{FB}, s_E is ultimately a loop-gain fraction of s_I:

$$s_E = s_I - s_{FB} = s_I - s_E A_{FW}\beta_{FB} = s_I - s_E A_{LG} = \frac{s_I}{1 + A_{LG}}. \tag{6.4}$$

This means that s_E is as low as A_{LG} is high. And when A_{LG} is much higher than one, s_E is so low that s_{FB} becomes a mirrored reflection of s_I and s_O a reverse β_{FB} translation of this reflection: s_I/β_{FB}. But generally, s_{FB} is

$$s_{FB} = s_E A_{FW}\beta_{FB} = (s_I - s_{FB})A_{LG} = \frac{s_I A_{LG}}{1 + A_{LG}}. \tag{6.5}$$

Since s_E is $s_I - s_{FB}$, A_{FW} translates s_E to s_O, and β_{FB} translates s_O to s_{FB}, s_O is ultimately a loop-gain fraction of $s_I A_{FW}$:

$$\begin{aligned} s_O &= s_E A_{FW} = (s_I - s_{FB})A_{FW} \\ &= (s_I - s_O\beta_{FB})A_{FW} = \frac{s_I A_{FW}}{1 + A_{FW}\beta_{FB}} = \frac{s_I A_{FW}}{1 + A_{LG}}. \end{aligned} \tag{6.6}$$

So the *closed-loop gain* A_{CL} from s_I to s_O is a loop-gain fraction of A_{FW}:

$$A_{CL} \equiv \frac{s_O}{s_I} = \frac{A_{FW}}{1 + A_{LG}} = \frac{A_{FW}}{1 + A_{FW}\beta_{FB}} = A_{FW}||\frac{1}{\beta_{FB}}. \tag{6.7}$$

When A_{LG} is much greater than one, A_{FW}'s cancel and A_{CL} reduces to $1/\beta_{FB}$.

A_{CL} is like the parallel combination of two forward translations: A_{FW} and $1/\beta_{FB}$. Using this analogy, A_{CL} follows the lowest forward translation. This is useful when determining the effects of feedback on A_{CL}, which diminish as A_{FW} falls and disappear when A_{FW} drops below $1/\beta_{FB}$, which happens when A_{LG} is less than one.

6.1.3 Frequency Response

A_{CL} follows the frequency response of the lowest forward translation. In Fig. 6.3, for example, A_{CL} (the boundary of the gray region) follows A_{FW} up to p_{X1} and past p_{X2} because A_{FW} is lower than $1/\beta_{FB}$ below p_{X1} and above p_{X2}. A_{CL} in Fig. 6.4 similarly follows A_{FW} between p_{X1} and p_{X2} and past p_{X34}. $1/\beta_{FB}$ sets A_{CL} otherwise.

So zeros and poles in A_{FW} are in A_{CL} when A_{FW} is lower. z_{FW1} and p_{FW4} in Fig. 6.3, for instance, are z_1 and p_3 in A_{CL}. And z_{FW1} in Fig. 6.4 is z_2 in A_{CL}.

Zeros and poles in β_{FB} are poles and zeros in $1/\beta_{FB}$. In Fig. 6.4, p_{FB1} and p_{FB2} in β_{FB} are zeros in $1/\beta_{FB}$ and z_{FB1} in β_{FB} is a pole in $1/\beta_{FB}$. Zeros and poles in $1/\beta_{FB}$ are also in A_{CL} when $1/\beta_{FB}$ is lower than A_{FW}. In Fig. 6.4, for example, p_{FB1} and p_{FB2} are zeros in $1/\beta_{FB}$ that set z_1 and z_3 in A_{CL}.

Poles and zeros also appear or disappear at gain crossings. In Fig. 6.3, z_{FW1} disappears at p_{X1} and p_{FW3} appears at p_{X2}. In Fig. 6.4, z_1 disappears at p_{X1}, z_2 disappears at p_{X2}, and z_3 disappears and p_{FW2} appears at p_{X34}. Reversing z_3 and falling with p_{FW2} at p_{X34} produces a double pole in A_{CL}.

Each pole and zero raises and lowers gain by the same amount that frequency climbs: 10× or 20 dB for every 10× or decade rise in frequency. In other words, gain and frequency scale. p_{X1}, for example, is $(1/\beta_{FB})/A_{FWO}$ times greater than z_1 in

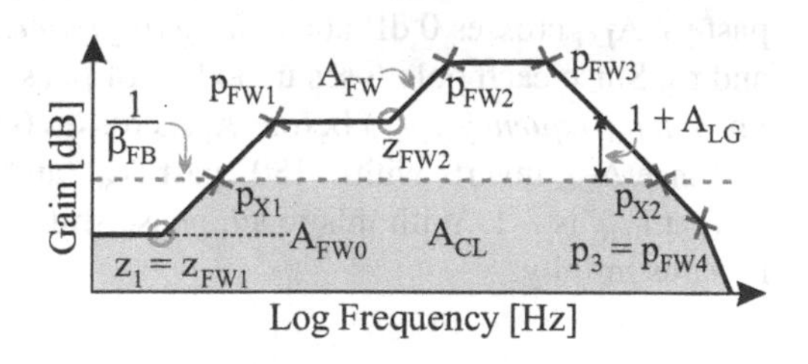

Fig. 6.3 Closed-loop response with constant β_{FB}

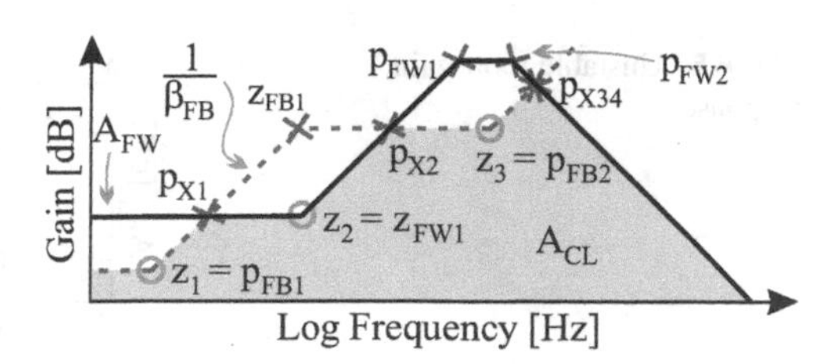

Fig. 6.4 Closed-loop response with variable β_{FB}

Fig. 6.3 and $A_{FW0}/(1/\beta_{FB0})$ times greater than z_1 in Fig. 6.4, where "0" in subscripts refers to gain at low frequency. Similarly, A_{CL} past p_{X2} in Fig. 6.4 is $1/\beta_{FB}$ past z_{FB1}, which is z_{FB1}/p_{FB1} times greater than $1/\beta_{FB0}$.

Like with gain, A_{CL}'s phase follows the phase of the lowest forward translation. So $\angle A_{CL}$ in Figs. 6.3 and 6.4 climbs up to 90° with z_{FW1} in A_{FW} and loses up to as much past p_{X1} in Fig. 6.3 and p_{X2} in Fig. 6.4 when $1/\beta_{FB}$ begins to dominate A_{CL}. $\angle A_{CL}$ in Fig. 6.4 also climbs up to 90° with the zero that p_{FB2} in β_{FB} sets and loses up to as much plus up to another 90° past p_{X34} when p_{FW2} in A_{FW} reduces A_{CL}.

6.1.4 Stability

A. Gain Objective

The principal aim of a feedback loop is to set an s_O that is a reverse β_{FB} translation of s_I's mirrored reflection. For A_{CL} to follow this translation, $1/\beta_{FB}$ should be lower than A_{FW}. But since gain is another goal, this $1/\beta_{FB}$ should be one or greater. So in practice, A_{FW} is usually higher than $1/\beta_{FB}$ across frequencies of interest and β_{FB} is lower than or equal to one.

B. Stability Criterion

High A_{LG} is desirable in feedback systems because amplifying s_E reduces the mismatch between s_I and s_{FB}. Translating s_O to s_{FB}, comparing s_{FB} to s_I, and amplifying the resulting s_E so this A_{LG} is high and s_O is accurate usually requires two or more stages in the loop. Since each stage incorporates one or more poles, finding two or more poles in A_{LG} is not uncommon.

In Fig. 6.5, just to cite an example, A_{LG}'s *zero-* or *low-frequency gain* A_{LG0} is well over 1 or 0 dB. A_{LG} falls 20 dB per decade past p_1 and another 20 dB per decade past p_2. A_{LG} crosses 0 dB at a *unity-gain frequency* f_{0dB} that is much higher than p_1 and p_2. Since each pole loses up to 90° of phase shift, $\angle A_{LG}$ reaches −180° (at the *inversion frequency* $f_{180°}$) before A_{LG} crosses 0 dB.

Since A_{LG} inverts with −180° past $f_{180°}$ and this inversion happens below f_{0dB}, A_{LG} at f_{0dB} is −1. With this much phase shift, positive feedback peaks A_{CL} at f_{0dB} towards infinity:

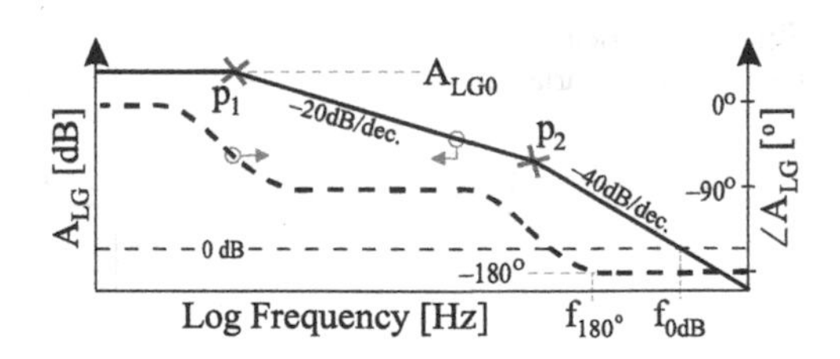

Fig. 6.5 Unstable loop-gain response

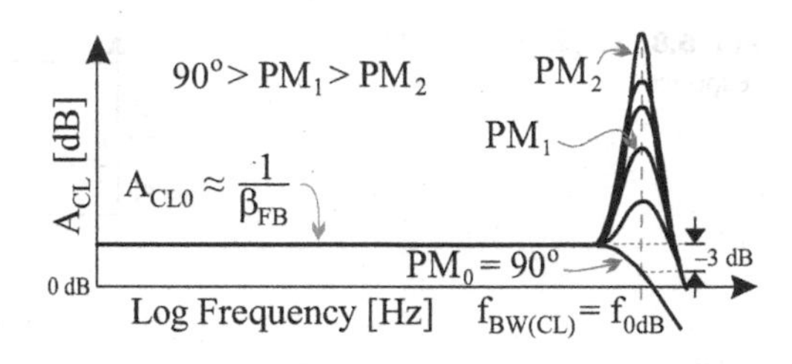

Fig. 6.6 Closed-loop response

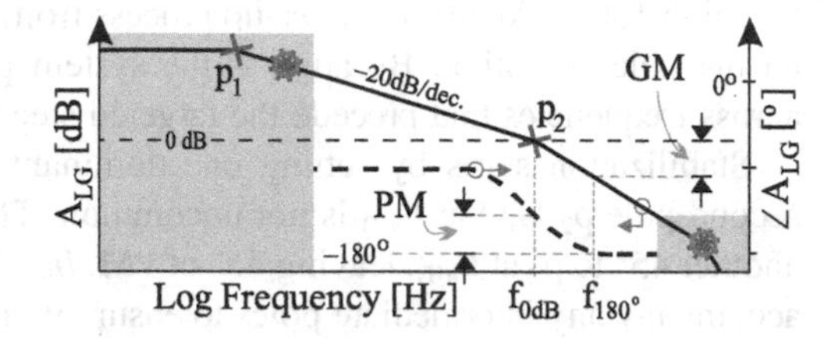

Fig. 6.7 Stable loop-gain response

$$A_{CL} = A_{FW} || \frac{1}{\beta_{FB}} = \frac{A_{FW}}{1 + A_{LG}}\bigg|_{A_{LG}=1\angle 180^\circ} = \frac{A_{FW}}{1-1} \rightarrow \infty. \tag{6.8}$$

To limit this peak, A_{LG} should reach f_{0dB} with less than 180° of phase shift. In other words, A_{LG} should reach f_{0dB} before f_{180°. This is the *stability criterion.*

So A_{CL} follows $1/\beta_{FB}$ in Fig. 6.6 and peaks at f_{0dB} to the extent that $\angle A_{LG}$ allows. The peak diminishes when the guard-band of unused phase-shift allowance, which is known as *phase margin* (PM), increases:

$$PM = 180^\circ + \angle A_{LG(0dB)} > 0^\circ. \tag{6.9}$$

The peak fades when this PM is 90°, which happens when p_2 is absent. Without p_2, p_1 produces the effect of a typical pole in A_{CL} at f_{0dB}.

f_{0dB} in A_{LG} is a gain crossing in $A_{FW} \parallel 1/\beta_{FB}$, where the effects of p_1 in A_{FW} appear in A_{CL} when A_{FW} falls below $1/\beta_{FB}$. So this f_{0dB} sets A_{CL}'s *closed-loop bandwidth* $f_{BW(CL)}$. And A_{CL} is stable (with a finite peak) when this f_{0dB} in Fig. 6.7 precedes f_{180°. A_{LG} at f_{180° should therefore be less than 1 or 0 dB. The guard-band this gain sets, known as *gain margin* (GM), is another measure of stability:

$$GM = 0\text{ dB} - A_{LG(180^\circ)} > 0\text{ dB}. \tag{6.10}$$

C. **Stabilization**

A_{CL} is stable when A_{LG} crosses 0 dB with less than 180° of phase shift, like Fig. 6.7 shows. Feedback stability is largely independent of what happens well below f_{0dB}. So A_{LG} can incorporate any combination of poles and zeros that keep $\angle A_{LG}$ from reaching 180° at f_{0dB}.

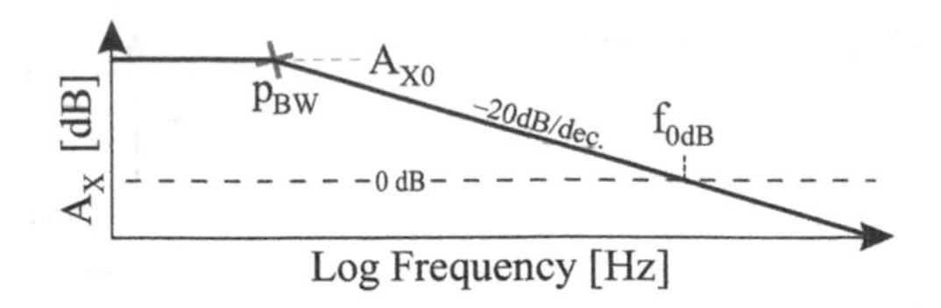

Fig. 6.8 Single-pole response

Even if $\angle A_{LG}$ dips to 180° before reaching f_{0dB}, A_{CL} remains stable as long as $\angle A_{LG}$ recovers PM at f_{0dB}. Engineers, however, normally keep $\angle A_{LG}$ from ever reaching 180° to keep the power-up process from creating and latching the system to an unstable condition. Because as the system powers, A_{LG0} rises and shifts f_{0dB} across frequencies that precede the targeted steady-state f_{0dB}.

Stabilization starts by setting one dominant low-frequency pole p_1. Letting a second pole p_2 land at f_{0dB} is not uncommon. This way, $\angle A_{LG}$ loses 90° to p_1 and another 45° to p_2 at f_{0dB}, leaving 45° of PM. *In-phase (left-hand-plane) zeros* should accompany any intermediate poles to ensure their combined contributions keep PM above 25°–30°.

Out-of-phase (right-half-plane) zeros should be ten times or more than ten times greater than f_{0dB}. This is because they invert and add gain. So they not only subtract phase but also extend f_{0dB} to higher frequency, where parasitic poles can deduct more phase.

D. **Gain–Bandwidth Product**

Gain A_X falls 20 dB per decade towards 0 dB like Fig. 6.8 shows when A_X incorporates one pole. The *gain–bandwidth product* (GBW) is A_X's low-frequency gain A_{X0} times this *bandwidth-setting pole* p_{BW}:

$$GBW \equiv A_{X0}p_{BW} \approx A_X f_{BW}\big|_{f_{BW} \geq p_{BW}} = f_{0dB} = \text{Constant.} \tag{6.11}$$

Since A_X drops as much as *operating frequency* f_O climbs past p_{BW}, GBW is also the product of A_X and the *bandwidth frequency* f_{BW} that f_O sets past p_{BW}, which as a result, is also f_{0dB} at 0 dB. This means that $A_X f_{BW}$ past p_{BW} is not only constant but also f_{0dB}. Projecting f_{0dB} this way: from the GBW that one pole sets, is useful when predicting and managing frequency response.

6.1.5 Loop Variations

A. **Pre-amplifier**

Amplifying the input of a feedback loop is equivalent to amplifying all forward translations. This is because a *pre-amplifier* A_{PRE}, or *pre-amp* for short, amplifies the forward translation that dominates:

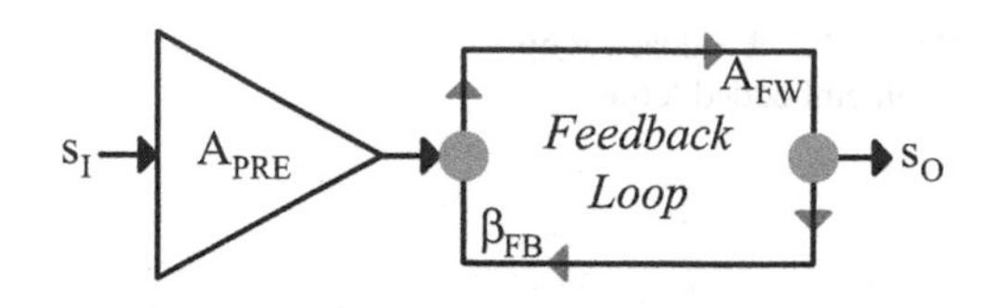

Fig. 6.9 Pre-amplified feedback loop

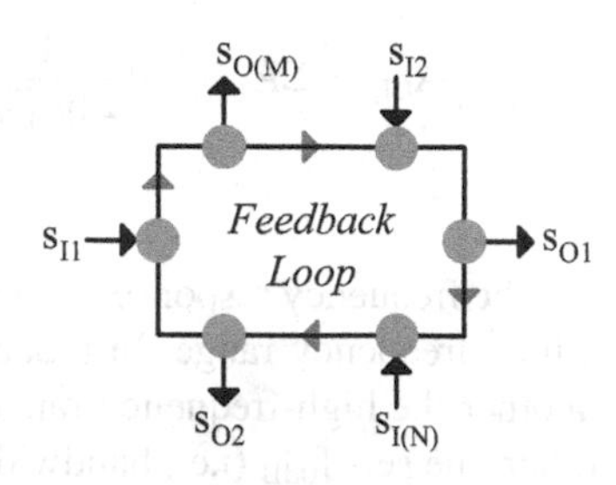

Fig. 6.10 Feedback loop with multiple inputs and outputs

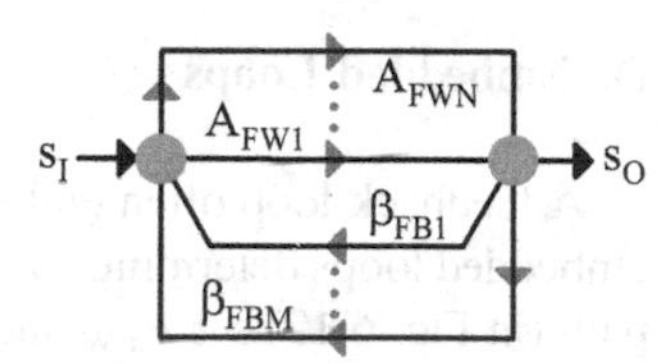

Fig. 6.11 Feedback loop with parallel paths

$$A_X \equiv \frac{s_O}{s_I} = A_{PRE}\left(A_{FW}||\frac{1}{\beta_{FB}}\right) = (A_{PRE}A_{FW})||\frac{A_{PRE}}{\beta_{FB}} \equiv A_F||A_\beta. \tag{6.12}$$

So the overall gain in Fig. 6.9 follows the pre-amplified lowest forward translation to s_O: the pre-amplified *overall forward gain* A_F that $A_{PRE}A_{FW}$ sets or the pre-amplified *feedback gain* A_β that A_{PRE}/β_{FB} sets.

B. Multiple Taps

A feedback loop can mix and sample multiple inputs and outputs. Since the loop is linear near its operating point, loop signals superimpose linear translations of the inputs. So an output in Fig. 6.10, which can be any signal in the loop, is the sum of individual closed-loop translations to that point:

$$s_{O(X)} = \sum_{K=1}^{N} s_{I(K)}A_{CL(K)} = \sum_{K=1}^{N} s_{I(K)}\left(A_{FW(K)}||\frac{1}{\beta_{FB(K)}}\right). \tag{6.13}$$

Each output is ultimately a unique closed-loop translation of each input.

C. Parallel Paths

Parallel paths in Fig. 6.11 can also amplify the error and translate the output. In these cases, the mixer and sampler add parallel A_{FW}'s and β_{FB}'s. This way, the highest-gain paths can dominate and set A_{CL}, and individual paths win dominance when their gains overwhelm the others:

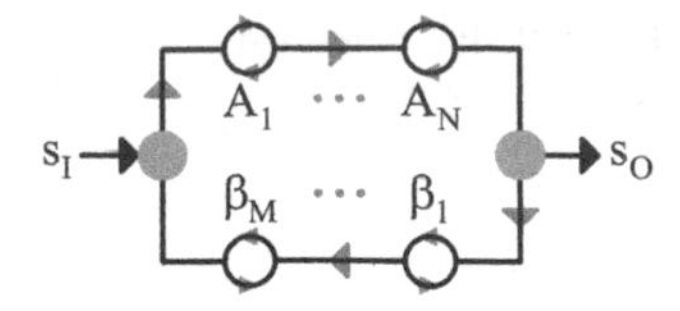

Fig. 6.12 Feedback loop with embedded loops

$$A_{CL} = \Sigma A_{FW(X)} || \frac{1}{\Sigma \beta_{FB(X)}} \approx \text{Max}\{A_{FW(X)}\} || \frac{1}{\text{Max}\{\beta_{FB(X)}\}}. \tag{6.14}$$

The frequency response of each parallel path dictates which translation dominates which frequency range. In practice, one path can set the low-frequency range and another the high-frequency range. This way, one sets the static translation and the other one sets f_{0dB} (i.e., bandwidth and stability criterion).

D. **Embedded Loops**

A feedback loop often embeds inner loops. The closed-loop translations of these embedded loops determine A_{FW}, β_{FB}, and as a result, A_{LG}. Inner loops in the forward path (in Fig. 6.12) set A_{FW} and inner loops in the feedback path set β_{FB}:

$$A_{CL} = A_{FW} || \frac{1}{\beta_{FB}} = \Pi_{X=1}^{N} A_X || \frac{1}{\Pi_{X=1}^{M} \beta_X}. \tag{6.15}$$

Designing feedback systems with embedded loops is usually a recursive process that starts with the outer loop. Determining A_{LG} and β_{FB} requirements for the outer loop is the first step. Setting and stabilizing individual inner loops is next. This way, with bandwidth-limited closed-loop translations of the inner loops in hand, formulating A_{FW} and β_{FB} and stabilizing A_{LG} is more straightforward.

6.2 Op-Amp Translations

6.2.1 Operational Amplifier

The *operational amplifier* (OA) in Fig. 6.13, or *op amp* for short, is useful in feedback systems because it can sense, amplify, and translate voltages. Its basic function is to amplify the difference between two input voltages, which is commonly known as the *differential input voltage* v_{ID}. Two other important features are high *input resistance* R_{IN} and low *output resistance* R_O.

R_{IN} is so much higher than other resistances that almost no current flows into the op amp. And R_O is so much lower than other resistances that R_O drops negligibly

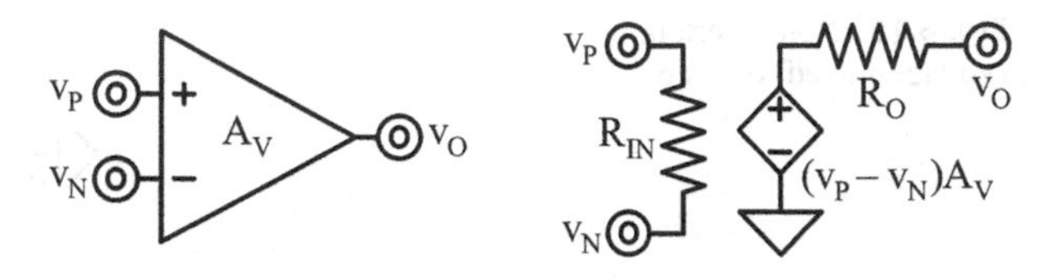

Fig. 6.13 Operational amplifier

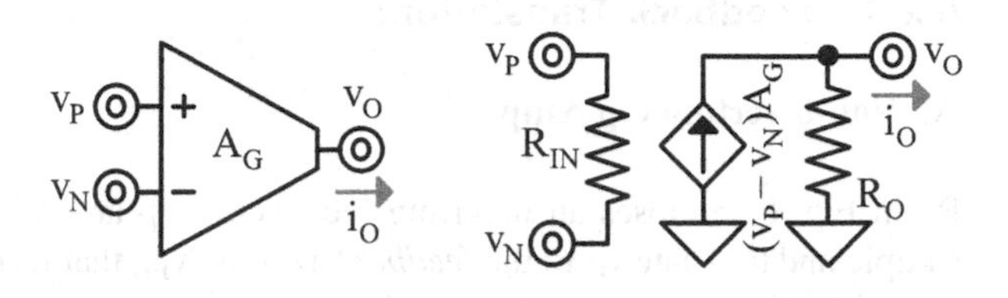

Fig. 6.14 Operational transconductance amplifier

low voltages. This way, R_{IN} does not load the input and the gain to the *output voltage* v_O is insensitive to loads that connect to the output.

The *amplifier voltage gain* A_V amplifies the v_{ID} between the *positive* and *negative inputs* v_P and v_N. A_V is normally constant up to the *amplifier pole* p_A. Past this p_A, A_V falls 20 dB per decade of frequency:

$$A_V \equiv \frac{v_O}{v_P - v_N} \approx \frac{A_{V0}}{1 + s/2\pi p_A}, \tag{6.16}$$

where A_{V0} is A_V's low-frequency gain. Zeros and other poles, if any, are so high, by design, that they do not affect frequencies of interest.

6.2.2 Operational Transconductance Amplifier

The *operational transconductance amplifier* (OTA) in Fig. 6.14 is basically an op amp with high R_O. R_O is so much higher than other resistances that the OTA is practically a current source. So the gain to the *output current* i_O is largely the *transconductance gain* A_G that amplifies v_P minus v_N:

$$A_G \equiv \frac{i_O}{v_P - v_N} \approx A_{G0}. \tag{6.17}$$

This A_G in OTAs is, also by design, largely independent of frequency. So A_G excludes the p_A that reduces A_V in op amps. Still, *parasitic capacitance* C_X across the *load resistance* R_{LD} at v_O reduces the gain to v_O past the *output pole* p_O that C_X sets when C_X shunts R_O and R_{LD}:

$$\left.\frac{1}{sC_X}\right|_{f_O \geq \frac{1}{2\pi(R_O||R_{LD})C_X} = p_O \approx \frac{1}{2\pi R_{LD} C_X}} \leq R_O||R_{LD} \approx R_{LD}. \tag{6.18}$$

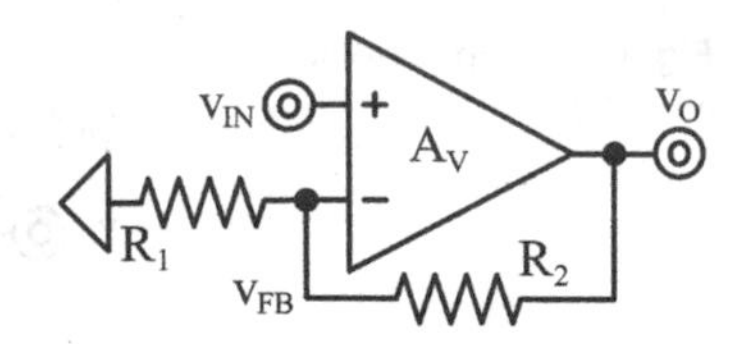

Fig. 6.15 Non-inverting (voltage-mixed) op amp

6.2.3 Feedback Translations

A. Non-inverting Op Amp

R_2 in Fig. 6.15 closes an inverting feedback loop around the op amp. R_2 with R_1 sample and translate v_O to the *feedback voltage* v_{FB} that feeds the op amp. This way, A_V mixes the *input voltage* v_{IN} and v_{FB}.

A_{FW} is nearly A_V because A_V amplifies the *error voltage* v_E between v_{IN} and v_{FB} to v_O and A_V's R_O is very low:

$$A_{FW} \equiv \frac{v_O}{v_E} = \frac{v_O}{v_{IN} - v_{FB}} \approx A_V = \frac{A_{V0}}{1 + s/2\pi p_A}. \tag{6.19}$$

Since A_V's R_{IN} is very high, β_{FB} is the v_O fraction that R_2 sets across R_1:

$$\beta_{FB} \equiv \frac{v_{FB}}{v_O} \approx \frac{R_1}{R_1 + R_2}. \tag{6.20}$$

So A_{LG} is $A_{FW}\beta_{FB}$ and A_{LG} reaches 0 dB at $A_{LG0}p_A$ or $A_{FW0}\beta_{FB}p_A$:

$$A_{LG} = A_{FW}\beta_{FB} \approx \left(\frac{A_{V0}}{1 + s/2\pi p_A}\right)\left(\frac{R_1}{R_1 + R_2}\right) \tag{6.21}$$

$$f_{0dB} \approx A_{LG0}p_A = A_{FW0}\beta_{FB}p_A \approx A_{V0}\left(\frac{R_1}{R_1 + R_2}\right)p_A. \tag{6.22}$$

And the voltage gain A_{VO} to v_O is A_{CL}'s $A_{V0} \parallel 1/\beta_{FB}$ up to f_{0dB}:

$$A_{VO} \equiv \frac{v_O}{v_{IN}} = A_{FW}||\frac{1}{\beta_{FB}} \approx \left(A_{V0}||\frac{R_1 + R_2}{R_1}\right)\left(\frac{1}{1 + s/2\pi f_{0dB}}\right). \tag{6.23}$$

This A_{VO} reduces to $1/\beta_{FB}$'s $(R_1 + R_2)/R_1$ up to f_{0dB} when A_{FW}'s A_{V0} is much greater than $1/\beta_{FB}$.

Example 1: Determine A_{FW0}, β_{FB}, A_{LG0}, f_{0dB}, A_{VO0}, and $f_{BW(CL)}$ when A_{V0} is 100 V/V, p_A is 10 kHz, R_1 is 10 kΩ, and R_2 is 90 kΩ.

Solution:

$$A_{FW0} \approx A_{V0} = 100\ V/V = 40\ dB$$

$$\beta_{FB} \approx \frac{R_1}{R_1 + R_2} = \frac{10k}{10k + 90k} = 100\ mV/V = -20\ dB$$

$$A_{LG0} = A_{FW0}\beta_{FB} \approx (100)(100m) = 10\ V/V = 20\ dB$$

$$f_{0dB} \approx A_{LG0}p_A \approx (10)(10k) = 100\ kHz$$

$$A_{VO0} = A_{FW0}||\frac{1}{\beta_{FB}} = 100||\frac{1}{100m} = 9.1\ V/V = 19\ dB$$

$$f_{BW(CL)} = f_{0dB} \approx 100\ kHz$$

Note: A_{VO0}'s 9.1 V/V is slightly lower than $1/\beta_{FB}$'s 10 V/V because A_{FW0}'s A_{V0} is only ten times greater than $1/\beta_{FB}$.

Explore with SPICE:
See Appendix A for notes on SPICE simulations.

```
* Non-inverting Op Amp
vi vi 0 dc=0 ac=1
eav0 va 0 vi vn 100
rpa va vb 1
cpa vb 0 15.92u
eavo vo 0 vb 0 1
r1 0 vn 10k
r2 vn vo 90k
.ac dec 1000 1k 10e6
.end
```

Tip: Plot v(vo) in dB to inspect A_{VO}.

B. **Inverting Op Amp**

R_2 in Figs. 6.15 and 6.16 close the same inverting feedback loop around the op amp. In Fig. 6.16, however, A_V's inputs do not mix v_{IN} and v_N. In this case, R_1 translates v_{IN} into an *input current* i_I, R_2 samples and translates v_O into a *feedback current* i_{FB}, and v_N mixes i_I and i_{FB}.

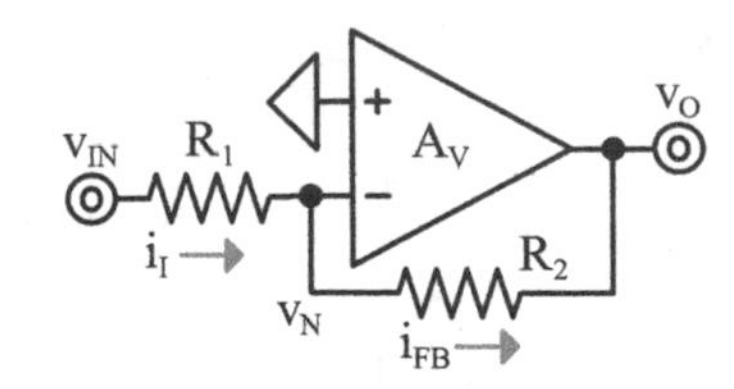

Fig. 6.16 Inverting (current-mixed) op amp

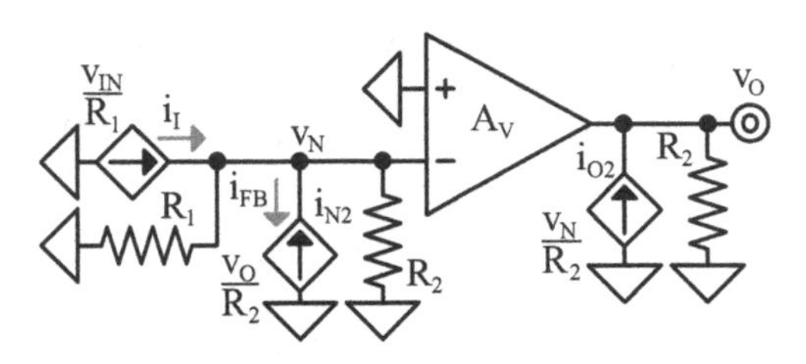

Fig. 6.17 Inverting (current-mixed) op-amp model

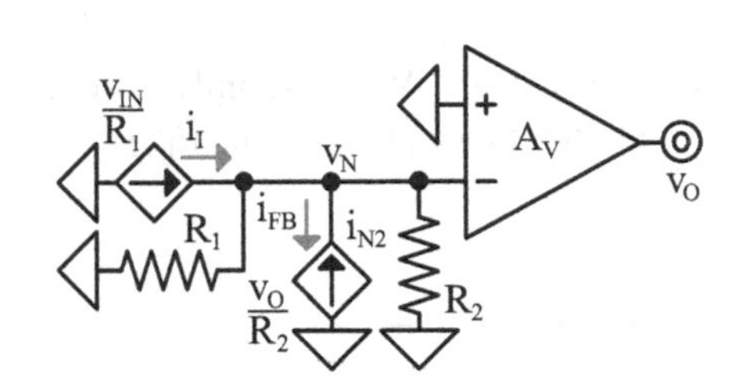

Fig. 6.18 Approximate inverting (current-mixed) op-amp model

To visualize and quantify how v_N mixes i_I and i_{FB}, consider R_1's and R_2's two-port models in Fig. 6.17. i_I–R_1 and i_{N2}–R_2 are Norton translations of v_{IN}–R_1 and v_O–R_2 where i_I and i_{N2} feed v_{IN}/R_1 and v_O/R_2 when a grounded v_N disables R_1 and R_2. i_{O2}–R_2 is a Norton translation of v_N–R_2 where i_{O2} feeds v_N/R_2 when a grounded v_O disables R_2.

v_N is therefore the voltage that i_I and i_{N2} drop across R_1 and R_2. But since i_{N2} is an inverted feedback translation of v_O, v_N is an Ohmic reflection of i_I minus i_{FB}. This is how v_N mixes i_I and i_{FB}.

When A_V's R_O is much lower than R_2, R_O sets the parallel resistance that R_O and R_2 establish at v_O. When A_V's equivalent current source $v_{ID}A_V/R_O$, which reduces to $-v_NA_V/R_O$ in Fig. 6.17, is similarly much greater than i_{O2}'s v_N/R_2, A_V sets the combined gain that A_V and R_2 establish. In other words, the effects of R_2 on v_O are negligible when A_V is a good op amp. In these cases, the simplified model in Fig. 6.18 is a good approximation.

Since R_{IN} is very high and R_O is very low in A_V, A_{FW} is the Ohmic and inverting translations that R_1 and R_2 establish at v_N and A_V sets across the op amp:

$$A_{FW} \equiv \frac{v_O}{i_E} = \frac{v_O}{i_I - i_{FB}} \approx (R_1||R_2)(-A_V) = -\frac{(R_1||R_2)A_{V0}}{1 + s/2\pi p_A}. \tag{6.24}$$

β_{FB} is an inverting reflection of the i_{N2} that v_O into R_2 sets:

$$\beta_{FB} \equiv \frac{i_{FB}}{v_O} = -\frac{i_{N2}}{v_O} = -\frac{1}{R_2}. \tag{6.25}$$

So A_{LG} is $A_{FW}\beta_{FB}$ and A_{LG} reaches 0 dB at $A_{LG0}p_A$ or $A_{FW0}\beta_{FB}p_A$:

$$\begin{aligned} A_{LG} &= A_{FW}\beta_{FB} \\ &= \left[-\frac{(R_1||R_2)A_{V0}}{1+s/2\pi p_A}\right]\left(-\frac{1}{R_2}\right) = \left(\frac{R_1}{R_1+R_2}\right)\left(\frac{A_{V0}}{1+s/2\pi p_A}\right) \end{aligned} \tag{6.26}$$

$$f_{0dB} \approx A_{LG0}p_A = A_{FW0}\beta_{FB}p_A \approx \left(\frac{R_1||R_2}{R_2}\right)A_{V0}p_A = \left(\frac{R_1}{R_1+R_2}\right)A_{V0}p_A. \tag{6.27}$$

And A_{VO} is $1/R_1$'s i_I translation of v_{IN} times A_{CL}'s translation $A_{FW0} \parallel 1/\beta_{FB}$ of i_I up to f_{0dB}:

$$\begin{aligned} A_{VO} &\equiv \frac{v_O}{v_{IN}} = \left(\frac{i_I}{v_{IN}}\right)\left(A_{FW}||\frac{1}{\beta_{FB}}\right) \\ &= A_F||A_\beta \approx \left(-\frac{R_2 A_{V0}}{R_1+R_2}\,||\,-\frac{R_2}{R_1}\right)\left(\frac{1}{1+s/2\pi f_{0dB}}\right). \end{aligned} \tag{6.28}$$

A_{VO} follows the lowest forward translation to v_O. A_F is the voltage-divided fraction that R_1 sets into R_2 times $-A_V$ and A_β is an Ohmic R_1 translation into $1/\beta_{FB}$'s $-R_2$. So when $A_V R_2/(R_1 + R_2)$ in A_F is greater than R_2/R_1 in A_β, A_{VO} follows A_β's $-R_2/R_1$ up to f_{0dB}.

Example 2: Determine A_{FW0}, β_{FB}, A_{LG0}, f_{0dB}, A_{VO0}, and $f_{BW(CL)}$ with the A_{V0}, p_A, R_1, and R_2 specified in Example 1.

Solution:

$$A_{FW0} \approx -(R_1 \,||\, R_2)A_{V0} = -(10k \,||\, 90k)(100) = -900\ kV/A = 120\ dB$$

$$\beta_{FB} = -\frac{1}{R_2} = \frac{1}{90k} = 11\ \mu A/V = -99\ dB$$

$$A_{LG0} = A_{FW0}\beta_{FB} \approx (-900k)(11\mu) = 9.9 = 20\ dB$$

$$f_{BW(CL)} = f_{0dB} \approx A_{LG0}p_A \approx (9.9)(10k) = 99\ kHz$$

$$A_{VO0} = -\frac{R_2 A_{V0}}{R_1 + R_2} || -\frac{R_2}{R_1}$$
$$= -\frac{(90k)(100)}{10k + 90k} || -\frac{90k}{10k} = -90|| -9 = -8.2\ V/V = 18\ dB$$

Note: Round-off errors reduce A_{LG0} and f_{0dB} to 9.9 and 99 kHz. Without these errors, A_{LG0} and f_{0dB} would be 10 and 100 kHz.

Explore with SPICE:
See Appendix A for notes on SPICE simulations.

```
* Inverting Op Amp
vi vi 0 dc=0 ac=1
r1 vi vn 10k
eav0 va 0 0 vn 100
rpa va vb 1
cpa vb 0 15.92u
eavo vo 0 vb 0 1
r2 vn vo 90k
.ac dec 1000 1k 10e6
.end
```

Tip: Plot v(vo) in dB to inspect A_{VO}.

C. **Differential Op Amp**

The op amp in Fig. 6.19 translates voltage- and current-mixed inputs into one v_O. Since A_V mixes v_I with v_{FB} and v_{FB} mixes i_I with i_{FB}, v_{FB} matches v_I and i_{FB} matches i_I to the extent A_{LG} determines. And v_O superimposes voltage- and current-mixed forward translations of v_{IN1} and v_{IN2}:

$$v_O = v_{IN1}\left(A_{F1}||A_{\beta1}\right) + v_{IN2}\left(A_{F2}||A_{\beta2}\right). \tag{6.29}$$

A_{F1} translates v_{IN1} to v_I and v_I – v_{FB} to v_O in the absence of v_{IN2} and A_{F2} translates v_{IN2} to i_I and i_I – i_{FB} to v_O in the absence of v_{IN1}:

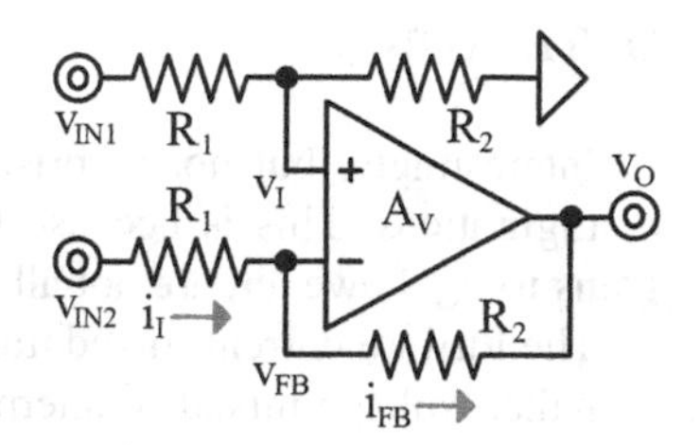

Fig. 6.19 Differential op amp

$$A_{F1} = \left(\frac{v_I}{v_{IN1}}\right)\left(\frac{v_O}{v_I - v_{FB}}\right)\Bigg|_{v_{IN2}=0} \approx \left(\frac{R_2}{R_1 + R_2}\right)A_V \quad (6.30)$$

$$A_{F2} = \left(\frac{i_I}{v_{IN2}}\right)\left(\frac{v_O}{i_I - i_{FB}}\right)\Bigg|_{v_{IN1}=0} \approx \left(\frac{R_2}{R_1 + R_2}\right)(-A_V) = -A_{F1}. \quad (6.31)$$

$A_{\beta 1}$ translates v_{IN1} to v_I and v_I to v_O in the absence of v_{IN2} and $A_{\beta 2}$ translates v_{IN2} to i_I and i_I to v_O in the absence of v_{I1}. So $A_{\beta 1}$'s $1/\beta_1$ and $A_{\beta 2}$'s $1/\beta_2$ are

$$\begin{aligned} A_{\beta 1} &= \frac{1}{\beta_1} = \left(\frac{v_I}{v_{IN1}}\right)\left(\frac{v_O}{v_I}\right)\Bigg|_{v_{IN2}=0} = \left(\frac{v_I}{v_{IN1}}\right)\left(\frac{1}{\beta_{FB1}}\right)\Bigg|_{v_{IN2}=0} \\ &= \left(\frac{R_2}{R_1 + R_2}\right)\left(\frac{R_1 + R_2}{R_1}\right) = \frac{R_2}{R_1} \end{aligned} \quad (6.32)$$

$$A_{\beta 2} = \frac{1}{\beta_2} = \left(\frac{i_I}{v_{IN2}}\right)\left(\frac{v_O}{i_I}\right)\Bigg|_{v_{IN1}=0} = \left(\frac{i_I}{v_{IN2}}\right)\left(\frac{1}{\beta_{FB2}}\right)\Bigg|_{v_{IN1}=0} = \left(\frac{1}{R_1}\right)(-R_2) = -A_{\beta 1}. \quad (6.33)$$

Since A_{F2} and $A_{\beta 2}$ are mirrored inversions of A_{F1} and $A_{\beta 1}$, the differential gain to v_O balances and matches v_{IN1}'s forward translations:

$$A_{VO} \equiv \frac{v_O}{v_{IN1} - v_{IN2}} = A_{F1}||A_{\beta 1} \approx \frac{R_2 A_V}{R_1 + R_2}||\frac{R_2}{R_1}. \quad (6.34)$$

When $A_{\beta 1}$'s R_2/R_1 is much lower than A_{F1}'s low-frequency gain, A_{VO} follows $A_{\beta 1}$ until p_A in A_V reduces A_{F1} below $A_{\beta 1}$:

$$\begin{aligned} A_{F1} &\approx \left(\frac{R_2}{R_1 + R_2}\right)\left(\frac{A_{V0}}{1 + s/2\pi p_A}\right)\Bigg|_{f_O > p_A} \\ &\approx \left(\frac{R_2}{R_1 + R_2}\right)A_{V0}\left(\frac{p_A}{f_O}\right)\Bigg|_{f_O \geq A_{V0}\left(\frac{R_1}{R_1 + R_2}\right)p_A = f_{0dB}} \leq A_{\beta 1} = \frac{R_2}{R_1}, \end{aligned} \quad (6.35)$$

where the "s" term overwhelms 1 above p_A. In other words, A_{VO} is nearly R_2/R_1 up to the f_{0dB} of the feedback loop that R_2 closes.

D. **Tradeoffs**

Interestingly, but not surprisingly, A_{LG} and f_{0dB} are the same for all op-amp configurations. This is because R_2 closes the same feedback loop. Their voltage gains to v_O, however, are not all the same.

The forward current-mixed translations to v_O (A_F and $A_β$) are lower in magnitude than their voltage-mixed counterparts. So for the same bandwidth, the voltage-mixed A_{VO} amplifies more than the current-mixed A_{VO}. Or for the same gain, the voltage-mixed A_{VO} amplifies to a higher f_{0dB} than the current-mixed A_{VO}.

But when inverting is necessary or convenient, or a differential translation is desirable, the sacrifice may be acceptable. For equal and opposite gains to v_O, for example, the voltage divider that feeds the voltage-mixed input of the multi-mixed op amp reduces the voltage-mixed gain A_{VO}. This way, the differential gain to v_O balances and matches the lower current-mixed gain A_{VO}.

6.3 Stabilizers

Feedback systems are stable when their loop gains reach 0 dB with less than 180° of phase shift. When needed, *stabilizers* add poles and zeros to ensure this happens. Although not always generalized this way, stabilizers normally adopt one of three basic strategies: Type I, II, or III.

6.3.1 Strategies

A. **Type I: Dominant Pole**

Type I adds one low-frequency pole p_{S1} that alone reduces A_{LG} in Fig. 6.20 to 0 dB. This way, after losing 90° of phase to p_{S1}, A_{LG} reaches f_{0dB} with 90° of phase margin. Even with a second pole p_{S2} at f_{0dB}, which loses another 45° at f_{0dB}, the margin reduces to 45°, which is still stable.

The *stabilizer gain* A_S is the part of A_{LG} that adds p_{S1}. Since no other pole or zero alters A_{LG} below f_{0dB}, A_{LG} drops with A_S past p_{S1}. And if A_S includes p_{S2}, A_{LG} falls faster with A_S past this p_{S2}. Although often off by a constant gain factor, A_{LG} with Type I stabilization usually follows A_S up to at least f_{0dB}.

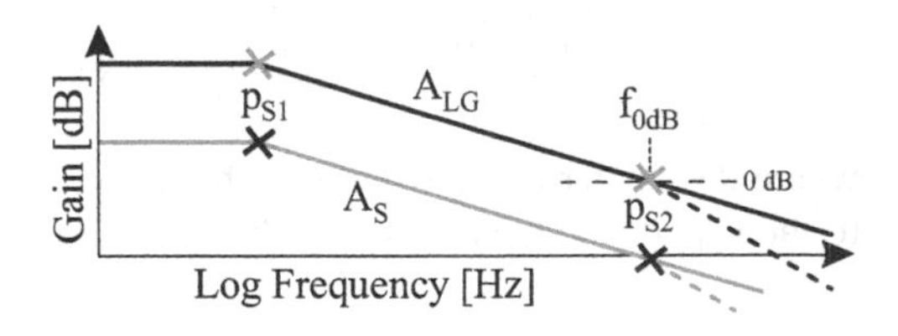

Fig. 6.20 Dominant-pole stabilization

B. Type II: Pole–Zero Pair

Type II adds p_{S1} and a zero z_{S1} that recovers the phase an intermediate pole p_1 in A_{LG} loses. This way, A_{LG} in Fig. 6.21 loses 180° of phase to p_{S1} and p_1, recovers 90° with z_{S1}, and reaches f_{0dB} with 90° of margin. Even with the loss of another pole p_{S2} at f_{0dB}, the margin is still 45°, which is stable.

A_S is the part of A_{LG} that adds p_{S1} and z_{S1}. z_{S1} should be within a decade of p_1 to keep A_{LG} from shifting 180°. Although often off by a constant gain factor, A_{LG} with Type II stabilization usually follows A_S up to p_1. Past p_1 and z_{S1}, A_{LG} falls and A_S flattens. And if A_S includes p_{S2}, A_{LG} falls faster than A_S past this p_{S2}.

C. Type III: Pole–Zero–Zero Triplet

Type III adds p_{S1} and two zeros z_{S1} and z_{S2} that recover the phase two intermediate poles p_1 and p_2 in A_{LG} lose. This way, A_{LG} in Fig. 6.22 loses 270° of phase to p_{S1}, p_1, and p_2 and recovers 180° with z_{S1} and z_{S2}, reaching f_{0dB} with 90° of margin. Even with the loss of another pole p_{S2} at f_{0dB}, the margin is still 45°, which is stable. Additional poles in A_S should be a decade or more higher than f_{0dB} to keep phase margin near 40°.

A_S is the part of A_{LG} that adds p_{S1}, z_{S1}, and z_{S2}. z_{S1} and z_{S2} should be above p_{S1} to let p_{S1} reduce A_{LG}, but within a decade of p_1 and p_2 to keep phase from shifting 180°. This way, A_{LG} follows A_S up to p_1 and continues to fall after z_{S1} and z_{S2} in A_S counter the effects of p_1 and p_2 in A_{LG}.

Parasitic poles in A_S eventually limit A_S's bandwidth. So after z_{S1} and z_{S2}, A_S flattens with p_{S2} and falls with p_{S3}. Although p_{S2} and p_{S3} are not always apart, only one of these poles can be close to f_{0dB} to keep PM from dipping below 30°–40°.

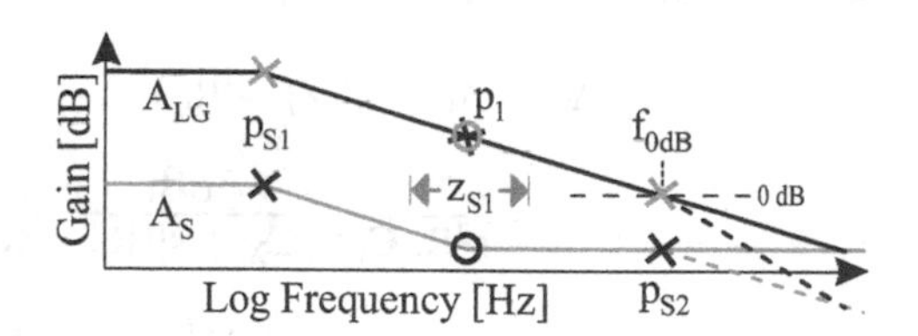

Fig. 6.21 Pole–zero stabilization

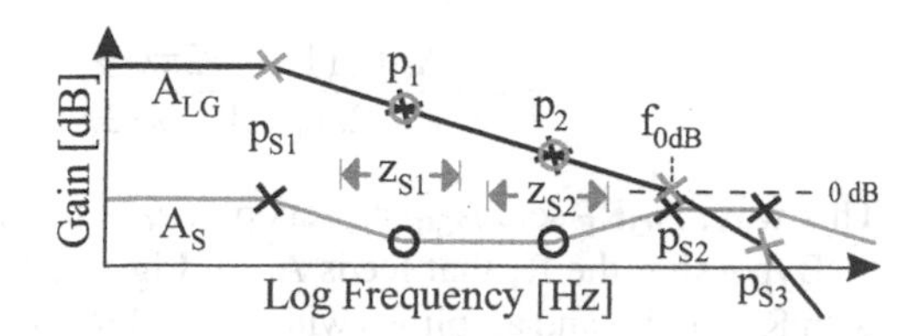

Fig. 6.22 Pole–zero–zero stabilization

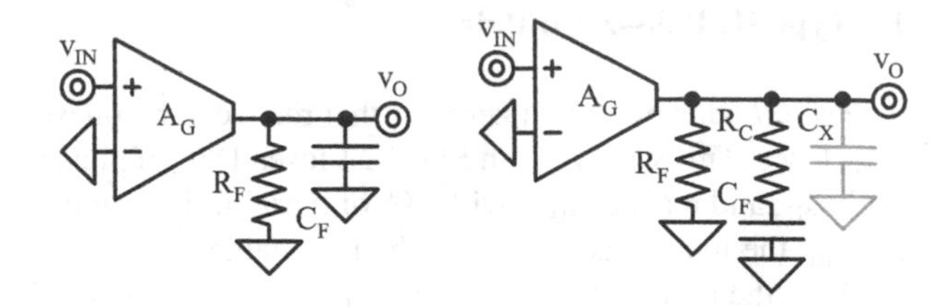

Fig. 6.23 Dominant-pole and pole–zero OTAs

6.3.2 Amplifier Translations

An op amp can add p_{S1}. This op amp, however, cannot be *any* op amp. This is because the low-frequency gain A_{S0} and p_{S1} that A_{V0} and p_A set should establish an f_{0dB} that keeps the feedback system stable:

$$A_S \approx \frac{A_{V0}}{1 + s/2\pi p_A}. \tag{6.36}$$

The OTAs in Fig. 6.23 can also add p_{S1}. A_{S0} is the gain that A_G sets across R_F. In the first implementation, A_S falls past p_F when C_F shunts R_F:

$$A_S \approx A_G\left(R_F || \frac{1}{sC_F}\right) = \frac{A_G R_F}{1 + sR_F C_F} = \frac{A_G R_F}{1 + s/2\pi p_F}. \tag{6.37}$$

Current-limiting C_F with R_C adds z_{S1}. With R_C, A_S falls past p_C when C_F and R_C shunt R_F before parasitic capacitance C_X at v_O shunts R_F. p_C eventually fades past z_{CX} when R_C current-limits C_F. Once C_F shorts, A_S flattens to $A_G(R_F \,||\, R_C)$ and later falls past p_O when C_X shunts R_F and R_C:

$$\begin{aligned} A_S &= A_G[R_F||(Z_F + R_C)||Z_X] \\ &= \frac{A_G R_F(1 + sC_F R_C)}{s^2 R_C C_F R_F C_X + s[(R_F + R_C)C_F + R_F C_X] + 1} \\ &\approx \frac{A_G R_F(1 + sR_C C_F)}{[1 + s(R_C + R_F)C_F][1 + s(R_F||R_C)C_X]} \\ &= \frac{A_G R_F(1 + s/2\pi z_{CX})}{(1 + s/2\pi p_C)(1 + s/2\pi p_O)}. \end{aligned} \tag{6.38}$$

This way, $A_G R_F$ sets A_{S0}, p_C sets p_{S1}, z_{CX} sets z_{S1}, and p_O sets p_{S2}.

Bypassing the R_1 that feeds A_G in Fig. 6.24 adds z_{S2}. Here, A_{S0} voltage-divides with R_1 and R_2 and amplifies with A_G and R_F. z_B raises A_S when C_B bypasses R_1 and p_{BX} reverses z_B when the parallel resistance R_1 and R_2 set current-limits C_B:

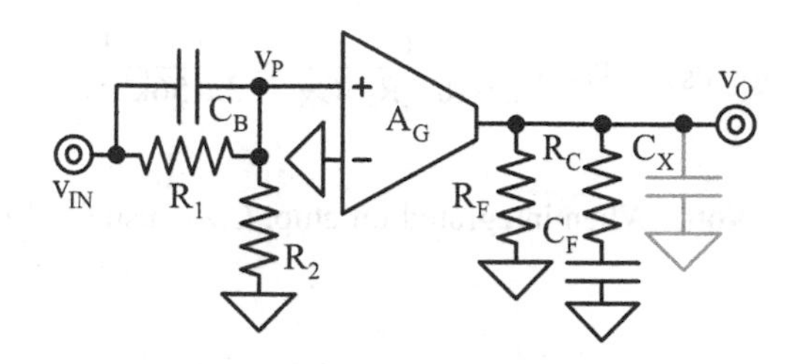

Fig. 6.24 Pole–zero–zero OTA

$$\begin{aligned}
A_S &\approx \left[\frac{R_2}{(R_1||Z_B) + R_2}\right] A_G[R_F||(Z_C + R_C)||Z_X] \\
&= \left(\frac{R_2 A_G R_F}{R_1 + R_2}\right)\left[\frac{(1 + R_1 C_B s)(1 + s/2\pi z_{CX})}{[1 + (R_1||R_2)C_B s](1 + s/2\pi p_C)(1 + s/2\pi p_O)}\right] \\
&= \left(\frac{R_2 A_G R_F}{R_1 + R_2}\right)\left[\frac{(1 + s/2\pi z_B)(1 + s/2\pi z_{CX})}{(1 + s/2\pi p_{BX})(1 + s/2\pi p_C)(1 + s/2\pi p_O)}\right].
\end{aligned} \quad (6.39)$$

This way, p_{S1} is p_C, z_{S1} and z_{S2} are z_{CX} and z_B, p_{S2} is p_{BX}, and p_{S3} is p_O.

Example 3: Determine R_2, C_B, A_G, C_F, R_C, and C_X so A_{S0} is 40 V/V, p_{S1} is 1 kHz, z_{S1} and z_{S2} are 10 kHz, p_{S2} exceeds 100 kHz, and p_{S3} matches or exceeds 1 MHz when R_1 and R_F are 500 kΩ.

Solution:

$$p_{S1} \equiv p_C < z_{S1} \equiv z_{CX} = z_{S2} \equiv z_B < p_{S2} \equiv p_{BX} < p_{S3} = p_O$$

$$\frac{z_{S1}}{p_{S1}} \equiv \frac{z_{CX}}{p_C} \approx \frac{R_C + R_F}{R_C} = \frac{R_C + 500k}{R_C} \equiv \frac{10\ \text{kHz}}{1\ \text{kHz}} \quad \therefore \quad R_C = 56\ \text{k}\Omega$$

$$\frac{p_{S2}}{z_{S2}} \equiv \frac{p_{BX}}{z_B} \approx \frac{R_1}{R_1||R_2} = \frac{500k}{500k||R_2} \equiv \frac{100\ \text{kHz}}{10\ \text{kHz}} \quad \therefore \quad R_2 = 56\ \text{k}\Omega$$

$$A_{S0} = \frac{R_2 A_G R_F}{R_1 + R_2} = \frac{(56k)A_G(500k)}{500k + 56k} \equiv 40\ \text{V/V} = 32\ \text{dB} \quad \therefore \quad A_G = 790\ \mu\text{A/V}$$

$$p_{S1} \equiv p_C \approx \frac{1}{2\pi(R_C + R_F)C_F} = \frac{1}{2\pi(56k + 500k)C_F} \equiv 1\ \text{kHz} \quad \therefore \quad C_F = 290\ \text{pF}$$

$$z_{S2} \equiv z_B \approx \frac{1}{2\pi R_1 C_B} = \frac{1}{2\pi(500k)C_B} \equiv 10\ \text{kHz} \quad \therefore \quad C_B = 32\ \text{pF}$$

$$p_{S3} = p_O \approx \frac{1}{2\pi(R_C||R_F)C_X} = \frac{1}{2\pi(56k||500k)C_X} \geq 1\ \text{MHz} \quad \therefore \quad C_X \leq 3.2\ \text{pF}$$

Note: When integrated on chip, C_X is usually lower than 3.2 pF.

Explore with SPICE:
See Appendix A for notes on SPICE simulations.

```
* Type III OTA
vi vi 0 dc=0 ac=1
r1 vi vp 500k
cb vi vp 32p
r2 vp 0 56k
gag 0 vo vp 0 790u
rf vo 0 500k
cf vo vc 0 290p
rc vc 0 56k
cx vo 0 3.2p
.ac dec 1000 10 10e6
.end
```

Tip: Plot v(vp) and v(vo) in dB to inspect z_B and p_{BX} and A_S.

6.3.3 Feedback Translations

A_S for feedback implementations of the stabilizer follows the lowest forward translation to v_O. The feedback translation component of A_S turns zeros and poles in the feedback path β_{FB} into poles and zeros in the forward path $1/\beta_{FB}$. So when the overall forward gain A_F exceeds the feedback gain A_β, A_S follows A_β, turning zeros and poles in β_{FB} into poles and zeros in A_β and A_S.

A. **Non-inverting**

The op amp in Fig. 6.25 turns the zero–pole pair in β_{FB} into p_{S1} and z_{S1}. $1/\beta_{FB}$ starts at $(R_F + R_B)/R_B$ at low f_O (when C_F opens), falls past p_F when C_F shunts R_F, and flattens to one past z_{FX} when R_B || R_F current-limit C_F. A_S follows this $1/\beta_{FB}$ until p_A reduces A_F's A_V below $1/\beta_{FB}$'s one. So p_{S1} is p_F, z_{S1} is z_{FX}, and p_{S2} is $A_{VO}p_A$'s projection to p_X, but only when A_{VO} and p_X exceed $A_{\beta 0}$ and z_{FX}:

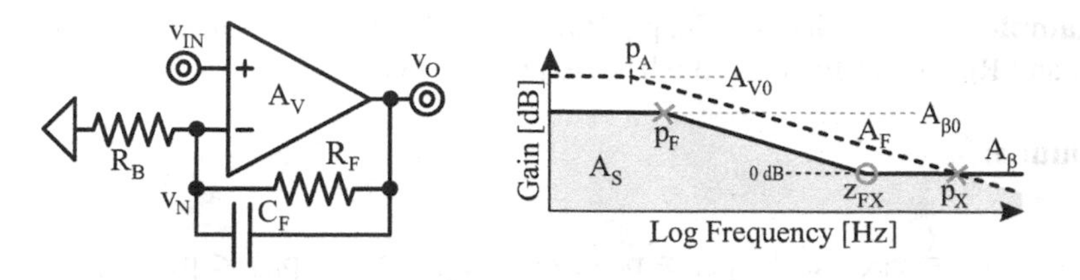

Fig. 6.25 Pole–zero non-inverting feedback translation

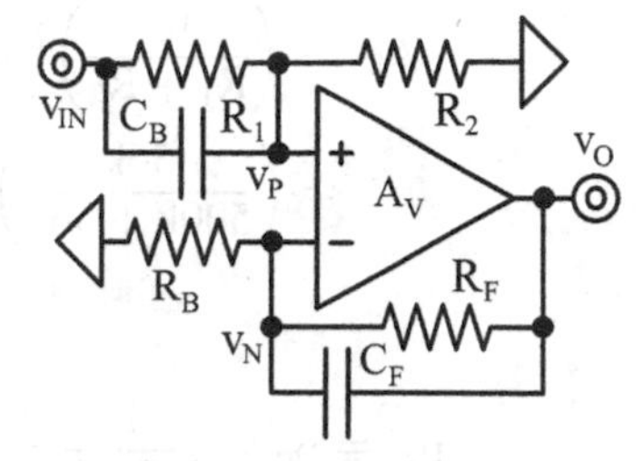

Fig. 6.26 Pole–zero–zero non-inverting feedback translation

$$A_{FW} \approx \frac{A_{V0}}{1 + s/2\pi p_A}\bigg|_{f_O > p_A} \approx A_{V0}\left(\frac{p_A}{f_O}\right)\bigg|_{f_O \geq A_{V0}p_A \approx p_X} \leq \frac{1}{\beta_{FB}}\bigg|_{f_O > z_{FX}} \approx 1 \tag{6.40}$$

$$\begin{aligned} A_S &= A_{FW}||\frac{1}{\beta_{FB}} \\ &\approx \left(\frac{R_F + R_B}{R_B}\right)\left[\frac{1 + s(R_B||R_F)C_F}{(1 + sR_FC_F)(1 + s/2\pi A_{V0}p_A)}\right] \\ &= \frac{A_{\beta 0}(1 + s/2\pi z_{FX})}{(1 + s/2\pi p_F)(1 + s/2\pi p_X)}. \end{aligned} \tag{6.41}$$

Bypassing the R_1 that feeds A_V in Fig. 6.26 adds z_{S2}. Here, A_{S0} voltage-divides with R_1 and R_2 and amplifies with A_{FW} and $1/\beta_{FB}$. z_B raises A_S when C_B bypasses R_1 and p_{BX} reverses z_B when $R_1 \parallel R_2$ current-limits C_B. This way, p_{S1} is p_F, z_{S1} and z_{S2} are z_{FX} and z_B, p_{S2} is p_{BX}, and p_{S3} is $A_{V0}p_A$'s projection to p_X, but only when A_{V0} and p_X exceed $A_{\beta 0}$ and z_{FX}:

$$\begin{aligned} A_S &= A_F||A_\beta \\ &= \left[\frac{R_2}{(R_1||Z_B) + R_2}\right]\left(A_{FW}||\frac{1}{\beta_{FB}}\right) \\ &\approx \frac{R_2(A_{V0}||A_{\beta 0})(1 + s/2\pi z_B)(1 + s/2\pi z_{FX})}{(R_1 + R_2)(1 + s/2\pi p_{BX})(1 + s/2\pi p_F)(1 + s/2\pi p_X)}. \end{aligned} \tag{6.42}$$

Example 4: Determine R_2, C_B, p_A, R_B, C_F, and z_{S1} with the A_{SO}, p_{S1}, z_{S2}, p_{S2}, p_{S3}, R_1, and R_F used in Example 3 when A_{VO} is 1 kV/V.

Solution:

$$z_{S1} \equiv z_{FX} \quad \text{and} \quad p_{S1} \equiv p_F < z_{S2} \equiv z_B < p_{S2} \equiv p_{BX} < p_{S3} = p_X$$

$$R_2 = 56\ \text{k}\Omega \text{ and } C_B = 32\ \text{pF from Example 3}$$

$$A_{SO} \approx \left(\frac{R_2}{R_1 + R_2}\right)\left(A_{VO} \middle\| \frac{R_F + R_B}{R_B}\right)$$
$$= \left(\frac{56\text{k}}{500\text{k} + 56\text{k}}\right)\left[1\text{k} \middle\| \left(1 + \frac{500\text{k}}{R_B}\right)\right] \equiv 40\ \text{V/V} = 32\ \text{dB}$$

$$\therefore \quad R_B = 760\ \Omega$$

$$p_{S1} \equiv p_F \approx \frac{1}{2\pi R_F C_F} = \frac{1}{2\pi(500\text{k})C_F} \equiv 1\ \text{kHz} \quad \therefore \quad C_F = 320\ \text{pF}$$

$$z_{S1} \equiv z_{FX} \approx \frac{1}{2\pi(R_B||R_F)C_F} = \frac{1}{2\pi(760||500\text{k})(320\text{p})} = 660\ \text{kHz}$$

$$p_{S3} = p_X \approx A_{VO}p_A = (1\text{k})p_A \geq 1\ \text{MHz} \quad \therefore \quad p_A \geq 1\ \text{kHz}$$

Note: Setting A_{SO} and p_{S1} automatically sets z_{S1} (with $1/\beta_{FBO}$). $1/\beta_{FBO}$ is high because R_1 and R_2 reduce A_{SO}, so R_B sets a high z_{S1}.

Explore with SPICE:
See Appendix A for notes on SPICE simulations.

```
* Type III Non-inverting Feedback Translation
vi vi 0 dc=0 ac=1
r1 vi vp 500k
cb vi vp 32p
r2 vp 0 56k
eav0 va 0 vp vn 1k
rpa va vb 1
cpa vb 0 159.2u
eavo vo 0 vb 0 1
```

(continued)

```
cf vo vn 320p
rf vo vn 500k
rb vn 0 760
.ac dec 1000 10 10e6
.end
```

Tip: Plot v(vp) and v(vo) in dB to inspect z_B and p_{BX} and A_S.

B. Inverting

The inverting op amp in Fig. 6.27 turns a zero in β_{FB} into p_{S1}. A_β starts with $-R_F/R_B$ and falls past p_F when C_F shunts R_F. A_S follows this A_β until A_F drops below A_β at p_X. So when A_{F0} and A_F's projection to p_X exceed $A_{\beta 0}$ and p_F, A_{S0} is $-R_F/R_B$, p_{S1} is p_F, and p_{S2} is p_X:

$$A_\beta \approx -\frac{R_F/R_B}{1 + sR_FC_F} = \frac{A_{\beta 0}}{1 + s/2\pi p_F} \tag{6.43}$$

$$A_S = A_F||A_\beta \approx \frac{-R_F/R_B}{(1 + s/2\pi p_F)(1 + s/2\pi p_X)}. \tag{6.44}$$

A_F voltage-divides v_{IN} with R_B into R_F–C_F and amplifies v_N with $-A_V$. As a result, A_F falls with p_A in A_V and p_C' when C_F shunts $R_F \parallel R_B$:

$$A_F \approx \left[\frac{R_F||Z_F}{R_B + (R_F||Z_F)}\right](-A_V) \approx \frac{-[R_F/(R_B + R_F)]A_{V0}}{(1 + s/2\pi p_A)\left(1 + s/2\pi p_C'\right)}. \tag{6.45}$$

With two poles, A_F falls faster than A_β. So A_F eventually drops below A_β:

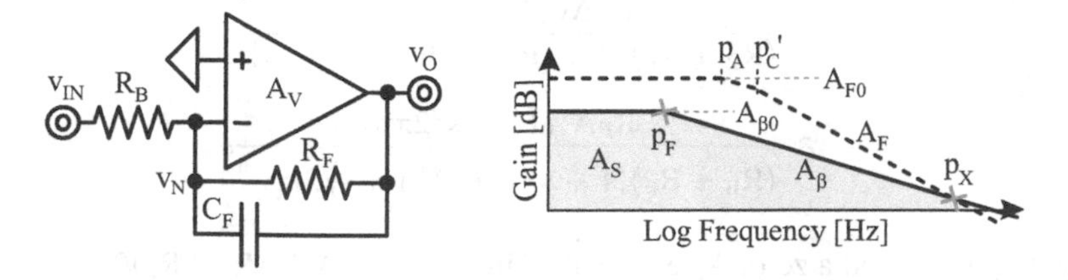

Fig. 6.27 Dominant-pole inverting feedback translation

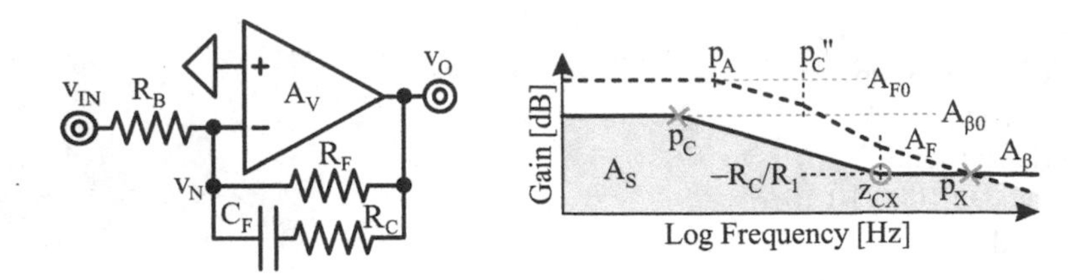

Fig. 6.28 Pole–zero inverting feedback translation

$$\left|A_F\right|_{f_O>p_A,p_{C'}} \approx \left|\frac{R_F A_{V0} p_A p_C'}{(R_B+R_F){f_O}^2}\right|_{f_O \ge \frac{A_{V0}R_B p_A p_C'}{(R_B+R_F)p_F} \approx p_X} \le \left|A_\beta\right|_{f_O>p_F} \approx \frac{R_F p_F}{R_B f_O}. \tag{6.46}$$

Current-limiting C_F with R_C in Fig. 6.28 reverses C_F's pole in A_β and A_F. This way, A_β falls past p_C when C_F and R_C shunt R_F and flattens to $-(R_C \parallel R_F)/R_B$ past z_{CX} when C_F shorts with respect to R_C. A_S follows this A_β until A_F drops below A_β at p_X. So when A_{F0} and A_F's projection to p_X exceed $A_{\beta0}$ and z_{CX}, A_{S0} is $-R_F/R_B$, p_{S1} is p_C, z_{S1} is z_{CX}, and p_{S2} is p_X:

$$\begin{aligned} A_\beta &\approx -\frac{R_F||(Z_F+R_C)}{R_B} \\ &= -\left(\frac{R_F}{R_B}\right)\left[\frac{1+R_C C_F s}{1+(R_C+R_F)C_F s}\right] \\ &= -\left(\frac{R_F}{R_B}\right)\left(\frac{1+s/2\pi z_{CX}}{1+s/2\pi p_C}\right) \end{aligned} \tag{6.47}$$

$$A_S = A_F||A_\beta \approx -\frac{(R_F/R_B)(1+s/2\pi z_{CX})}{(1+s/2\pi p_C)(1+s/2\pi p_X)}. \tag{6.48}$$

A_F voltage-divides v_{IN} with R_B into R_F–C_F–R_C and amplifies v_N with $-A_V$. So A_F falls with p_A in A_V and p_C'' when C_F and R_C shunt $R_F \parallel R_B$ and z_{CX} reverses p_C'' when R_C current-limits C_F:

$$\begin{aligned} A_F &\approx \frac{[R_F||(Z_F+R_C)](-A_V)}{R_B+[R_F||(Z_F+R_C)]} \\ &\approx \frac{-R_F A_V(1+R_C C_F s)}{(R_B+R_F)\{1+[R_C+(R_F||R_B)]C_F s\}} \\ &\approx \frac{-R_F A_{V0}(1+s/2\pi z_{CX})}{(R_B+R_F)(1+s/2\pi p_A)\left(1+s/2\pi p_C''\right)}. \end{aligned} \tag{6.49}$$

With two poles and a zero, A_F eventually drops below A_β's $(R_C \parallel R_F)/R_B$:

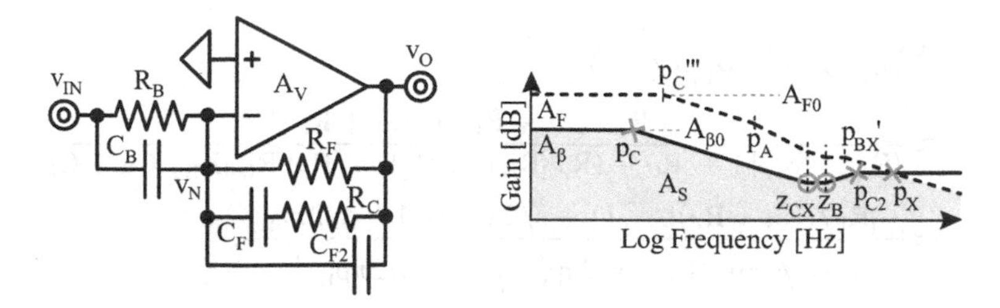

Fig. 6.29 Pole–zero–zero inverting feedback translation

$$|A_F|_{f_O>z_{CX}} \approx \frac{R_F A_{V0} p_A p_C''}{(R_B + R_F) z_{CX} f_O} \leq |A_\beta|_{f_O>z_{CX}} \approx \frac{R_F p_C}{R_B z_{CX}} = \frac{R_C||R_F}{R_B} \tag{6.50}$$

when

$$f_O > \left(\frac{R_B A_{V0}}{R_B + R_F}\right)\left(\frac{p_A p_C''}{p_C}\right) \approx p_X. \tag{6.51}$$

Bypassing R_B with C_B in Fig. 6.29 raises A_β and A_F. This way, A_β starts with $-R_F/R_B$ at low f_O, falls past p_C when C_F and R_C shunt R_F, and flattens and rises past z_{CX} and z_B when R_C current-limits C_F and C_B bypasses R_B. Adding C_{F2} flattens A_β past p_{C2} when C_{F2} shunts R_C || R_F, past which A_β remains $-C_B/C_{F2}$. A_S follows this A_β until A_F drops below A_β at p_X. So when A_{F0} and A_F's projection to p_X exceed $A_{\beta 0}$ and p_{C2}, p_{S1} is p_C, z_{S1} and z_{S2} are z_{CX} and z_B, p_{S2} is p_{C2}, and p_{S3} is p_X:

$$\begin{aligned} A_\beta &\approx -\left(\frac{R_F}{R_B}\right)\left\{\frac{(1 + R_C C_F s)(1 + R_B C_B s)}{[1 + (R_C + R_F)C_F s][1 + (R_C||R_F)C_{F2} s]}\right\} \\ &\approx -\left(\frac{R_F}{R_B}\right)\left[\frac{(1 + s/2\pi z_{CX})(1 + s/2\pi z_B)}{(1 + s/2\pi p_C)(1 + s/2\pi p_{C2})}\right] \end{aligned} \tag{6.52}$$

$$\begin{aligned} A_S &= A_F||A_\beta \\ &\approx -\left(\frac{R_F A_{V0}}{R_B + R_F}\,||\,\frac{R_F}{R_B}\right)\left[\frac{(1 + s/2\pi z_{CX})(1 + s/2\pi z_B)}{(1 + s/2\pi p_C)(1 + s/2\pi p_{C2})(1 + s/2\pi p_X)}\right]. \end{aligned} \tag{6.53}$$

A_F voltage-divides v_{IN} with R_B–C_B into R_F–C_F–R_C–C_{F2} and amplifies v_N with $-A_V$. So A_F falls with p_A in A_V and p_C''' when C_F and R_C shunt R_F || R_B before p_{BX}' later reduces A_F when R_F || R_C || R_B current-limits C_B and C_{F2} (which also corresponds to C_{F2} and C_B shunting R_F || R_C || R_B) and climbs with z_{CX} and z_B when R_C current-limits C_F and C_B bypasses R_B:

$$
\begin{aligned}
A_F &\approx \frac{[R_F||(Z_F + R_C)||Z_{F2}](-A_V)}{(R_B||Z_B) + [R_F||(Z_F + R_C)||Z_{F2}]} \\
&\approx \frac{-R_F A_V(1 + R_C C_F s)(1 + R_B C_B s)}{(R_B + R_F)\{1 + [R_C + (R_F||R_B)]C_F s\}[1 + (R_F||R_C||R_B)(C_B + C_{F2})s]} \\
&\approx \frac{-[R_F/(R_B + R_F)]A_{V0}(1 + s/2\pi z_{CX})(1 + s/2\pi z_B)}{(1 + s/2\pi p_A)\left(1 + s/2\pi p_C{'''}\right)\left(1 + s/2\pi p_{BX}{'}\right)}.
\end{aligned}
\tag{6.54}
$$

With three poles and two zeros, A_F eventually drops below the A_β that C_B/C_{F2} sets after C_B shunts R_B and C_{F2} shunts $R_F \parallel R_C$:

$$
\begin{aligned}
|A_F|_{f_O > p_C{'''}, p_A, z_{CX}, z_B, p_{BX}{'}} &\approx \frac{R_F A_{V0} p_A p_C{'''} p_{BX}{'}}{(R_B + R_F) z_{CX} z_B f_O} \\
&\le |A_\beta|_{f_O > p_C, z_{CX}, z_B, p_{C2}} \approx \frac{R_F p_C p_{C2}}{R_B z_{CX} z_B} = \frac{C_B}{C_{F2}}
\end{aligned}
\tag{6.55}
$$

when

$$
f_O \ge \frac{R_B A_{V0} p_A p_C{'''} p_{BX}{'}}{(R_B + R_F) p_C p_{C2}} \approx p_X. \tag{6.56}
$$

Without C_{F2}, A_S would rise past z_{CX} and z_B with A_β and later fall with A_F. Falling after rising this way is the effect of a double pole p_{S23}: the disappearance of a zero (z_B) and appearance of a pole ($p_{BX}{'}$). The purpose of C_{F2} is to split this p_{S23} into p_{S2} and p_{S3}. This way, p_{S2} appears at an f_{0dB} that keeps the feedback system stable without losing much phase shift to p_{S3}.

Example 5: Determine p_A, R_B, C_B, C_F, R_C, and C_{F2} with the $|A_{S0}|$, p_{S1}, z_{S1}, z_{S2}, p_{S3}, and R_F used in Example 3 so p_{S2} matches or exceeds 100 kHz when A_{V0} is 1 kV/V.

Solution:

$$p_{S1} \equiv p_C < z_{S1} \equiv z_{CX} = z_{S2} \equiv z_B < p_{S2} \equiv p_{C2} < p_{S3} = p_X$$

$$R_C = 56\ \text{k}\Omega \text{ and } C_F = 290\ \text{pF from Example 3}$$

$$A_{S0} = -\left(\frac{R_F A_{V0}}{R_B + R_F} || \frac{R_F}{R_B}\right) = -\left(\frac{(500k)(1k)}{R_B + (500k)} || \frac{500k}{R_B}\right)$$
$$\equiv -40\ V/V = 32dB \quad \therefore \quad R_B = 12\ k\Omega$$

$$z_{S2} \equiv z_B \approx \frac{1}{2\pi R_B C_B} = \frac{1}{2\pi(12k)C_B} \equiv 10\ kHz \quad \therefore \quad C_B = 1.3\ nF$$

$$p_{S2} \equiv p_{C2} \approx \frac{1}{2\pi(R_C||R_F)C_{F2}} = \frac{1}{2\pi(56k||500k)C_{F2}} \geq 100\ kHz \quad \therefore \quad C_{F2} \leq 32\ pF$$

$$p_{S3} \equiv p_X \approx \frac{R_B A_{V0} p_A p_C' p_{BX}'}{(R_B + R_F)p_C p_{C2}}$$
$$\approx \frac{[A_{V0} p_A R_B/(R_B + R_F)](R_C + R_F)(R_F||R_C)C_{F2}}{[R_C + (R_B||R_F)](R_F||R_C||R_B)(C_B + C_{F2})}$$
$$\approx \left[\frac{A_{V0} p_A (12k)}{12k + 500k}\right]\left[\frac{56k + 500k}{56k + 12k}\right]\left[\frac{(500k||56k)(32p)}{(9.7k)(1.3n + 32p)}\right]$$
$$= (24m)A_{V0} p_A = (24m)(1k)p_A \geq 1\ MHz \quad \therefore \quad p_A \geq 42\ kHz$$

Explore with SPICE:
See Appendix A for notes on SPICE simulations.

```
* Type III Inverting Feedback Translation
vi vi 0 dc=0 ac=1
rb vi vn 12k
cb vi vn 1.3n
eav0 va 0 0 vn 1000
rpa va vb 1
cpa vb 0 3.791u
eavo vo 0 vb 0 1
rf vn vo 500k
cf vn vx 290p
rc vx vo 56k
cf2 vn vo 32p
.ac dec 1000 10 10e6
.end
```

Tip: Plot v(vo) in dB to inspect A_S.

6.3.4 Mixed Translations

Forward gains can also play a dominant role in A_S. In these cases, A_S follows A_F before transitioning to $A_β$ and ultimately succumbing again to A_F. This way, A_F and $A_β$ set A_{S0} and add p_{S1}, z_{S1}, and z_{S2} before A_F limits A_S at high frequency.

A. **Non-inverting**

A_F in the non-inverting op amp in Fig. 6.30 amplifies v_{IN} with A_V and $A_β$ with R_F and C_F. A_F starts at A_{V0} and falls past A_V's p_A. $A_β$ starts high with Z_C/R_F, falls as C_F shorts, and flattens to one past z_{FX} when C_F shorts with respect to R_F:

$$A_F \approx A_V = \frac{A_{V0}}{1 + s/2\pi p_A} \tag{6.57}$$

$$\begin{aligned} A_\beta &= \frac{Z_C + R_F}{R_F} = \frac{1 + sR_FC_F}{sR_FC_F} = \frac{1 + s/2\pi z_{FX}}{s/2\pi p_F} \\ &= \left(\frac{Z_C}{R_F}\right)\left(1 + \frac{s}{2\pi z_{FX}}\right) = \left(\frac{p_F}{f_{Oi}}\right)\left(1 + \frac{s}{2\pi z_{FX}}\right). \end{aligned} \tag{6.58}$$

So A_S follows A_F's A_{V0} until $A_β$ drops below A_F's A_{V0} at p_{X1}. A_S then falls and flattens with $A_β$ past z_{FX} until A_F falls below $A_β$'s one at p_{X2}. This way, A_{S0} is A_{V0}, p_{S1} is p_{X1}, z_{S1} is z_{FX}, and p_{S2} is p_{X2}, but only when $A_β$'s projection to p_{X1} precedes p_A and z_{FX} and A_F's projection to p_{X2} exceeds z_{FX}:

$$A_\beta\Big|_{f_O<z_F} \approx \frac{Z_C}{R_F} = \frac{1}{sR_FC_F}\Bigg|_{f_O \geq \frac{1}{2\pi A_{V0}R_FC_F} = \frac{z_{FX}}{A_{V0}} = \frac{p_F}{A_{V0}} \approx p_{X1}} \leq A_F\Big|_{f_O<p_A} \approx A_{V0}, \tag{6.59}$$

$$A_F\Big|_{f_O>p_A} \approx A_{V0}\left(\frac{p_A}{f_O}\right)\Bigg|_{f_O \geq A_{V0}p_A \approx p_{X2}} \leq A_\beta\Big|_{f_O>z_{FX}} \approx 1, \tag{6.60}$$

$$A_S = A_F || A_\beta \approx \frac{A_{V0}(1 + s/2\pi z_{FX})}{(1 + s/2\pi p_{X1})(1 + s/2\pi p_{X2})}. \tag{6.61}$$

Bypassing the R_1 that feeds A_V in Fig. 6.31 adds z_{S2}. Here, R_1 and R_2 voltage-divide and A_F and $A_β$ amplify. z_B raises A_S when C_B bypasses R_1 and p_{BX} reverses

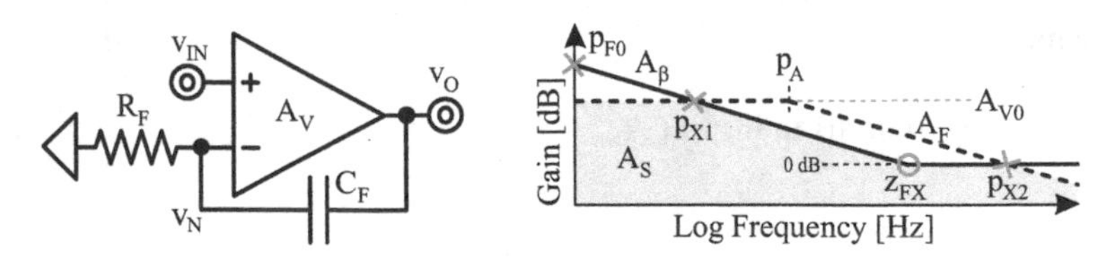

Fig. 6.30 Pole–zero non-inverting mixed translation

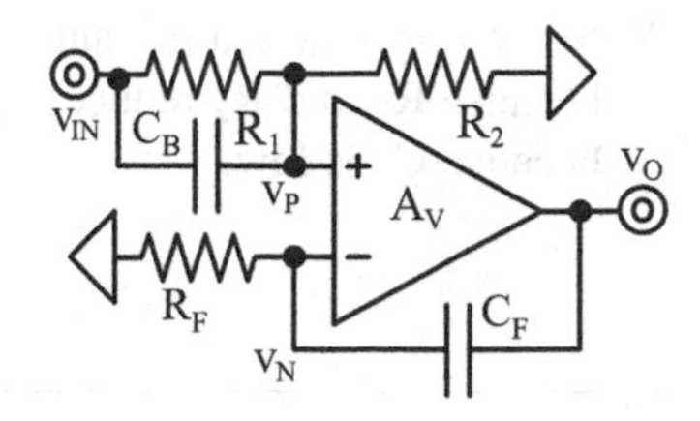

Fig. 6.31 Pole–zero–zero non-inverting mixed translation

z_B when $R_1 \parallel R_2$ current-limits C_B. So A_{S0} is $A_{V0}R_2/(R_1 + R_2)$, p_{S1} is p_{X1}, z_{S1} and z_{S2} are z_{FX} and z_B, p_{S2} is p_{BX}, and p_{S3} is p_{X2}:

$$A_S = \left[\frac{R_2}{(R_1||Z_B) + R_2}\right](A_F||A_\beta)$$
$$= \frac{R_2 A_{V0}(1 + s/2\pi z_B)(1 + s/2\pi z_{FX})}{(R_1 + R_2)(1 + s/2\pi p_{BX})(1 + s/2\pi p_{X1})(1 + s/2\pi p_{X2})}. \quad (6.62)$$

Example 6: Determine C_B, R_2, A_{V0}, p_A, C_F, and z_{S1} with the A_{S0}, p_{S1}, z_{S2}, p_{S2}, p_{S3}, R_1, and R_F used in Example 3.

Solution:

$$R_2 = 56\ \text{k}\Omega \text{ and } C_B = 32\ \text{pF from Example 3}$$

$$A_{S0} = \frac{R_2 A_{V0}}{R_1 + R_2} = \frac{(56\text{k})A_{V0}}{500\text{k} + 56\text{k}} \equiv 40\ \text{V/V} = 32\ \text{dB} \quad \therefore \quad A_{V0} = 400\ \text{V/V} = 52\ \text{dB}$$

$$p_{S1} \equiv p_{X1} < z_{S1} \equiv z_{FX} = z_{S2} \equiv z_B < p_{S2} \equiv p_{BX} < p_{S3} = p_{X2}$$

$$p_{S1} \equiv p_{X1} \approx \frac{1}{2\pi R_F C_F A_{V0}} = \frac{1}{2\pi(500\text{k})C_F(400)} \equiv 1\ \text{kHz} \quad \therefore \quad C_F = 800\ \text{fF}$$

$$z_{S1} \equiv z_{FX} \approx \frac{1}{2\pi R_F C_F} = \frac{1}{2\pi(500\text{k})(800\text{f})} = 400\ \text{kHz}$$

$$p_{S3} = p_{X2} \approx A_{V0} p_A = (400)p_A \geq 1\ \text{MHz} \quad \therefore \quad p_A \geq 2.5\ \text{kHz} \approx p_{S1}$$

Note: Setting A_{SO} and p_{S1} automatically sets z_{S1} (with A_{VO}). A_{VO} needs to be high because R_1 and R_2 reduce A_{SO}, C_F is low because A_{VO} is high, and z_{S1} is high because C_F is low.

Explore with SPICE:

See Appendix A for notes on SPICE simulations.

```
* Type III Non-inverting Mixed Translation
vi vi 0 dc=0 ac=1
r1 vi vp 500k
cb vi vp 32p
r2 vp 0 56k
eav0 va 0 vp vn 400
rpa va vb 1
cpa vb 0 63.69u
eavo vo 0 vb 0 1
cf vo vn 800f
rf vn 0 500k
.ac dec 1000 10 10e6
.end
```

Tip: Plot v(vp) and v(vo) in dB to inspect z_B and p_{BX} and A_S.

B. Inverting

A_F in the inverting op amp in Fig. 6.32 voltage-divides v_{IN} with R_F into C_F and amplifies with $-A_V$. A_F starts with $-A_{VO}$ and falls past A_V's p_A and p_F when C_F shunts R_F. A_β starts high with $-Z_C/R_F$ and falls as C_F shorts:

$$A_F \approx \left(\frac{Z_C}{R_F + Z_C}\right)(-A_V) = \frac{-A_V}{1 + sR_FC_F} = \frac{-A_{VO}}{(1 + s/2\pi p_A)(1 + s/2\pi p_F)} \tag{6.63}$$

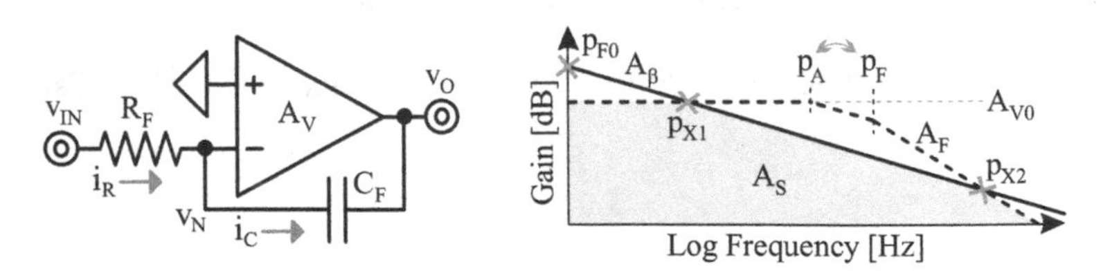

Fig. 6.32 Dominant-pole inverting mixed translation

$$A_\beta = -\frac{Z_C}{R_F} = -\frac{1}{sR_FC_F} = -\frac{2\pi p_F}{s} = -\frac{p_F}{f_{Oi}}. \tag{6.64}$$

A_S follows A_F's A_{V0} until A_β drops below A_{V0} at p_{X1}. With two poles in A_F and one in A_β, A_F falls faster than A_β. As a result, A_S falls with A_β past p_{X1} until A_F falls below A_β at p_{X2}. This way, A_{S0} is $-A_{V0}$, p_{S1} is p_{X1}, and p_{S2} is p_{X2}, but only when A_β's projection to p_{X1} precedes p_A and p_F and A_F's projection to p_{X2} exceeds p_{X1}:

$$|A_\beta| \approx \frac{Z_C}{R_F} = \frac{1}{sR_FC_F}\bigg|_{f_O \geq \frac{1}{2\pi R_FC_FA_{V0}} = \frac{p_F}{A_{V0}} \approx p_{X1}} \leq |A_F|_{f_O < p_A} \approx A_{V0}, \tag{6.65}$$

$$|A_F|_{f_O > p_A, p_F} \approx \frac{A_{V0}p_Ap_F}{f_O^2}\bigg|_{f_O \geq A_{V0}p_A \approx p_{X2}} \leq |A_\beta| = \frac{p_{F0}}{f_O}, \tag{6.66}$$

$$A_S = A_F || A_\beta \approx \frac{-A_{V0}}{(1 + s/2\pi p_{X1})(1 + s/2\pi p_{X2})}. \tag{6.67}$$

Current-limiting C_F with R_C in Fig. 6.33 reverses C_F's pole in A_F and A_β. So A_F starts with $-A_{V0}$ and falls past A_V's p_A and p_C when C_F and R_C shunt R_F and z_{CX} reverses p_C when R_C current-limits C_F. A_β starts high at $-Z_C/R_F$, falls as C_F shorts, and flattens to $-R_C/R_F$ past z_{CX} when C_F shorts with respect to R_C:

$$A_F \approx \left(\frac{R_C + Z_C}{R_F + R_C + Z_C}\right)(-A_V) = \frac{-A_{V0}(1 + s/2\pi z_{CX})}{(1 + s/2\pi p_A)(1 + s/2\pi p_C)} \tag{6.68}$$

$$A_\beta = -\frac{Z_C + R_C}{R_F} = -\frac{1 + sR_CC_F}{sR_FC_F} = -\frac{1 + s/2\pi z_{CX}}{s/2\pi p_F} = -\left(\frac{Z_C}{R_F}\right)\left(1 + \frac{s}{2\pi z_{CX}}\right) = -\left(\frac{p_F}{f_{Oi}}\right)\left(1 + \frac{s}{2\pi z_{CX}}\right). \tag{6.69}$$

A_S follows A_F's A_{V0} until A_β drops below A_{V0} at p_{X1}. A_S falls and flattens with A_β past z_{CX} until A_F falls below A_β's $-R_C/R_F$ at p_{X2}. This way, A_{S0} is $-A_{V0}$, p_{S1} is p_{X1}, z_{S1} is z_{CX}, and p_{S2} is p_{X2}, but only when A_β's projection to p_{X1} precedes p_A and p_C and A_F's projection to p_{X2} exceeds z_{CX}:

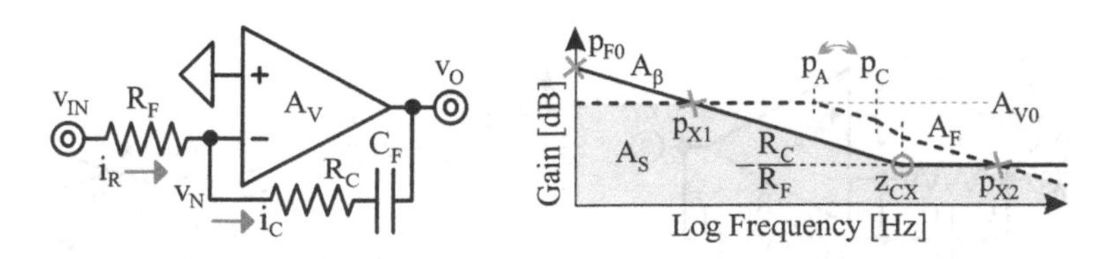

Fig. 6.33 Pole–zero inverting mixed translation

$$\left| A_\beta \right|_{f_O<z_{CX}} \approx \frac{Z_C}{R_F} = \frac{1}{sR_FC_F}\Bigg|_{f_O \geq \frac{1}{2\pi R_F C_F A_{V0}} = \frac{p_F}{A_{V0}} \approx p_{X1}} \leq \left|A_F\right|_{f_O<p_A} \approx A_{V0}, \tag{6.70}$$

$$\left| A_F \right|_{f_O>z_{CX}} \approx \frac{A_{V0}p_Ap_C}{z_{CX}f_O}\Bigg|_{f_O \geq A_{V0}\left(\frac{p_Ap_C}{p_F}\right) \approx p_{X2}} \leq \left|A_\beta\right|_{f_O>z_{CX}} = \frac{R_C}{R_F} = \frac{p_F}{z_{CX}}, \tag{6.71}$$

and

$$A_S = A_F||A_\beta \approx \frac{-A_{V0}(1 + s/2\pi z_{CX})}{(1 + s/2\pi p_{X1})(1 + s/2\pi p_{X2})}. \tag{6.72}$$

Bypassing R_B with C_B in Fig. 6.34 raises A_F and A_β. This way, A_F falls past A_V's p_A, p_C when C_F and R_C shunt R_F before C_B shunts R_F, and p_{BX}'' when $R_C \parallel R_F$ current-limits C_B. A_F climbs past z_{CX} when R_C current-limits C_F and past z_B when C_B bypasses R_F. A_β starts high with $-Z_C/R_F$, falls as C_F shorts, flattens past z_{CX} when R_C current-limits C_F, and climbs past z_B when C_B bypasses R_B:

$$\begin{aligned} A_F &\approx \left[\frac{R_C + Z_C}{(R_F||Z_B) + R_C + Z_C}\right](-A_V) \\ &= \frac{-A_V(1 + sR_FC_B)(1 + sR_CC_F)}{s^2R_FR_CC_BC_F + s[(R_F + R_C)C_F + R_FC_B] + 1} \\ &\approx (1)\left[\frac{(1 + s/2\pi z_{CX})(1 + s/2\pi z_B)}{(1 + s/2\pi p_C)\left(1 + s/2\pi p_{BX}''\right)}\right]\left(\frac{-A_{V0}}{1 + s/2\pi p_A}\right) \end{aligned} \tag{6.73}$$

$$\begin{aligned} A_\beta = -\frac{Z_C + R_C}{R_F||Z_B} &= -\left(\frac{Z_C}{R_F}\right)\left(1 + \frac{s}{2\pi z_{CX}}\right)\left(1 + \frac{s}{2\pi z_B}\right) \\ &= -\left(\frac{p_F}{f_{Oi}}\right)\left(1 + \frac{s}{2\pi z_{CX}}\right)\left(1 + \frac{s}{2\pi z_B}\right). \end{aligned} \tag{6.74}$$

A_S follows A_F's A_{V0} until A_β drops below A_{V0} at p_{X1}. A_S falls and flattens and rises with A_β past z_{CX} and z_B until A_β rises above A_F at p_{X2}. A_S then follows and falls with A_F past p_A. As a result, A_{S0} is $-A_{V0}$, p_{S1} is p_{X1}, z_{S1} and z_{S2} are z_{CX} and z_B, p_{S2} is p_{X2}, and p_{S3} is p_A, but only when A_β's projection to p_{X1} precedes and A_F's projection to p_{X2} exceeds p_C, z_{CX}, z_B, and p_{BX}'':

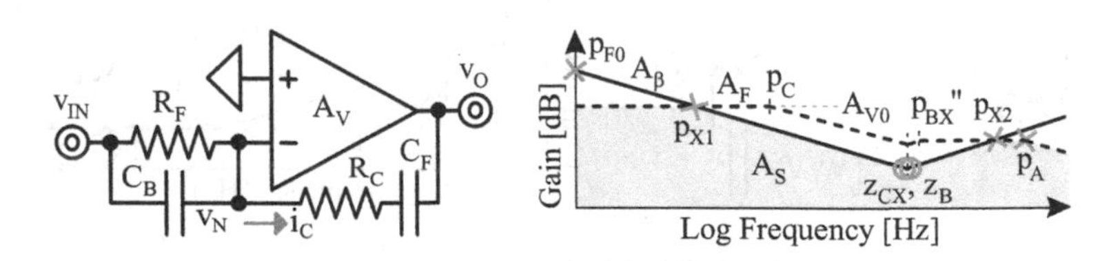

Fig. 6.34 Pole–zero–zero inverting mixed translation

$$|A_{\beta 0}|_{f_O < z_{CX}, z_B} \approx \frac{Z_C}{R_F} = \frac{1}{sR_FC_F}\bigg|_{f_O \geq \frac{1}{2\pi R_F C_F A_{V0}} = \frac{p_F}{A_{V0}} \approx p_{X1}} \leq A_{F0} \approx A_{V0}, \quad (6.75)$$

$$|A_\beta|_{f_O > z_{CX}, z_B} \approx \left(\frac{Z_C}{R_F}\right)\left(\frac{f_O{}^2}{z_{CX}z_B}\right) \approx \frac{p_F f_O}{z_{CX}z_B}\bigg|_{f_O \geq A_{V0}\left(\frac{p_C p_{BX}''}{p_F}\right) \approx p_{X2}}$$
$$\geq |A_F|_{f_O > p_C, z_{CX}, z_B, p_{BX''}} \approx A_{V0}\left(\frac{p_C p_{BX}''}{z_{CX}z_B}\right), \quad (6.76)$$

and

$$A_S = A_F || A_\beta \approx \frac{-A_{V0}(1 + s/2\pi z_{CX})(1 + s/2\pi z_B)}{(1 + s/2\pi p_{X1})(1 + s/2\pi p_{X2})(1 + s/2\pi p_A)}. \quad (6.77)$$

Example 7: Determine C_B, A_{V0}, p_A, C_F, R_C, and z_{S2} with the $|A_{S0}|$, p_{S1}, z_{S1}, p_{S2}, and p_{S3} used in Example 5.

Solution:

$$A_{S0} = -A_{V0} \equiv -40\ \text{V/V} = 32\ \text{dB}$$

$$p_{S1} \equiv p_{X1} < z_{S1} \equiv z_{CX} = z_{S2} \equiv z_B < p_{S2} \equiv p_{X2} < p_{S3} \approx p_A$$

$$p_{S1} \equiv p_{X1} \approx \frac{p_F}{A_{V0}} = \frac{1}{2\pi R_F C_F A_{V0}} = \frac{1}{2\pi(500k)C_F(40)} \equiv 1\ \text{kHz} \quad \therefore \quad C_F = 8.0\ \text{pF}$$

$$z_{S1} \equiv z_{CX} \approx \frac{1}{2\pi R_C C_F} = \frac{1}{2\pi R_C(8.0p)} \equiv 10\ \text{kHz} \quad \therefore \quad R_C = 2.0\ \text{M}\Omega$$

$$p_{S2} \equiv p_{X2} \approx A_{V0}\left(\frac{p_C p_{BX}''}{p_F}\right) = \frac{A_{V0}}{2\pi R_C C_B} = \frac{40}{2\pi(2M)C_B} \geq 100\ \text{kHz} \quad \therefore \quad C_B \leq 32\ \text{pF}$$

$$z_{S2} \equiv z_B \approx \frac{1}{2\pi R_F C_B} \geq \frac{1}{2\pi(500k)(32p)} = 10\ \text{kHz}$$

$$p_{S3} \approx p_A \geq 1\ \text{MHz}$$

Note: Setting A_{S0} and p_{S2} automatically sets z_{S2}.

Explore with SPICE:
See Appendix A for notes on SPICE simulations.

```
* Type III Inverting Mixed Translation
vi vi 0 dc=0 ac=1
rf vi vn 500k
cb vi vn 32p
eav0 va 0 0 vn 40
rpa va vb 1
cpa vb 0 159.2n
eavo vo 0 vb 0 1
rc vn vx 2e6
cf vx vo 8p
.ac dec 1000 10 10e6
.end
```

Tip: Plot v(vo) in dB to inspect A_S.

6.3.5 Tradeoffs

Tolerance, bandwidth, capacitance, and number of components are important considerations for stabilizers. Low variation reduces *field failure rate*. Higher bandwidth (which implies higher p_{S2} and p_{S3}) increases phase margin. Lower capacitance saves silicon area. And fewer components reduce space and cost.

Feedback translations are usually reliable and predictable because resistor ratios set their gain. The bandwidth of on-chip OTAs is normally wide because they exclude p_A, whose projection limits the bandwidth of feedback implementations. The disadvantage of feedback translations and OTAs is that they need considerable capacitance to establish low-to-moderate frequency poles and zeros.

Mixed translations need less capacitance because they effectively magnify the shunting effect of the filter capacitor. Inverting mixed translations use fewer components because they do not need a resistor to set the gain or an input filter to add a second zero. The drawback of feedback implementations is bandwidth, because p_A reduces and projects A_V to a frequency past which the feedback gain no longer sets gain.

6.4 Voltage Control

The basic aim of voltage regulators is to supply current at predetermined voltages. Engineers close feedback loops to v_O for this purpose. This way, the system supplies the *load current* i_{LD} needed to set and keep v_O steady.

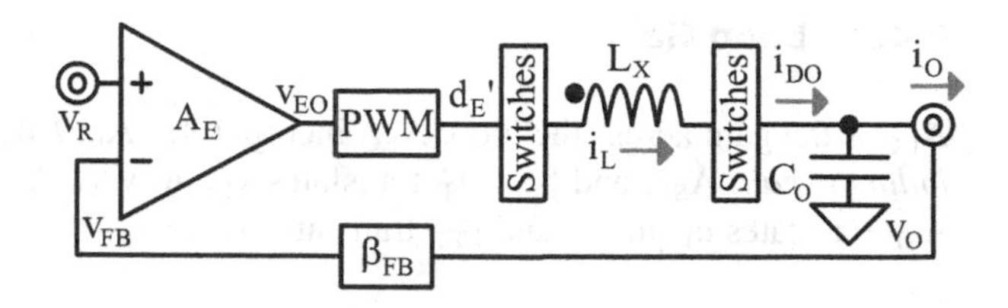

Fig. 6.35 Voltage-mode voltage controller

6.4.1 Controller

A. Composition

The switched inductor, *output capacitor* C_O, *feedback scaler* β_{FB}, *error amplifier* or *amp* A_E, and *pulse-width modulator* (PWM) in Fig. 6.35 close the inverting feedback loop that sets and keeps v_O steady. L_X and C_O are the *power stage*. L_X transfers input power to v_O and C_O supplies and absorbs dynamic mismatches between i_L and the i_{LD} that i_O supplies.

β_{FB}, A_E, and the PWM are the *feedback controller*. β_{FB} senses and scales v_O to the v_{FB} that A_E mixes and compares with a *reference voltage* v_R. A_E mixes, amplifies, and when needed, stabilizes the loop. The PWM converts the *amplified error* v_{EO} that A_E outputs into an *energize duty-cycle command* d_E'. This d_E' sets the *inductor current* i_L that ultimately feeds v_O.

As a whole, A_E amplifies the error that adjusts d_E' and i_L so v_{FB} mirrors v_R. As a result, v_O is a reverse β_{FB} translation of v_R's mirrored reflection:

$$v_O = \frac{v_{FB}}{\beta_{FB}} \approx \frac{v_R}{\beta_{FB}}. \tag{6.78}$$

So when v_O deviates from this v_R/β_{FB}, the loop adjusts i_L until v_O recovers.

B. Feedback Objectives

High A_{LG} is desirable because v_{FB} matches v_R to the extent A_{LG} exceeds one. High f_{0dB} is also desirable because v_{FB} matches v_R up to the $f_{BW(CL)}$ that f_{0dB} sets. This is important because $f_{BW(CL)}$ determines the *response time* t_R needed to adjust i_O when responding to sudden *load dumps* Δi_{LD}. In short, higher A_{LG} improves v_O accuracy and higher f_{0dB} shortens t_R.

6.4.2 Loop Gain

A_{LG} is the gain across the loop. A_{LG} incorporates A_E, *PWM gain* A_{PWM}, *switched-inductor gain* A_{SL}, and β_{FB}. A_E translates v_{FB} to v_{EO}, A_{PWM} translates v_{EO} to d_E', A_{SL} translates d_E' to v_O, and β_{FB} translates v_O to v_{FB}:

$$A_{LG} \equiv \frac{v_{FB}}{v_E} = \frac{v_{FB}}{v_R - v_{FB}} = A_E A_{PWM} A_{SL} \beta_{FB}. \tag{6.79}$$

A_E is the gain engineers use to stabilize the feedback loop. Since β_{FB} is the translation that sets v_O and v_R is an independent voltage that v_{FB} mirrors, v_R and v_O's target set β_{FB} to v_R/v_O. The operating mechanics of the PWM and power stage determine the other gain translations.

PWM d_E' in Figs. 6.35 and 6.36 is a digital signal that pulses once every *switching period* t_{SW}. v_{EO} sets the pulse width, which the switched inductor uses to set L_X's *energize time* t_E. So d_E' is ultimately the t_E fraction of t_{SW} across which L_X energizes.

d_E' ranges from close to zero to nearly one or 100%. The v_{EO} variation $v_{EO(PP)}$ that sweeps d_E' across this maximum range $\Delta d_{E(MAX)}'$ determines the gain of the PWM. But since the PWM delays this translation, A_{PWM} is roughly $1/v_{EO(PP)}$ up to the *PWM pole* p_{PWM} that the *PWM delay* t_{PWM} determines:

$$A_{PWM} \equiv \frac{d_e'}{v_{eo}} \approx \left(\frac{\Delta d_{E(MAX)}'}{v_{EO(PP)}}\right)\left(\frac{1}{1 + s/2\pi p_{PWM}}\right) \approx \frac{1/v_{EO(PP)}}{1 + s/2\pi p_{PWM}}. \tag{6.80}$$

This delay should be a small fraction of t_{SW}. So p_{PWM} is normally much higher than the *switching pole* p_{SW} that a t_{SW} delay establishes.

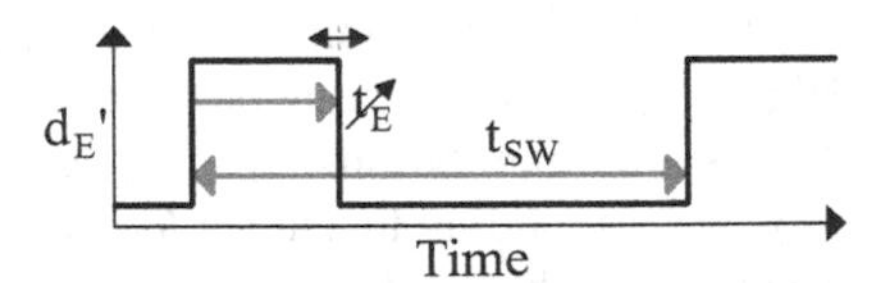

Fig. 6.36 Energize duty-cycle command

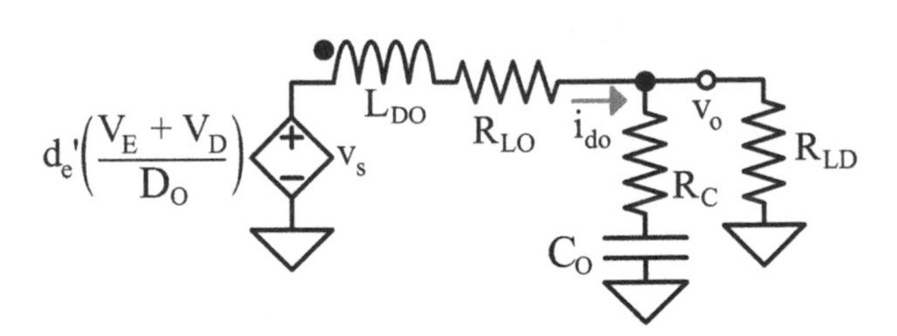

Fig. 6.37 Small-signal model of the switched inductor in CCM

SL The switched inductor in *continuous-conduction mode* (CCM) is basically a voltage-sourced inductor. The *small-signal voltage source* v_s is d_e' times an *output duty-cycle* d_O translation of the *voltages* v_E and v_D that *energize* and *drain* L_X. The *duty-cycled inductance* L_{DO} is a d_O translation of L_X with a *duty-cycled inductor resistance* R_{LO} that is usually negligibly lower than R_{LD}. So the static components of d_O, v_E, and v_D set L_{DO} and R_{LO} in Fig. 6.37 to L_X/D_O^2 and R_L/D_O and A_{SL} to

$$A_{SL(CCM)} \equiv \frac{v_O}{d_e'} \approx \left(\frac{V_E + V_D}{D_O}\right)\left(\frac{R_{LD}}{R_{LO} + R_{LD}}\right) \left\{\frac{(1 + s/2\pi z_C)(1 - s/2\pi z_{DO})}{\left[(s/2\pi p_{LC})^2 + s/2\pi p_{LC}Q_{LC} + 1\right](1 + s/2\pi p_{SW})}\right\}. \tag{6.81}$$

This gain drops as L_X opens with frequency because L_X feeds v_O less current. A_{SL} also falls as C_O shorts and pulls current away from v_O. The resulting *inductor* and *capacitor poles* p_L and p_C appear together as a double pole p_{LC} at the *transitional LC resonant frequency* f_{LC}, which happens when L_{DO}'s impedance sL_{DO} overcomes C_O's $1/sC_O$. p_C eventually fades past z_C when the *capacitor resistance* R_C current-limits C_O.

Duty-cycled outputs connect L_X to v_O only when draining L_X. So when the *switching frequency* f_{SW} is constant, extending t_E shortens L_X's *drain time* t_D. Reducing drain current this way produces an inverting (out-of-phase) zero when the loss outpaces the gain. This *duty-cycled zero* z_{DO} normally appears above p_{LC}, but not by far. When present, z_{DO} is usually below p_{SW}.

p_{LC} is challenging because it shifts phase 180° and peaks the gain. Since L_{DO}'s and C_O's impedances cancel at f_{LC}, *inductor resistance* R_L and R_C impose a *series resistance* R_S that current-limits this peak. R_{LD} dampens it below this level because R_{LD} adds to the resistance that limits the LC current. But since R_L and R_C are usually low and R_{LD} is variable, A_{SL} in CCM can still peak 20–30 dB over its f_{LC} projection. z_{DO} is more problematic because z_{DO} inverts in addition to increasing the gain.

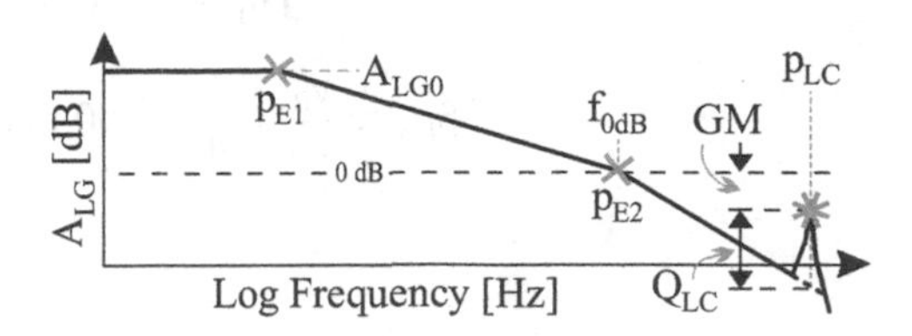

Fig. 6.38 Dominant-pole CCM stabilization

6.4.3 Voltage Mode

A. Type I: Dominant Pole

When stabilizing with one pole, A_{LG} should reach 0 dB at an f_{0dB} in Fig. 6.38 that precedes and keeps the peak at p_{LC} from reaching 0 dB. A_E should therefore incorporate gain for accuracy, p_{E1} so f_{0dB} is a decade or more below p_{LC}, and p_{E2} at f_{0dB} to suppress the peak at f_{LC}. This way, phase margin can near 45° and gain margin can be up to 40 dB.

A_{LG} peaks at p_{LC} to the gain that the *LC quality factor* Q_{LC} of the power stage sets. When targeting 45° of PM, p_{E2} should near f_{0dB}, which $A_{LG0}p_{E1}$ sets. This way, p_{E1} and p_{E2} project A_{LG} to a gain at p_{LC} that Q_{LC} peaks to:

$$A_{LG}\big|_{f_{LC}>p_{E1},p_{E2}} \approx A_{LG0}\left(\frac{p_{E1}p_{E2}}{f_{LC}^2}\right)Q_{LC} \approx \left(\frac{f_{0dB}^2}{p_{LC}^2}\right)Q_{LC} = \frac{1}{GM}, \tag{6.82}$$

where R_S current-limits Q_{LC} to C_O's impedance $Z_{C(LC)}$ at f_{LC} over R_S:

$$Q_{LC} = \frac{Z_{C(LC)}}{R_S} = \frac{f_{SC}}{f_{LC}} = \frac{1}{2\pi R_S C_O f_{LC}} \approx \frac{1}{2\pi(R_{LO}+R_C)C_O p_{LC}}. \tag{6.83}$$

This peaked projection, which R_{LD} also dampens, should be lower than one by the desired gain margin.

Example 8: Determine A_{E0}, p_{E1}, and p_{E2} so A_{LG0} is 100 V/V, PM nears 45°, and GM exceeds 10 V/V when v_O is 4 V, v_R is 1 V, v_E is 2 V, v_D is 4 V, d_O is 33%, $v_{EO(PP)}$ is 500 mV, L_X is 10 μH, R_L is 50 mΩ, C_O is 5 μH, R_C is 10 mΩ, R_{LD} is 100 Ω, and f_{SW} is 1 MHz.

Solution:

$$A_{PWM0} \approx \frac{1}{v_{EO(PP)}} = \frac{1}{500m} = 2\ V^{-1} = 6\ dB$$

$$A_{SL0} \approx \frac{V_E + V_D}{D_O} = \frac{2+4}{33\%} = 18\ V = 25\ dB$$

$$\beta_{FB} \approx \frac{v_R}{v_O} = \frac{1}{4} = 250\ mV/V = -12\ dB$$

$$A_{LG0} = A_{E0}A_{PWM0}A_{SL0}\beta_{FB} \approx A_{E0}(2)(18)(250m) \equiv 100\ V/V = 40\ dB$$

$$\therefore \quad A_{E0} = 11\ V/V = 21\ dB$$

$$L_{DO} = \frac{L_X}{D_O{}^2} = \frac{10\mu}{33\%^2} = 92\ \mu H$$

$$R_{LO} = \frac{R_L}{D_O} = \frac{50m}{33\%} = 150\ m\Omega$$

$$p_{LC} = \frac{1}{2\pi\sqrt{L_{DO}C_O}} = \frac{1}{2\pi\sqrt{(92\mu)(5\mu)}} = 7.4\ kHz$$

$$z_{DO} = \left(\frac{V_E + V_D}{2\pi L_X}\right)\left(\frac{D_O}{I_{L(HI)}}\right)$$

$$\approx \left(\frac{V_E + V_D}{2\pi L_X}\right)\left(\frac{D_O{}^2}{v_O/R_{LD}}\right) = \left[\frac{2+4}{2\pi(10\mu)}\right]\left(\frac{33\%^2}{4/100}\right) = 260\ kHz$$

$$z_C = \frac{1}{2\pi R_C C_O} = \frac{1}{2\pi(10m)(5\mu)} = 3.2\ MHz$$

$$PM \approx 45^\circ \quad \therefore \quad p_{E2} \equiv f_{0dB} = A_{LG0}p_{E1} = (100)p_{E1}$$

$$A_{LG}|_{p_{LC}} \approx A_{LG0}p_{E1}\left(\frac{p_{E2}}{p_{LC}{}^2}\right)\left[\frac{1}{2\pi(R_{LO} + R_C)C_O p_{LC}}\right]$$

$$= \left(\frac{f_{0dB}{}^2}{7.4k^2}\right)\left[\frac{1}{2\pi(150m + 10m)(5\mu)(7.4k)}\right]$$

$$< \frac{1}{GM} = \frac{1}{10} = -20\ dB$$

$$\therefore \quad p_{E2} \equiv f_{0dB} = 450\ Hz << z_{DO} < p_{SW} = f_{SW} \quad \text{and} \quad p_{E1} = 4.5\ Hz$$

Note: R_{LD} reduces A_{LG} at p_{LC}, so f_{0dB} and p_{E2} can be slightly higher.

Example 9: Determine A_{VO}, p_A, and C_F for an inverting mixed translation so A_{E0}, p_{E1}, and p_{E2} satisfy the requirements specified in Example 8 when R_F is 500 kΩ.

Solution:

$$A_{V0} = A_{E0} \equiv 11 \text{ V/V} = 21 \text{ dB}$$

$$p_{E1} \approx \frac{p_F}{A_{V0}} = \frac{1}{2\pi R_F C_F A_{V0}} = \frac{1}{2\pi(500k)C_F(11)} \equiv 4.5 \text{ Hz} \quad \therefore \quad C_F = 6.4 \text{ nF}$$

$$p_{E2} = p_X \approx A_{V0} p_A = 11 p_A \equiv 450 \text{ Hz} \quad \therefore \quad p_A = 41 \text{ Hz}$$

Explore with SPICE:
See Appendix A for notes on SPICE simulations.

```
* Type I CCM Stabilization with an Inverting Mixed Translation
vi vi 0 dc=0 ac=1
rf vi vn 500k
eav0 va 0 0 vn 11
rpa va vb 1
cpa vb 0 3.884m
eavo veo 0 vb 0 1
cf vn veo 6.4n
epwm de 0 veo 0 2
elv v1 0 de 0 3.84
rlv v1 v2 1
rpsw v2 0 352m
ezdo v3 0 0 de 3.84
clv v3 v2 612n
elv0 vs 0 v2 0 18
ldo vs vl 92u
rl vl vo 150m
co vo vc 5u
rc vc 0 10m
*rld vo 0 100
efb vfb 0 vo 0 250m
.ac dec 1000 0.1 10e6
.end
```

Tip: Plot v(veo) and v(vfb) in dB to inspect A_E and A_{LG} and un-comment rld to see effects on A_{LG}.

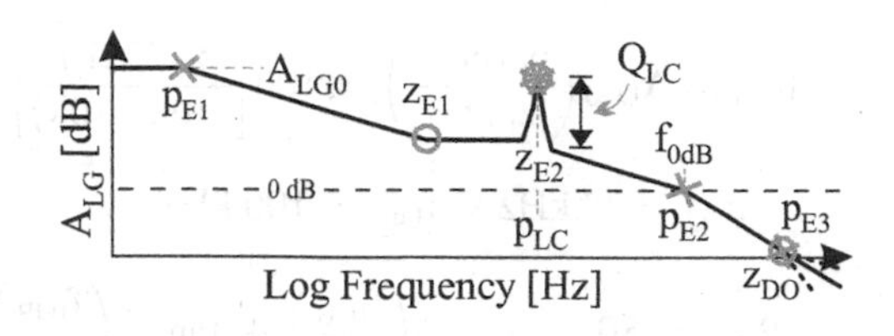

Fig. 6.39 Pole–zero–zero stabilization in CCM

B. **Type III: Pole–Zero–Zero Triplet**

When z_{DO} is absent or well above p_{LC}, two zeros can recover the phase p_{LC} loses before A_{LG} in Fig. 6.39 reaches f_{0dB}. This way, f_{0dB} can exceed p_{LC}. For this, A_E should add p_{E1}, z_{E1} and z_{E2} near p_{LC}, p_{E2} near f_{0dB}, and p_{E3} over f_{0dB}.

p_{E1}, p_{LC}, z_{E1}, and z_{E2} should project A_{LG} to an f_{0dB} that is close to or below p_{E2} and a decade or more below p_{E3} and z_{DO}. To keep phase from shifting too much, z_{E1} and z_{E2} should be near or below p_{LC}. This way, A_{LG} loses 270° to p_{E1} and p_{LC} and recovers up to 180° with z_{E1} and z_{E2}, leaving 90° of margin at f_{0dB} for p_{E2}, p_{E3}, and z_{DO}:

$$A_{LG}\Big|_{f_{0dB}>p_{LC}\geq z_{E1},z_{E2}>p_{E1}} \approx \frac{A_{LG0}p_{E1}p_{LC}^2}{z_{E1}z_{E2}f_{0dB}}\Bigg|_{f_{0dB}\approx A_{LG0}\left(\frac{p_{E1}p_{LC}^2}{z_{E1}z_{E2}}\right)\leq p_{E2}<\frac{p_{E3}}{10},\frac{z_{DO}}{10}} = 1. \quad (6.84)$$

Example 10: Determine A_{E0}, p_{E1}, and PM so f_{0dB} is 100 kHz when z_{E1} and z_{E2} are 10% of f_{0dB}, p_{E2} nears f_{0dB}, p_{E3} is at f_{SW}, and d_O is 100% with the other parameters used in Example 8.

Solution:

$$A_{PWM0} = 2\ V^{-1}, \beta_{FB} \approx 250\ mV/V, f_{SW} = 1\ MHz,$$
$$\text{and } z_C = 3.2\ MHz \text{ from Example 8}$$

$$A_{SL0} \approx V_E + V_D = 2 + 4 = 6\ V = 16\ dB$$

$$A_{LG0} = A_{E0}A_{PWM0}A_{SL0}\beta_{FB} \approx A_{E0}(2)(6)(250m) \equiv 100\ V/V = 40\ dB$$

$$\therefore \quad A_{E0} = 33\ V/V = 30\ dB$$

$$d_O = 1 \quad \therefore \quad L_{DO} = L_X, R_{LO} = R_L, \text{and no } z_{DO}$$

$$p_{LC} = \frac{1}{2\pi\sqrt{L_X C_O}} = \frac{1}{2\pi\sqrt{(10\mu)(5\mu)}} = 22\ kHz$$

$$f_{0dB} \approx A_{LG0}\left(\frac{p_{E1}p_{LC}{}^2}{z_{E1}z_{E2}}\right) = 100\left[\frac{p_{E1}(22k)^2}{(10k)(10k)}\right] \equiv 100\ \text{kHz} \quad \therefore \quad p_{E1} = 210\ \text{Hz}$$

$$p_{LC} = 22\ \text{kHz} < f_{0dB} = 100\ \text{kHz} \quad \therefore \quad \angle A_{LC(0dB)} \equiv \angle A_{LC}\ \text{at}\ f_{0dB} \approx -180°$$

$$PM = 180° - \tan^{-1}\left(\frac{f_{0dB}}{p_{E1}}\right) + \tan^{-1}\left(\frac{f_{0dB}}{z_{E1}}\right) + \tan^{-1}\left(\frac{f_{0dB}}{z_{E2}}\right) + \angle A_{LC(0dB)}$$

$$- \tan^{-1}\left(\frac{f_{0dB}}{p_{E2}}\right) - \tan^{-1}\left(\frac{f_{0dB}}{p_{E3}}\right) - \tan^{-1}\left(\frac{f_{0dB}}{p_{SW}}\right) + \tan^{-1}\left(\frac{f_{0dB}}{z_C}\right)$$

$$\approx 180° - 90° + 84° + 84° - 180° - 45° - 6° - 6° + 2° = 23°$$

Note: PM is less than 45° because z_{E2} is within a decade of f_{0dB}, so z_{E2} saves less than 90°, and p_{E3} and p_{SW} lose 12° when they are a decade above f_{0dB}.

Example 11: Determine A_{V0}, p_A, C_F, and R_C for an inverting mixed translation so A_{E0}, p_{E1}, p_{E2}, and p_{E3} satisfy the requirements in Example 8 when R_F is 500 kΩ.

Solution:

$$A_{V0} = A_{E0} \equiv 33\ \text{V/V} = 30\ \text{dB}$$

$$p_{E1} \approx \frac{p_F}{A_{V0}} = \frac{1}{2\pi R_F C_F A_{V0}} = \frac{1}{2\pi(500k)C_F(33)} \equiv 210\ \text{Hz} \quad \therefore \quad C_F = 46\ \text{pF}$$

$$z_{E1} = \frac{1}{2\pi R_F C_B} = \frac{1}{2\pi(500k)C_B} \equiv 10\ \text{kHz} \quad \therefore \quad C_B = 32\ \text{pF}$$

$$z_{E2} = \frac{1}{2\pi R_C C_F} = \frac{1}{2\pi R_C(46p)} \equiv 10\ \text{kHz} \quad \therefore \quad R_C = 350\ \text{k}\Omega$$

$$p_{E2} \approx A_{V0}\left(\frac{p_C p_{BX}{}''}{p_F}\right) = \frac{A_{V0}}{2\pi R_C C_B} = \frac{33}{2\pi(350k)(32p)} = 470\ \text{kHz}$$
$$> f_{0dB} = 100\ \text{kHz}$$

$$p_A = p_{E3} \equiv f_{SW} = 1\ \text{MHz}$$

Explore with SPICE:
See Appendix A for notes on SPICE simulations.

```
* Type III CCM Stabilization with an Inverting Mixed Translation
vi vi 0 dc=0 ac=1
rf vi vn 500k
cb vi vn 32p
eav0 va 0 0 vn 33
rpa va vb 1
cpa vb 0 159.2n
eavo veo 0 vb 0 1
rcf vn vx 350k
cf vx veo 46p
epwm de 0 veo 0 2
epsw vy 0 de 0 1
rpsw vy vz 1
cpsw vz 0 159.2n
esl0 vs 0 vz 0 6
lx vs vl 10u
rl vl vo 50m
co vo vc 5u
rc vc 0 10m
rld vo 0 100
efb vfb 0 vo 0 250m
.ac dec 1000 10 10e6
.end
```

Tip: Plot v(veo) and v(vfb) in dB to inspect A_E and A_{LG}.

C. **Design Notes**

The drawback of one dominant pole is low bandwidth, because f_{0dB} should precede p_{LC}. Adding zeros extends f_{0dB} beyond p_{LC}, but only to the extent unintended poles and z_{DO} allow. R_C sometimes reverses the effects of p_C in p_{LC} when R_C current-limits C_O below f_{0dB}. But R_C is usually very low, and adding R_C for the sake of reducing z_C burns more Ohmic power and increases dynamic fluctuations in v_O.

6.4.4 Current Mode

One way of eliminating p_{LC} is by regulating i_L. This way, the feedback translation that determines i_L is largely independent of sL_X. Removing this dependence to sL_X eliminates the LC interaction that produces p_{LC}.

A. **Current Loop**

A_{IE}, the PWM, the switched inductor, and β_{IFB} in Fig. 6.40 close an inverting feedback loop that sets i_L. A_{IE} senses and amplifies the error that adjusts d_E' and i_L so v_{IFB} nears v_{EO}. This way, i_L is a reverse β_{IFB} translation of v_{EO}'s mirrored reflection, which is independent of L_X's impedance sL_X:

$$i_L = \frac{v_{IFB}}{\beta_{IFB}} \approx \frac{v_{EO}}{\beta_{IFB}}. \tag{6.85}$$

This is like removing L_X from the circuit.

B. **Loop Gain**

When the overall forward gain A_{IF} surpasses the feedback gain $A_{I\beta}$, the gain A_G to i_L follows $A_{I\beta}$'s $1/\beta_{IFB}$ up to the p_G that the loop's f_{I0dB} sets:

$$\begin{aligned} A_G &\equiv \frac{i_L}{v_{EO}} = A_{IF}||A_{I\beta} \\ &\approx \frac{1/\beta_{IFB}}{(1 + s/2\pi f_{I0dB})(1 + s/2\pi p_{SW})} = \frac{1/\beta_{IFB}}{(1 + s/2\pi p_G)(1 + s/2\pi p_{SW})}. \end{aligned} \tag{6.86}$$

A_G drops faster past p_{SW} when f_O surpasses f_{SW}. This β_{IFB} is usually constant. So the loop that sets i_L in Fig. 6.40 is basically a bandwidth-limited transconductor that d_O in Fig. 6.41 duty-cycles.

A_{LG} is the gain across A_E, A_G, d_O into C_O with R_C and R_{LD}, and β_{FB}. A_{LG} starts with $A_{E0}A_{G0}D_OR_{LD}\beta_{FB}$. A_{LG} falls past p_G, p_{CP}, and p_{SW} when f_{I0dB} bandwidth-limits A_G, C_O and R_C shunt R_{LD}, and f_O surpasses f_{SW}. p_{CP} eventually fades (past z_C)

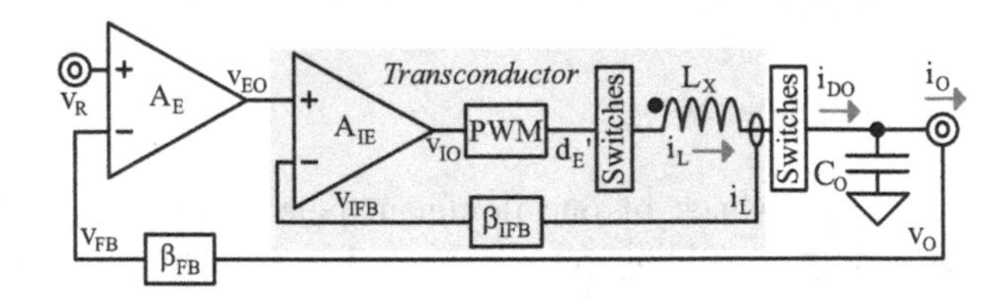

Fig. 6.40 Current-mode voltage controller

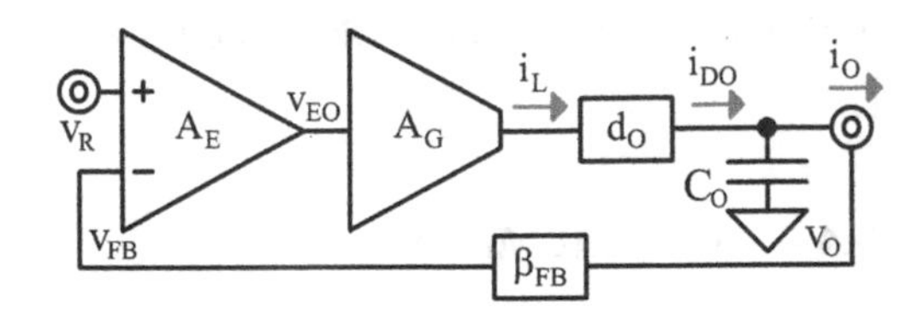

Fig. 6.41 Equivalent current-mode voltage controller

when R_C current-limits C_O:

$$A_{LG} \equiv \frac{v_{FB}}{v_E} = \frac{v_{FB}}{v_R - v_{FB}} = A_E A_G D_O[(Z_C + R_C)||R_{LD}]\beta_{FB}$$
$$\approx \frac{A_E A_{G0} D_O R_{LD} \beta_{FB}(1 + s/2\pi z_C)}{(1 + s/2\pi p_G)(1 + s/2\pi p_{SW})(1 + s/2\pi p_{CP})}. \quad (6.87)$$

p_{CP} normally precedes p_{LC}, and although not always, p_G can surpass p_{LC}. When this happens: when p_{CP} precedes p_G, the net effect of current-mode controllers is to split p_{LC} into p_{CP} and p_G. z_C is usually higher than these poles because R_C is often very low.

C. Stabilization

p_{CP} is typically so low that p_{CP} alone can reduce A_{LG0} to 0 dB. In these cases, A_{E0} can raise A_{LG0} to a level that extends f_{0dB} to p_G with a p_{E1} that is a decade or more above p_G. The system is Type I and inherently stable this way.

Since R_C is usually low, R_{LD} in p_{CP} overwhelms R_C. So R_{LD}'s in A_{LG0} and p_{CP} practically cancel in f_{0dB}. This independence to R_{LD} is desirable because R_{LD} is variable and largely unpredictable:

$$f_{0dB} \approx A_{LG0} p_{CP} \approx \frac{A_{E0} A_{DO0} R_{LD} \beta_{FB}}{2\pi(R_C + R_{LD})C_O} \approx \frac{A_{E0} A_{G0} D_O \beta_{FB}}{2\pi C_O} \approx \frac{A_{E0} D_O \beta_{FB}}{2\pi C_O \beta_{IFB}} \leq p_G \leq \frac{p_{E1}}{10}, \frac{f_{SW}}{10}. \quad (6.88)$$

The challenge with this approach is keeping p_{E1} well above p_G.

Adding a zero can ease A_E's GBW requirement $A_{E0}p_{E1}$. In this case, with a Type II stabilizer, p_{E1}, z_{E1}, and p_{CP} can project A_{LG} to an f_{0dB} that nears p_G. p_{E2} should be a decade or more over p_G:

$$A_{LG}\Big|_{f_{0dB} > p_{E1}, z_{E1}, p_{CP}} \approx A_{LG0}\left(\frac{p_{E1} p_{CP}}{z_{E1} f_{0dB}}\right)\Big|_{f_{0dB} \approx A_{LG0}\left(\frac{p_{E1} p_{CP}}{z_{E1}}\right) \leq p_G \leq \frac{p_{E2}}{10}, \frac{f_{SW}}{10}} = 1. \quad (6.89)$$

With or without z_{E1}, f_{0dB} can reach p_G. Since this p_G can be higher than p_{LC}, current-mode voltage controllers in CCM respond faster than Type I voltage-mode controllers. This is the advantage of current-mode control.

Example 12: Determine A_{E0}, p_{CP}, p_{E1}, and PM so f_{0dB} is at p_G when β_{IFB} is 1 Ω and p_G is 100 kHz with the other parameters used in Example 8.

Solution:

$$\beta_{FB} \approx 250\ \text{mV/V}, f_{SW} = 1\ \text{MHz, and } z_C = 3.2\ \text{MHz from Example 8}$$

$$p_{CP} = \frac{1}{2\pi(R_C + R_{LD})C_O} = \frac{1}{2\pi(10m + 100)(5\mu)} = 320\ \text{Hz}$$

$$f_{0dB} \approx \frac{A_{E0}A_{G0}D_O\beta_{FB}}{2\pi C_O} \approx \frac{A_{E0}D_O\beta_{FB}}{2\pi C_O\beta_{IFB}} = \frac{A_{E0}(33\%)(250m)}{2\pi(5\mu)(1)} \equiv p_G = 100\ \text{kHz} << z_C$$

$$\therefore\quad A_{E0} = 38\ \text{V/V} = 32\ \text{dB} \quad \text{and} \quad p_{E1} \geq 10f_{0dB} = 1\ \text{MHz}$$

$$PM = 180^\circ - \tan^{-1}\left(\frac{f_{0dB}}{p_{CP}}\right) - \tan^{-1}\left(\frac{f_{0dB}}{p_G}\right) - \tan^{-1}\left(\frac{f_{0dB}}{p_{SW}}\right) - \tan^{-1}\left(\frac{f_{0dB}}{p_{E1}}\right)$$

$$\approx 180^\circ - 90^\circ - 45^\circ - 6^\circ - 6^\circ = 33^\circ$$

Note: A_E's GBW projection $A_{E0}p_{E1}$ should reach or exceed 38 MHz.

Example 13: Determine A_{G0} and C_F for an OTA so A_{E0} and p_{E1} satisfy the requirements in Example 12 when R_F is 500 kΩ.

Solution:

$$A_{V0} = A_{G0}R_F = A_{G0}(500k) = A_{E0} \equiv 38\ \text{V/V} = 32\ \text{dB} \ \therefore\ A_{G0} = 76\ \mu\text{S} = -82\ \text{dB}$$

$$p_{E1} = \frac{1}{2\pi R_F C_F} = \frac{1}{2\pi(500k)C_F} \geq 1\ \text{MHz} \quad \therefore \quad C_F \leq 320\ \text{fF}$$

Explore with SPICE:
See Appendix A for notes on SPICE simulations.

```
* Type I Current-Mode Stabilization with an OTA
vi vi 0 dc=0 ac=1
ge0 0 veo 0 vi 76u
rf veo 0 500k
cf veo 0 320f
epg vw 0 veo 0 1
rpg vw vx 1
cpg vx 0 1.592u
epsw vy 0 vx 0 1


rpsw vy vz 1
cpsw vz 0 159.2n
gido 0 vo vy 0 330m
co vo vc 5u
rc vc 0 10m
rld vo 0 100
efb vfb 0 vo 0 250m
.ac dec 1000 1 10e6
.end
```

Tip: Plot v(veo), i(Gido)/v(veo), and v(vfb) in dB to inspect A_E, A_{DO}, and A_{LG}.

Example 14: Determine A_E's GBW $A_{E0}p_{E1}$ so z_{E1} is at p_{CP} with the other parameters used in Examples 8 and 12.

Solution:

$$\beta_{FB} \approx 250\ \text{mV/V and } z_C = 3.2\ \text{MHz from Example 8}$$

$$z_{E1} \equiv p_{CP} = 320\ \text{Hz from Example 12}$$

$$A_{LG0} = A_{E0}\left(\frac{1}{\beta_{IFB}}\right)D_O R_{LD}\beta_{FB} \approx A_{E0}(1)(33\%)(100)(250\text{m}) = 8.2A_{E0}$$

$$f_{0dB} \approx A_{LG0}p_{E1}\left(\frac{p_{CP}}{z_{E1}}\right) = 8.2A_{E0}p_{E1}\left(\frac{320}{320}\right) \equiv p_G = 100\ \text{kHz} \quad \therefore\ A_{E0}p_{E1} = 12\ \text{kHz}$$

6.4.5 Discontinuous Conduction

A. Switched Inductor

The switched inductor in *discontinuous-conduction mode* (DCM) is basically a current-sourced resistor. The *small-signal current source* i_s in Fig. 6.42 is d_e' times an *energize duty-cycle* d_E translation of L_X's *peak inductor current* $i_{L(PK)}$. The *duty-cycled resistance* R_{DO} is a timed d_O translation of L_X with an R_L/D_O that is usually negligibly lower.

i_s into R_{DO}, C_O with R_C, and R_{LD} sets the switched-inductor gain to v_O. A_{SL} falls past p_{CS} when C_O and R_C shunt $R_{DO} \parallel R_{LD}$ and past p_{SW} when f_O surpasses f_{SW}. p_{CS} eventually fades past z_C when R_C current-limits C_O:

$$A_{SL(DCM)} \equiv \frac{v_O}{d_e'} = \left(\frac{I_{L(PK)}}{D_E}\right)\left(\frac{R_{DO}||R_{LD}}{1 + s/2\pi p_{SW}}\right)\left(\frac{1 + s/2\pi z_C}{1 + s/2\pi p_{CS}}\right), \tag{6.90}$$

where

$$R_{DO} = 2\left(\frac{L_X}{{D_O}^2}\right)\left(\frac{T_{SW}}{{T_C}^2}\right) + \frac{R_L}{D_O} = 2L_{DO}\left(\frac{T_{SW}}{{T_C}^2}\right) + R_{LO}, \tag{6.91}$$

t_C is L_X's *conduction time*, and uppercase variables are static components.

A notable difference between DCM and CCM is the absence of p_L and z_{DO}. L_X's small-signal Ohmic current does not change in DCM because L_X's average voltage across a cycle does not vary. And since L_X depletes before every cycle ends in DCM, d_O does not sacrifice i_L. p_{CS} is also variable and lower than p_{LC} because T_C falls with i_O to levels that keep R_{DO} high.

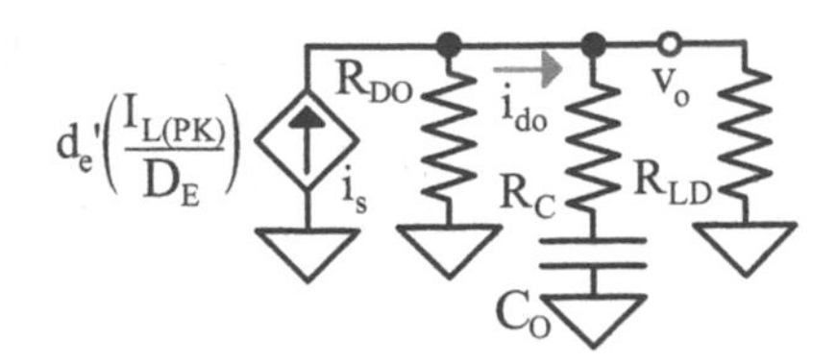

Fig. 6.42 Small-signal model of the switched inductor in DCM

B. Stabilization

p_{CS} is usually so low that p_{CS} alone can reduce A_{LG0} to 0 dB. In these cases, A_{E0} can raise A_{LG0} to a level that extends f_{0dB} to a decade below f_{SW} with a p_{E1} that nears f_{SW}. The system is Type I and inherently stable this way.

Since R_C is usually low, $R_{DO} \parallel R_{LD}$ in p_{CS} overwhelms R_C. So $R_{DO} \parallel R_{LD}$'s in A_{LG0} and p_{CS} practically cancel in f_{0dB}. This independence to R_{LD} is desirable because R_{LD} is variable and unpredictable:

$$f_{0dB} \approx A_{LG0}p_{CS} = \frac{A_{E0}A_{PWM0}A_{LI}(R_{DO}||R_{LD})\beta_{FB}}{2\pi[R_C + (R_{DO}||R_{LD})]C_O} \approx \frac{A_{E0}A_{PWM0}I_{L(PK)}\beta_{FB}}{2\pi C_O D_E}$$
$$\leq p_{E1}, \frac{f_{SW}}{10}. \tag{6.92}$$

p_{E1} should near or exceed f_{0dB} to keep PM near or above 45°.

Example 15: Determine A_{E0}, p_{E1}, p_{CS}, and f_{0dB} so f_{0dB} is 10% of f_{SW} and PM is 45° when $I_{L(PK)}$ is 57 mA, D_E is 67%, R_{DO} is 560 Ω, and R_{LD} is 500 Ω with the other parameters used in Example 8.

Solution:

$$\beta_{FB} \approx 250\text{ mV/V}, A_{PWM0} = 2\text{ V}^{-1}, f_{SW} = 1\text{ MHz, and } z_C = 3.2\text{ MHz from Example 8}$$

$$p_{CS} = \frac{1}{2\pi[R_C + (R_{DO}||R_{LD})]C_O} = \frac{1}{2\pi[10m + (560||500)](5\mu)} = 120\text{ Hz}$$

$$f_{0dB} \approx \frac{A_{E0}A_{PWM0}I_{L(PK)}\beta_{FB}}{2\pi D_E C_O} = \frac{A_{E0}(2)(57m)(250m)}{2\pi(67\%)(5\mu)} \equiv \frac{f_{SW}}{10} = 100\text{ kHz} << z_C$$

$$\therefore \quad A_{E0} = 74\text{ V/V} = 37\text{ dB}$$

$$PM = 180° - \tan^{-1}\left(\frac{f_{0dB}}{p_{CS}}\right) - \tan^{-1}\left(\frac{f_{0dB}}{p_{E1}}\right) - \tan^{-1}\left(\frac{f_{0dB}}{p_{SW}}\right) + \tan^{-1}\left(\frac{f_{0dB}}{z_C}\right)$$

$$= 180° - 90° - \tan^{-1}\left(\frac{f_{0dB}}{p_{E1}}\right) - 6° + 2° \equiv 45° \quad \therefore \quad p_{E1} \geq 120\text{ kHz}$$

Note: A_E's GBW projection $A_{E0}p_{E1}$ should reach or exceed 8.9 MHz.

Example 16: Determine A_{G0} and C_F for an OTA so A_{E0} and p_{E1} satisfy the requirements in Example 15 when R_F is 500 kΩ.

Solution:

$$A_{V0} = A_{G0}R_F = A_{G0}(500k) = A_{E0} \equiv 74\ V/V = 37\ dB\ \therefore\ A_{G0} = 150\ \mu S = -76\ dB$$

$$p_{E1} = \frac{1}{2\pi R_F C_F} = \frac{1}{2\pi(500k)C_F} \geq 120\ kHz \quad \therefore \quad C_F \leq 2.6\ pF$$

Explore with SPICE:
See Appendix A for notes on SPICE simulations.

```
* Type I DCM Stabilization with an OTA
vi vi 0 dc=0 ac=1
ge0 0 veo 0 vi 150u
rf veo 0 500k
cf veo 0 2.6p
epwm de 0 veo 0 2
epsw vx 0 de 0 1
rpsw vx vy 1
cpsw vy 0 159.2n
gis 0 vo vy 0 85m
rdo vo 0 560
co vo vc 5u
rc vc 0 10m
rld vo 0 500
efb vfb 0 vo 0 250m
.ac dec 1000 0.1 10e6
.end
```

Tip: Plot v(veo), i(Gis)/v(de), and v(vfb) in dB to inspect A_E, A_{SL}, and A_{LG}.

6.5 Current Control

Current-mode voltage regulators, chargers, and LED drivers use current controllers to regulate current. Their basic aim is to output a current that is insensitive to v_O. They enclose feedback loops that sense and set i_L, i_{DO}, or i_O for this purpose. This way, v_O variations do not alter the current.

6.5.1 Controller

β_{IFB}, A_{IE}, and the PWM in Fig. 6.43 comprise the feedback controller that sets i_L. β_{IFB} senses and translates i_L to the v_{IFB} that A_{IE} mixes with the *input* v_I. A_{IE} senses and amplifies the error between v_{IFB} and v_I and the PWM converts the amplified error into the duty-cycle command d_E' that sets i_L.

Together, A_{IE} amplifies the error that adjusts d_E' and i_L so v_{IFB} nears v_I. This way, i_L is a reverse β_{IFB} translation of v_I's mirrored reflection:

$$i_L = \frac{v_{IFB}}{\beta_{IFB}} \approx \frac{v_I}{\beta_{IFB}}. \tag{6.93}$$

So when i_L deviates from this v_I/β_{IFB}, the loop adjusts d_E' until i_L recovers.

6.5.2 Transconductance Gain

The switched inductor in continuous conduction is basically a voltage-sourced inductor. v_s in Fig. 6.37 is d_e' times a d_O translation of v_E and v_D and L_{DO} and R_{LO} are d_O translations of L_X and R_L. The gain A_{DO} to i_{DO} is an Ohmic translation of L_{DO} with R_{LO}, C_O with R_C, and R_{LD}, where R_{LO} and R_C are usually much lower than R_{LD}. Since d_O duty-cycles i_L to i_{DO} and induces z_{DO}, the gain A_{IL} to i_L is a reverse d_O translation of A_{DO} without z_{DO}:

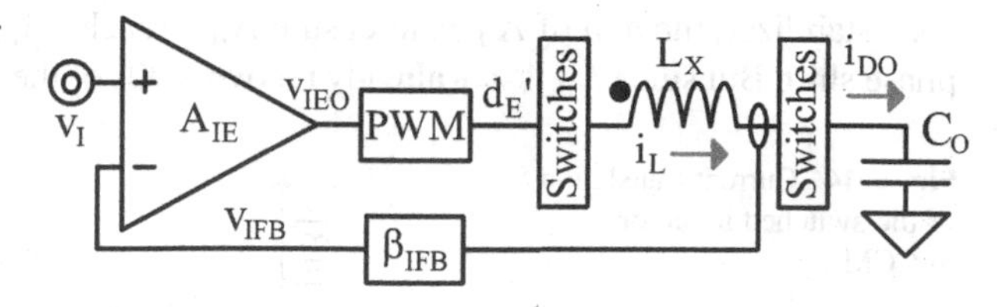

Fig. 6.43 Current controller

$$A_{IL(CCM)} \equiv \frac{i_l}{d_e'} = \frac{i_{do}/D_O}{d_e'}\bigg|_{No\ z_{DO}}$$
$$= \frac{A_{DO}}{D_O}\bigg|_{No\ z_{DO}} = \frac{A_{LV(CCM)}/D_O}{Z_{DO} + R_{LO} + \left[(Z_C + R_C)||R_{LD}\right]}\bigg|_{No\ z_{DO}}$$
$$\approx \left[\frac{V_E + V_D}{D_O^2(R_{LO} + R_{LD})}\right]\left\{\frac{1 + s/2\pi z_{CP}}{\left[(s/2\pi p_{LC})^2 + s/2\pi p_{LC}Q_{LC} + 1\right](1 + s/2\pi p_{SW})}\right\}. \tag{6.94}$$

At low frequency, when L_{DO} shorts and C_O opens, v_s into R_{LO} and R_{LD} sets A_{IL0} in Fig. 6.44 to $(V_E + V_D)/[D_O^2(R_{LO} + R_{LD})]$ or $K_V/(R_{LO} + R_{LD})$, which reduces to K_V/R_{LD} when R_{LO} is much lower than R_{LD}. A_{IL} rises past z_{CP} with K_V/Z_C's sK_VC_O when C_O and R_C shunt R_{LD}. A_{IL} falls past p_{LC} with K_V/Z_L's K_V/sL_X when Z_L overcomes Z_C. At p_{LC}, A_{IL} peaks Q_{LC} over K_V/Z_C's projection. A_{IL} eventually falls faster past p_{SW} when f_O exceeds f_{SW}.

The gain across the loop A_{ILG} is $A_{IE}A_{PWM}A_{IL}\beta_{IFB}$:

$$A_{ILG} \equiv \frac{v_{IFB}}{v_{IE}} = \frac{v_{IFB}}{v_{IR} - v_{IFB}} = A_{IE}A_{PWM}A_{IL}\beta_{FB}. \tag{6.95}$$

The closed-loop gain A_G to i_L follows the lowest forward translation to i_L. A_{IE}, A_{PWM}, and the switched-inductor gain A_{IL} to i_L set the overall forward gain A_{IF} and $1/\beta_{IFB}$ sets the feedback gain $A_{I\beta}$. So A_G is

$$A_G \equiv \frac{i_L}{v_I} = A_{IF}||A_{I\beta} = (A_{IE}A_{PWM}A_{IL})||\frac{1}{\beta_{IFB}}. \tag{6.96}$$

$A_{IE}A_{PWM}A_{IL}$ is the part of A_{ILG} that determines feedback accuracy. This is because A_G follows $A_{I\beta}$ to the extent A_{IF}'s $A_{IE}A_{PWM}A_{IL}$ exceeds $A_{I\beta}$. In other words, A_G approaches $1/\beta_{IFB}$ as A_{IF} increases. This means, regulation accuracy improves with higher A_{IF}.

6.5.3 Type I: Inherent Stability

As a stabilizer, the aim of A_{IE} is to ensure A_{ILG} reaches f_{I0dB} with less than 180° of phase shift. But since A_{IL}'s z_{CP} already recovers 90° of the 180° that p_{LC} loses, A_{IE}'s

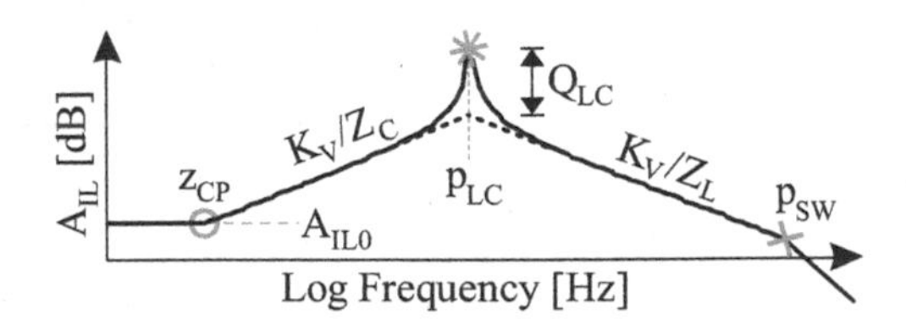

Fig. 6.44 Current translation of the switched inductor in CCM

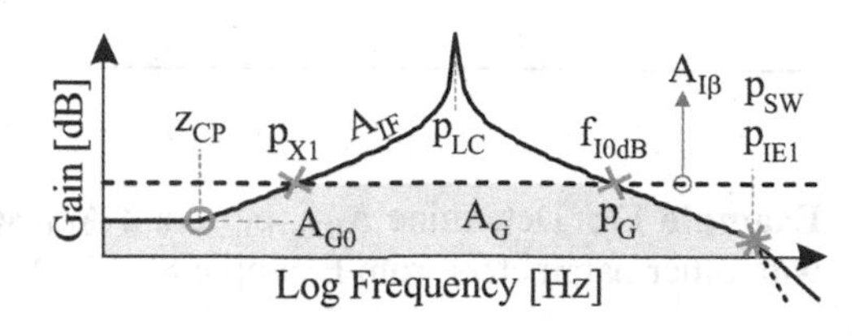

Fig. 6.45 Inherent transconductance in CCM

role can be to increase gain, and that way, extend f_{I0dB}. For 45° or so of PM, f_{I0dB} should be at or below A_{IE}'s bandwidth p_{IE1} and a decade or more below f_{SW}:

$$A_{ILG}\Big|_{f_{I0dB} > p_{LC} > z_{CP}} \approx \frac{A_{ILG0}{p_{LC}}^2}{z_{CP}f_{I0dB}}\Bigg|_{f_{I0dB}=A_{ILG0}\left(\frac{p_{LC}^2}{z_{CP}}\right)=p_G\le p_{IE1},\frac{f_{SW}}{10}} = 1. \qquad (6.97)$$

Since A_{ILG} rises and falls to 0 dB, A_{ILG} usually starts low, which means A_{IF0} is also low. So A_{IF} in Fig. 6.45 starts low, climbs past z_{CP}, falls past p_{LC}, and falls faster past p_{SW}. Although not always, A_{IF0}'s $A_{IE0}A_{PWM0}A_{IL0}$ is often lower than $A_{I\beta}$'s $1/\beta_{IFB}$. So A_G often starts with A_{IF0}.

A_G climbs with A_{IF} past z_{CP} until A_{IF} surpasses $A_{I\beta}$. This z_{CP} is usually low because C_O is high and R_{LD} is moderate. Since A_{IF} is the part of A_{ILG} that excludes β_{IFB}, A_{ILG0} is less than one after A_{IF} surpasses $A_{I\beta}$'s $1/\beta_{IFB}$. So A_{IF} crosses $A_{I\beta}$ at a p_{X1} that is $1/A_{ILG0}$ times greater than z_{CP}:

$$A_{IF}\Big|_{z_{CP} < f_O < p_{LC}} = A_{IE0}A_{PWM0}A_{IL0}\left(\frac{f_O}{z_{CP}}\right)\Bigg|_{f_O \ge \frac{z_{CP}}{A_{ILG0}} \approx p_{X1} > z_{CP}} \ge A_{I\beta} = \frac{1}{\beta_{IFB}}. \qquad (6.98)$$

A_G follows $A_{I\beta}$'s $1/\beta_{IFB}$ past p_{X1} until A_{IF} falls below $1/\beta_{IFB}$, which happens at the p_G that f_{I0dB} sets. A_G falls faster with A_{IF} past p_{IE1} and p_{SW}:

$$\begin{aligned} A_G &\approx \frac{A_{IE}A_{PWM0}A_{IL0}(1+s/2\pi z_{CP})}{(1+s/2\pi p_{X1})(1+s/2\pi p_G)(1+s/2\pi p_{SW})} \\ &\approx \frac{A_{IE0}A_{PWM0}A_{IL0}(1+s/2\pi z_{CP})}{(1+s/2\pi p_{X1})(1+s/2\pi p_G)(1+s/2\pi p_{IE1})(1+s/2\pi p_{SW})}\Bigg|_{f_O > p_{X1}} \\ &\approx \frac{1/\beta_{IFB}}{(1+s/2\pi p_G)(1+s/2\pi p_{IE1})(1+s/2\pi p_{SW})}. \end{aligned} \qquad (6.99)$$

So A_G reaches and follows $1/\beta_{IFB}$ past p_{X1} up to p_G.

Since A_{IF0} is low, A_{G0} is lower than the targeted $1/\beta_{IFB}$, which means dc accuracy is low. This accuracy, however, is not always important. In current-mode voltage regulators, for example, the feedback loop that sets v_O adjusts v_I until v_{FB} matches v_R, irrespective of v_I-to-i_L's dc accuracy.

Example 17: Determine A_{IE0}, p_{IE1}, and A_{G0} so f_{I0dB} is $10\%f_{SW}$ when β_{IFB} is 1 Ω with other parameters from Example 8.

Solution:

$$R_{LO} = 150\ m\Omega, A_{PWM0} = 2\ V^{-1}, \text{and } p_{LC} = 7.4\ kHz \text{ from Example 8}$$

$$A_{IL0} \approx \frac{V_E + V_D}{{D_O}^2(R_{LO} + R_{LD})} = \frac{2+4}{33\%^2(150m + 100)} = 550\ mA = -5.2\ dB$$

$$z_{CP} = \frac{1}{2\pi(R_C + R_{LD})C_O} = \frac{1}{2\pi(10m + 100)(5\mu)} = 320\ Hz$$

$$f_{I0dB} \approx A_{ILG0}\left(\frac{{p_{LC}}^2}{z_{CP}}\right) = A_{ILG0}\left(\frac{7.4k^2}{320}\right) \equiv \frac{f_{SW}}{10} = 100\ kHz \ \therefore\ A_{ILG0} = 580\ mV/V = -4.7\ dB$$

$$A_{ILG0} = A_{IE0}A_{PWM0}A_{IL0}\beta_{IFB} \approx A_{IE0}(2)(550m)(1) = 580\ mV/V = -4.7\ dB$$

$$\therefore \quad A_{IE0} = 530\ mV/V \quad \text{and} \quad p_{IE1} \geq 10f_{I0dB} = 1\ MHz$$

$$A_{G0} = (A_{IE0}A_{PWM0}A_{IL0})||\frac{1}{\beta_{IFB}}$$
$$\approx [(530m)(2)(550m)]||1 = 370\ mV/A = -8.6\ dB$$

Note: A_{G0} is less than the one that $A_{I\beta}$ sets with β_{IFB} because A_{IF0} is less than $A_{I\beta}$.

Explore with SPICE:
See Appendix A for notes on SPICE simulations.

```
* Type I Stabilization for Current Loop
vi vi 0 dc=0 ac=1
eaie0 va 0 vi 0 530m
epie vb 0 va 0 1
rpie vb vieo 1
cpie vieo 0 159.2n
epwm de 0 vieo 0 2
epsw v1 0 de 0 1
```

(continued)

```
rpsw v1 v2 1
cpsw v2 0 159.2n
esl vs 0 v2 0 55
ldo vs vz 92u
rl vz vo 150m
co vo vc 5u
rc vc 0 10m
rld vo 0 100
fifb vifb 0 esl 1
rifb vifb 0 1
.ac dec 1000 10 10e6
.end
```

Tip: Plot v(vieo), i(Esl)/v(de), and v(vifb) in dB to inspect A_{IE}, A_{IL}, and A_{ILG}.

6.5.4 Type II: Pole–Zero Stabilization

When dc accuracy is important, A_{IE} can add gain, p_{IE1} near z_{CP} to counter z_{CP}, and z_{IE1} near p_{LC} to recover the phase lost to p_{IE1}. This way, A_{IF} in Fig. 6.46 starts high, falls and flattens past p_{IE1} and z_{CP}, and climbs and falls past z_{IE1} and p_{LC}. Since A_{IF0} exceeds $A_{I\beta}$, A_{G0} nears $1/\beta_{IFB}$.

A_G follows $A_{I\beta}$'s $1/\beta_{IFB}$ until A_{IF} drops below $1/\beta_{IFB}$. For 45° or so of PM, f_{I0dB} should be at or below A_{IE}'s p_{IE2} and a decade or more below f_{SW}. This way, A_G follows $1/\beta_{IFB}$ up to p_G, drops faster past p_{IE2}, and even faster past p_{SW}:

$$A_{IF}\big|_{f_O>p_{LC}\geq z_{IE1}>z_{CP}\geq p_{IE1}} = A_{IE0}A_{PWM0}A_{IL0}\left(\frac{p_{IE1}{p_{LC}}^2}{z_{IE1}z_{CP}f_O}\right) \leq A_{I\beta} = \frac{1}{\beta_{IFB}} \quad (6.100)$$

when

$$f_O \geq A_{ILG0}\left(\frac{p_{IE1}{p_{LC}}^2}{z_{IE1}z_{CP}}\right) \approx p_G = f_{I0dB} \leq p_{IE2}, \frac{f_{SW}}{10} \quad (6.101)$$

and

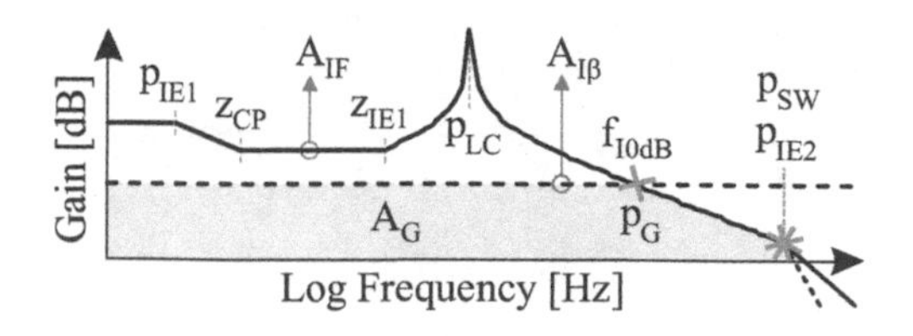

Fig. 6.46 Pole–zero transconductance in CCM

$$A_G \approx \frac{1/\beta_{IFB}}{(1 + s/2\pi p_G)(1 + s/2\pi p_{IE2})(1 + s/2\pi p_{SW})}. \quad (6.102)$$

Example 18: Determine A_{IE0}, p_{IE2}, and A_{G0} so f_{0dB} is 10% of f_{SW} when p_{IE1} is at z_{CP} and z_{IE1} is at p_{LC} with the other parameters used in Examples 8 and 17.

Solution:

$$A_{PWM0} = 2\ V^{-1}, f_{SW} = 1\ \text{MHz, and } z_{IE1} \equiv p_{LC} = 7.4\ \text{kHz from Example 8}$$

$$A_{IL0} = 550\ \text{mA and } p_{IE1} \equiv z_{CP} = 320\ \text{Hz from Example 17}$$

$$f_{I0dB} \approx A_{ILG0}\left(\frac{p_{IE1}p_{LC}^{2}}{z_{IE1}z_{CP}}\right) = A_{ILG0}\left[\frac{(320)(7.4k)^{2}}{(7.4k)(320)}\right] = A_{ILG0}(7.4k)$$

$$\equiv \frac{f_{SW}}{10} = 100\ \text{kHz}$$

$$\therefore \quad A_{ILG0} = 14\ V/V = 23\ \text{dB} \quad \text{and} \quad p_{IE2} \geq f_{I0dB} = 100\ \text{kHz}$$

$$A_{ILG0} = A_{IE0}A_{PWM0}A_{IL0}\beta_{IFB} \approx A_{IE0}(2)(550m)(1) \equiv 14\ V/V = 23\ \text{dB}$$

$$\therefore \quad A_{IE0} = 13\ V/V = 22\ \text{dB}$$

$$A_{G0} = (A_{IE0}A_{PWM0}A_{IL0})||\frac{1}{\beta_{IFB}} \approx [(13)(2)(550m)]||1 = 930\ \text{mA/V} = -0.63\ \text{dB}$$

Note: A_{G0} is closer to the one $A_{I\beta}$ sets with β_{IFB} than in Example 17 because A_{IF0} is greater than $A_{I\beta}$.

Example 19: Determine A_{V0}, p_A, C_F, and R_C for the inverting mixed translation so A_{IE0}, p_{IE1}, z_{IE1}, and p_{E2} satisfy the requirements in Example 18 when R_F is 500 kΩ.

Solution:

$$A_{V0} = A_{IE0} \equiv 13\ \text{V/V} = 22\ \text{dB}$$

$$p_{IE1} \approx \frac{p_F}{A_{V0}} = \frac{1}{2\pi A_{V0} R_F C_F} = \frac{1}{2\pi(13)(500k)C_F} \equiv 320\ \text{Hz} \quad \therefore \quad C_F = 77\ \text{pF}$$

$$z_{IE1} = \frac{1}{2\pi R_C C_F} = \frac{1}{2\pi R_C (77p)} \equiv p_{LC} = 7.4\ \text{kHz} \quad \therefore \quad R_C = 280\ \text{k}\Omega$$

$$p_{IE2} \approx A_{V0} p_A \left(\frac{p_C}{p_F}\right) = A_{V0} p_A \left(\frac{R_F}{R_F + R_C}\right) = (13) p_A \left(\frac{500k}{500k + 280k}\right)$$
$$\geq f_{0dB} = 100\ \text{kHz} \quad \therefore \quad p_A \geq 12\ \text{kHz}$$

Explore with SPICE:
See Appendix A for notes on SPICE simulations.

```
* Type II Stabilization for Current Loop with Inverting Mixed Translation
vi vi 0 dc=0 ac=1
rf vi vn 500k
eav va 0 0 vn 13
rpa va vb 1
cpa vb 0 13.27u
eavo vieo 0 vb 0 1
rcf vn vv 280k
cf vv vieo 77p
epwm de 0 vieo 0 2
epsw v1 0 de 0 1
rpsw v1 v2 1
cpsw v2 0 159.2n
esl vs 0 v2 0 55
ldo vs vz 92u
rl vz vo 150m
```

(continued)

```
co vo vc 5u
rc vc 0 10m
rld vo 0 100
fifb vifb 0 esl 1
rifb vifb 0 1
.ac dec 1000 10 10e6
.end
```

Tip: Plot v(vieo), i(Esl)/v(de), and v(vifb) in dB to inspect A_{IE}, A_{IL}, and A_{ILG}.

6.5.5 Discontinuous Conduction

The switched inductor in discontinuous conduction is basically a current-sourced resistor. i_s in Fig. 6.42 is d_e' times a d_E translation of $i_{L(PK)}$ and R_{DO} is a timed d_O translation of L_X with a d_O translation of R_L. v_o is an Ohmic translation of i_s into R_{DO}, C_O with R_C, and R_{LD} and i_{do} is an Ohmic translation of v_o into C_O with R_C and R_{LD}. And since d_O duty-cycles i_L to i_{DO}, the small-signal switched-inductor gain A_{IL} to i_L is a reverse d_O translation of the gain A_{DO} to i_{DO}:

$$\begin{aligned}A_{IL} &\equiv \frac{i_l}{d_e'} = \left(\frac{i_s}{d_e'}\right)\left(\frac{v_o}{i_s}\right)\left(\frac{i_{do}}{v_o}\right)\left(\frac{i_l}{i_{do}}\right)\\ &= \frac{i_{do}/D_O}{d_e'} = \frac{A_{DO}}{D_O} \approx \left(\frac{A_{LI}}{D_O}\right)\left[\frac{R_{DO}||(Z_C + R_C)||R_{LD}}{(Z_C + R_C)||R_{LD}}\right]\\ &= \left(\frac{I_{L(PK)}}{D_E D_O}\right)\left(\frac{R_{DO}||R_{LD}}{R_{LD}}\right)\left[\frac{1 + s/2\pi z_{CP}}{(1 + s/2\pi p_{CS})(1 + s/2\pi p_{SW})}\right].\end{aligned} \tag{6.103}$$

A_{IL} starts at low f_O (when C_O opens) with i_s into $R_{DO} \parallel R_{LD}$ over R_{LD}. A_{IL} climbs past z_{CP} when C_O and R_C shunt R_{LD} and flattens past p_{CS} when C_O shorts with respect to R_C and $R_{DO} \parallel R_{LD}$. A_{IL} eventually falls past p_{SW} when f_O surpasses f_{SW}.

Since the system cannot respond within one t_{SW}, A_{ILG} should reach 0 dB below f_{SW}. A_{IE} should therefore add gain and p_{IE1} so A_{ILG} drops to 0 dB at an f_{I0dB} that is a decade or more below f_{SW}. This way, A_{IF} in Fig. 6.47 rises and flattens with A_{IL} past z_{CP} and p_{CS} and falls with A_{IE} past p_{IE1}.

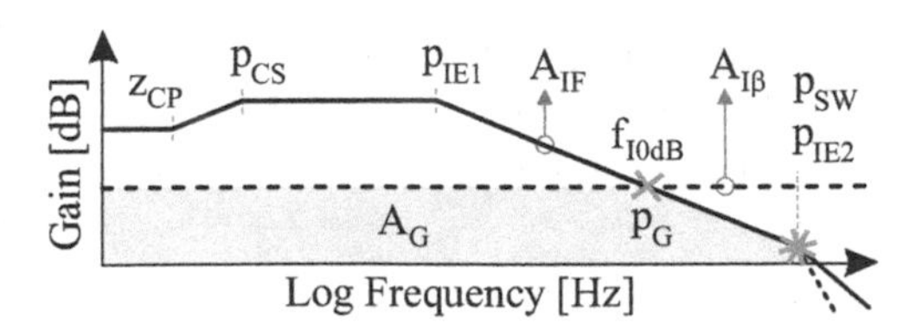

Fig. 6.47 Dominant-pole transconductance in DCM

With A_{IF} higher than $A_{I\beta}$, A_G follows $1/\beta_{IFB}$ until A_{IF} falls below $1/\beta_{IFB}$. This happens at the p_G that f_{I0dB} sets. To keep 45° or so of PM, this f_{I0dB} can be at or below A_{IE}'s p_{IE2} and a decade or more below f_{SW}:

$$A_{ILG0}\Big|_{f_{I0dB}>p_{IE1}>p_{CS}>z_{CP}} \approx A_{ILG0}\left(\frac{p_{CS}p_{IE1}}{z_{CP}f_{I0dB}}\right)\Bigg|_{f_{I0dB}\approx A_{ILG0}p_{IE1}\left(\frac{p_{CS}}{z_{CP}}\right)=p_G\le p_{IE2},\frac{f_{SW}}{10}} = 1. \quad (6.104)$$

This way, A_G follows $1/\beta_{IFB}$ up to a p_G that nears p_{IE2} and is well below f_{SW}:

$$A_G \approx \frac{1/\beta_{IFB}}{(1+s/2\pi p_G)(1+s/2\pi p_{IE2})(1+s/2\pi p_{SW})}. \quad (6.105)$$

Example 20: Determine A_{IE0}, p_{IE1}, p_{IE2}, and A_{G0} so A_{ILG0} is 100 V/V when β_{IFB} is 1 Ω with the other parameters used in Examples 8 and 15.

Solution:

$$A_{PWM0} = 2\ \text{V}^{-1} \text{ from Example 8 and } p_{CS} = 120 \text{ Hz from Example 15}$$

$$A_{IL0} \approx \frac{I_{L(PK)}(R_{DO}||R_{LD})}{D_E D_O R_{LD}} = \frac{(57\text{m})(560||500)}{(67\%)(33\%)(500)} = 140\text{ mA} = -17\text{ dB}$$

$$A_{ILG0} = A_{IE0}A_{PWM0}A_{IL0}\beta_{IFB} = A_{IE0}(2)(140\text{m})(1) \equiv 100\text{ V/V} = 40\text{ dB}$$

$$\therefore \quad A_{IE0} = 360\text{ V/V} = 51\text{ dB}$$

$$z_{CP} = \frac{1}{2\pi(R_C+R_{LD})C_O} = \frac{1}{2\pi(10\text{m}+500)(5\mu)} = 64\text{ Hz}$$

$$f_{I0dB} \approx A_{ILG0}\left(\frac{p_{CS}}{z_{CP}}\right)p_{IE1} = (100)\left(\frac{120}{64}\right)p_{IE1} \equiv \frac{f_{SW}}{10} = 100\text{ kHz}$$

$$\therefore \quad p_{IE1} = 530\text{ Hz} \quad \text{and} \quad p_{IE2} \ge f_{I0dB} = 100\text{ kHz}$$

$$A_{G0} = (A_{IE0}A_{PWM0}A_{IL0})||\frac{1}{\beta_{IFB}} \approx [(190)(2)(140\text{m})]||1 = 980\text{ mA/V} = -0.18\text{ dB}$$

Example 21: Determine A_{V0}, p_A, and C_F for the inverting mixed translation so A_{IE0}, p_{IE1}, and p_{E2} satisfy the requirements in Example 18 when R_F is 500 kΩ.

Solution:

$$A_{V0} = A_{IE0} \equiv 360\ \text{V/V} = 51\ \text{dB}$$

$$p_{E1} \approx \frac{p_F}{A_{V0}} = \frac{1}{2\pi R_F C_F A_{V0}} = \frac{1}{2\pi(500k)C_F(360)} \equiv 530\ \text{Hz} \quad \therefore \quad C_F = 1.7\ \text{pF}$$

$$p_{IE2} \approx A_{V0}p_A = (360)p_A \geq f_{I0dB} = 100\ \text{kHz} \quad \therefore \quad p_A \geq 280\ \text{Hz}$$

Explore with SPICE:
See Appendix A for notes on SPICE simulations.

```
* Type I DCM Stabilization with an Inverting Mixed Translation
vi vi 0 dc=0 ac=1
rf vi vn 500k
eav va 0 0 vn 360
rpa va vb 1
cpa vb 0 568.7u
eavo vieo 0 vb 0 1
cf vn vieo 1.7p
epwm de 0 vieo 0 2
epsw v1 0 de 0 1
rpsw v1 v2 1
cpsw v2 0 159.2n
gis 0 vs v2 0 260m
rdo vs 0 560
vio vs vo dc=0
co vo vc 5u
rc vc 0 10m
rld vo 0 500
fifb 0 vifb vio 1
rifb vifb 0 1
.ac dec 1000 10 10e6
.end
```

Tip: Plot v(vieo), i(Vio)/v(de), and v(vifb) in dB to inspect A_{IE}, A_{IL}, and A_{ILG}.

6.6 Digital Control

Feedback controllers use the voltage or current they sense to generate a pulsing command. From this perspective, feedback controllers are *analog–digital converters* (ADC). Mostly *analog controllers* mix, amplify, and stabilize the feedback system in the analog domain and mostly *digital controllers* in the digital domain.

Conventional ADCs digitize the voltage or current that digital controllers sense. Clocked *digital-signal processors* (DSP) use this digital word to mix, amplify, stabilize, and drive the switched inductor. Like analog controllers, digital controllers set loop gains that reach 0 dB with less than 180° of phase shift, if possible, at the highest manageable f_{0dB}.

6.6.1 Voltage Controller

Digital voltage-mode voltage controllers translate v_O in Fig. 6.48 to v_{FB} with β_{FB} and v_{FB} into an N-bit digital word d_{1-N} with ADCs. DSPs mix and compare this word d_{1-N} with a reference word d_R and use the difference to output the pulsing command d_E' that adjusts i_L. This way, DSPs sense and amplify the error that adjusts d_E' and i_L so v_{FB} mirrors the v_R that d_R represents and v_O nears a β_{FB} translation of v_R.

Although possible, digital current-mode voltage controllers are not popular. This is because they need another ADC to convert i_L into a digital word. And this second ADC consumes power and requires silicon area.

6.6.2 Current Controller

Current controllers translate i_{DO} or i_O in Fig. 6.49 to v_{IFB} with β_{IFB}. ADCs turn v_{IFB} into d_{1-N} and DSPs mix and compare d_{1-N} with d_{IR} to output the d_E' that adjusts i_L. This way, DSPs sense and amplify the error that adjusts i_L so v_{IFB} mirrors the v_{IR} that d_{IR} represents and i_{DO} or i_O nears a β_{IFB} translation of v_{IR}'s mirrored reflection.

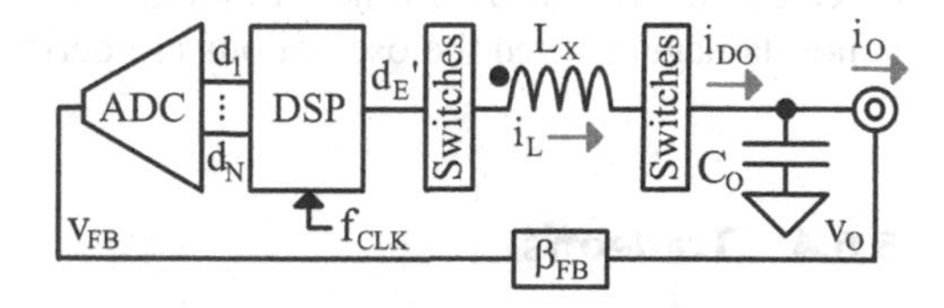

Fig. 6.48 Digital voltage-mode voltage controller

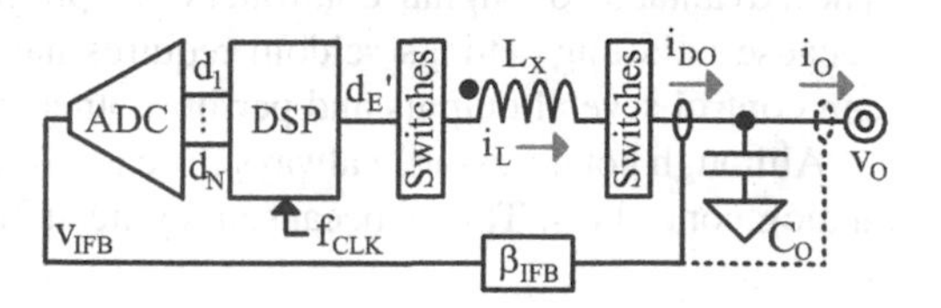

Fig. 6.49 Digital current controller

6.6.3 Digital Response

A. Digital Gain

Like PWMs, the ADC and DSP convert a voltage into a pulsing d_E'. So like the PWM, the v_{FB} variation Δv_{FB} that sweeps d_E' across its maximum range $\Delta d_{E(MAX)}'$ sets the *digital gain* A_{DIG}. N_{LSB} is the *number of least-significant bits* (LSB) needed for this Δv_{FB}, v_{LSB} is the *LSB voltage*, and Δv_{FB} is the sum of v_{LSB}'s the DSP uses to swing d_E' across 1 or 100%. So N_{SLB} and v_{LSB} set the low-frequency gain A_{DIG0} to $1/N_{LSB}v_{LSB}$:

$$\begin{aligned} A_{DIG} \equiv \frac{d_e'}{v_{fb}} &= \frac{A_{DIG0}}{1 + s/2\pi p_{DIG}} \\ &\approx \frac{\Delta d_{E(MAX)}'/\Delta v_{FB}}{1 + s/2\pi f_{BW}} \\ &\approx \frac{1/(N_{LSB} v_{LSB})}{1 + s/2\pi(f_{CLK}/N_{CLK})} = \frac{1/(N_{LSB} v_{LSB})}{1 + s/2\pi(N_{CLK} t_{CLK})^{-1}}. \end{aligned} \tag{6.106}$$

B. Bandwidth

The ADC and DSP require several clock cycles to sense and process a d_E' adjustment. The *number of clock cycles* N_{CLK} needed and the *clock period* t_{CLK} limit the bandwidth to $(t_{CLK}N_{CLK})^{-1}$ or f_{CLK}/N_{CLK}. This can be the Type-I dominant pole that reduces A_{LG} to 0 dB so f_{0dB} is below p_{LC}. The DSP can also add one or two intervening zeros in the digital domain, and that way, implement Type-II or -III stabilization.

C. Limit Cycling

Digital controllers often alternate d_E' between nearest states. This is because the d_E' needed to keep v_{FB} or v_{IFB} near v_R or v_{IR} can be between two d_E' settings. So the higher state prompts the controller to revert to the lower state, which induces reversals that repeat over time. Although not always acceptable, this *limit cycling* is not damaging when the oscillations are periodic and stable like just described.

6.6.4 Tradeoffs

The advantages of digital controllers are programmability and flexibility. This is because adjusting settings seldom requires hardware modifications. And one DSP can control several outputs and perform other functions.

Although not necessarily always the case, digital controllers are often slower than analog controllers. This is because they need N_{CLK} clock cycles to adjust d_E'. This drawback fades when t_{CLK} is shorter, which is possible when faster (thinner-oxide) transistors implement the DSP. Adding the manufacturing steps needed to integrate these faster transistors into the same substrate often represents an additional cost.

Another disadvantage of digital control is silicon real estate. One DSP usually requires considerably more silicon area than the two or three analog blocks needed for analog control. So digital controllers often cost more than their analog counterparts. This cost tradeoff diminishes when the DSP manages more outputs and functions.

6.7 Summary

Power supplies use inverting feedback loops to set and regulate their output. These loops sense, translate, and amplify the error that adjusts the output. This way, the output becomes a feedback translation of a dependable reference.

The translation is accurate when the forward gain is high and stable when the loop gain reaches 0 dB with less than 180° of phase shift. When stable, the output follows the lowest forward translation. These translations can embed internal loops that, on their own, should also be stable.

Op amps help translate signals across the loop. Closing local feedback loops around them desensitizes their translations. When looped and limited to the same bandwidth, non-inverting op amps amplify more than their inverting counterparts.

Stabilizers use these op amps to add gain, poles, and zeros that ensure a feedback system is stable. Un-looped on-chip translations are faster (wider bandwidth), feedback translations are more reliable and predictable, and mixed translations need lower capacitance. Of these, inverting mixed translations use fewer components.

Voltage controllers that add one pole and two zeros or embed a current loop can reach a bandwidth that exceeds the double LC pole. Parasitic poles in the stabilizer and the switching frequency ultimately limit this bandwidth. Stabilizing the switched-inductor system in discontinuous-conduction mode is simpler because the double pole reduces to one pole.

Current controllers are inherently stable when their low-frequency loop gain (and dc accuracy) is low. Adding a low-frequency pole and a higher-frequency zero can counter the zero that keeps gain down and recover the phase shift that the pole loses. This way, loop gain and accuracy can be higher.

Digital controllers mix, amplify, and stabilize in the digital domain. This way, adjusting the settings does not require hardware modifications. And the same DSP can control several outputs and perform other functions. The main drawback is silicon area, because DSPs usually occupy more area than the analog blocks analog controllers need. But this tradeoff fades when the DSP manages more functions.

Control Loops

7

Abbreviations

LED	Light-emitting diode
OCP	Over-current protection
PWM	Pulse-width modulator
SL	Switched inductor
SR	Set–reset flip flop
A_E	Error amplifier
A_{IE}	Current-error amplifier
A_G	Transconductance gain
A_{G0}	Static low/zero-frequency transconductance gain
A_{LG}	Loop gain
A_{PWM}	PWM gain translation
A_{V0}	Static low/zero-frequency voltage gain
β_{FB}	Feedback translation
β_{IFB}	Current-feedback translation
C_{LDC}	Load-compensation capacitor
CP_E	Error comparator
CP_{HYS}	Hysteretic comparator
CP_{PWM}	PWM comparator
CP_T	Time-loop comparator
d_D	Drain duty cycle
d_E	Energize duty cycle
d_E'	Energize duty-cycle command
d_O	Output duty cycle
d_{OFF}	Off duty cycle
d_{ON}	On duty cycle
Δi_{LD}	Load-dump current
Δv_T	Hysteresis voltage

G. A. Rincón-Mora, *Switched Inductor Power IC Design*,
https://doi.org/10.1007/978-3-030-95899-2_7

f_{0dB}	Unity-gain frequency
f_{CLK}	Clock frequency
f_{I0dB}	Current loop's unity-gain frequency
f_{SW}	Switching frequency
i_L	Inductor current
i_{LD}	Load current
i_O	Output current
K_O	Overdrive factor
L_C	Capacitor's equivalent series inductance
L_X	Switched transfer inductor
p_{BW}	Bandwidth-setting pole
p_C	Capacitor pole
p_{HYS}	Hysteretic pole
p_{IBW}	Current loop's bandwidth
p_{LC}	LC double pole
p_{LD}	Load-compensation pole
Q	Current state
Q^{-1}	Previous state
$\overline{Q}$	Opposite/complementary state
R	Reset flip-flop terminal
R_C	Capacitor's equivalent series resistance
R_L	Inductor's equivalent series resistance
R_{LDC}	Load-compensation resistor
S	Set flip-flop terminal
SR_{OFF}	Off-time flip flop
SR_{ON}	On-time flip flop
SR_{PK}	Peak-current flip flop
SR_{VL}	Valley-current flip flop
Σv_{ID}	Differential sum
t_{CLK}	Clock period
t_E	Energize time
t_{LC}	LC's resonant period
t_O	Output pulse width
t_{OFF}	Off time
t_{ON}	On time
t_{OSC}	Oscillating period
t_P	Propagation delay
$t_{P(BW)}$	Bandwidth propagation delay
$t_{P(SR)}$	Slew-rate propagation delay
t_R	Response time
t_{SR}	SR flip flop's propagation delay
t_{SW}	Switching period
τ_{BW}	Bandwidth-setting time constant
τ_{HYS}	Hysteretic time constant

v_D	Drain voltage
v_{DD}	Positive power supply
v_E	Energize voltage
v_{EO}	Amplified error voltage
v_{FB}	Feedback voltage
v_{HOS}	Hysteretic offset voltage
v_I	Input voltage
v_{ID}	Differential input voltage
v_{IDI}	Differential current-mode input voltage
v_{IDV}	Differential voltage-mode input voltage
v_{IFB}	Current-feedback voltage
v_{IN}	Input/input voltage/input power supply
v_{IO}	Amplified current-error voltage
v_{IOS}	Current-loop offset
v_{IR}	Current-reference voltage
v_L	Inductor voltage
v_{LD}	Loading-effect voltage
v_{LOS}	Loop-offset voltage
v_N	Negative input
v_O	Output/output voltage
V_{OH}	Output high
V_{OL}	Output low
v_P	Positive input
v_{POS}	Projection offset voltage
v_R	Reference voltage
v_S	Sawtooth voltage
v_{SOS}	Slope compensation offset voltage
v_{SS}	Negative power supply
v_{SW}	Switch-node voltage
$v_{T(HI)}$	Rising trip point
$v_{T(LO)}$	Falling trip point
v_{VOS}	Voltage-loop offset
z_C	Capacitor zero
z_{DO}	Duty-cycled zero

Switched-inductor (SL) power supplies normally close feedback loops that sense the output, amplify the error, and adjust inductor current so the error is low. With this feedback action, they can feed any current the load demands or controller commands. This is how *voltage regulators*, *battery chargers*, and *light-emitting diode* (LED) *drivers* supply power.

The most distinguishing feature in their implementations is how the error adjusts the inductor current. In this respect, although often described in unique and stand-alone terms, most switched-inductor power-supply systems evolve from two basic primitives: pulse-width-modulated and hysteretic loops. Variations on these then germinate features and restrictions that ultimately set them apart.

7.1 Primitives

7.1.1 PWM Loop

A. **Comparator**

A *comparator* compares two analog voltages and outputs a digital voltage to indicate which is higher. The *output* v_O reaches the *output high* V_{OH} in Fig. 7.1 when the *positive input* v_P surpasses the *negative input* v_N and reaches the *output low* V_{OL} when the opposite happens: when v_N exceeds v_P. In CMOS implementations, V_{OH} and V_{OL} usually near the *positive* and *negative power supplies* v_{DD} and v_{SS}.

The *differential input voltage* v_{ID} is the difference between v_P and v_N. From this perspective, v_O reaches V_{OH} when v_{ID} is positive and V_{OL} when v_{ID} is negative. So comparators also dual as *polarity detectors*.

In practice, v_{ID} should overcome a *minimum threshold* $v_{ID(MIN)}$ to assert a clear digital state. Below this level, v_O is between V_{OL} and V_{OH}. $v_{ID(MIN)}$ is therefore a function of v_{ID}'s gain translation to v_O and v_O's total swing. This A_{V0} is the *static (low/zero-frequency) voltage gain* of the comparator:

$$A_{V0} \equiv \frac{\Delta v_O}{\Delta v_{ID}} \approx \frac{\Delta v_{O(MAX)}}{\Delta v_{ID(MIN)}} = \frac{V_{OH} - V_{OL}}{2v_{ID(MIN)}}. \tag{7.1}$$

The comparator also requires time to react. As small variations in v_{ID} grow to large variations in v_O, poles and slew rate delay v_O's response to v_{ID}. The *bandwidth-setting pole* p_{BW} in A_V is to blame for small-signal delays. Or to be more precise, p_{BW}'s *bandwidth-setting time constant* τ_{BW} delays v_{ID}'s translation to the final output voltage $v_{O(F)}$ that $v_{ID}A_{V0}$ sets:

$$A_V = \frac{A_{V0}}{1 + s/2\pi p_{BW}} = \frac{A_{V0}}{1 + \tau_{BW}s} \tag{7.2}$$

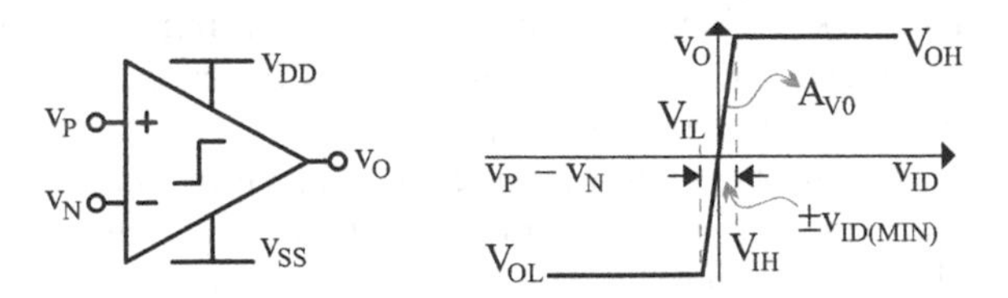

Fig. 7.1 Comparator

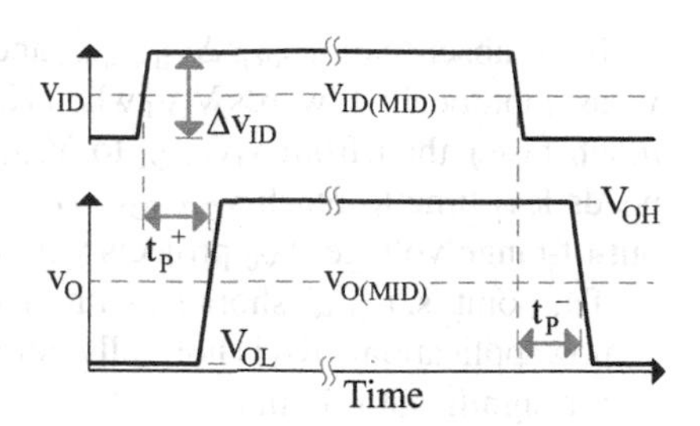

Fig. 7.2 Propagation delay

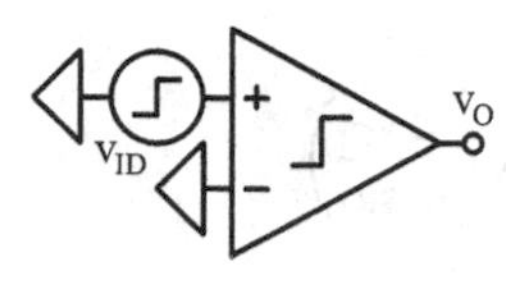

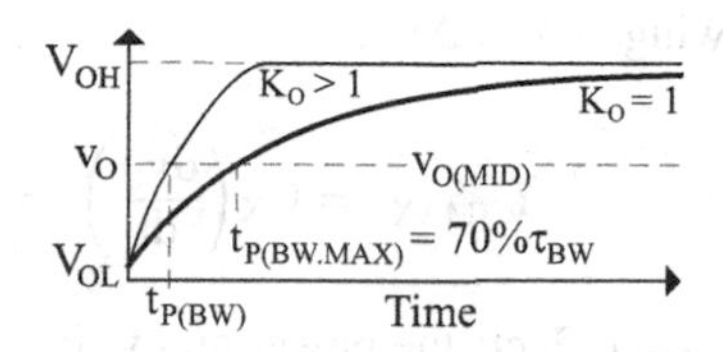

Fig. 7.3 Step response

$$v_O = v_{O(F)}\left[1 - \exp\left(\frac{-t}{\tau_{BW}}\right)\right] = v_{ID}A_{V0}\left[1 - \exp\left(\frac{-t}{\tau_{BW}}\right)\right]. \tag{7.3}$$

Propagation delay t_P is the delay between v_{ID}'s and v_O's halfway points in Fig. 7.2. Since p_{BW} delays small signals before slew rate delays larger transitions, *bandwidth* and *slew-rate delays* $t_{P(BW)}$ and $t_{P(SR)}$ tend to add:

$$t_P \approx t_{P(BW)} + t_{P(SR)}. \tag{7.4}$$

Although not always, rising and falling delays t_P^+ and t_P^- often match (by design).

When v_{ID}'s transition is much shorter than t_P, the input variation Δv_{ID} is practically an instantaneous step. When this happens, v_O in Fig. 7.3 requires $t_{P(BW)}$ to rise halfway across v_O's swing: from V_{OL} to $v_{O(MID)}$ across $\Delta v_{O(MID)}$. The *overdrive factor* K_O is the factor by which Δv_{ID} overcomes the $\Delta v_{ID(MIN)}$ that $\pm v_{ID(MIN)}$ determines. Since K_O should match or exceed one to assert a clear digital state, v_O needs less than 70% of τ_{BW} to reach $v_{O(MID)}$:

$$\Delta v_{O(MID)} = \frac{\Delta v_{O(MAX)}}{2} = \frac{V_{OH} - V_{OL}}{2} = \Delta v_{ID}A_{V0}\left[1 - \exp\left(\frac{-t_{P(BW)}}{\tau_{BW}}\right)\right], \tag{7.5}$$

$$K_O \equiv \frac{\Delta v_{ID}}{\Delta v_{ID(MIN)}} = \frac{\Delta v_{ID}}{2v_{ID(MIN)}} = \Delta v_{ID}\left(\frac{A_{V0}}{V_{OH} - V_{OH}}\right) \geq 1, \tag{7.6}$$

$$t_{P(BW)} = \tau_{BW}\ln\left(1 - \frac{1}{2K_O}\right)^{-1} \leq 70\%\tau_{BW}. \tag{7.7}$$

In the absence of $t_{P(SR)}$, $\Delta v_{ID(MIN)}$ and the negative exponential that p_{BW} sets raise v_O asymptotically towards V_{OH} when K_O is one. Notice v_O rises from V_{OL} to $v_{O(MID)}$ much faster than from $v_{O(MID)}$ to V_{OH}. And when Δv_{ID} surpasses $\Delta v_{ID(MIN)}$, v_O needs less time to reach $v_{O(MID)}$. V_{OH}, however, clamps v_O before v_O reaches the out-of-range voltage A_{V0} projects with a higher Δv_{ID}.

The point is, $t_{P(BW)}$ shortens with increasing K_O's. But since v_{ID} is usually low in analog applications, K_O is normally not very high. So bandwidth and overdrive often play a significant role in t_P.

Capacitances, current drive, and voltage swings determine slew-rate delays. This is because the maximum current that a circuit avails with $i_{C(MAX)}$ sets the $t_{P(SR)}$ that C_X needs to swing across Δv_C:

$$i_{C(MAX)} = C_X\left(\frac{dv_C}{dt}\right) = C_X\left(\frac{\Delta v_C}{t_{P(SR)}}\right). \tag{7.8}$$

So in a way, $t_{P(SR)}$ reflects the power and voltage an engineer allows the comparator to consume and swing.

B. Pulse-Width Modulator

The purpose of *pulse-width modulators* (PWM) is to scale the *output pulse width* t_O with an *input voltage* v_I across a constant *clock period* t_{CLK}. t_O's fraction of t_{CLK} is the *output duty-cycle* d_O. So t_O and d_O rise and fall with v_I:

$$d_O \equiv \frac{t_O}{t_{CLK}} \propto v_I. \tag{7.9}$$

The *PWM comparator* CP_{PWM} in Fig. 7.4 scales t_O with v_I by comparing v_I to a *sawtooth voltage* v_S. v_S ramps from $v_{S(LO)}$ to $v_{S(HI)}$ and resets back to $v_{S(LO)}$ with the *clock frequency* f_{CLK} after every t_{CLK}. When v_I is between $v_{S(LO)}$ and $v_{S(HI)}$, v_O rises when f_{CLK} resets v_S below v_I and falls when v_S surpasses v_I. This way, t_O ends when v_S overcomes v_I, v_I sets t_O and d_O, and static variations in v_I modify t_O and d_O.

t_O and d_O are zero when v_I is below $v_{S(LO)}$ and they are t_{CLK} and one when v_I surpasses $v_{S(HI)}$. Between these, t_O and d_O scale linearly with v_I because v_S ramps

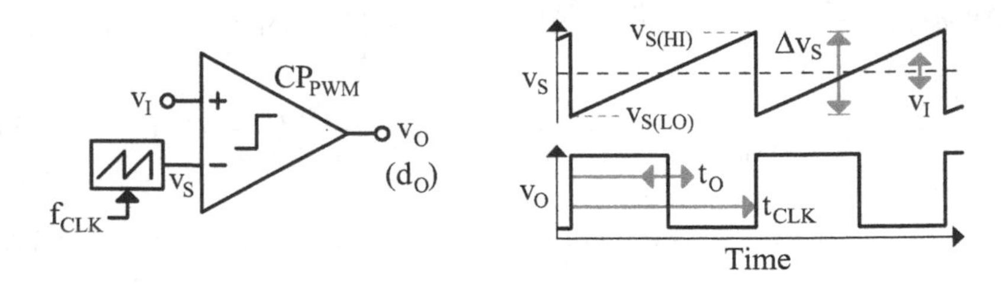

Fig. 7.4 Non-inverting pulse-width modulator

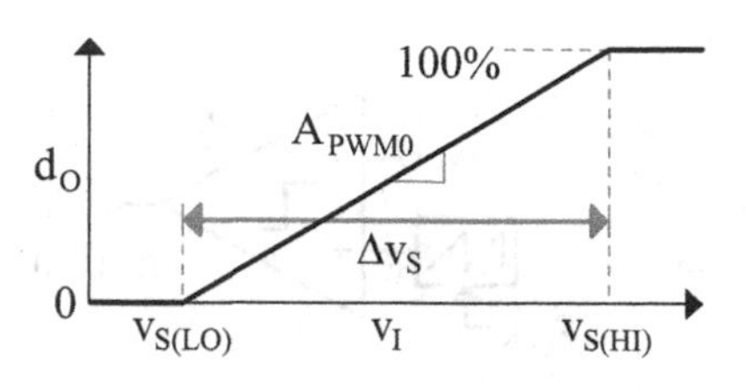

Fig. 7.5 Static PWM translation

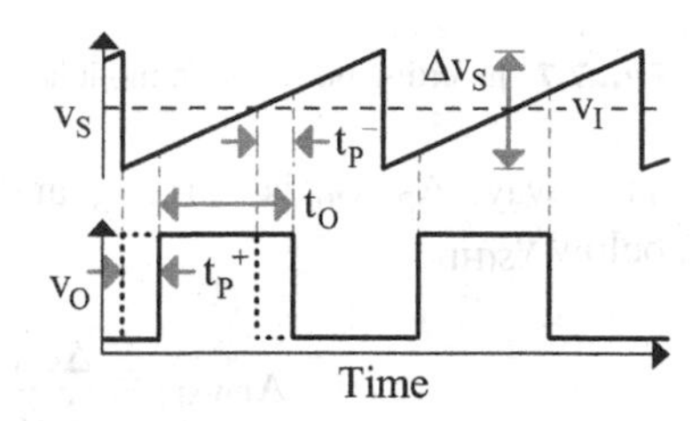

Fig. 7.6 Dynamic PWM translation

linearly across t_{CLK}. Since d_O traverses 0–1 (zero to 100%) when v_I sweeps across v_S's range Δv_S, v_I's static *PWM gain translation* A_{PWM0} to d_O in Fig. 7.5 is $1/\Delta v_S$:

$$A_{PWM0} \equiv \frac{\Delta d_O}{\Delta v_I} = \frac{1-0}{v_{S(HI)} - v_{S(LO)}} = \frac{1}{\Delta v_S}. \tag{7.10}$$

In practice, t_P delays v_O's transitions. v_O in Fig. 7.6 rises t_P^+ after v_S resets and falls t_P^- after v_S overcomes v_I. When t_P's match, t_P^+ shortens t_O by the same amount t_P^- extends t_O. So t_P does not alter t_O.

v_S ramps $\Delta v_S/t_{CLK}$ across t_P^+ and t_O when setting t_O. The v_I that sets t_O is the v_S that ramps across t_P^+ and the part of t_O that excludes t_P^-. But when t_P's match, t_P^- cuts what t_P^+ adds. So v_I is a d_O fraction of Δv_S over $v_{S(LO)}$:

$$v_I = v_{S(LO)} + \left(t_P^+ + t_O - t_P^-\right)\left(\frac{dv_S}{dt}\right) \approx v_{S(LO)} + t_O\left(\frac{\Delta v_S}{t_{CLK}}\right) = v_{S(LO)} + \Delta v_S d_O. \tag{7.11}$$

Since t_O cannot end before t_P^- elapses, t_P^- is also the lowest t_O possible. This means that d_O's minimum is a t_P^- fraction of t_{CLK}:

$$d_{O(MIN)} = \frac{t_{O(MIN)}}{t_{CLK}} = \frac{t_P^-}{t_{CLK}} \approx \frac{t_P}{t_{CLK}}. \tag{7.12}$$

In short, the only real effect of t_P on operation is limiting $d_{O(MIN)}$.

Swapping inputs and flipping the ramp like Fig. 7.7 shows inverts the translation to d_O. This is because v_O rises to start t_O a t_P^+ after v_S resets high and v_O falls to end t_O a t_P^- after v_S falls below v_I'. So increasing v_I' shortens t_O and d_O and *vice versa*.

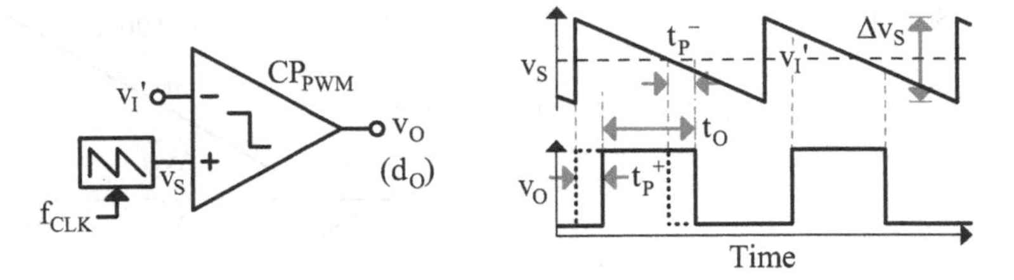

Fig. 7.7 Inverting pulse-width modulator

This way, A_{PWM0} is $-1/\Delta v_S$ and the v_I' that sets t_O is a d_O fraction of Δv_S below $v_{S(HI)}$:

$$A_{PWM0} \equiv \frac{\Delta d_O}{\Delta v_I} = \frac{1-0}{v_{S(LO)} - v_{S(HI)}} = -\frac{1}{\Delta v_S} \tag{7.13}$$

$$\begin{aligned} v_I' &\approx v_{S(HI)} + (t_P^+ + t_O - t_P^-)\left(\frac{dv_S}{dt}\right) \\ &\approx v_{S(HI)} - t_O\left(\frac{\Delta v_S}{t_{CLK}}\right) = v_{S(HI)} - \Delta v_S d_O. \end{aligned} \tag{7.14}$$

Example 1: Determine v_I, v_I', and $d_{O(MIN)}$ when $v_{S(LO)}$ and $v_{S(HI)}$ are 200 and 500 mV, d_O is 45%, t_P is 100 ns, and t_{CLK} is 1 µs.

Solution:

$$\Delta v_S = v_{S(HI)} - v_{S(LO)} = 500\text{m} - 200\text{m} = 300 \text{ mV}$$

$$v_I = v_{S(LO)} + \Delta v_S d_O = 200\text{m} + (300\text{m})(45\%) = 335 \text{ mV}$$

$$v_I' = v_{S(HI)} - \Delta v_S d_O = 500\text{m} - (300\text{m})(45\%) = 365 \text{ mV}$$

$$d_{O(MIN)} = \frac{t_P}{t_{CLK}} = \frac{100\text{n}}{1\mu} = 10\%$$

Explore with SPICE:
See Appendix A for notes on SPICE simulations.

```
*Non-inverting Pulse-Width Modulator
v1v v1v 0 dc=1
vi vi 0 dc=335m
vs vs 0 dc=200m pulse 200m 500m 0998n 1n 1n 1u
xpwm vi vs vo v1v 0 cp
.lib lib.txt
.tran 5u
.end
```

Tip: Plot v(vi), v(vs), and v(vo).

C. PWM Loops

The *PWM voltage loop* in Fig. 7.8 is the classic *PWM voltage-mode regulator*. CP_{PWM} and v_S translate the *amplified error voltage* v_{EO} into the *energize duty-cycle command* d_E' that energizes the *switched inductor* L_X. f_{CLK} determines the *switching frequency* and *period* f_{SW} and t_{SW} of L_X.

The *feedback translation* β_{FB} scales v_O to the *feedback voltage* v_{FB} that the *error amplifier* A_E compares to the *reference voltage* v_R. So A_E outputs the v_{EO} that sets and adjusts d_E' so L_X's *inductor current* i_L supplies the *output current* i_O needed to keep the error between v_R and v_{FB} low. This way, v_{FB} nears v_R and v_O is close to a reverse β_{FB} translation of v_R.

The *PWM current loop* in Fig. 7.9 is a direct translation of the voltage loop in Fig. 7.8. This loop senses and regulates i_L or i_O. So the error that feeds CP_{PWM} and adjusts d_E' is the *amplified current-error voltage* v_{IO}.

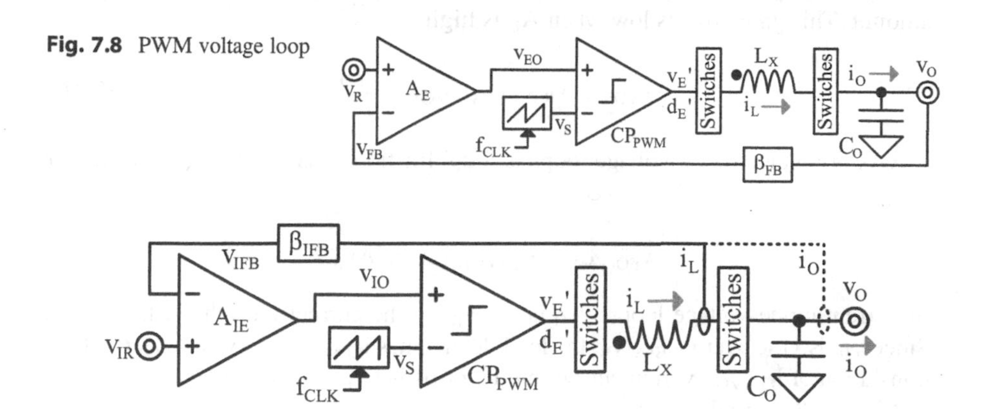

Fig. 7.8 PWM voltage loop

Fig. 7.9 PWM current loop

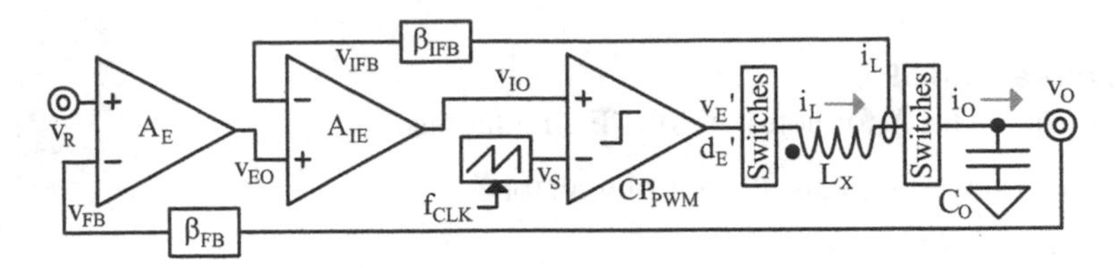

Fig. 7.10 PWM current-mode voltage loop

The *current-feedback translation* β_{IFB} scales i_L or i_O to the *current-feedback voltage* v_{IFB} that the *current-error amplifier* A_{IE} compares to the *current-reference voltage* v_{IR}. A_{IE} outputs the v_{IO} that sets and adjusts d_E' so the error between v_{IFB} and v_{IR} is low. This way, v_{IFB} nears v_{IR} and i_L or i_O is close to a reverse β_{IFB} translation of v_{IR}.

The *PWM current-mode voltage loop* in Fig. 7.10 uses the PWM current loop to control i_L. In this implementation, β_{FB} scales v_O to v_{FB}, A_E compares v_{FB} to v_R, and the current loop converts v_{EO} to i_L. So v_{EO} adjusts the i_L that sets i_O so v_{FB} nears v_R and v_O is close to a reverse β_{FB} translation of v_R.

D. **Offsets**

v_{IR} and $v_{IFB(AVG)}$ set the steady v_{IO} that determines d_E'. This is the *offset* v_{IOS} the *current loop* needs to control $i_{L(AVG)}$. This means, v_{IR} should surpass the targeted $v_{IFB(AVG)}$ by this amount. This gain error is low when A_{IE} is high. Since v_{IO} in PWM loops is a d_E' fraction of Δv_S over $v_{S(LO)}$ or under $v_{S(HI)}$, this v_{IOS} is

$$v_{IOS} \equiv v_{IR} - v_{IFB(AVG)} = \frac{v_{IO}}{A_{IE}} = \frac{v_{S(LO/HI)} \pm d_E'\Delta v_S}{A_{IE}}. \tag{7.15}$$

v_R and $v_{FB(AVG)}$ similarly set the steady v_{EO} that ultimately determines d_E'. This is the *offset* v_{VOS} the *voltage loop* needs to control v_O. $v_{FB(AVG)}$ is below v_R by this amount. This gain error is low when A_E is high:

$$v_{VOS} \equiv v_R - v_{FB(AVG)} = \frac{v_{EO}}{A_E}. \tag{7.16}$$

v_{EO} in voltage-mode voltage loops is a d_E' fraction of Δv_S over $v_{S(LO)}$ or under $v_{S(HI)}$:

$$v_{EO(VM)} = v_{S(LO/HI)} \pm d_E'\Delta v_S. \tag{7.17}$$

In current-mode voltage loops, v_{EO} is the v_{IR} in the current loop that sets $i_{L(AVG)}$. Since v_{IR} is v_{IOS} over $v_{IFB(AVG)}$, the load determines $i_{L(AVG)}$, and $v_{IFB(AVG)}$ is a β_{IFB} translation of $i_{L(AVG)}$, v_{EO} in current-mode loops is v_{IOS} over the *loading effect* v_{LD} that $i_{L(AVG)}\beta_{IFB}$ establishes:

$$v_{EO(IM)} = v_{LD} + v_{IOS} = i_{L(AVG)}\beta_{IFB} + \frac{v_{IO}}{A_{IE}}. \quad (7.18)$$

β_{FBI} and β_{FB} are the feedback translations needed to set $i_{L(AVG)}$ or i_O and $v_{O(AVG)}$. They can account for v_{IOS} and v_{VOS}. But since the i_O and *energize* and *drain voltages* v_E and v_D that set $i_{L(AVG)}$ and d_E' vary with the *input supply* v_{IN}, v_O, and load, v_{IR} and v_R and v_{IOS}'s and v_{VOS}'s statistical mean can set β_{FBI} and β_{FB}. This way, β_{FBI} and β_{FB} can center their outputs $i_{L(AVG)}$ or i_O and $v_{O(AVG)}$ about their targets:

$$\beta_{IFB} \equiv \frac{\overline{v_{IFB(AVG)}}}{i_{L(AVG)/O}} = \frac{v_{IR} - \overline{v_{IOS}}}{i_{L(AVG)/O}} \quad (7.19)$$

$$\beta_{FB} \equiv \frac{\overline{v_{FB(AVG)}}}{v_{O(AVG)}} = \frac{v_R - \overline{v_{VOS}}}{v_{O(AVG)}}. \quad (7.20)$$

Example 2: Determine v_{IO}, v_{IOS}, v_{VOS}, $v_{FB(AVG)}$, β_{FB}, and v_O's error for the PWM current-mode voltage loop with the PWM in Example 1 so v_O is 1.8 V when v_R is 1.2 V, β_{IFB} is 1 Ω, A_E and A_{IE} are 10 V/V, and $i_{L(AVG)}$ is 100–500 mA.

Solution:

$$v_{IO} = 340 \text{ mV from Example 1}$$

$$v_{IOS} = \frac{v_{IO}}{A_{IE}} = \frac{340\text{m}}{10} = 34 \text{ mV}$$

$$v_{VOS} = \frac{i_{L(AVG)}\beta_{IFB} + v_{IOS}}{A_E} = \frac{i_{L(AVG)}(1) + 34\text{m}}{10} = 33 \pm 20 \text{ mV}$$

$$v_{FB(AVG)} = v_R - v_{VOS} = 1.2 - v_{VOS} = 1.17 \text{ V} \pm 20 \text{ mV}$$

$$\beta_{FB} \equiv \frac{\overline{v_{FB(AVG)}}}{v_{O(AVG)}} = \frac{1.17}{1.80} = 65\%$$

$$v_{O(AVG)} = \frac{v_{FB(AVG)}}{\beta_{FB}} = \frac{v_{FB(AVG)}}{65\%} = 1.80 \text{ V} \pm 31 \text{ mV}$$

Note: v_O's loading effect is ±31 mV or ±1.7%.

Explore with SPICE:
See Appendix A for notes on SPICE simulations.

```
*Current-Mode PWM Voltage Buck
vin vin 0 dc=4
sei vin vsw de 0 sw1v
ddg 0 vsw idiode
lx vsw vl 10u
vi vl vo 0
co vo 0 5u
ro vo 0 18
io vo 0 pwl 0 0 1.2m 0 1.3m 400m
vr vr 0 dc=1.2 pwl 0 0 1m 1.2
efb vfb 0 vo 0 0.65
eae veo 0 vr vfb 10
fbifb 0 vifb vi 1
rbifb vifb 0 1
cbifb vifb 0 1p
gaie 0 vio veo vifb 10
rieo vio 0 1
cpie vio x 88.5u
rzie x 0 81.8 m
v1v v1v 0 dc=1
vs vs 0 dc=200 m pulse 200m 500m 0998n 1n 1n 1u
xpwm vio vs de v1v 0 cp
.lib lib.txt
.tran 1.4m
.end
```

Tip: Wait for simulation to finish (because it may require some time to complete) and plot v(vo), v(vfb), v(vifb), v(veo), v(vio), v(vs), and i(Lx) across 1.4 ms, between 1.100 and 1.105 ms, and across last 5 μs.

E. Design Notes

Stabilizers in PWM loops ensure the loop gain falls below one past a unity-gain frequency that is lower than f_{SW}. This way, PWM loops suppress frequency components near and above f_{SW}, averaging the v_O, i_L, or i_O they control and set. This is why literature often calls them *average loops*.

t_P and v_S are fundamental limitations in PWM loops. d_E' cannot fall below t_P/t_{CLK}. This bounds v_{IN} and v_O because v_{IN} and v_O in v_E and v_D set d_E'. v_S offsets the output, but only to the extent A_E, A_{IE}, β_{IFB}, and β_{FB} allow.

PWM loops can saturate. Since feedback loops require time to respond, sudden variations in load, v_R, or v_{IR} shift v_{FB} or v_{IFB} away from their steady-state points.

When this error is high, A_E or A_{IE} can push v_{EO} or v_{IO} outside v_S's range, which would saturate d_E' to zero or one.

When d_E' rails this way, L_X can energize or drain across consecutive cycles. This can be a problem because energizing L_X across extended periods can grow i_L to a level that overloads v_{IN}. And draining L_X into v_{IN} can reverse i_L back into v_{IN}, which might not be capable of receiving much charge.

The stabilizers in A_E and A_{IE} can suppress v_{EO}'s and v_{IO}'s excursions, but not always by enough. So engineers often add guardrails that keep d_E' from saturating. *Over-current protection* (OCP), for example, can force the switcher to drain L_X when i_L reaches a maximum threshold. A_E and A_{IE} can also clamp their outputs so they do not swing too far.

Example 3: Determine $v_{O(MIN)}$ for a buck–boost in *continuous conduction* when v_{IN} is 2 V, t_P is 100 ns, and t_{CLK} is 1 μs.

Solution:

$$d_E = \frac{v_D}{v_E + v_D} = \frac{v_O}{2 + v_O} \geq \frac{t_P}{t_{CLK}} = \frac{100n}{1\mu} \quad \therefore \quad v_O \geq 220\ \text{mV}$$

7.1.2 Hysteretic Loop

A. Hysteretic Comparator

The *hysteretic comparator* CP_{HYS} in Fig. 7.11 is a comparator that decouples and shifts v_{ID}'s *rising* and *falling trip points* $v_{T(HI)}$ and $v_{T(LO)}$. This way, v_O rises to V_{OH} after v_{ID} rises over $v_{T(HI)}$ and falls to V_{OL} after v_{ID} falls under $v_{T(LO)}$. But v_O does not trip when v_{ID} rises over $v_{T(LO)}$ or when v_{ID} falls below $v_{T(HI)}$. The difference between these trip points is the *hysteresis* Δv_T of the comparator.

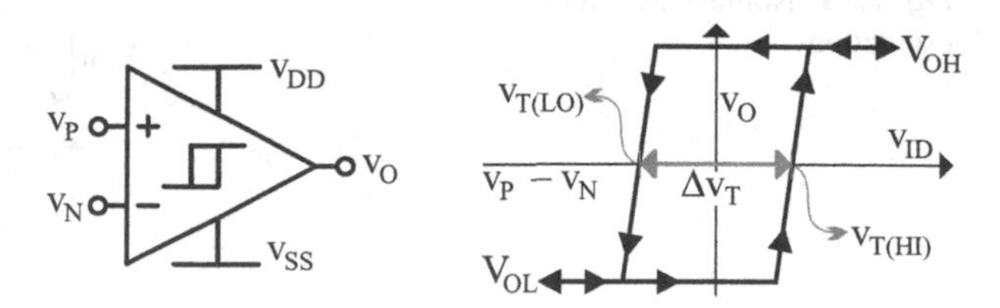

Fig. 7.11 Hysteretic comparator

v_N's trip points oppose v_P's in v_{ID}. This is because v_O trips low when v_N rises $-v_{T(LO)}$ over v_P and trips high when v_N falls $-v_{T(HI)}$ below v_P. This is like saying v_N's $v_{T(HI)}$ and $v_{T(LO)}$ are v_P's $-v_{T(LO)}$ and $-v_{T(HI)}$.

B. Hysteretic Loops

The *hysteretic current loop* in Fig. 7.12 is a *relaxation oscillator* that hinges on the hysteretic comparator and slewing action of i_L. β_{IFB} is usually a resistor that converts i_L to v_{IFB}. This way, v_{IFB} rises and falls with i_L.

Oscillators oscillate when the gain of a non-inverting feedback loop is one. In the case of *ring oscillators*, delay inverts a negative feedback loop and swing limits keep the cycle-to-cycle gain at one. Hysteretic current loops oscillate the same way because i_L reverses when CP_{HYS} trips (negative feedback), i_L slews across trip points (delay), and trip points keep $i_{L(HI)}$ and $i_{L(LO)}$ steady (cycle-to-cycle gain is one).

CP_{HYS} waits for the v_E the switcher impresses across L_X to slew i_L and v_{IFB} to v_{IFB}'s $v_{T(HI)}$ in Fig. 7.13. CP_{HYS} trips v_E' low when v_{IFB} overcomes $v_{T(HI)}$. This prompts the switcher to apply v_D across L_X, which slews i_L and v_{IFB} down. CP_{HYS} "relaxes" across this time, until v_{IFB} reaches v_{IFB}'s $v_{T(LO)}$. At this point, CP_{HYS} trips high and v_{IFB} again climbs towards $v_{T(HI)}$. In short, CP_{HYS} relaxes between transitions and activates (to reverse i_L) when v_{IFB} reaches CP_{HYS}'s hysteretic thresholds.

v_{IFB} and $i_L\beta_{IFB}$ oscillate this way between v_{IFB}'s $v_{T(LO)}$ and $v_{T(HI)}$. Since these v_T's are usually constant and symmetrical about v_{IR}, v_{IFB}'s average and v_{IR} are halfway between v_T's. v_{IR} determines v_{IFB}'s average $v_{IFB(AVG)}$ this way. And with β_{IFB}, v_{IR} sets i_L's average $i_{L(AVG)}$ and Δv_T sets i_L's ripple Δi_L.

The *hysteretic current-mode voltage loop* in Fig. 7.14 uses the hysteretic current loop to control $i_{L(AVG)}$. Here, β_{FB} scales v_O to v_{FB}, A_E compares v_{FB} to v_R, and the current loop converts v_{EO} to $i_{L(AVG)}$. So v_{EO} adjusts the $i_{L(AVG)}$ that sets i_O so v_{FB} nears v_R and v_O is close to a reverse β_{FB} translation of v_R.

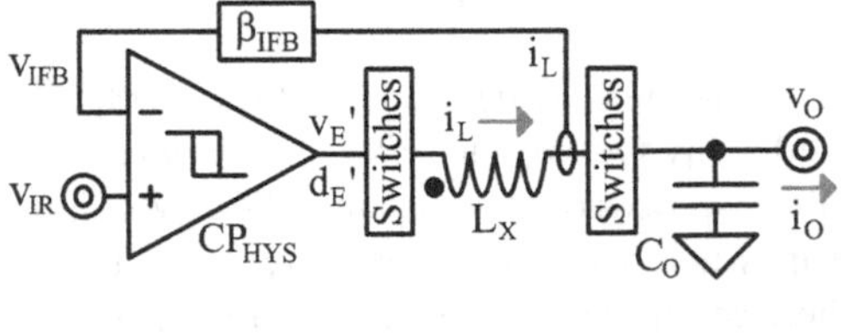

Fig. 7.12 Hysteretic current loop

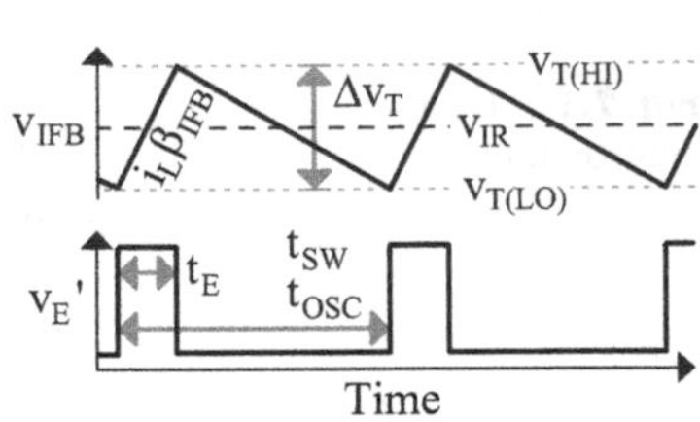

Fig. 7.13 Nominal hysteretic oscillation

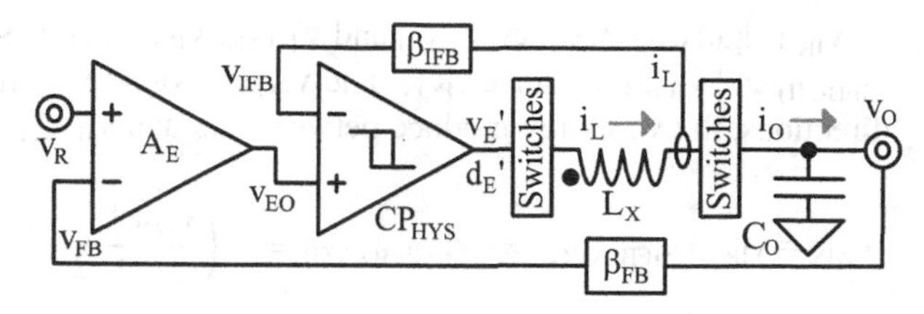

Fig. 7.14 Hysteretic current-mode voltage loop

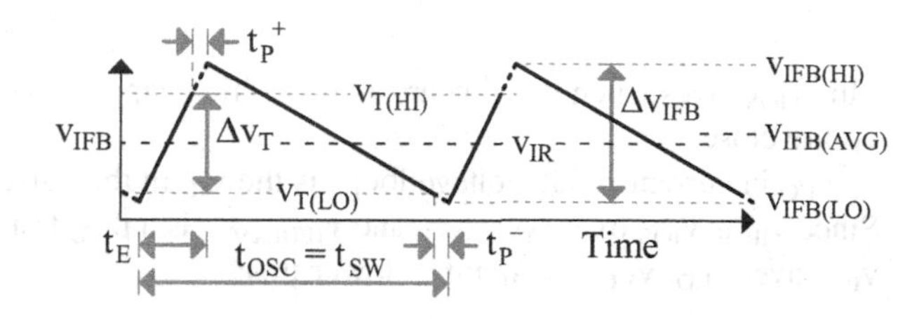

Fig. 7.15 Actual hysteretic oscillation

C. **Offsets**

In practice, t_P's across CP_{HYS} and the switcher delay transitions. So v_{IFB} in Fig. 7.15 does not stop rising or falling past $v_{T(HI)}$ or $v_{T(LO)}$ until after t_P elapses. Letting v_{IFB} rise and fall across t_P's extends v_{IFB}'s swing Δv_{IFB}.

t_P^+ and t_P^- induce rising and falling *projection offset voltages* v_{POS}^+ and v_{POS}^- that raise $v_{IFB(HI)}$ above $v_{T(HI)}$, reduce $v_{IFB(LO)}$ below $v_{T(LO)}$, and expand Δv_{IFB} beyond Δv_T. i_L therefore ripples across the Δi_L that the expanded Δv_{IFB} and β_{IFB} set.

Since v_E energizes L_X across t_P^+ and v_D drains L_X across t_P^-, the rising and falling dv_{IFB}/dt rates that project v_{IFB} across t_P^+ and t_P^- are β_{IFB} translations of i_L's v_E/L_X and v_D/L_X. When v_T's are symmetrical, $v_{T(HI)}$ is $0.5\Delta v_T$ over v_{IR}, $v_{T(LO)}$ is $0.5\Delta v_T$ below v_{IR}, and v_{IR} is halfway across Δv_T. So rising and falling v_{POS}'s, high and low v_{IFB} points, and the ripple that Δv_T and v_{POS}'s induce in v_{IFB} are

$$v_{POS}^{\pm} = t_P^{\pm}\left(\frac{dv_{IFB}}{dt_{E/D}}\right) = t_P^{\pm}\left(\frac{di_L\beta_{IFB}}{dt_{E/D}}\right) = t_P^{\pm}\left(\frac{v_{E/D}}{L_X}\right)\beta_{IFB}, \tag{7.21}$$

$$v_{IFB(HI)} = v_{T(HI)} + v_{POS}^{+} = v_{IR} + \frac{\Delta v_T}{2} + t_P^{+}\left(\frac{v_E}{L_X}\right)\beta_{IFB}, \tag{7.22}$$

$$v_{IFB(LO)} = v_{T(LO)} - v_{POS}^{-} = v_{IR} - \frac{\Delta v_T}{2} - t_P^{-}\left(\frac{v_D}{L_X}\right)\beta_{IFB}, \tag{7.23}$$

and

$$\Delta v_{IFB} = \Delta i_L\beta_{IFB} = \Delta v_T + v_{POS}^{+} + v_{POS}^{-} \approx \Delta v_T + t_P\left(\frac{v_E + v_D}{L_X}\right)\beta_{IFB}. \tag{7.24}$$

v_{IR} is halfway between $v_{T(HI)}$ and $v_{T(LO)}$. $v_{IFB(AVG)}$ is similarly halfway between $v_{IFB(HI)}$ and $v_{IFB(LO)}$. Since v_{POS}^+ and v_{POS}^- extend $v_{IFB(HI)}$ and $v_{IFB(LO)}$ in opposite directions, the v_{IOS} they produce between v_{IR} and $v_{IFB(AVG)}$ is half their difference:

$$v_{IOS} \equiv v_{IR} - v_{IFB(AVG)} = -\Delta v_{IFB(AVG)} = -\left(\frac{v_{POS}^+ - v_{POS}^-}{2}\right) \approx t_P\left(\frac{v_D - v_E}{2L_X}\right)\beta_{IFB}. \tag{7.25}$$

This v_{IOS} fades when v_{POS}^+ nears v_{POS}^-, which happens when t_P's match and v_E and v_D are close.

v_{EO} in current-mode voltage loops is the v_{IR} in the current loop that sets $i_{L(AVG)}$. Since v_{IR} is v_{IOS} over $v_{IFB(AVG)}$ and $v_{IFB(AVG)}$ is a β_{IFB} translation of $i_{L(AVG)}$, v_{EO} is v_{IOS} over v_{LD}. v_{VOS} is A_E times lower this:

$$v_{VOS} \equiv v_R - v_{FB(AVG)} = \frac{v_{EO}}{A_E} = \frac{v_{LD} + v_{IOS}}{A_E} \approx \left[i_{L(AVG)} + t_P\left(\frac{v_D - v_E}{2L_X}\right)\right]\left(\frac{\beta_{IFB}}{A_E}\right). \tag{7.26}$$

β_{FBI} and β_{FB} can account for v_{IOS} and v_{VOS}. But since $i_{L(AVG)}$, v_E, and v_D vary with load, v_{IN}, and v_O, v_{IR} and v_R and v_{IOS}'s and v_{VOS}'s statistical means can set β_{FBI} and β_{FB}. This way, β_{FBI} and β_{FB} can center $i_{L(AVG)}$ or i_O and $v_{O(AVG)}$ about their targets.

Example 4: Determine Δv_{IFB}, v_{IOS}, v_{VOS}, $v_{FB(AVG)}$, β_{FB}, and v_O's error for the hysteretic current-mode voltage loop so v_O is 1.8 V when v_R is 1.2 V, A_E is 10 V/V, β_{IFB} is 1 Ω, Δv_T is 50 mV, t_P is 100 ns, v_E is 2.2 V, v_D is 1.8 V, L_X is 10 μH, and $i_{L(AVG)}$ is 100–500 mA.

Solution:

$$\Delta v_{IFB} = \Delta v_T + t_P\left(\frac{v_E + v_D}{L_X}\right)\beta_{IFB} = 50\text{m} + (100\text{n})\left(\frac{2.2 + 1.8}{10\mu}\right)(1) = 90\text{ mV}$$

$$v_{IOS} = t_P\left(\frac{v_D - v_E}{2L_X}\right)\beta_{IFB} = (100\text{n})\left[\frac{1.8 - 2.2}{2(10\mu)}\right](1) = -2\text{ mV}$$

$$v_{VOS} = \frac{i_{L(AVG)}\beta_{IFB} + v_{IOS}}{A_E} = \frac{i_{L(AVG)}(1) - 2\text{m}}{10} = 30 \pm 20\text{ mV}$$

$$v_{FB(AVG)} = v_R - v_{VOS} = 1.2 - v_{VOS} = 1.17\ V \pm 20\ mV$$

$$\beta_{FB} \equiv \frac{\overline{v_{FB(AVG)}}}{v_{O(AVG)}} = \frac{1.17}{1.80} = 65\%$$

$$v_{O(AVG)} = \frac{v_{FB(AVG)}}{\beta_{FB}} = \frac{v_{FB(AVG)}}{65\%} = 1.8\ V \pm 31\ mV$$

Note: v_O's ±31 mV or ±1.7% loading effect equals that of Example 2 because $i_{L(AVG)}$, β_{IFB}, and A_E in both examples match, so A_E suppresses the same loading offset.

Explore with SPICE:
See Appendix A for notes on SPICE simulations.

```
*Hysteretic Current-Mode Voltage Buck
vin vin 0 dc=4
sei vin vsw de 0 sw1v
ddg 0 vsw idiode
lx vsw vl 10u
vi vl vo 0
co vo 0 5u
ro vo 0 18
io vo 0 pwl 0 0 1.2m 0 1.3m 400m
vr vr 0 dc=1.2 pwl 0 0 1m 1.2
efb vfb 0 vo 0 0.65
eae veo 0 vr vfb 10
fbifb 0 vifb vi 1
rbifb vifb 0 1
cifb vifb 0 1p
v1v v1v 0 dc=1
xhys veo vifb de v1v 0 cphys50m
.lib lib.txt
.tran 1.4m
.end
```

Tip: Wait for simulation to finish (because it may require some time to complete) and plot v(vo), v(vfb), v(vifb), v(veo), and i(Lx) across 1.4 ms, between 1.100 and 1.105 ms, and across last 5 μs.

D. **Oscillating Period**

v_E' and the *oscillation period* t_{OSC} set the *energize time* t_E and t_{SW} that determine d_E in the switched inductor. t_{SW} is therefore an *energize duty-cycle* d_E translation (which v_E and v_D set) of the t_E that v_{IFB}'s $i_L\beta_{IFB}$ requires to rise across Δv_{IFB} when v_E energizes L_X:

$$\begin{aligned} t_{SW} &= \frac{t_E}{d_E} = \left(\frac{\Delta v_{IFB}}{d_E}\right)\left(\frac{dt_E}{di_L\beta_{IFB}}\right) \\ &= \left(\frac{\Delta v_{IFB}}{d_E}\right)\left(\frac{L_X}{v_E}\right)\left(\frac{1}{\beta_{IFB}}\right) \\ &= \left(\frac{\Delta v_{IFB}}{\beta_{IFB}}\right)\left(\frac{v_E + v_D}{v_D v_E}\right)L_X \\ &= \left(\frac{\Delta v_{IFB}}{\beta_{IFB}}\right)\left(\frac{L_X}{v_E||v_D}\right) = \Delta i_L\left(\frac{L_X}{v_E||v_D}\right). \end{aligned} \quad (7.27)$$

So v_E and v_D with L_X translate the Δi_L that Δv_{IFB} and β_{IFB} set into t_{SW}. Note t_{SW}'s sensitivity to v_T's and t_P in Δv_{IFB}, v_E, v_D, and L_X.

v_{IFB} can slew this way across Δv_{IFB} only when v_E and v_D are steady. This is usually true for battery chargers and LED drivers because batteries and LEDs drop steady v_O's. In voltage regulators, v_O is steady without *LC* oscillations when t_{SW} is shorter than L_X and C_O's *resonant period* t_{LC}.

Example 5: Determine t_{SW} for Example 4.

Solution:

$$\Delta v_{IFB} = 90 \text{ mV from Example 4}$$

$$t_{SW} = \left(\frac{\Delta v_{IFB}}{\beta_{IFB}}\right)\left(\frac{L_X}{v_E||v_D}\right) = \left(\frac{90m}{1}\right)\left(\frac{10\mu}{2.2||1.8}\right) = 910 \text{ ns}$$

Note: f_{SW}'s $1/t_{SW}$ is about 1.1 MHz.

E. Response Time (Bandwidth)

The aim of the hysteretic loop is to set $i_{L(AVG)}$ with v_{IR}. β_{IFB} sets v_{IR}'s *static transconductance gain* A_{G0} to $i_{L(AVG)}$. The *hysteretic pole* p_{HYS} is the bandwidth past which the *transconductance gain* A_G falls:

$$A_G \equiv \frac{i_{L(AVG)}}{v_{IR}} = \frac{A_{G0}}{1 + s/2\pi p_{HYS}} = \frac{1/\beta_{IFB}}{1 + \tau_{HYS} s}. \quad (7.28)$$

This p_{HYS} determines the *response time* t_R of the loop.

When v_{IR} in Fig. 7.12 rises suddenly, CP_{HYS} trips and keeps v_E' high until i_L raises v_{IFB} to the new v_{IR}. So v_{IR}/β_{IFB}'s rise in Fig. 7.16 prompts i_L to slew across the $\Delta i_{L(AVG)}$ that $\Delta v_{IR}/\beta_{IFB}$ sets. t_R is the time i_L requires to traverse across $\Delta v_{IR}/\beta_{IFB}$, which L_X and L_X's *inductor voltage* v_L determine:

$$t_R = \frac{\Delta i_{L(AVG)}}{di_L/dt} = \Delta i_{L(AVG)}\left(\frac{L_X}{v_L}\right) = \left(\frac{\Delta v_{IR}}{\beta_{IFB}}\right)\left(\frac{L_X}{v_L}\right). \quad (7.29)$$

Before and after this t_R, i_L oscillates normally across Δi_L.

An exponential model of i_L can approximate this response. The *hysteretic time constant* τ_{HYS} that matches the model i_L' to i_L when i_L is 78% of its target (after 78% of t_R elapses) splits the maximum error between i_L' and i_L across t_R so positive and negative errors are the same. This τ_{HYS} is roughly half of t_R:

$$78\%\Delta i_{L(AVG)} = \Delta i_{L(AVG)}\left[1 - \exp\left(\frac{-78\% t_R}{\tau_{HYS}}\right)\right], \quad (7.30)$$

so

$$\tau_{HYS} = \frac{78\% t_R}{\ln(1 - 78\%)} = \frac{t_R}{1.9} = 52\% t_R \quad (7.31)$$

and

$$\text{Error} = i_L' - i_L = \Delta i_{L(AVG)}\left[1 - \exp\left(\frac{-t_X}{\tau_{HYS}}\right)\right] - t_X\left(\frac{\Delta i_{L(AVG)}}{t_R}\right). \quad (7.32)$$

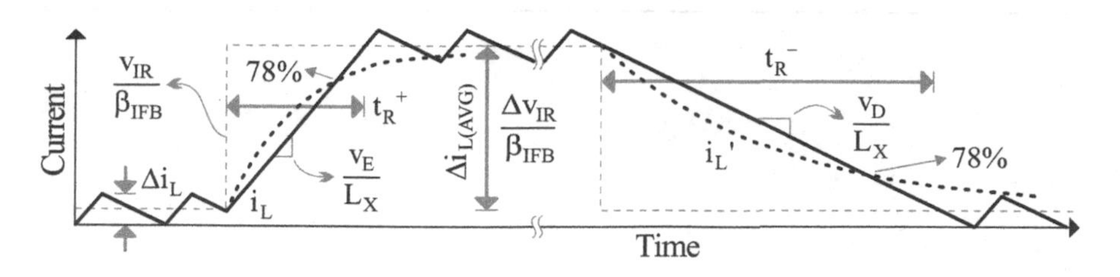

Fig. 7.16 Hysteretic response

The τ_{HYS} that produces this exponential response sets p_{HYS} to

$$\begin{aligned} p_{HYS} &= \frac{1}{2\pi\tau_{HYS}} \approx \left(\frac{1}{2\pi}\right)\left(\frac{1.9}{t_R}\right) \\ &= \left(\frac{1.9}{2\pi}\right)\left(\frac{1}{\Delta i_{L(AVG)}}\right)\left(\frac{v_L}{L_X}\right) = \left(\frac{1.9}{2\pi}\right)\left(\frac{\beta_{IFB}}{\Delta v_{IR}}\right)\left(\frac{v_{E/D}}{L_X}\right). \end{aligned} \tag{7.33}$$

Note p_{HYS} is low when v_L is low and Δv_{IR} and $\Delta i_{L(AVG)}$ are high, where v_L is v_E or v_D. And relative to f_{SW}, p_{HYS} for the slower transition (with lower v_L) nears f_{SW} when Δv_{IR} is close to or less than $\Delta v_T'$:

$$\frac{p_{HYS}}{f_{SW}} = p_{HYS} t_{SW} = \left(\frac{1.9}{2\pi}\right)\left(\frac{\Delta v_T'}{\Delta v_{IR}}\right)\left(\frac{v_{E/D}}{v_E || v_D}\right). \tag{7.34}$$

Example 6: Determine t_R and p_{HYS} for Example 4 when $\Delta i_{L(AVG)}$ is $\pm$400 mA.

Solution:

$$t_R{}^+ = \Delta i_{L(AVG)}{}^+\left(\frac{L_X}{v_E}\right) = (+400\text{m})\left(\frac{10\mu}{2.2}\right) = 1.8\ \mu\text{s}$$

$$t_R{}^- = \Delta i_{L(AVG)}{}^-\left(\frac{L_X}{v_D}\right) = (400\text{m})\left(\frac{10\mu}{1.8}\right) = 2.2\ \mu\text{s}$$

$$p_{HYS} \approx \frac{1.9}{2\pi t_R} = 140\ \text{kHz and } 170\ \text{kHz}$$

Note: p_{HYS} is roughly a decade below f_{SW}'s 1.1 MHz.

Example 7: Determine f_{SW} for Example 4 when Δv_T is 40–60 mV, t_P is 50–150 ns, and L_X is 7–13 µH.

Solution:

$$v_E = 2.2\text{ V and } v_D = 1.8\text{ V from Example 4}$$

$$\Delta v_{IFB} = \Delta v_T + t_P\left(\frac{v_E + v_D}{L_X}\right)\beta_{IFB} = 55-150\text{ mV}$$

$$f_{SW} = \frac{1}{t_{SW}} = \left(\frac{\beta_{IFB}}{\Delta v_{IFB}}\right)\left(\frac{v_E \| v_D}{L_X}\right) = 510\text{ kHz to }2.6\text{ MHz}$$

Note: f_{SW} shifts with Δv_T, t_P, and L_X variations.

Load Dumps The basic purpose of C_O is to supply and pull dynamic mismatches between i_L and the *load* i_{LD}. Capacitance is usually high because C_O's higher aim is to suppress transient variations in v_O. This way, C_O can supply and pull large mismatches without suffering significant v_O fluctuations.

Higher capacitance is especially important in buck–boosts and boosts because they duty-cycle their outputs. So when L_X disconnects from v_O, C_O supplies all of i_{LD}. In bucks, C_O only conducts i_L's ripple because $i_{L(AVG)}$ supplies the entire static load. This is why bucks typically need lower capacitance than boosts and buck–boosts.

Dynamic loads induce another current mismatch. Feedback is to blame for how v_O responds to these i_{LD} fluctuations. Bucks, boosts, and buck–boosts are all the same in this respect: they all depend on feedback to suppress the effects of i_{LD} variations on v_O.

v_{LD} in v_{VOS}, which represents the gain error of the feedback loop, accounts for the loading effect of static variations in i_{LD}. t_R across the loop determines the impact of dynamic i_{LD} fluctuations. This is because, as the feedback loop reacts, C_O supplies the part of i_L that i_{LD} cannot pull or pulls the part of i_{LD} that i_L cannot supply.

When i_{LD} rises suddenly, for example, C_O supplies the additional load Δi_{LD}. C_O similarly pulls the excess Δi_{LD} that i_L supplies when i_{LD} falls suddenly. C_O supplies and pulls these rising and falling *load dumps* across the t_R the voltage loop needs to adjust i_L.

v_{EO} in current-mode voltage regulators is the v_{IR} that controls i_L so i_O supplies i_{LD}. t_R across the voltage loop is the time v_{EO} requires to respond to Δi_{LD}'s plus the time i_L needs to reach the $i_{L(AVG)}$ needed to satisfy i_{LD}. This last part is the t_R of the current loop: the time i_L requires to respond to variations in v_{IR}.

F. **Design Notes**

Short t_R is the fundamental advantage of the hysteretic current loop. This is because the loop responds within one t_{SW} and slews without interruptions to its target. So minimum bandwidth is only a function of i_L's maximum variation and minimum slew rate, which the application (via v_E and v_D) and L_X set. This is the shortest worst-case t_R that a switched inductor needs to respond.

The drawback is t_{SW}'s sensitivity to v_T's, t_P, v_E, v_D, and L_X, which vary with fabrication runs and operating conditions. Predicting and suppressing the switching noise that a variable f_{SW} produces is not always easy. This is why synchronizing f_{SW} to a clock is often desirable.

7.2 Summing Contractions

7.2.1 Summing Comparator

The *summing comparator* in Fig. 7.17 is a comparator that adds inputs. So v_O trips high when v_P's overcome v_N's and low when v_N's surpass v_P's. This way, as a polarity detector, v_O reaches V_{OH} when the *differential sum* Σv_{ID} is positive and V_{OL} when Σv_{ID} is negative.

Summing v_P's and v_N's is like adding v_{P2} and v_{N2}'s v_{ID2} to v_{P1}:

$$\Sigma v_{ID} = (v_{P1} + v_{P2}) - (v_{N1} + v_{N2}) = (v_{P1} + v_{ID2}) - v_{N1} = v_{ID1} + v_{ID2}. \quad (7.35)$$

This is equivalent to adding a v_{ID2} offset to v_{P1} in Fig. 7.18, or adding v_{P1} and v_{N1}'s v_{ID1} to v_{ID2}, which is adding differential voltages. In other words, summing comparators are analog summers that can add and subtract voltages.

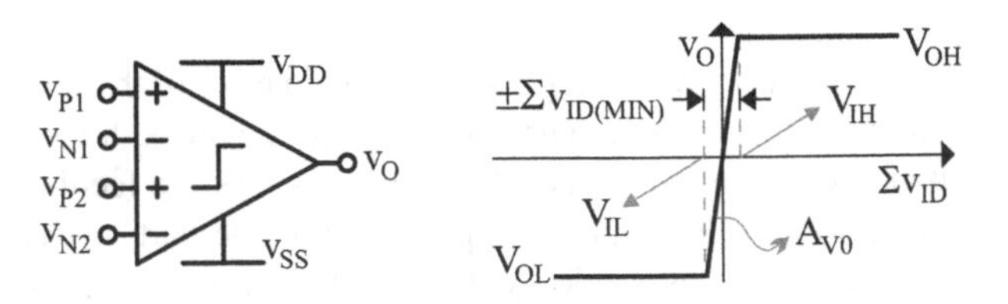

Fig. 7.17 Summing comparator

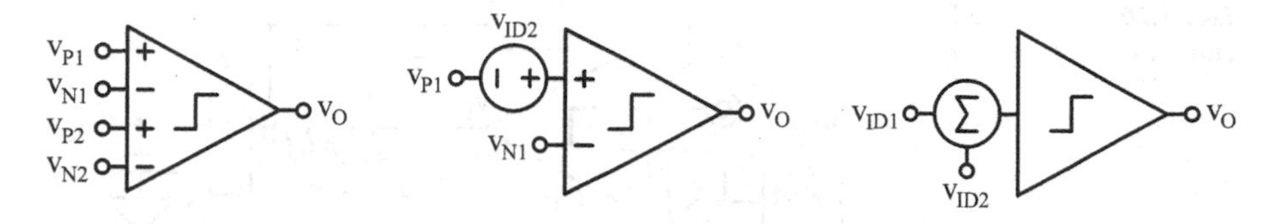

Fig. 7.18 Summing equivalents

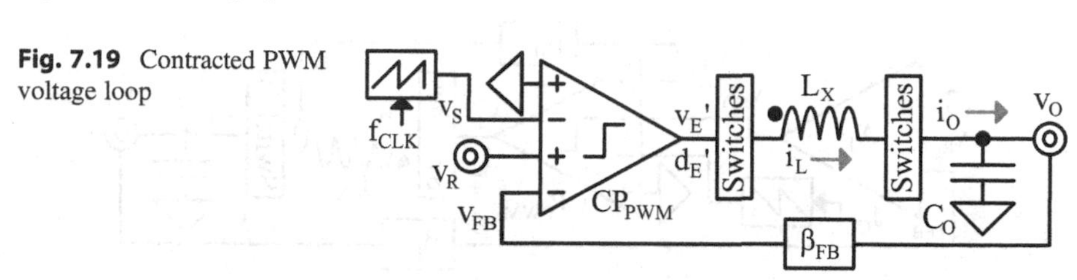

Fig. 7.19 Contracted PWM voltage loop

7.2.2 PWM Contractions

A. **Voltage Loop**

CP_{PWM} in the PWM voltage loop in Fig. 7.8 trips when v_{EO} and v_S's v_{ID} crosses zero. Since v_{EO} is the amplified error A_E outputs, v_{ID} in CP_{PWM} subtracts v_S from the v_{EO} that $(v_R - v_{FB})A_E$ sets. When A_E is one, this v_{ID} effectively subtracts v_S and v_{FB} from v_R like CP_{PWM} in the *contracted PWM voltage loop* in Fig. 7.19:

$$\Sigma v_{ID} = v_{EO} - v_S = (v_R - v_{FB})A_E - v_S|_{A_E \equiv 1} = v_R - v_{FB} - v_S. \tag{7.36}$$

Without the stabilizer normally embedded in A_E, the loop is stable only under certain conditions. A buck voltage regulator with a resistive C_O is a popular example. This is because the out-of-phase *duty-cycled zero* z_{DO} is absent in bucks. And C_O's *capacitor zero* z_C recovers some of the phase lost to L_X and C_O's *LC double pole* p_{LC} before the *loop gain* A_{LG} reaches the *unity-gain frequency* f_{0dB}.

B. **Current Loop**

A_{IE} in the PWM current loop in Fig. 7.9 amplifies the error between v_{IR} and v_{IFB}. v_{ID} in CP_{PWM} subtracts v_S from this amplified error v_{IO}. When A_{IE} is one, this v_{ID} effectively subtracts v_{IFB} and v_S from v_{IR} like CP_{PWM} in Fig. 7.20:

$$\Sigma v_{ID} = v_{IO} - v_S = (v_{IR} - v_{IFB})A_{IE} - v_S|_{A_{IE} \equiv 1} = v_{IR} - v_{IFB} - v_S. \tag{7.37}$$

Stability is often not a concern for this contraction, especially when z_{DO} is absent. This is because current loops without z_{DO}'s are usually inherently stable. *Contracted PWM current loops* are flexible this way.

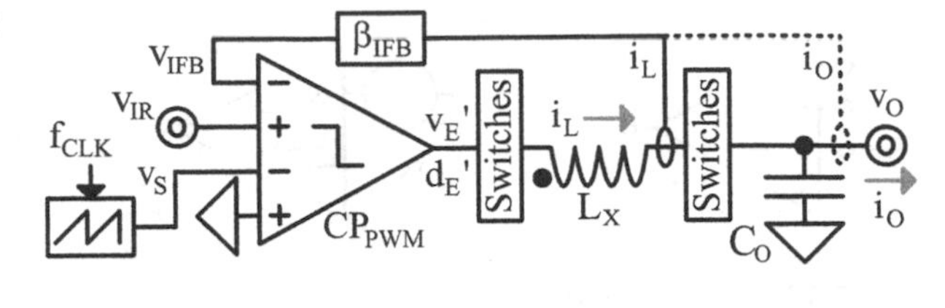

Fig. 7.20 Contracted PWM current loop

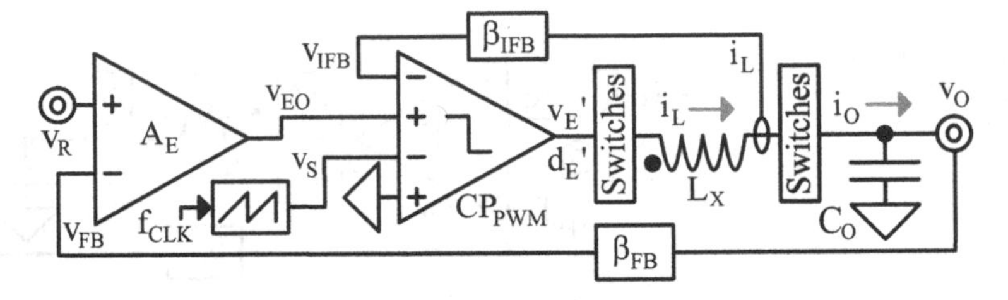

Fig. 7.21 Contracted PWM current-mode voltage loop

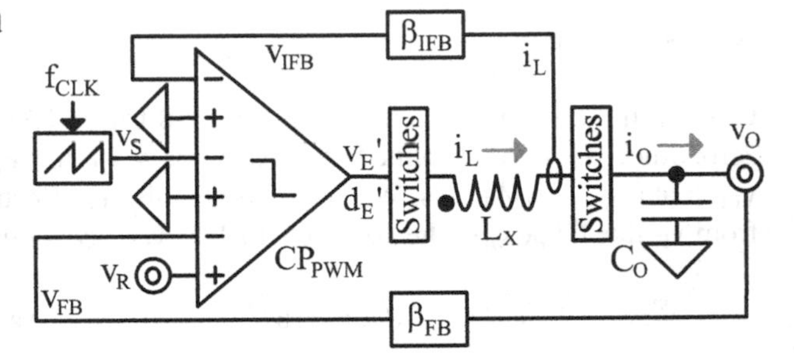

Fig. 7.22 Doubly contracted PWM current-mode voltage loop

C. Current-Mode Voltage Loop

The *contracted PWM current-mode voltage loop* in Fig. 7.21 uses the contracted current loop to control i_L. β_{FB} scales v_O to v_{FB}, A_E compares v_{FB} to v_R, and the current loop translates v_{EO} to i_L. So v_{EO} adjusts the i_L that sets i_O so v_{FB} nears v_R and v_O is close to a reverse β_{FB} translation of v_R.

A_E amplifies the error between v_R and v_{FB}. Σv_{ID} in CP_{PWM} subtracts v_{IFB} and v_S from the amplified error v_{EO}. When A_E is one, this Σv_{ID} effectively subtracts v_{FB}, v_{IFB}, and v_S from v_R like CP_{PWM} in Fig. 7.22:

$$\begin{aligned}\Sigma v_{ID} &= v_{EO} - v_{IFB} - v_S \\ &= (v_R - v_{FB})A_E - v_{IFB} - v_S\big|_{A_E \equiv 1} \\ &= v_R - v_{FB} - v_{IFB} - v_S \\ &= v_R + (-v_S) - v_{FB} - v_{IFB}\end{aligned} \tag{7.38}$$

This is like subtracting v_{FB} and v_{IFB} from v_R and $-v_S$. With v_S inverted this way, Σv_{ID} adds two v_P's and two v_N's like CP_{PWM} in Fig. 7.23. Either way, incorporating A_E and A_{IE} into CP_{PWM} is a *double contraction.*

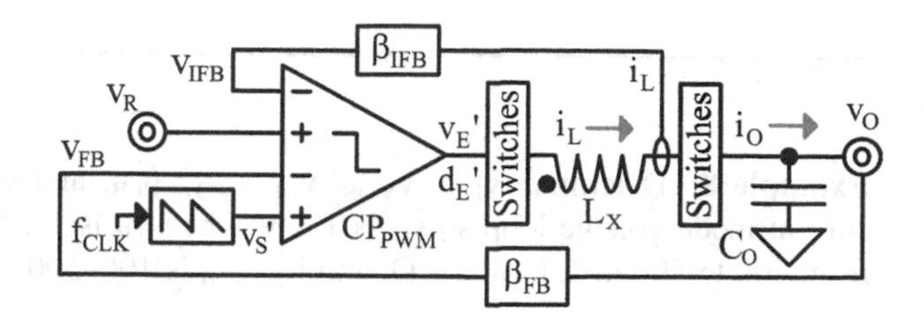

Fig. 7.23 Compact PWM current-mode voltage loop

These double contractions are stable without z_{DO} when the *capacitor pole* p_C that C_O produces reduces A_{LG} to f_{0dB} at or below the *current loop's bandwidth* p_{IBW} that the *current loop's unity-gain frequency* f_{I0dB} sets. It is also stable when the roles of p_C and p_{IBW} reverse. p_C and p_{IBW} can also precede f_{0dB} when z_C reverses p_C at or below f_{0dB}. With z_{DO}, stable operating conditions are more elusive.

D. **Offsets**

The difference between conventional and contracted PWM loops is the absence of A_{IE} and A_E. Without these, v_{IO} and v_{EO} in v_{IOS} and v_{VOS} are unsuppressed. So v_{IOS} and v_{VOS} are v_{IO} and v_{EO}, where v_{IO} and v_{EO} in voltage-mode loops are a d_E' fraction of Δv_S over $v_{S(LO)}$ or below $v_{S(HI)}$ and v_{EO} in current-mode loops is v_{IOS} over v_{LD}:

$$\begin{aligned} v_{IOS} &\equiv v_{IR} - v_{IFB(AVG)} = v_{IO} \\ &= v_{S(LO/HI)} \pm d_E'\Delta v_S \end{aligned} \tag{7.39}$$

$$v_{VOS} \equiv v_R - v_{FB(AVG)} = v_{EO}. \tag{7.40}$$

Unsuppressed this way, v_{VOS} can be so high that it alters v_O in the v_L that sets d_E', so it also affects d_E'.

In voltage loops, v_S adds v_{IOS} to the v_{EO} that sets v_{VOS} when v_S connects to a terminal whose polarity matches that of v_R. When inverting and feeding v_S to an opposing terminal, v_S subtracts v_{IOS} from the v_{EO} that sets v_{VOS}. So v_{VOS} is $+v_{IOS}$ or $-v_{IOS}$ in voltage-mode loops and $v_{LD} \pm v_{IOS}$ in current-mode loops:

$$v_{EO(VM)} = \pm v_{IOS} \tag{7.41}$$

$$v_{EO(IM)} = v_{LD} \pm v_{IOS}. \tag{7.42}$$

Regulating i_L introduces a peculiarity that is usually absent otherwise. This is because i_L normally ripples across a noticeable Δi_L fraction of i_L. i_O and v_O are steadier because C_O is, by design, high enough to keep them steady. So the v_{IFB} that feeds and sets the voltage that drives the PWM is usually a ripple.

This means that v_{IR} and v_R control the point in v_{IFB}'s ripple that sets the d_E' that v_E and v_D need, not v_{IFB}'s halfway point $v_{IFB(AVG)}$ or $i_{L(AVG)}\beta_{IFB}$. Instead, they set the point in i_L with which v_{IFB}'s $i_L\beta_{IFB}$ crosses v_S. A capacitor C_{IFB} in β_{IFB} that suppresses v_{IFB}'s ripple reduces the difference between this point in i_L and $i_{L(AVG)}$. But to keep the loop stable, the pole this C_{IFB} adds should match or surpass f_{I0dB}.

Example 8: Determine v_{IOS}, v_{VOS}, $v_{FB(AVG)}$, β_{FB}, and v_O's error for the PWM current-mode voltage loop so v_O is 1.8 V when v_R is 1.2 V, d_E' is 45%, v_S ramps from 200 to 500 mV, β_{IFB} is 1 Ω, and $i_{L(AVG)}$ is 100–500 mA.

Solution:

$$\Delta v_S = v_{S(HI)} - v_{S(LO)} = 500m - 200m = 300 \text{ mV}$$

$$v_{IOS} = v_{S(LO)} + \Delta v_S d_E' = 200m + (300m)(45\%) = 340 \text{ mV}$$

$$v_{VOS} \approx i_{L(AVG)}\beta_{IFB} + v_{IOS} = i_{L(AVG)}(1) + 340m = 640 \pm 200 \text{ mV}$$

$$v_{FB(AVG)} = v_R - v_{VOS} = 1.2 - v_{VOS} \approx 560 \pm 200 \text{ mV}$$

$$\beta_{FB} \equiv \frac{\overline{v_{FB(AVG)}}}{v_{O(AVG)}} = \frac{560m}{1.80} = 31\%$$

$$v_{O(AVG)} = \frac{v_{FB(AVG)}}{\beta_{FB}} = \frac{v_{FB(AVG)}}{31\%} = 1.81 \text{ V} \pm 640 \text{ mV}$$

Note: v_O's ±640 mV or ±36% variation is so high that d_E' and v_{IOS} also vary, so this loading effect is an approximation.

7.2.3 Hysteretic Contraction

A. **Current-Mode Voltage Loop**

A_E in the hysteretic current-mode voltage loop in Fig. 7.14 amplifies the error between v_R and v_{FB}. v_{ID} in CP_{HYS} subtracts v_{IFB} from this amplified error v_{EO}. When A_E is one, this v_{ID} effectively subtracts v_{FB} and v_{IFB} from v_R like CP_{HYS} in the *contracted hysteretic current-mode voltage loop* in Fig. 7.24:

$$\Sigma v_{ID} = v_{EO} - v_{IFB} = (v_R - v_{FB})A_E - v_{IFB}\big|_{A_E \equiv 1} = v_R - v_{FB} - v_{IFB}. \tag{7.43}$$

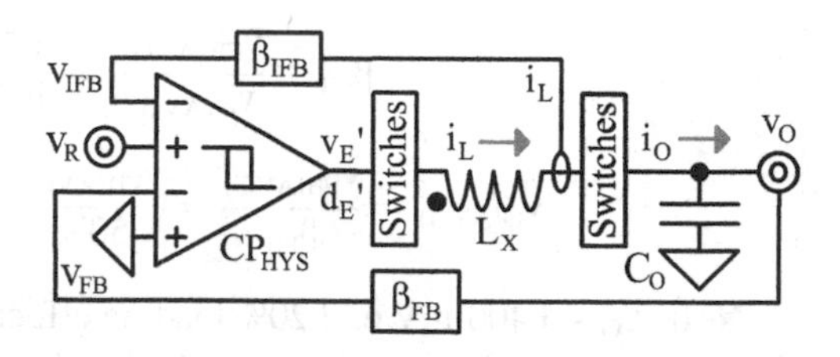

Fig. 7.24 Contracted hysteretic current-mode voltage loop

The loop is stable without z_{DO} when p_C reduces A_{LG} to f_{0dB} at or below p_{IBW}, or when p_{IBW} reduces A_{LG} to f_{0dB} at or below p_C. p_C and p_{IBW} can also precede f_{0dB} when z_C reverses p_C at or below f_{0dB}. With z_{DO}, stable conditions are more elusive.

B. Offset

The difference between conventional and contracted hysteretic voltage loops is the absence of A_E. Without A_E, v_{EO} in v_{VOS} is unsuppressed. So v_{VOS} is the v_{EO} that v_{IOS} over v_{LD} sets:

$$v_{VOS} \equiv v_R - v_{FB(AVG)}$$
$$= v_{LD} + v_{IOS} = v_{LD} + v_{POS}^- - v_{POS}^+ \approx \left[i_{L(AVG)} + t_P\left(\frac{v_D - v_E}{2L_X}\right)\right]\beta_{IFB}. \tag{7.44}$$

Although v_{VOS} in the v_{FB} that sets v_O also alters the v_L's that set v_{POS}'s in v_{IOS}, t_P is (numerically) usually a fraction of L_X, so $t_P/2L_X$ suppresses v_E and v_D variations in v_{IOS}.

Example 9: Determine v_{IOS}, v_{VOS}, $v_{FB(AVG)}$, β_{FB}, and v_O's error for the hysteretic current-mode voltage loop so v_O is 1.8 V when v_R is 1.2 V, v_E is 2.2 V, v_D is 1.8 V, L_X is 10 μH, Δv_T is 50 mV, t_P is 100 ns, β_{IFB} is 1 Ω, and $i_{L(AVG)}$ is 100–500 mA.

Solution:

$$v_{IOS} = t_P\left(\frac{v_D - v_E}{2L_X}\right)\beta_{IFB} = (100\text{n})\left[\frac{1.8 - 2.2}{2(10\mu)}\right](1) = -2\text{ mV}$$

$$v_{VOS} = i_{L(AVG)}\beta_{IFB} + v_{IOS} = i_{L(AVG)} - 2\text{m} = 298 \pm 200\text{ mV}$$

$$v_{FB(AVG)} = v_R - v_{VOS} = 1.2 - v_{VOS} = 902 \pm 200\text{ mV}$$

$$\beta_{FB} \equiv \frac{\overline{v_{FB(AVG)}}}{v_{O(AVG)}} = \frac{902m}{1.80} = 50\%$$

$$v_{O(AVG)} = \frac{v_{FB(AVG)}}{\beta_{FB}} = \frac{v_{FB(AVG)}}{50\%} = 1.80 \text{ V} \pm 400 \text{ mV}$$

Note: v_O's ±400-mV or ±20% loading effect is lower than in Example 8 because the v_{VOS} that sets β_{FB} is higher here. And with a higher β_{FB}, v_{LD}'s reverse translation to v_O is lower.

7.2.4 Load Compensation

Offsets in v_{FB} determine v_O's *static accuracy*. Since β_{FB} can compensate for v_{IOS}, but only center the effect of v_{LD}, v_O can vary no less than $\pm 0.5 v_{LD}/\beta_{FB}$. This can be problematic when the i_L that sets i_O varies widely.

The effect of loading on v_{EO} is essentially the need for adding v_{LD}, which is a β_{IFB} translation of $i_{L(AVG)}$. Luckily, this information is embedded in v_{IFB}. So averaging v_{IFB} and adding $v_{IFB(AVG)}$ to v_{EO} satisfies the need (i.e., offset) v_{LD} produces. This way, without this need, v_{VOS} excludes v_{LD}.

The low-pass filter and summing comparator in Fig. 7.25 do this for v_{EO}. R_{LDC} and C_{LDC} average v_{IFB} so v_{LD}' steadies about $v_{IFB(AVG)}$ and CP_E adds the resulting average to the v_{ID} that produces v_{EO} (which is $v_R - v_{FB}$) when A_E is one. Since v_{LD}' excludes higher-frequency components, v_{LD}' satisfies the offset $i_{L(AVG)}\beta_{IFB}$ produces without altering the feedback dynamics near f_{0dB}, which ensure the loop is stable.

The *load-compensation resistor* and *capacitor* produce the low-frequency pole that averages v_{IFB}. C_{LDC} shunts to ground the dynamics v_{IFB} injects into v_{LD}' via R_{LDC}. So when the *load-compensation pole* p_{LDC} is much lower than f_{0dB}, v_{LD}' is largely free of dynamics near f_{0dB}:

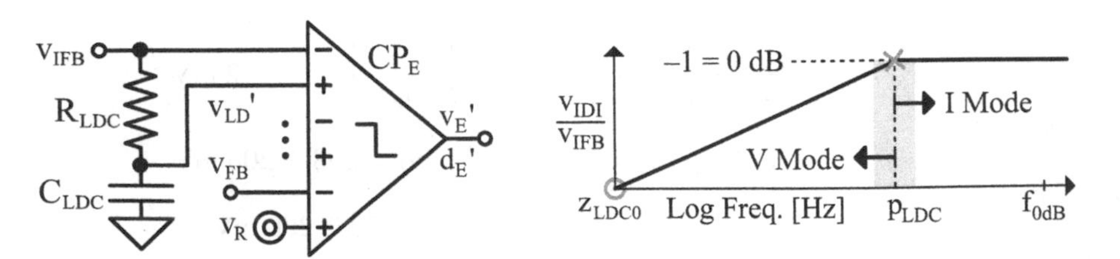

Fig. 7.25 Load compensation

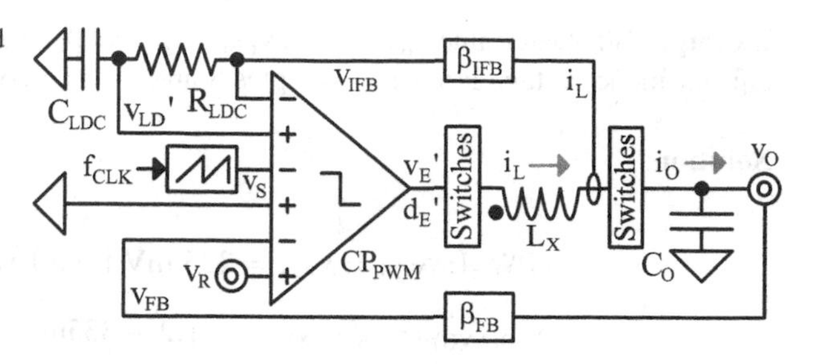

Fig. 7.26 Load-compensated PWM current-mode voltage loop

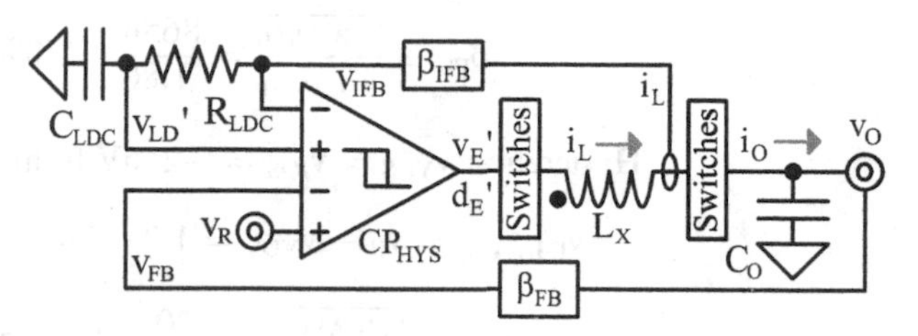

Fig. 7.27 Load-compensated hysteretic current-mode voltage loop

$$p_{LDC} = \frac{1}{2\pi R_{LDC} C_{LDC}} << f_{0dB}. \tag{7.45}$$

In practice, p_{LDC} is below p_{LC} or near p_{LC} when z_{DO} is absent and f_{0dB} is above p_{LC}.

The resulting *differential current-mode voltage* v_{IDI} that v_{IFB} and v_{LD}' produce is zero at dc, increases with frequency as C_{LDC} shunts current, and nears $-v_{IFB}$ past p_{LDC} when C_{LDC} shorts v_{LD}' to ground:

$$v_{IDI} = v_{LD}' - v_{IFB} = \frac{v_{IFB}}{1 + s/2\pi p_{LDC}} - v_{IFB} = -v_{IFB}\left(\frac{s/2\pi p_{LD}}{1 + s/2\pi p_{LD}}\right). \tag{7.46}$$

So the current loop is fully active only above p_{LDC}. This way, the system behaves more like a voltage-mode loop below p_{LDC} (when v_{IDI} excludes v_{IFB}: v_{IDI} is very low) and like a current-mode loop above p_{LDC} (when v_{IDI} carries v_{IFB}: v_{IDI} is $-v_{IFB}$). v_{IDI} excludes v_{IFB} in Fig. 7.25 when v_{IFB}'s translation to v_{IDI} is very low and carries v_{IFB} when v_{IDI}/v_{IFB} approaches -1.

The PWM and hysteretic systems in Figs. 7.26 and 7.27 add this filtered v_{LD}' to a v_P in CP_{PWM} and CP_{HYS}. This adds the v_{LD} that v_{EO} needs when setting $i_{L(AVG)}$. v_O's can be very close to their targets this way because, without v_{LD} in v_{VOS}, β_{FB}'s can account for the rest.

Example 10: Determine v_{IOS}, v_{VOS}, $v_{FB(AVG)}$, and β_{FB} for the PWM and hysteretic current-mode voltage loops in Examples 8 and 9 when compensated.

Solution:

$$\text{PWM}: v_{VOS} = v_{IOS} = 335\text{ mV from Example 8}$$

$$v_{FB(AVG)} = v_R - v_{VOS} = 1.2 - 335m = 865\text{ mV}$$

$$\beta_{FB} \equiv \frac{\overline{v_{FB(AVG)}}}{v_{O(AVG)}} = \frac{865m}{1.80} = 48\%$$

$$\text{Hysteretic.}: v_{VOS} = v_{IOS} = -2\text{ mV from Example 9}$$

$$v_{FB(AVG)} = v_R - v_{VOS} = 1.2 - 2m = 1.20\text{ V}$$

$$\beta_{FB} \equiv \frac{\overline{v_{FB(AVG)}}}{v_{O(AVG)}} = \frac{1.20}{1.80} = \frac{2}{3} = 67\%$$

Note: v_O is 1.8 V without loading effect.

Explore with SPICE:
See Appendix A for notes on SPICE simulations.

```
*Contracted PWM Buck
vin vin 0 dc=4
sei vin vsw de 0 sw1v
ddg 0 vsw idiode
lx vsw vl 10u
vi vl vo 0
co vo 0 5u
ro vo 0 18
io vo 0 pwl 0 0 1m 0 1.2m 0 1.3m 400m
vr vr 0 dc = 0 pwl 0 0 1 m 1.2
vs vs 0 dc = 200m pulse 200m 500m 0998n 1n 1n 1u
efb vfb 0 vo 0 0.48
fbifb 0 vifb vi 1
rbifb vifb 0 2
```

(continued)

```
cbifb vifb 0 1u
rldc vifb vld 2
cldc vld 0 1.81u
v1v v1v 0 dc=1
xpwm vr vifb 0 vfb vld vs de v1v 0 cp3vid
.lib lib.txt
.tran 1.4m
.end
```

Tip: Wait for simulation to finish (because it may require some time to complete) and plot v(vo), v(vfb), v(vifb), v(vld), and i(Lx) across 1.4 ms, between 1.100 and 1.105 ms, and across last 5 µs.

Explore with SPICE:

See Appendix A for notes on SPICE simulations.

```
*Contracted Hysteretic Buck
vin vin 0 dc=4
si vin vsw de 0 sw1v
dg 0 vsw idiode
lx vsw vswx 10u
vi vswx vo 0
co vo 0 5u
ro vo 0 18
io vo 0 pwl 0 0 1.2m 0 1.4m 400m 1.6m 400m 1.6001m 0 1.7m 0
+1.7001m 400m
vr vr 0 dc=0 pwl 0 0 1m 1.2
efb vfb 0 vo 0 0.667
fbifb 0 vifb vi 1
rbifb vifb 0 2
cbifb vifb 0 1p
rldc vifb vld 2
cldc vld 0 1.81u
v1v v1v 0 dc=1
xhys vr vfb vld vifb 0 0 de v1v 0 cphys3vid50m
.lib lib.txt
.tran 1.8m
.end
```

Tip: Wait for simulation to finish (because it may require some time to complete) and plot v(vo), v(vfb), v(vifb), v(vld), and i(Lx) across 1.8 ms, between 1.100 and 1.105 ms, and between 1.500 and 1.505 ms.

7.2.5 Design Notes

Contractions are good because they integrate A_E and A_{IE} into one summing comparator. The benefits of removing A_E and A_{IE} are lower power, smaller silicon area, and lower cost. Plus, the summing comparator can also cancel the loading effect that offsets v_O.

Without a stabilizer, stability for current-mode voltage loops can be elusive, especially for buck–boosts and boosts, because they incorporate z_{DO}'s. Adding a stabilizing A_E can solve this problem. This way, many of the benefits (like load compensation) can remain.

7.3 Constant-Time Peak/Valley Loops

7.3.1 SR Flip Flop

Flip flops are bistable *state machines* that flip-flop between states. They are *one-bit memory* devices because they can "remember" their previous digital state. They are also *sequential circuits* because their outputs depend on this previous state. And they are *latches* because they latch or clamp their outputs when the inputs are static.

Set–reset (SR) *flip flops* "set" or "reset" their *current state* "Q" in Fig. 7.28 high or low when their "S" or "R" input is high. They hold their *previous state* Q^{-1} when both inputs are low. When both inputs are high, *set-dominant flip flops* output one and *reset-dominant flip flops* output zero. SR circuits normally output Q and Q's *opposite* or *complementary state* $\overline{Q}$.

SR flip flops are useful in feedback controllers because they decouple on and off commands. With flip flops, one circuit or loop can control the on state with S and another can control the off state with R. In constant-time loops, a feedback loop often controls one and a timer commands the other.

7.3.2 Pulse Generator

A. **Set-Enabled**

The *set-enabled pulse generator* in Fig. 7.29 is an interruptible relaxation oscillator. v_O and v_R close a feedback loop that oscillates when v_S is high. Grounding v_S stops

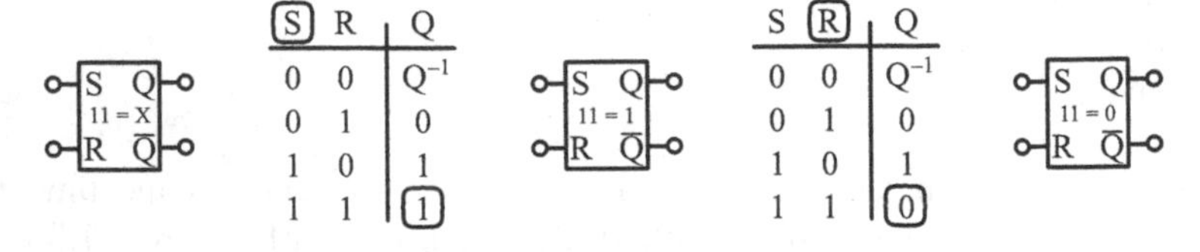

Fig. 7.28 Set- and reset-dominant SR flip flops

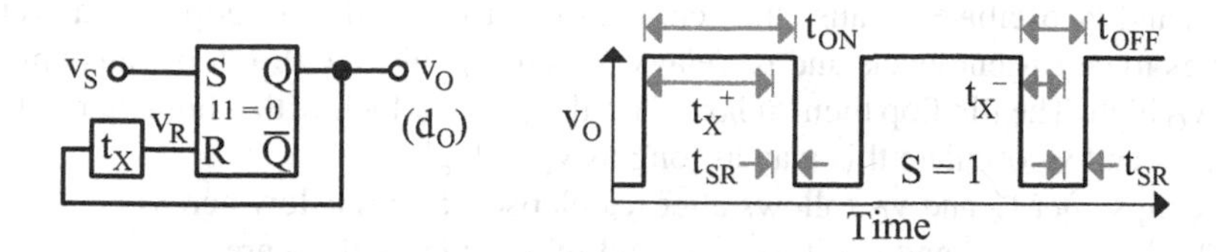

Fig. 7.29 Set-enabled pulse generator

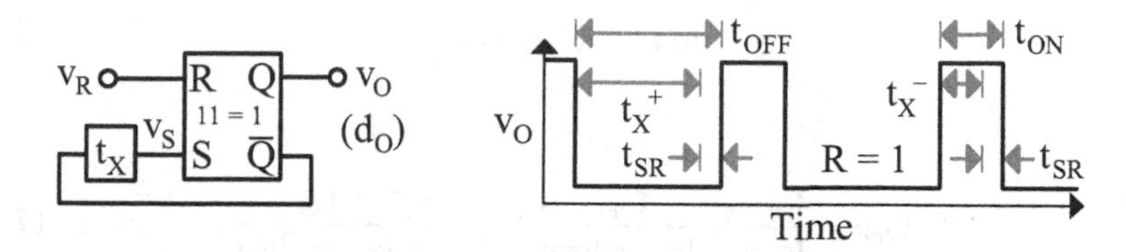

Fig. 7.30 Reset-enabled pulse generator

the oscillations because a reset-dominant flip flop cannot trip v_O high without a high v_S. This is how v_S enables the oscillator. Enabled pulse generators are also *one-shot pulses* (or *one shots* for short) because they pulse once each time their enabling inputs pulse.

v_O and v_R oscillate because, like a *ring oscillator*, they close an inverting feedback loop with a v_O that swings to consistent levels and t_X delays. When v_O rises, the flip flop waits until v_R resets v_O low. The flip flop then "relaxes" until v_R falls, which with a high v_S sets v_O high. v_O and v_R oscillate this way as long as v_S is high.

v_R flips after t_X and v_O follows after *SR's propagation delay* t_{SR} elapses. So v_O is high across t_X's rising delay $t_X{}^+$ and t_{SR} and low across t_X's falling delay $t_X{}^-$ and t_{SR}. *On/off times* $t_{ON/OFF}$ and *on duty-cycle* d_{ON} are therefore

$$t_{ON} = t_X{}^+ + t_{SR}, \tag{7.47}$$

$$t_{OFF} = t_X{}^- + t_{SR}, \tag{7.48}$$

and

$$d_{ON} = \frac{t_{ON}}{t_{OSC}} = \frac{t_{ON}}{t_{ON/OFF} + t_{OFF}} = \frac{t_X{}^\pm + t_{SR}}{t_X{}^+ + t_X{}^- + 2t_{SR}}. \tag{7.49}$$

B. **Reset-Enabled**

The *reset-enabled pulse generator* in Fig. 7.30 is the complement of the set-enabled circuit. v_O and v_R close a feedback loop that oscillates when v_R is high. Grounding v_R stops the oscillations because a set-dominant flip flop cannot trip v_O low without a high v_R. This is how v_R enables the oscillator.

v_O and v_S oscillate because they close an inverting feedback loop with a v_O that swings to consistent levels and t_X delays. When v_O falls, the flip flop waits until v_S sets v_O high. The flip flop then "relaxes" until v_S falls, which with a high v_R resets v_O low. v_O and v_S oscillate this way as long as v_R is high.

v_S flips after t_X and v_O follows after t_{SR} elapses. So v_O is low across t_X^+ and t_{SR} and high across t_X^- and t_{SR}. t_{OFF}, t_{ON}, and *off duty-cycle* d_{OFF} are

$$t_{OFF} = t_X^+ + t_{SR}, \tag{7.50}$$

$$t_{ON} = t_X^- + t_{SR}, \tag{7.51}$$

and

$$d_{OFF} = \frac{t_{OFF}}{t_{OSC}} = \frac{t_{OFF}}{t_{ON} + t_{OFF}} = \frac{t_X^+ + t_{SR}}{t_X^+ + t_X^- + 2t_{SR}}. \tag{7.52}$$

7.3.3 Constant On-Time Valley Loop

A. Current Loop

The *constant on-time valley current loop* in Fig. 7.31 uses a set-enabled pulse generator to start the t_{ON} that sets t_E. β_{IFB} scales i_L to v_{IFB} and the *time-loop comparator* CP_T triggers a t_{ON} pulse whenever v_{IFB} falls below v_{IR}. This way, v_{IR} sets $v_{IFB(LO)}$ and t_{ON} fixes t_E, which (with v_E and L_X) projects v_{IFB} to $v_{IFB(HI)}$. Setting $v_{IFB(LO)}$ is a form of *valley control*.

In practice, t_P in CP_T and t_{SR} in the *on flip flop* SR_{ON} delay transitions. So L_X energizes across the t_{ON} that t_X^+ and t_{SR} set. Across this t_E in Fig. 7.32, di_L/dt_E projects v_{IFB} across Δv_{IFB}. And with this t_E, v_E and v_D in d_E set t_{SW}:

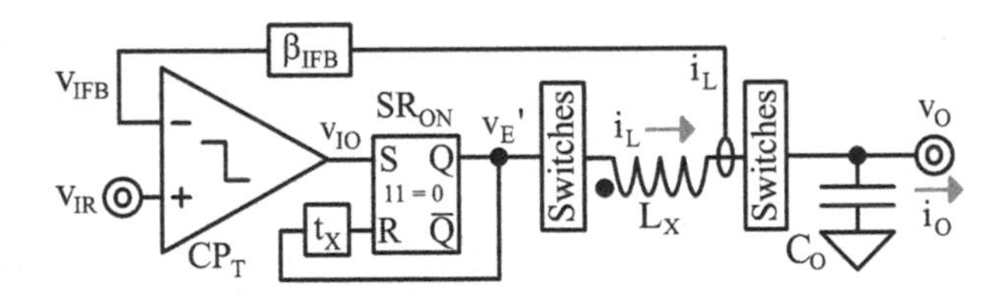

Fig. 7.31 Constant on-time valley current loop

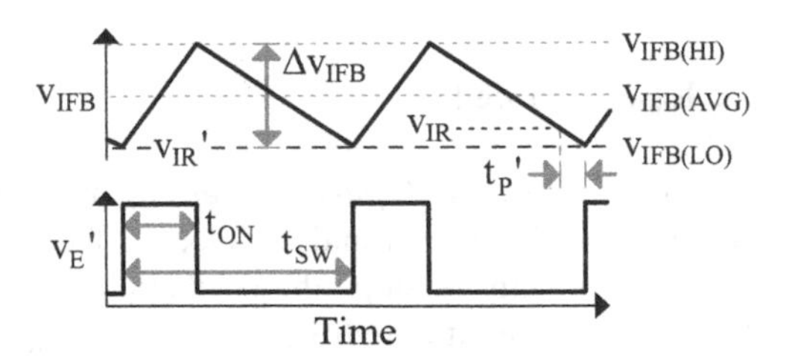

Fig. 7.32 Constant on-time valley oscillation

$$\Delta v_{IFB} = t_E\left(\frac{dv_{IFB}}{dt_E}\right) = t_E\left(\frac{di_L}{dt_E}\right)\beta_{IFB} = t_{ON}\left(\frac{v_E}{L_X}\right)\beta_{IFB} \quad (7.53)$$

$$t_{SW} = \frac{t_E}{d_E} = t_{ON}\left(\frac{v_E + v_D}{v_D}\right). \quad (7.54)$$

B. Offsets

t_P^- and t_{SR} in t_P' in Fig. 7.32 delay t_D's termination (i.e., t_E's onset). So v_{IFB} falls v_{POS}^- below v_{IR} to v_{IR}'. This v_{POS}^- is the offset di_L/dt_D's v_D/L_X projects across t_P'. $v_{IFB(AVG)}$ is halfway between $v_{IFB(HI)}$ and $v_{IFB(LO)}$, which is also half Δv_{IFB} over v_{IR}'. In other words, v_{IR}' is half Δv_{IFB} below $v_{IFB(AVG)}$, v_{IR} is v_{POS}^- above v_{IR}', and $v_{IFB(AVG)}$ is half Δv_{IFB} over the v_{IR}' that v_{POS}^- subtracts from v_{IR}:

$$\begin{aligned} v_{IOS} \equiv v_{IR} - v_{IFB(AVG)} &= -\left(\frac{\Delta v_{IFB}}{2} - v_{POS}^-\right) \\ &= \left[t_P'\left(\frac{v_D}{L_X}\right) - \left(\frac{t_{ON}}{2}\right)\left(\frac{v_E}{L_X}\right)\right]\beta_{IFB}. \end{aligned} \quad (7.55)$$

v_{EO} in current-mode voltage loops is the v_{IR} that sets $i_{L(AVG)}$ in current loops. Since v_{IR} is v_{IOS} over $v_{IFB(AVG)}$ and $v_{IFB(AVG)}$ is a β_{IFB} translation of $i_{L(AVG)}$, v_{EO} is v_{IOS} over v_{LD}. And v_{VOS} is A_E times lower:

$$\begin{aligned} v_{VOS} \equiv v_{IR} - v_{FB(AVG)} &= \frac{v_{EO}}{A_E} = \frac{v_{LD} + v_{IOS}}{A_E} \\ &= \left[i_{L(AVG)} + t_P'\left(\frac{v_D}{L_X}\right) - \left(\frac{t_{ON}}{2}\right)\left(\frac{v_E}{L_X}\right)\right]\left(\frac{\beta_{IFB}}{A_E}\right). \end{aligned} \quad (7.56)$$

Example 11: Determine t_{SW}, Δv_{IFB}, v_{IOS}, v_{VOS}, and $v_{FB(AVG)}$ for the constant on-time current-mode voltage loop when v_R is 1.2 V, A_E is 10 V/V, β_{IFB} is 1 Ω, t_P is 100 ns, t_{SR} is 10 ns, t_X^+ is 450 ns, v_E is 2.2 V, v_D is 1.8 V, L_X is 10 μH, and $i_{L(AVG)}$ is 100–500 mA.

Solution:

$$t_E = t_{ON} = t_X^+ + t_{SR} = 450n + 10n = 460 \text{ ns}$$

$$t_{SW} = t_E\left(\frac{v_E + v_D}{v_D}\right) = (460n)\left(\frac{2.2 + 1.8}{1.8}\right) = 1.0\ \mu s$$

$$\Delta v_{IFB} = t_E\left(\frac{v_E}{L_X}\right)\beta_{IFB} = (460n)\left(\frac{2.2}{10\mu}\right)(1) = 100\ mV$$

$$t_P' = t_P + t_{SR} = 100n + 10n = 110\ ns$$

$$v_{IOS} = t_P'\left(\frac{v_D}{L_X}\right)\beta_{IFB} - \frac{\Delta v_{IFB}}{2} = (110n)\left(\frac{1.8}{10\mu}\right)(1) - \frac{100m}{2} = -30\ mV$$

$$v_{VOS} = \frac{i_{L(AVG)}\beta_{IFB} + v_{IOS}}{A_E} = \frac{i_{L(AVG)}(1) - 30m}{10} = 27 \pm 20\ mV$$

$$v_{FB(AVG)} = v_R - v_{VOS} = 1.2 - v_{VOS} = 1.17\ V \pm 20\ mV = 1.17\ V \pm 1.7\%$$

Note: Contracting the system increases v_{FB}'s loading effect and load compensation cancels it.

Explore with SPICE:

See Appendix A for notes on SPICE simulations.

```
*Constant On-Time Valley Current-Mode Voltage Buck
vin vin 0 dc=4
sei vin vsw de 0 sw1v
ddg 0 vsw idiode
lx vsw vl 10u
vi vl vo 0
co vo 0 5u
ro vo 0 18
io vo 0 pwl 0 0 1.2m 0 1.3m 400m
vr vr 0 dc=0 pwl 0 0 1m 1.2
efb vfb 0 vo 0 0.65
cfb vfb 0 1p
eae veo 0 vr vfb 10
fbifb 0 vifb vi 1
rbifb vifb 0 1
cbifb vifb 0 1p
v1v v1v 0 dc = 1
```

(continued)

```
xcp veo vifb vio v1v 0 cp
xsrr vio vtx de vqn srr
xdly de vtx dly
.lib lib.txt
.tran 1.4m
.end
```

Tip: Wait for simulation to finish (because it may require some time to complete) and plot v(vo), v(vfb), v(veo), v(vifb), and i(Lx) across 1.4 ms, between 1.100 and 1.105 ms, and across last 5 μs.

C. Response Time

When v_{IR} rises suddenly, v_{IR} surpasses v_{IFB}, so v_{IO} rises and stays high. This keeps v_E' oscillating. t_X^+ is, by design, much greater than t_X^-, so d_{ON} is fairly high (near or above 90%). This way, when $t_{ON}v_E$ exceeds $t_{OFF}v_D$, i_L across t_R^+ in Fig. 7.33 rises di_L^+ more than i_L falls di_L^-:

$$di_L{}^+ = t_E\left(\frac{di_L}{dt_E}\right) = t_{ON}\left(\frac{v_E}{L_X}\right) > di_L{}^- = t_D\left(\frac{di_L}{dt_D}\right) = t_{OFF}\left(\frac{v_D}{L_X}\right). \tag{7.57}$$

This is the only way i_L can rise and reach a higher level (across t_R^+).

When v_{IR} falls suddenly, v_{IR} falls below v_{IFB}, so v_{IO} falls and stays low. This stops SR_{ON} from pulsing. So L_X drains across t_R^- without interruptions (like in the hysteretic current loop) until i_L reduces v_{IFB} to the new target.

t_R^+ is longer than in the hysteretic current loop because t_{OFF} interrupts i_L's ascent. But when t_{OFF} is very short, t_R^+ is nearly the same. So this system can respond nearly as fast as the hysteretic current loop.

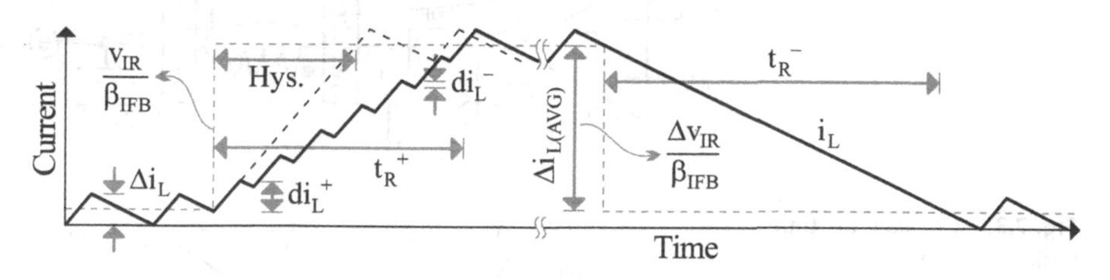

Fig. 7.33 Constant on-time valley response

7.3.4 Constant Off-Time Peak Loop

A. Current Loop

The *constant off-time peak current loop* in Fig. 7.34 is the complement of the on-time loop. It uses a reset-enabled pulse generator to start the t_{OFF} that sets t_D. In this case, CP_T triggers t_{OFF} pulses when v_{IFB} rises over v_{IR}. This way, v_{IR} sets $v_{IFB(HI)}$; t_{OFF} fixes t_D, which (with v_E and L_X) projects v_{IFB} to $v_{IFB(LO)}$; and v_E' pulses when v_{IO} is high. Setting $v_{IFB(HI)}$ is a form of *peak control.*

In practice, t_P in CP_T and t_{SR} in the *off flip flop* SR_{OFF} delay transitions. So L_X drains across the t_{OFF} that t_X^+ and t_{SR} set. Across this t_D in Fig. 7.35, di_L/dt_D projects v_{IFB} across Δv_{IFB}. And with this t_D, v_E and v_D in d_D set t_{SW}:

$$\Delta v_{IFB} = t_D\left(\frac{dv_{IFB}}{dt_D}\right) = t_D\left(\frac{di_L}{dt_D}\right)\beta_{IFB} = t_{OFF}\left(\frac{v_D}{L_X}\right)\beta_{IFB} \tag{7.58}$$

and

$$t_{SW} = \frac{t_D}{d_D} = \frac{t_D}{1 - d_E} = t_{OFF}\left(\frac{v_E + v_D}{v_E}\right), \tag{7.59}$$

where t_D is a *drain duty-cycle* d_D fraction of t_{SW} and d_D is $1 - d_E$.

B. Offsets

In practice, t_P^+ and t_{SR} in t_P' in Fig. 7.35 delay t_E's termination. So v_{IFB} rises v_{POS}^+ over v_{IR} to v_{IR}'. This v_{POS}^+ is the offset di_L/dt_E's v_E/L_X projects across t_P'. $v_{IFB(AVG)}$ is halfway between $v_{IFB(HI)}$ and $v_{IFB(LO)}$, which is half Δv_{IFB} below this v_{IR}'. In other words, v_{IR} is half Δv_{IFB} over $v_{IFB(AVG)}$ and v_{POS}^+ below v_{IR}':

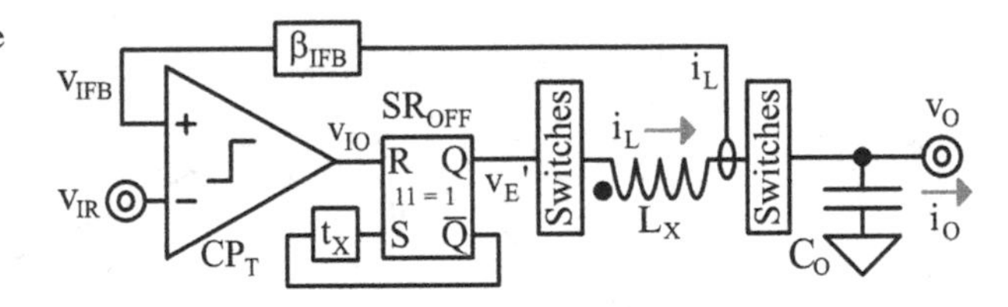

Fig. 7.34 Constant off-time peak current loop

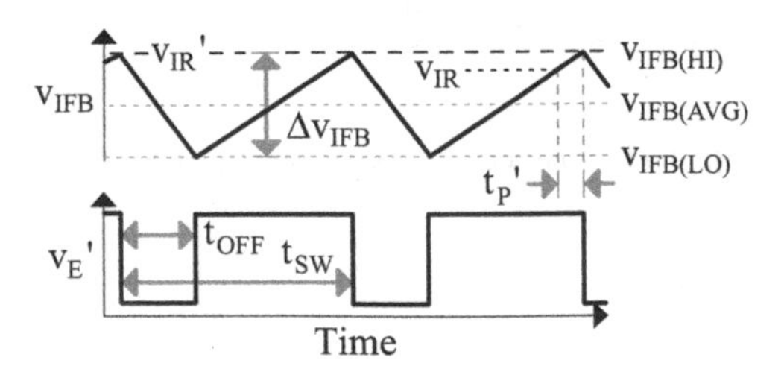

Fig. 7.35 Constant off-time peak oscillation

$$v_{IOS} \equiv v_{IR} - v_{IFB(AVG)} = \frac{\Delta v_{IFB}}{2} - v_{POS}{}^{+} = \left[\left(\frac{t_{OFF}}{2}\right)\left(\frac{v_D}{L_X}\right) - t_P{}'\left(\frac{v_E}{L_X}\right)\right]\beta_{IFB}. \tag{7.60}$$

v_{EO} in current-mode voltage loops is the v_{IR} that sets $i_{L(AVG)}$ in current loops. Since v_{IR} is v_{IOS} over $v_{IFB(AVG)}$ and $v_{IFB(AVG)}$ is a β_{IFB} translation of $i_{L(AVG)}$, v_{EO} is v_{IOS} over v_{LD}. And v_{VOS} is A_E times lower:

$$v_{VOS} \equiv v_R - v_{FB(AVG)} = \frac{v_{EO}}{A_E} = \frac{v_{LD} + v_{IOS}}{A_E} = \left[i_{L(AVG)} + \left(\frac{t_{OFF}}{2}\right)\left(\frac{v_D}{L_X}\right) - t_P{}'\left(\frac{v_E}{L_X}\right)\right]\left(\frac{\beta_{IFB}}{A_E}\right). \tag{7.61}$$

C. **Response Time**

When v_{IR} rises suddenly, v_{IR} surpasses v_{IFB}, so v_{IO} falls and stays low. This stops SR_{OFF} from pulsing. So L_X energizes across $t_R{}^+$ in Fig. 7.36 without interruptions (like in the hysteretic current loop) until v_{IFB} reaches the new target.

When v_{IR} falls suddenly, v_{IR} falls below v_{IFB}, so v_{IO} rises and stays high. This keeps $v_E{}'$ oscillating. $t_X{}^+$ is, by design, much greater than $t_X{}^-$, so d_{OFF} is fairly high (near or above 90%). This way, when $t_{OFF}v_D$ exceeds $t_{ON}v_D$, i_L across $t_R{}^-$ falls $di_L{}^-$ more than i_L rises $di_L{}^+$:

$$di_L{}^- = t_D\left(\frac{di_L}{dt_D}\right) = t_{OFF}\left(\frac{v_D}{L_X}\right) > di_L{}^+ = t_E\left(\frac{di_L}{dt_E}\right) = t_{ON}\left(\frac{v_E}{L_X}\right). \tag{7.62}$$

This is the only way i_L can fall and reach a lower target (across $t_R{}^-$).

$t_R{}^-$ is longer than in the hysteretic current loop because t_{ON} interrupts i_L's descent. But when t_{ON} is very short, $t_R{}^-$'s nearly match. So this system can respond nearly as fast as the hysteretic current loop.

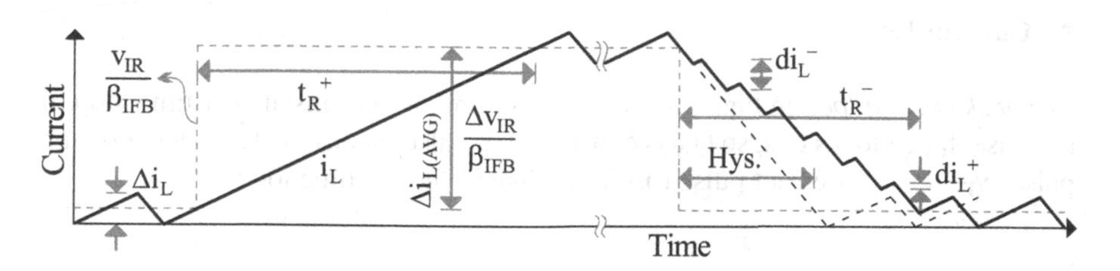

Fig. 7.36 Constant off-time peak response

Explore with SPICE:
See Appendix A for notes on SPICE simulations.

```
*Constant Off-Time Current-Mode Voltage Buck
vin vin 0 dc=4
sei vin vsw de 0 sw1v
ddg 0 vsw idiode
lx vsw vl 10u
vi vl vo 0
co vo 0 5u
ro vo 0 4.33
io vo 0 pwl 0 0 50u 0 50.001u 500m 75u 500m 75.001u 0
vr vr 0 dc=1.2
efb vfb 0 vo 0 0.53
eae veo 0 vr vfb 10
fbifb 0 vifb vi 1
rbifb vifb 0 1
v1v v1v 0 dc=1
xcp vifb veo vio v1v 0 cp
xsroff vtx vio de vqn srs
xdly vqn vtx dly
.ic i(lx)=0
.lib lib.txt
.tran 100u
.end
```

Tip: Plot v(vo), v(vfb), v(veo), v(vifb), and i(Lx) across 100 μs, between 40 and 45 μs, and between 65 and 70 μs.

7.3.5 Constant-Period Peak/Valley Loops

A. Current Loop

The *peak current loop* in Fig. 7.37 is a close sibling of the constant off-time loop. In this case, f_{CLK} clocks t_E's, so t_{SW} is constant. f_{CLK} also breaks the feedback loop that pulses v_E', so v_E' does not pulse (oscillate) like in the off-time loop.

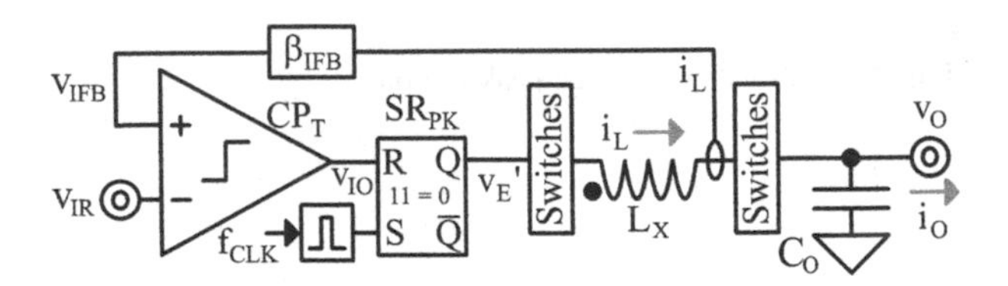

Fig. 7.37 Constant-period peak current loop

The static oscillation is largely the same for both loops. CP_T starts t_D when v_{IFB} overcomes v_{IR} and v_{IFB} falls until, in this case, f_{CLK} ends t_D and starts another t_E. This way, v_{IR} sets $v_{IFB(HI)}$, t_{CLK} fixes t_{SW}, and t_D is a d_D fraction of t_{SW}:

$$t_D = d_D t_{SW} = (1 - d_E)t_{CLK} = \left(\frac{v_E}{v_E + v_D}\right)t_{CLK}. \tag{7.63}$$

t_R^+ is also the same. When v_{IR} rises suddenly, v_{IR} surpasses v_{IFB}, so v_{IO} falls and stays low. This raises and keeps v_E' high because, once f_{CLK} sets v_E', subsequent lows in the S of the reset-dominant flip flop keep v_E' high. In other words, v_E' cannot reset when v_{IO} stays low. So i_L rises across t_R^+ without interruptions.

t_R^- is shorter, however. When v_{IR} falls suddenly, v_{IR} falls below v_{IFB}, so v_{IO} rises and stays high. This keeps v_E' low because the *peak flip flop* SR_{PK} is reset-dominant. In other words, v_E' cannot set when v_{IO} stays high. This way, i_L falls across t_R^- without interruptions.

The *valley current loop* in Fig. 7.38 is the complement of the peak loop. CP_T starts t_E when v_{IFB} falls below v_{IR}, v_{IFB} rises until f_{CLK} ends t_E, and v_E' stays low and high with v_{IO} across t_R's. So v_{IR} sets $v_{IFB(LO)}$, t_{CLK} fixes t_{SW}, t_E is a d_E fraction of t_{SW}, and i_L rises and falls across t_R^+ and t_R^- without interruptions:

$$t_E = d_E t_{SW} = \left(\frac{v_D}{v_E + v_D}\right)t_{CLK}. \tag{7.64}$$

The *valley flip flop* SR_{VL} is set-dominant to ensure f_{CLK} does not reset Q when v_{IO} is high.

B. Sub-harmonic Oscillation

v_{IN} noise can change the di_L/dt that projects i_L across $\Delta v_{IFB}/\beta_{IFB}$, which in turn, can alter t_E or t_D. A temporary rise in v_E's v_{IN}, for example, raises di_L/dt_E in Fig. 7.39. So i_L reaches v_{IR}'/β_{IFB} sooner, falls across a longer t_D, and reaches di_{LO} below the nominal $i_{L(LO)}$.

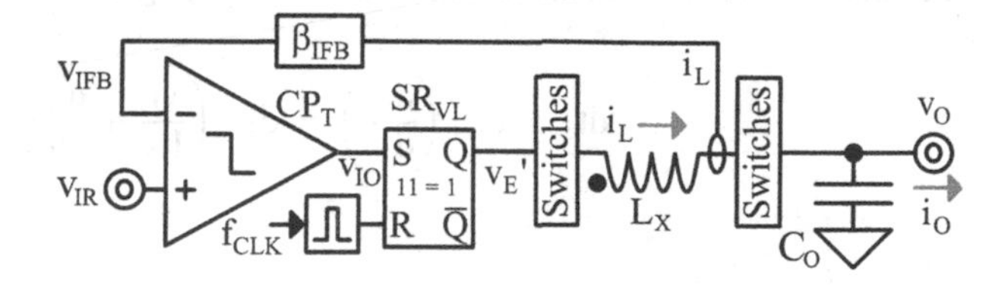

Fig. 7.38 Constant-period valley current loop

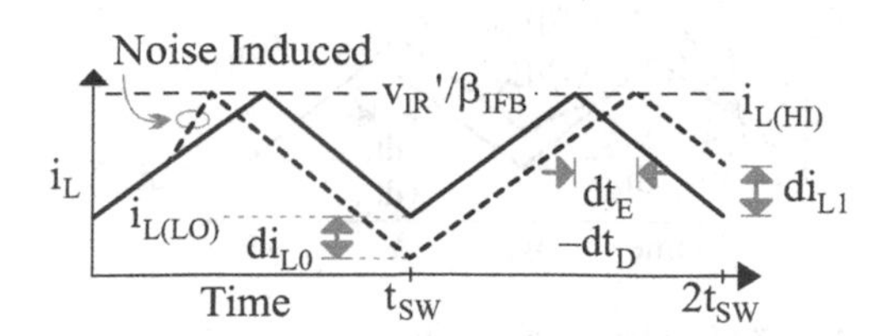

Fig. 7.39 Sub-harmonic oscillation

di_{L0} extends the next t_E that di_L/dt_E projects. This shortens the t_D that follows because t_{SW} is constant. So di_L/dt_D projects i_L above the nominal $i_{L(LO)}$. Since d_E/d_D is v_D/v_E, the new imbalance di_{L1} is an inverting d_E/d_D translation of di_{L0}:

$$\begin{aligned} di_{L1} &= di_{L0}\left(\frac{dt_E}{di_L}\right)\left(\frac{dt_D}{dt_E}\right)\left(\frac{di_L}{dt_D}\right) \\ &= di_{L0}\left(\frac{L_X}{v_E}\right)(-1)\left(\frac{v_D}{L_X}\right) = -di_{L0}\left(\frac{d_E}{d_D}\right) = -di_{L0}\left(\frac{d_E}{1-d_E}\right), \end{aligned} \tag{7.65}$$

where t_E prolongs (dt_E) as much as t_D shortens ($-dt_D$).

Note this imbalance in i_L inverts every cycle and repeats every other cycle. So the frequency of this *sub-harmonic oscillation* is half f_{SW}. Also note that the imbalance shrinks when d_E/d_D is less than one, repeats when d_E/d_D is one, and grows when d_E/d_D is greater than one. This means that oscillations persist and grow when d_E is at or over 50%. Sub-harmonic oscillations fade with time when d_E is below 50%.

C. Slope Compensation

Sloping v_{IR} so v_{IR}''/β_{IFB} in Fig. 7.40 falls with di_L/dt_D fixes the problem. This way, noise across t_E projects i_L to a v_{IR}''/β_{IFB} that aligns and projects i_L back to i_L's normal falling trajectory. And noise across t_D projects an imbalance di_{L0} that di_L/dt_E's v_E/L_X projects to v_{IR}''/β_{IFB}, which realigns i_L across t_D.

Since t_E extends as much as t_D shortens when t_{SW} is constant, the dt_E that di_{L0} extends is also $-dt_D$. So v_{IR}/β_{IFB}'s slope di_L^*/dt generates a t_E imbalance $di_{L(E)}$ that opposes the t_D imbalance $di_{L(D)}$ that di_L/dt_D produces. The resulting di_{L1} is zero when di_L^*/dt matches di_L/dt_D's v_D/L_X:

$$di_{L1} = di_{L(E)} + di_{L(D)} = dt_E\left(\frac{di_L^*}{dt_E}\right) + dt_D\left(\frac{di_L}{dt_D}\right) = dt_E\left(\frac{di_L^*}{dt_E} - \frac{di_L}{dt_D}\right). \tag{7.66}$$

The aggregate slope in di_L/dt_E projects di_{L0} to a variation dt_E in t_E. Interestingly, this dt_E, in turn, induces a $|di_{L1}|$ that is less than $|di_{L0}|$ when di_L^*/dt_E is half di_L/dt_D because $2(d_D/d_E) + 1$ in the resulting di_{L1} is always greater than one, irrespective of d_E:

$$dt_E = di_{L0}\left(\frac{dt_E}{di_L}\right) = di_{L0}\left(\frac{v_E}{L_X} + \frac{di_L^*}{dt_E}\right)^{-1} \tag{7.67}$$

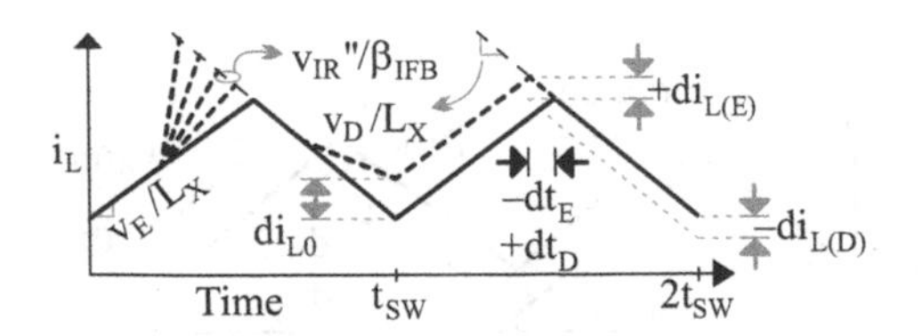

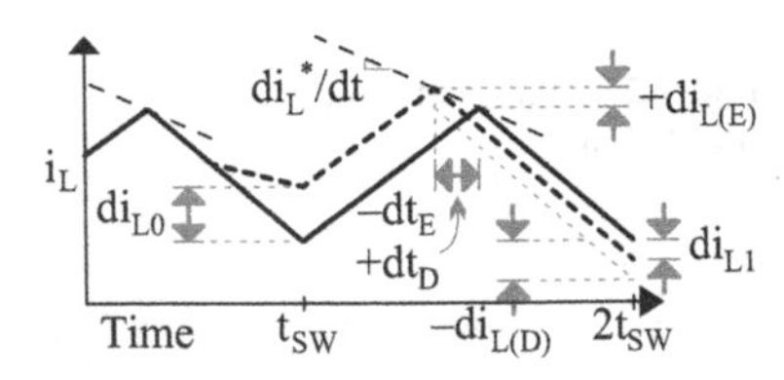

Fig. 7.40 Slope compensation

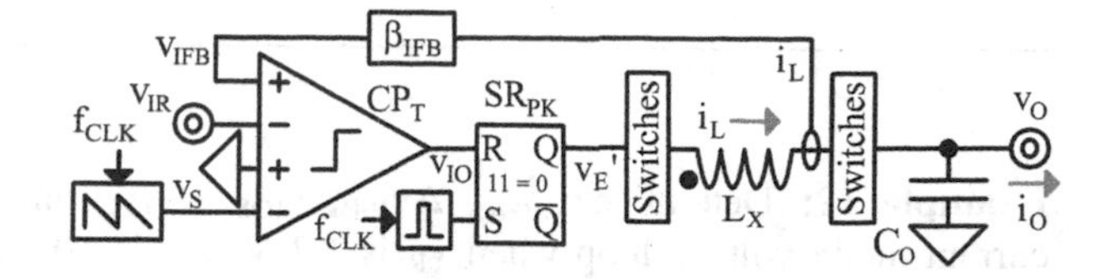

Fig. 7.41 Slope-compensated peak current loop

$$
\begin{aligned}
di_{L1} &= dt_E\left(\frac{di_L{}^*}{dt_E} - \frac{di_L}{dt_D}\right)\Bigg|_{\frac{di_L{}^*}{dt_E}=\frac{di_L}{2dt_D}} \\
&= di_{L0}\left(\frac{v_E + 0.5v_D}{L_X}\right)^{-1}\left(\frac{0.5v_D - v_D}{L_X}\right) \qquad (7.68)\\
&= di_{L0}\left(\frac{-0.5v_D}{v_E + 0.5v_D}\right) = \frac{-di_{L0}}{2(d_D/d_E)+1}.
\end{aligned}
$$

This means that oscillations shrink when $di_L{}^*/dt_E$ is $0.5(v_D/L_X)$. And since subtracting this $di_L{}^*/dt$ from v_{IR}/β_{IFB} when v_{IR} is a v_N in Fig. 7.37 is like adding $di_L{}^*/dt$ to i_L's v_E/L_X when v_{IFB} is a v_P, oscillations also shrink when dv_{IR}/dt is $-0.5(v_D/L_X)\beta_{IFB}$.

The sawtooth in the *slope-compensated peak current loop* in Fig. 7.41 adds this slope (Δv_S across t_{CLK}) to v_{IR}. v_S should fall when v_S and v_{IR} are both v_N's or both v_P's. v_S should rise otherwise, when their terminal polarities oppose.

D. **Offsets**

Since v_{IR} still sets i_L's valley or peak, v_{IOS} in constant-period loops incorporates the same components that set v_{IOS} in constant on/off loops. When included, *slope compensation* adds another offset v_{SOS}. This v_{SOS} is a d_E' fraction of Δv_S over $v_{S(LO)}$ or below $v_{S(HI)}$:

$$v_{SOS} = v_{S(LO/HI)} \pm \Delta v_S d_E'. \qquad (7.69)$$

When fed to a terminal whose polarity matches v_{IR}'s, v_{SOS} adds to the v_{IOS} that Δv_{IFB} and t_P in v_{POS}'s set in constant on/off loops. v_{SOS} subtracts from this v_{IOS} when v_S feeds a terminal whose polarity opposes v_{IR}'s:

$$
\begin{aligned}
v_{IOS} &\equiv v_{IR} - v_{IFB(AVG)} = v_{IOS(ON/OFF)} \pm v_{SOS} \\
&= \left(\mp\frac{\Delta v_{IFB}}{2} \pm v_{IOS\mp}\right) \pm \left(v_{S(LO/HI)} \pm \Delta v_S d_E{}'\right). \qquad (7.70)
\end{aligned}
$$

The polarity of v_{SOS} in v_{IOS} depends on connectivity, not the nature of the slope.

Example 12: Determine $v_{S(HI)}$, Δv_{IFB}, v_{IOS}, v_{VOS}, and $v_{FB(AVG)}$ for the peak current-mode voltage loop when v_R is 1.2 V, A_E is 10 V/V, β_{IFB} is 1 Ω, t_{CLK} is 1 μs, t_P is 100 ns, t_{SR} is 10 ns, v_S ramps from $v_{S(LO)}$, $v_{S(LO)}$ is 200 mV, v_E is 2.2 V, v_D is 1.8 V, L_X is 10 μH, and $i_{L(AVG)}$ is 100–500 mA.

Solution:

$$\Delta v_S = t_{CLK}\left(\frac{dv_{IFB}{}^*}{dt}\right) \equiv t_{CLK}\left(\frac{0.5v_D}{L_X}\right) = (1\mu)\left[\frac{0.5(1.8)}{10\mu}\right] = 90\text{ mV}$$

$$v_{S(HI)} = v_{S(LO)} + \Delta v_S = 200\text{m} + 90\text{m} = 290\text{ mV}$$

$$d_E = \frac{v_D}{v_E + v_D} = \frac{1.8}{2.2 + 1.8} = 45\%$$

$$t_E = d_E t_{CLK} = (45\%)(1\mu) = 450\text{ ns}$$

$$\Delta v_{IFB} = t_E\left(\frac{v_E}{L_X}\right)\beta_{IFB} = (450\text{n})\left(\frac{2.2}{10\mu}\right)(1) = 99\text{ mV}$$

$$t_P{}' = t_P + t_{SR} = 100\text{n} + 10\text{n} = 110\text{ ns}$$

$$v_{SOS} = v_{S(LO)} + d_E\Delta v_S = 200\text{m} + (45\%)(90\text{m}) = 240\text{ mV}$$

$$v_{IOS} = \frac{\Delta v_{IFB}}{2} - t_P{}'\left(\frac{v_E}{L_X}\right)\beta_{IFB} + v_{SOS}$$

$$= \frac{99\text{m}}{2} - (110\text{n})\left(\frac{2.2}{10\mu}\right)(1) + 240\text{ m} = 265\text{ mV}$$

$$v_{VOS} = \frac{i_{L(AVG)}\beta_{IFB} + v_{IOS}}{A_E} = \frac{i_{L(AVG)}(1) + 265\text{m}}{10} = 56 \pm 20\text{ mV}$$

$$v_{FB(AVG)} = v_R - v_{VOS} = 1.2 - v_{VOS} = 1.14\text{ V} \pm 20\text{ mV}$$

Note: Contracting the system increases the loading effect in v_{FB} and load compensation removes it.

Explore with SPICE:
See Appendix A for notes on SPICE simulations.

```
*Constant-Period Peak Current-Mode Voltage Buck
vin vin 0 dc=4
sei vin vsw de 0 sw1v
ddg 0 vsw idiode
lx vsw vl 10u
vi vl vo 0
co vo 0 5u
ro vo 0 18.3
io vo 0 pwl 0 0 1.2m 0 1.2001m 403m 1.4m 403m 1.4001m 0
vr vr 0 dc=1.2 pwl 0 0 1 m 1.2
efb vfb 0 vo 0 0.635
eae veo 0 vr vfb 10
fbifb 0 vifb vi 1
rbifb vifb 0 1
cbifb vifb 0 1p
v1v v1v 0 dc=1
vs vs 0 dc=200m pulse 200m 290m 0998n 1n 1n 1u
xcp vifb veo vs 0 0 0 vio v1v 0 cp3vid
xsrr vclk vio de vqn srr
vclk vclk 0 dc=0 pulse 0 1100n 1n 1n 50n 1u
.lib lib.txt
.tran 1.5m
.end
```

Tip: Wait for simulation to finish (because it may require some time to complete) and plot v(vo), v(vfb), v(veo), v(vs), v(vifb), and i(Lx) across 1.5 ms, between 1.100 and 1.105 ms, and between 1.300 and 1.305 ms.

7.3.6 Design Notes

Constant-time loops are derivatives of hysteretic loops. This is because, like hysteretic loops, they use a comparator to control $v_{IFB(HI)}$ or $v_{IFB(LO)}$. The difference is that an SR flip-flop timer sets the opposite peak. In other words, constant-time loops are flip flop-decoupled hysteretic loops.

The benefit of constant-period peak/valley loops is t_{SW}. They are as fast as hysteretic loops with a constant t_{SW}, which produces predictable f_{SW} noise. The drawback is slope compensation, which complicates the system.

Slope compensation is not always necessary, though. Still, even when d_E is under 50%, systems are more stable with slope compensation. Without it, sub-harmonic oscillations require time to fade.

Fortunately, constant on/off valley/peak loops do not need slope compensation. The trade-off is t_{SW} sensitivity, because t_{SW} varies with v_E and v_D in d_E. This variation, however, is lower than in hysteretic loops because t_{SW} in constant on/off-time loops does not scale with v_T's, t_P, or L_X.

Dynamic accuracy is an important consideration when choosing between constant on/off-time loops. The error between v_{IR} and v_{IFB} or v_R and v_{FB} is typically greater when t_R is longer. So between rising and falling t_R's, the slower di_L/dt transition is usually more limiting. Choosing the scheme that interrupts the faster di_L/dt transition normally sacrifices less accuracy.

Offset is a weakness for constant-time peak/valley loops. $v_{IFB(AVG)}$ is half Δv_{IFB} over or under v_{IR}. This offset, however, is not always a problem, especially when β_{IFB} or β_{FB} accounts for it.

7.4 Oscillating Voltage-Mode Bucks

7.4.1 Resistive Capacitor

A. **Output Voltage**

Bucks connect L_X directly to C_O and v_O. So i_L in Fig. 7.42 feeds i_O to the load and i_C to C_O. In steady state, when i_O is static, the load receives i_L's dc average $i_{L(DC)}$ or $i_{L(AVG)}$ and C_O receives i_L's alternating ripple $i_{L(AC)}$ or Δi_L.

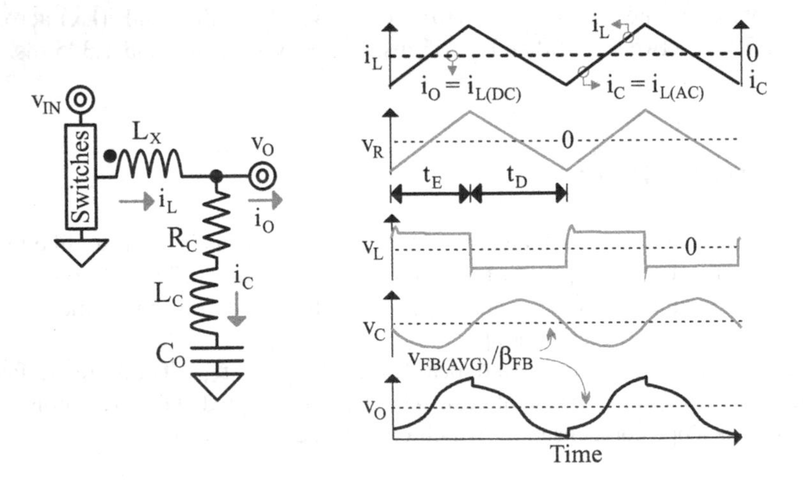

Fig. 7.42 Buck output

In practice, C_O incorporates unintended *equivalent series resistance* and *inductance* R_C and L_C. Since i_C carries $i_{L(AC)}$, R_C, L_C, and C_O drop ac voltages about C_O's dc average $v_{C(DC)}$ or $v_{C(AVG)}$. This $v_{C(DC)}$ sets v_O's dc average $v_{O(DC)}$ or $v_{O(AVG)}$.

v_O's alternating ripple Δv_O or $v_{O(AC)}$ is the superimposed sum of R_C's, L_C's, and C_O's ripples. Since i_C carries i_L's triangular ripple, v_R is triangular, v_L pulses positive and negative with di_L/dt_E and di_L/dt_D across t_E and t_D, and v_C rises and falls parabolically when $i_{L(AC)}$ is positive and negative:

$$\begin{aligned}\Delta v_O \equiv v_{O(AC)} &= v_R + v_L + v_{C(AC)} \\ &= i_{L(AC)}R_C + L_C\left(\frac{di_L}{dt_{E/D}}\right) + \int \frac{i_{L(AC)}}{C_O}dt.\end{aligned} \tag{7.71}$$

v_R in resistive C_O's overwhelms v_L and $v_{C(AC)}$. So v_O ripples with v_R about v_C's average. This way, Δi_L sets v_O's ripple to $\Delta i_L R_C$. This means, v_O in steady state rises and falls with i_L.

B. Comparator Loops

Hysteretic and constant-time current loops respond quickly because comparators generate the error that adjusts i_L. Using comparators for this purpose is possible because v_{IFB} rises and falls across t_E and t_D. This way, comparators can start and end t_E when v_{IFB} reaches $v_{IFB(HI)}$ and $v_{IFB(LO)}$.

Since resistive capacitors in bucks produce v_O's that rise and fall across t_E and t_D, comparators can close similar voltage loops. Hysteretic loops can start t_E when v_O reaches $v_{O(LO)}$ and end t_E when v_O reaches $v_{O(HI)}$. Valley loops can start t_E the same way and end t_E after t_{ON} or t_{CLK} elapses. And peak loops can end t_E the same way and start t_E after t_{OFF} or t_{CLK} elapses.

The *resistive voltage-mode buck* in Fig. 7.43 embodies this principle. β_{FB} scales v_O to v_{FB} and the *error comparator* CP_E compares v_{FB} to v_R. The v_{EO} that CP_E outputs adjusts i_L so v_{FB} and v_O ripple about v_R and v_R/β_{FB}, respectively. CP_E and the switcher can close hysteretic, valley, or peak loops this way.

In valley loops, v_R and v_{FB} connect to CP_E's v_P and v_N and v_{EO} sets a clocked or constant on-time flip flop. Peak loops connect v_R to v_N and v_{FB} to v_P and v_{EO} resets a clocked or constant off-time flip flop. And CP_E in hysteretic loops is a hysteretic comparator with v_R and v_{FB} connected to v_P and v_N.

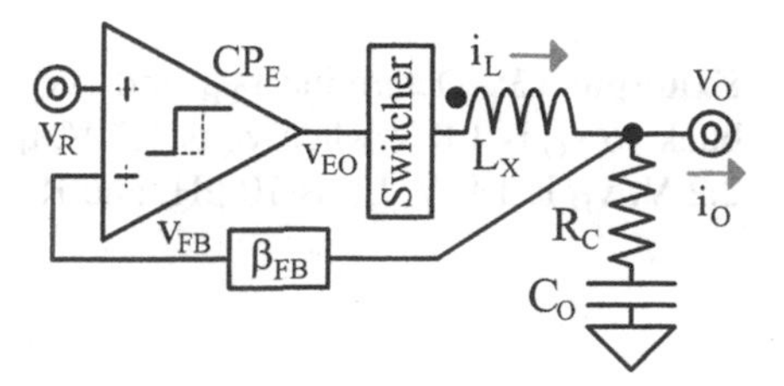

Fig. 7.43 Resistive voltage-mode buck

C. **Oscillating Period**

Constant-time t_{SW}'s are the same in current and voltage loops. t_{SW} in hysteretic loops is also the same, but with a different β_{IFB}. In the resistive buck, R_C and β_{FB} convert Δi_L to Δv_O and Δv_O to Δv_{FB}. So $R_C\beta_{FB}$ translates di_L/dt projections v_{POS}'s beyond CP_E's Δv_T in Δv_{FB} and Δi_L to Δv_{FB} in t_{SW}:

$$\Delta v_{FB} = \Delta v_O\beta_{FB} = \Delta v_T + v_{POS}^+ + v_{POS}^- \approx \Delta v_T + t_P\left(\frac{v_E + v_D}{L_X}\right)R_C\beta_{FB} \quad (7.72)$$

$$t_{SW} = \frac{t_E}{d_E} = \left(\frac{\Delta i_L}{d_E}\right)\left(\frac{dt_E}{di_L}\right) = \left(\frac{\Delta i_L}{d_E}\right)\left(\frac{L_X}{v_E}\right) = \left(\frac{\Delta v_{FB}}{\beta_{FB}R_C}\right)\left(\frac{L_X}{v_E||v_D}\right). \quad (7.73)$$

v_O's ripple Δv_O is a β_{FB} translation of this Δv_{FB}.

D. **Offsets**

v_{FB} in hysteretic loops ripples about v_R when di_L/dt's projections across t_P's match. $R_C\beta_{FB}$ translates asymmetric di_L/dt projections to a *hysteretic offset* v_{HOS} that shifts $v_{FB(AVG)}$ away from v_R. This v_{HOS} on v_{FB} is low when t_P's in v_{POS}'s are short:

$$v_{HOS} = \frac{v_{POS}^+ - v_{POS}^-}{2} \approx \left(\frac{t_P}{2}\right)\left(\frac{dv_{FB}}{dt_E} - \frac{dv_{FB}}{dt_D}\right) \approx t_P\left(\frac{v_E - v_D}{2L_X}\right)R_C\beta_{FB}. \quad (7.74)$$

v_{HOS} subtracts from v_{VOS} because, as defined, this offset raises v_{FB}.

In constant-time loops, v_R is half Δv_{FB} and a dv_{FB}/dt offset projection away from $v_{FB(AVG)}$. $R_C\beta_{FB}$ translates the di_L/dt projections that set Δv_{FB} and $v_{POS}^{+/-}$. So Δv_{FB} and v_{POS}'s are

$$\Delta v_{FB} = t_{E/D}\left(\frac{dv_{FB}}{dt_{E/D}}\right) = t_{E/D}\left(\frac{di_L}{dt_{E/D}}\right)R_C\beta_{FB} = t_{E/D}\left(\frac{v_{E/D}}{L_X}\right)R_C\beta_{FB} \quad (7.75)$$

$$v_{POS}^{\pm} = t_P'\left(\frac{dv_{FB}}{dt_{E/D}}\right) = t_P'\left(\frac{v_{E/D}}{L_X}\right)R_C\beta_{FB}. \quad (7.76)$$

Example 13: Determine β_{FB} and Δv_O for the constant off-time peak-controlled buck so v_O is 1.8 V when v_R is 1.2 V, t_P is 100 ns, t_{SR} is 10 ns, t_X^+ is 450 ns, v_E is 2.2 V, v_D is 1.8 V, L_X is 10 μH, and R_C is 1 Ω.

Solution:

$$t_D = t_{OFF} = t_X{}^+ + t_{SR} = 450n + 10n = 460\ ns$$

$$t_P' = t_P + t_{SR} = 100n + 10n = 110\ ns$$

$$\Delta v_{FB} = t_D\left(\frac{v_D}{L_X}\right)R_C\beta_{FB} = (460n)\left(\frac{1.8}{10\mu}\right)(1)\beta_{FB} = 8.3\%\beta_{FB}$$

$$v_{POS}{}^+ = t_P'\left(\frac{v_E}{L_X}\right)R_C\beta_{FB} = (110n)\left(\frac{2.2}{10\mu}\right)(1)\beta_{FB} = 2.4\%\beta_{FB}$$

$$v_{VOS} = \frac{\Delta v_{FB}}{2} - v_{POS}{}^+ = \left(\frac{8.3\%}{2} - 2.4\%\right)\beta_{FB} = 1.8\%\beta_{FB}$$

$$\beta_{FB} \equiv \frac{v_{FB(AVG)}}{v_{O(AVG)}} = \frac{v_R - v_{VOS}}{v_{O(AVG)}} = \frac{1.2 - 1.8\%\beta_{FB}}{1.80} = 66\%$$

$$\Delta v_O = \frac{\Delta v_{FB}}{\beta_{FB}} = \frac{8.3\%\beta_{FB}}{\beta_{FB}} = 83\ mV$$

Note: R_C sets v_{FB}'s and v_O's ripples Δv_{FB} and Δv_O.

Explore with SPICE:
See Appendix A for notes on SPICE simulations.

```
*RC-Sensed Voltage-Mode Constant Off-Time Buck
vin vin 0 dc=4
si vin vsw de 0 sw1v
dg 0 vsw idiode
lx vsw vo 10u
co vo vc 5u
rc vc 0 1
ro vo 0 18
io vo 0 pwl 0 0100u 0101u 400m 120u 400m 121u 0
vr vr 0 dc=1.2 pwl 0 0 50u 1.2
efb vfb 0 vo 0 0.66
cvfb vfb 0 1p
```

(continued)

```
v1v v1v 0 dc=1
xcp vfb vr veo v1v 0 cp
xsroff vtx veo de vqn srs
xdly vqn vtx dly
.ic i(lx)=0
.lib lib.txt
.tran 150u
.end
```

Tip: Plot v(vo), v(vfb), and i(Lx) across 150 μs, between 80 and 85 μs, and between 115 and 120 μs.

7.4.2 RC Filter

A. **Comparator Loops**

The *filtered voltage-mode buck* in Fig. 7.44 is a variation of the resistive buck. Since L_X shorts and C_F opens at low frequency, v_F's average follows that of the *switch-node voltage* v_{SW}, which in turn follows v_O's. So $v_{F(AVG)}$, $v_{SW(AVG)}$, and $v_{O(AVG)}$ match, $v_{FB(AVG)}$ nears v_R, and $v_{O(AVG)}$ is close to v_R/β_{FB}.

R_F and C_F ripple v_F like R_C in the resistive buck ripples v_O. When C_F shorts with respect to R_F past an f_F that is well below f_{SW}, v_F steadies across t_{SW}. So across t_{SW}, R_F drops $v_{SW(E/D)} - v_{F(AVG)}$, which nears the $v_{IN} - v_{O(AVG)}$ and $0 - v_{O(AVG)}$ that sets v_E and v_D. This means that v_E/R_F and v_D/R_F slew C_F up and down across t_E and t_D:

$$f_F = \frac{1}{2\pi R_F C_F} << f_{SW} \quad (7.77)$$

$$i_C{}^{\pm} = \frac{v_R{}^{\pm}}{R_F}\bigg|_{f_{SW}>>f_F} \approx \frac{v_{SW(E/D)} - v_{F(AVG)}}{R_F} \approx \frac{v_{SW(E/D)} - v_{O(AVG)}}{R_F} = \frac{v_{E/D}}{R_F}. \quad (7.78)$$

This way, with an i_C that v_E and v_D set, v_F rises across t_E and falls across t_D. This is what CP_E needs to oscillate i_L. Notice $v_{E/D}/R_FC_F$ ripples v_F like $(v_{E/D}/L_X)R_C$ ripples v_O in the resistive buck:

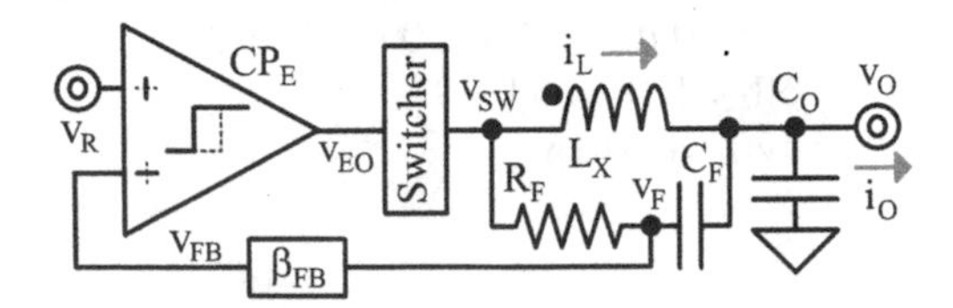

Fig. 7.44 Filtered voltage-mode buck

$$\Delta v_F = \left(\frac{dv_F}{dt_{E/D}}\right)t_{E/D} = \left(\frac{i_C^{\pm}}{C_F}\right)t_{E/D} \approx \left(\frac{v_{E/D}}{R_F C_F}\right)t_{E/D}. \quad (7.79)$$

R_F and C_F, however, decouple v_O's ripple from v_{FB}'s. Decoupled this way, Δv_O can be lower than the Δv_{FB} that CP_E needs to distinguish v_{FB} from noise. With high-quality (low-R_C) C_O's, Δv_O is the v_C that C_O sets with $i_{L(AC)}$, which can be very low. Since $i_{L(AC)}$ is positive for half of t_E and half of t_D and negative for the other halves, v_O rises and falls across

$$\Delta v_O \approx \int_0^{0.5t_E+0.5t_D} \frac{i_{L(AC)}}{C_O} dt = \int_0^{0.5t_E} \frac{v_E t}{L_X C_O} dt + \int_0^{0.5t_D} \frac{v_D t}{L_X C_O} dt = \frac{v_E {t_E}^2 + v_D {t_D}^2}{8 L_X C_O}. \quad (7.80)$$

B. Oscillation Period

Constant-time t_{SW}'s are the same in current and voltage loops. t_{SW} in hysteretic loops is also the same, but with a different β_{IFB}. In the filtered buck, $v_{E/D}/R_F C_F$ slews v_F and β_{FB} converts Δv_F to Δv_{FB}. So $v_{E/D}\beta_{FB}/R_F C_F$ projects v_{FB} over CP_E's Δv_T in Δv_{FB} and across Δv_{FB} in t_{SW}:

$$\Delta v_{FB} = \Delta v_F \beta_{FB} = \Delta v_T + {v_{POS}}^+ + {v_{POS}}^- \approx \Delta v_T + t_P\left(\frac{dv_F}{dt_E} + \frac{dv_F}{dt_D}\right)\beta_{FB} \approx \Delta v_T + t_P\left(\frac{v_E + v_D}{R_F C_F}\right)\beta_{FB} \quad (7.81)$$

$$t_{SW} = \frac{t_E}{d_E} = \left(\frac{v_E + v_D}{v_D}\right)\Delta v_F\left(\frac{dt_E}{dv_F}\right) \approx \left(\frac{\Delta v_{FB}}{\beta_{FB}}\right)\left(\frac{R_F C_F}{v_E || v_D}\right). \quad (7.82)$$

C. Offsets

v_{FB} in hysteretic loops ripples about v_R when dv_F/dt's projections across t_P's match. Asymmetric dv_F/dt projections produce a hysteretic offset v_{HOS} that shifts $v_{FB(AVG)}$ away from v_R. This v_{HOS} on v_{FB} is low when t_P in v_{POS}'s is short:

$$v_{HOS} = \frac{{v_{POS}}^+ - {v_{POS}}^-}{2} \approx \left(\frac{t_P}{2}\right)\left(\frac{dv_F}{dt_E} - \frac{dv_F}{dt_D}\right)\beta_{FB} \approx \left(\frac{t_P}{2}\right)\left(\frac{v_E - v_D}{R_F C_F}\right)\beta_{FB}. \quad (7.83)$$

v_{HOS} subtracts from v_{VOS} because, as defined, this offset raises v_{FB}.

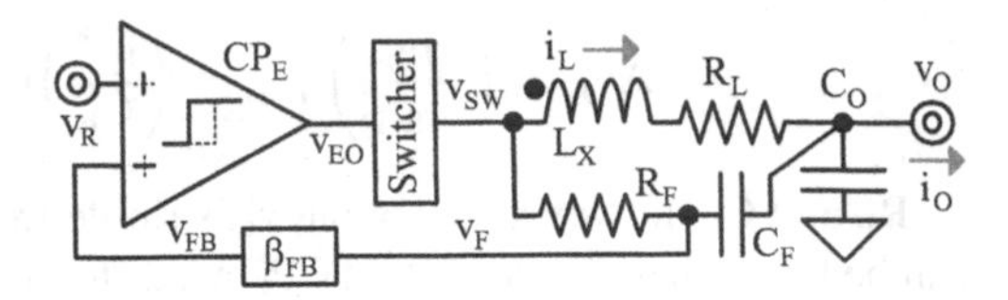

Fig. 7.45 Filtered voltage-mode buck with resistive inductor

In constant-time loops, v_R is half Δv_{FB} and a dv_F/dt offset projection away from $v_{FB(AVG)}$. β_{FB} translates the dv_F/dt projections that set Δv_{FB} and $v_{POS}{}^{+/-}$. These Δv_{FB} and $v_{POS}{}^{+/-}$ are

$$\Delta v_{FB} = \Delta v_F \beta_{FB} \approx t_{E/D}\left(\frac{v_{E/D}}{R_F C_F}\right)\beta_{FB} \tag{7.84}$$

$$v_{POS}{}^{\pm} = t_P\left(\frac{dv_F}{dt_{E/D}}\right)\beta_{FB} = t_P{}'\left(\frac{v_{E/D}}{R_F C_F}\right)\beta_{FB}. \tag{7.85}$$

In practice, L_X incorporates an unintended *equivalent series resistance* R_L. R_L's dc component $R_{L(DC)}$ in Fig. 7.45 reduces $v_{O(AVG)}$ below $v_{SW(AVG)}$ and $v_{F(AVG)}$ with the $i_{L(AVG)}$ that supplies i_O. This loading effect reduces $v_{FB(AVG)}$ by v_{LD}, which is the effect of subtracting v_{LD} from $v_{FB(AVG)}$ and adding v_{LD} to v_{VOS}:

$$v_{LD} = i_{L(AVG)} R_{L(DC)} \beta_{FB} = i_O R_{L(DC)} \beta_{FB}. \tag{7.86}$$

This offset is fairly low with high-quality (low-R_L) inductors.

Example 14: Determine R_F, Δv_T, β_{FB}, Δv_F, Δv_{FB}, Δv_O, and v_O's error for the filtered hysteretic buck so v_O is 1.8 V and f_{SW} is 1 MHz when v_R is 1.2 V, t_P is 100 ns, C_F is 20 pF, v_E is 2.2 V, v_D is 1.8 V, L_X is 10 μH, R_L is 100 mΩ, and $i_{L(AVG)}$ is 100–500 mA.

Solution:

$$f_S = \frac{1}{2\pi R_F C_F} = \frac{1}{2\pi R_F (20p)} \equiv \frac{f_{SW}}{100} = \frac{1M}{100} \quad \therefore \quad R_F = 800\ k\Omega$$

$$v_{HOS} \approx t_P\left(\frac{v_E - v_F}{2R_F C_F}\right)\beta_{FB} = \frac{(100n)(2.2 - 1.8)\beta_{FB}}{2(800k)(20p)} = 0.12\%\beta_{FB}$$

$$v_{LD} = i_O R_L \beta_{FB} = i_{L(AVG)}(100m)\beta_{FB} = 3.0\% \pm 2.0\%\ \text{of}\ \beta_{FB}$$

$$v_{VOS} = v_{LD} - v_{HOS} = (3.0\% \pm 2.0\% - 0.12\%)\beta_{FB} = 2.9\% \pm 2.0\% \text{ of } \beta_{FB}$$

$$\beta_{FB} \equiv \frac{\overline{v_{FB}}}{v_O} = \frac{v_R - \overline{v_{VOS}}}{v_O} = \frac{1.2 - 2.9\%\beta_{FB}}{1.80} = 66\%$$

$$\therefore v_{HOS} = 790\ \mu V, v_{LD} = 20 \pm 13\ mV, v_{VOS} = 19 \pm 13\ mV,$$
$$\text{and } v_O = 1.79V \pm 20mV$$

$$t_{SW} = \left(\frac{\Delta v_{FB}}{\beta_{FB}}\right)\left(\frac{R_F C_F}{v_E || v_D}\right) = \frac{\Delta v_{FB}(800k)(20p)}{66\%(2.2||1.8)} = \frac{1}{f_{SW}} \equiv 1\ \mu s$$

$$\therefore \Delta v_{FB} \approx \Delta v_T + t_P\left(\frac{v_E + v_D}{R_F C_F}\right)\beta_{FB}$$
$$= \Delta v_T + \frac{(100n)(2.2 + 1.8)(66\%)}{(800k)(20p)} = 41mV \rightarrow \Delta v_T = 24\ mV$$

$$\Delta v_F = \frac{\Delta v_{FB}}{\beta_{FB}} = \frac{41m}{66\%} = 62\ mV$$

$$t_E = d_E t_{SW} = \left(\frac{v_D}{v_E + v_D}\right)t_{SW} = \left(\frac{1.8}{2.2 + 1.8}\right)(1\mu) = 450\ ns$$

$$\Delta v_O \approx \frac{v_E t_E{}^2 + v_D t_D{}^2}{8L_X C_O} \approx \frac{v_E t_E{}^2 + v_D(t_{SW} - t_E)^2}{8L_X C_O}$$
$$= \frac{(2.2)(450n)^2 + (1.8)(1\mu - 450n)^2}{8(10\mu)(5\mu)} = 2.5\ mV$$

Note: Δv_O is much lower than Δv_F, which is large enough to overwhelm millivolt noise.

Explore with SPICE:
See Appendix A for notes on SPICE simulations.

```
*Filtered Voltage-Mode Hysteretic Buck
vin vin 0 dc=4
sei vin vsw de 0 sw1v
ddg 0 vsw idiode
lx vsw vl 10u
```

(continued)

```
rl vl vo 100m
co vo 0 5u
ro vo 0 18
io vo 0 pwl 0 0 70u 0 71u 400m 120u 400m 121u 0
vr vr 0 dc=1.2 pwl 0 0 20u 1.2
rf vsw vf 800k
cf vf vo 20p
efb vfb 0 vf 0 0.66
v1v v1v 0 dc=1
xhys vr vfb de v1v 0 cphys25m
.ic i(lx)=0
.lib lib.txt
.tran 150u
.end
```

Tip: Plot v(vo), v(vf), v(vfb), and i(Lx) across 150 μs, between 60 and 65 μs, and between 110 and 115 μs.

D. **Voltage-Mode Voltage Loop**

The *voltage-mode voltage-looped buck* in Fig. 7.46 uses the filtered buck like a current loop, but as a voltage buffer instead of a transconductor. In other words, β_{FB2} scales v_O to v_{FB2}, A_{E2} compares v_{FB2} to v_R, and the filtered buck translates v_{EO2} to $v_{O(AVG)}$ so $v_{SW(AVG)}$ nears v_{EO2}/β_{FB1}. This way, the feedback action of A_{E2} sets v_{FB2} near v_R and v_O close to v_R/β_{FB2}.

v_{FB2} and v_{EO2} are steady because C_O suppresses v_O's ripple. CP_{E1}, however, needs this v_{EO2} to carry a β_{FB1} translation of $v_{F(AVG)}$. v_R and v_{FB2} must generate this v_{EO2} plus the other offsets the filtered buck adds to v_{VOS}. Since $v_{F(AVG)}$ nears $v_{O(AVG)}$, this *loop offset* v_{LOS} is close to $v_{O(AVG)}\beta_{FB1}$:

$$\begin{aligned} v_{LOS} &= v_{EO2}\big|_{v_{LD}=v_{IOS}=0} \\ &\approx v_{FB1(AVG)} = v_{F(AVG)}\beta_{FB1} = v_{SW(AVG)}\beta_{FB1} \approx v_{O(AVG)}\beta_{FB1}. \end{aligned} \tag{7.87}$$

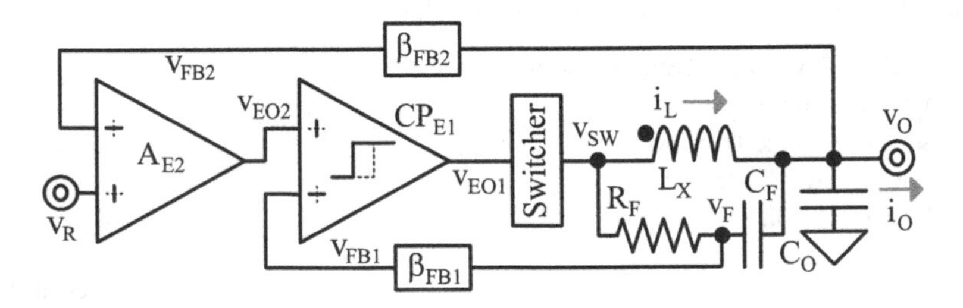

Fig. 7.46 Voltage-mode voltage-looped buck

v_{EO2} also carries the v_{IOS} and v_{LD} that t_P's and R_L induce. Since v_R and v_{FB2} with A_{E2} generate v_{EO2}, v_{VOS} is a reverse A_{E2} translation of v_{LD}, v_{IOS}, and v_{LOS}. The overall offset is low when A_{E2} is high:

$$v_{VOS} \equiv v_R - v_{FB(AVG)} = \frac{v_{EO2}}{A_{E2}} = \frac{v_{LD} + v_{IOS} + v_{LOS}}{A_{E2}}. \quad (7.88)$$

The ultimate aim of the second loop is to suppress the effects of v_{IOS} and v_{LD}. In practice, R_L is low and v_{LOS} is much greater than v_{OS} and v_{LD}. A_{E2} should therefore reduce v_{LOS} below v_{OS} and v_{LD} to reap the benefits of A_{E2}.

But since β_{FB1} and β_{FB2} can account for v_{VOS}'s mean, the real advantage of A_2 is to suppress v_{LD}'s variable loading effect. In this light, lower-quality (high-R_L) inductors subjected to heavy loads stand to benefit the most. A popular name for this double voltage-loop strategy is *voltage squared.*

Example 15: Determine β_{FB2} and v_O's error for a voltage-squared buck that uses the filtered hysteretic buck in Example 14 so v_O is 1.8 V when A_{E2} is 10 V/V.

Solution:

$$v_{HOS} = 790\ \mu V \text{ and } v_{LD} = 20 \pm 13 \text{ mV from Example 14}$$

$$v_{LOS} \approx v_{O(AVG)}\beta_{FB1} = (1.8)(66\%) = 1.2 \text{ V}$$

$$v_{VOS} = \frac{v_{LD} - v_{HOS} + v_{LOS}}{A_{E2}} \approx \frac{20m \pm 13m - 790\mu + 1.2}{10} = 120 \pm 1.3 \text{ mV}$$

$$\beta_{FB2} \equiv \frac{\overline{v_{FB2}}}{v_O} = \frac{v_R - \overline{v_{VOS}}}{v_O} = \frac{1.2 - 120m}{1.8} = 60\%$$

$$v_O = \frac{v_{FB2}}{\beta_{FB2}} = \frac{v_R - v_{VOS}}{\beta_{FB2}} = \frac{1.2 - v_{VOS}}{60\%} = 1.8 \text{ V} \pm 2 \text{ mV}$$

Explore with SPICE:
See Appendix A for notes on SPICE simulations.

```
*Voltage-Mode Voltage-Looped Hysteretic Buck
vin vin 0 dc=4
sei vin vsw de 0 sw1v
```

(continued)

```
ddg 0 vsw idiode
lx vsw vl 10u
rl vl vo 100m
co vo 0 5u
ro vo 0 18
io vo 0 pwl 0 0 70u 0 71u 400m 120u 400m 121u 0
vr vr 0 dc=1.2 pwl 0 0 20u 1.2
efb1 vfb1 0 vo 0 0.60
eae1 veo1 0 vr vfb1 10
rf vsw vf 800k
cf. vf vo 20p
efb2 vfb2 0 vf 0 0.66
v1v v1v 0 dc=1
xhys veo1 vfb2 de v1v 0 cphys25m
.ic i(lx)=0
.lib lib.txt
.tran 150u
.end
```

Tip: Plot v(vo), v(vf), v(vfb1), v(vfb2), and i(Lx) across 150 μs, between 60 and 65 μs, and between 110 and 115 μs.

7.4.3 Design Notes

Resistive bucks are fast, compact, and low cost. The drawback is R_C. This is because resistive C_O's burn more Ohmic power. They also drop higher voltages. v_O's ripple and dynamic excursions when v_R or load variations outpace the loop are therefore higher.

Filtered bucks are more efficient. Dynamic accuracy is also better, because they can use higher-quality C_O's. Their disadvantage is static accuracy, because v_O falls with the i_O that $i_{L(AVG)}$ feeds. Voltage-squared loops can reduce this variation with another amplifier. But the reduction in the loading effect i_O and R_L produce should justify the additional silicon area and power A_{E2} requires.

7.5 Summary

Most feedback controllers descend from PWM and hysteretic loops. PWM loops average the signals that set their outputs, so they require multiple switching cycles to respond. Hysteretic loops oscillate the i_L about a v_R that sets their outputs. These loops are faster because they slew i_L without interruptions to their new targets.

The drawback of hysteretic implementations is t_{SW}'s sensitivity to v_T's, t_P, L_X, v_E, and v_D. Constant on/off-time valley/peak loops are almost as fast as hysteretic loops without t_{SW}'s sensitivity to v_T's, t_P, and L_X. Constant-period loops are better: just as fast as hysteretic loops with a constant t_{SW}. The caveat is, they need slope compensation when d_E reaches or surpasses 50%, which requires a sawtooth.

Summing comparators can contract these loops when A_E is unnecessary. Although offsets are higher without this A_E, β_{FB} can account for their mean. And summing comparators can subtract loading effects from current-mode voltage loops. But removing A_E is possible only when loops do not need stabilizers to remain stable.

Resistive voltage-mode bucks are fast, compact, and low cost. The problem is higher R_C's sacrifice power and dynamic accuracy. Filtered bucks do not need high R_C's, but their R_L's drop v_O with the i_O they supply, which degrades static accuracy. Enclosing a second loop reduces this loading effect. The tradeoffs here are power and silicon area.

Oscillating voltage-mode bucks are possible because v_O and reflections of v_O rise and fall across t_E and t_D. This is how hysteretic, valley, and peak loops can start and/or end t_E. Since voltage-mode boosts and buck–boosts disconnect L_X from their outputs when L_X drains across t_D, deriving rising and falling components is less straightforward. Boosts and buck–boosts usually do not benefit from hysteretic, peak, or valley voltage-mode control for this reason.

Building Blocks

8

Abbreviations

ADC	Analog–digital converter
DCM	Discontinuous-conduction mode
EMI	Electromagnetic interference
ICMR	Input common-mode range
LED	Light-emitting diode
OCP	Over-current protection
PM	Phase margin
SL	Switched inductor
SR	Set–reset flip flop
ZCD	Zero-current detector
A_0	Low-frequency gain
A_β	Feedback gain
A_E	Error amplifier
A_F	Forward gain
A_{IE}	Current-error amplifier
A_{LG}	Loop gain
A_M	Mirror gain
A_S	Stabilizing filter response
β_{IFB}	Current-feedback translation
$\beta_{IFB(MF)}$	Moderate-frequency current-feedback translation
$\beta_{IFB(LF)}$	Low-frequency current-feedback translation
β_{FB}	Voltage-feedback translation
C_{CH}	Channel capacitance
C_D	Drain capacitance
C_{EI}	Error amplifier's input capacitance
C_G	Gate capacitance
C_{GI}	Input gate capacitance

G. A. Rincón-Mora, *Switched Inductor Power IC Design*,
https://doi.org/10.1007/978-3-030-95899-2_8

C_{GO}	Output gate capacitance
C_O	Output capacitance
C_{OX}''	Oxide capacitance (per unit area)
CP_{IE}	Current-error comparator
D_{DG}	Ground drain diode
D_{DO}	Output drain diode
d_E'	Duty-cycle command
D_H	High-side diode
D_L	Low-side diode
f_{0dB}	Unity-gain frequency
f_{LC}	Transitional LC (resonant) frequency
f_o	Interstage fan out
F_O	Total fan out
f_{RC}	RC frequency
f_{RL}	RL frequency
I_B	Bias current
i_C	Collector current
i_E	Error current
i_{FB}	Feedback current
i_L	Inductor current
$i_{L(D)}$	Inductor drain current
i_O	Output current
i_P	Power-transfer current
i_S	Sense current
I_S	Saturation current
i_{ST}	Shoot-through current
k_{SL}	Self-loading coefficient
L	Length
L_{OL}	Overlap length
L_X	Switched inductor
λ	Channel-length modulation parameter
M_H	High-side switch
M_L	Low-side switch
M_P	Power transistor
M_S	Sense transistor
p_A	Amplifier pole
p_{BW}	Bandwidth-setting pole
P_{DRV}	Gate-driver power
P_{IV}	i_{DS}–v_{DS} overlap power
P_G	Gate-charge power
P_Q	Quiescent power
P_{ST}	Shoot-through power
Q	Current state
$\overline{Q}$	Complementary/opposite state

Q^{-1}	Previous state
q_G	Gate charge
R_D	Drain resistance
R_{DS}	Drain–source resistance
R_L	Inductor's equivalent series resistance
R_{LD}	Load resistance
R_O	Output resistance
R_S	Series sense resistor
Σv_{ID}	Differential sum
t_{CLK}	Clock period
t_D	Drain time
t_{DT}	Dead time
t_I	Inverter-chain delay
t_E	Energize time
t_{LC}	LC period
t_{OSC}	Oscillation period
t_P	Propagation delay
t_{PW}	Pulse width
t_R	Response/reset time
t_S	Suppression time
t_{ST}	Shoot-through time
t_{SW}	Switching period
τ_{LC}	LC time constant
τ_{RC}	RC time constant
v_B	Base voltage
v_{BE}	Base–emitter voltage
V_{BG}	Band-gap voltage
v_{DD}	Positive power supply
$v_{DS(SAT)}$	Saturation voltage
v_E	Error voltage
v_{FB}	Feedback voltage
v_G	Gate voltage
v_H	High input
v_I	Input
v_{ID}	Differential input voltage
v_{IFB}	Current-feedback voltage
v_{IN}	Input voltage
v_L	Inductor voltage/low input
v_n	Noise voltage
v_N	Negative input
v_O	Output/output voltage
$V_{OS(S)}$	Systemic offset voltage
v_P	Positive input
v_R	Reference voltage
v_S	Sawtooth voltage

v_{SS}	Negative power supply
v_{SWI}	Input switching node/voltage
v_{SWO}	Output switching node/voltage
v_T	Trip point
V_t	Thermal voltage
V_{T0}	Zero-bias threshold voltage
$v_{T(HI)}$	Rising threshold
$v_{T(LO)}$	Falling threshold
v_{VOS}	Voltage-loop offset
W	Width
Z_{LD}	Load impedance

Switched-inductor (SL) dc–dc power supplies draw and deliver power. They are *regulators* when *feedback controllers* switch them so their outputs follow a target. *Voltage regulators* regulate output voltage, *battery chargers* and *light-emitting diode* (LED) *drivers* regulate output current, *battery-charging voltage regulators* regulate both, and *energy harvesters* regulate input power.

Good power supplies are *efficient* and *accurate*. Efficient supplies lose a small fraction of the power they draw. And accurate supplies deliver the rest with a voltage or current that is very close to a target. Supplying power at a particular voltage or with a particular current like this is a form of *power conditioning*.

Power conditioning requires several actions. Some of these are analog in nature, others are digital, and those in between are mixed-mode. The building blocks that power supplies use reflect this diversity.

8.1 1. Current Sensors

8.1.1 Series Resistance

A. Sense Resistor

Inserting a *series sense resistor* R_S into the conduction path is one way of sensing current. R_S in Fig. 8.1 can be in series with the *switched inductor* L_X, the *output* v_O,

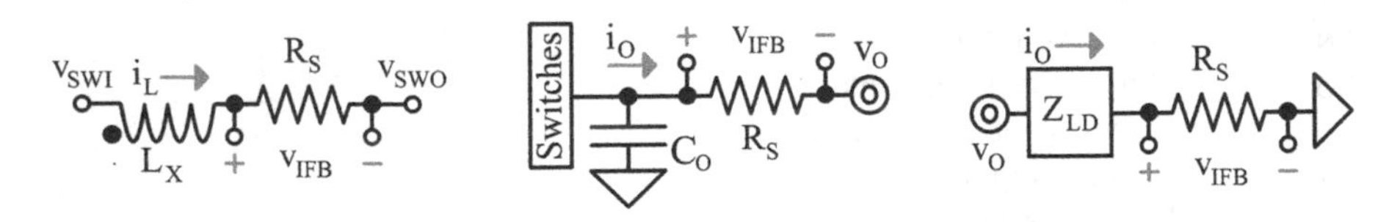

Fig. 8.1 Sense resistor

or the *load* Z_{LD}. In all cases, the *current-feedback translation* β_{IFB} that senses v_O's *output current* i_O or L_X's *inductor current* i_L is R_S:

$$\beta_{IFB} = \frac{v_{IFB}}{i_{L/O}} = \frac{i_{L/O}R_S}{i_{L/O}} = R_S. \tag{8.1}$$

The challenge with adding R_S is the Ohmic power R_S consumes. This is why R_S is usually about or below 1 Ω. The *current-feedback voltage* v_{IFB} that R_S generates is therefore very low. This is unfortunate because *current-error amplifiers* or *comparators* A_{IE} and CP_{IE} require additional *quiescent power* P_Q when distinguishing millivolt signals from the noise the switching network generates.

B. MOS Resistance

Sensing resistances already in the conduction path saves the Ohmic power that adding R_S burns. But the only resistances accessible are those of the switches, which conduct only when they close. So they can only sense part of i_L or i_O:

$$\beta_{IFB} = \left.\frac{v_{IFB}}{i_{L/O}}\right|_{t_{ON}} = \left.\frac{i_{L/O}R_{DS}}{i_{L/O}}\right|_{t_{ON}} = R_{DS}\big|_{t_{ON}}. \tag{8.2}$$

Since these MOS triode *drain–source resistances* R_{DS} are usually very low, they drop similarly low voltages. This is unfortunate because A_{IE}'s and CP_{IE}'s need more P_Q when distinguishing the v_{IFB} they produce in Fig. 8.2 from the switching noise they and other transistors generate.

Reconstructing the v_{IFB} that i_L should set across the *switching period* t_{SW} is possible by sensing switches that conduct across complementary periods. An energize switch can translate i_L across L_X's *energize time* t_E and a drain switch across L_X's *drain time* t_D:

$$\beta_{IFB} = \frac{v_{IFB}}{i_L} = \frac{v_{IFB}\big|_{t_E} + v_{IFB}\big|_{t_D}}{i_L} = R_{DS(E)}\big|_{t_E} + R_{DS(D)}\big|_{t_D}. \tag{8.3}$$

Superimposing their v_{IFB}'s, however, requires more processing, and as a result, more silicon area and power. Luckily, sensing all components of i_L is not always necessary.

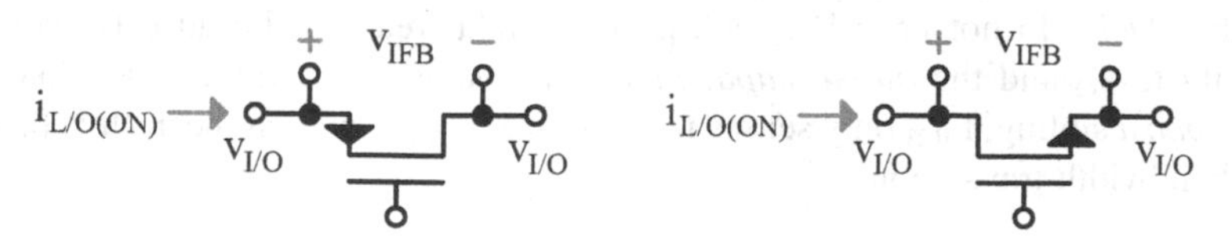

Fig. 8.2 MOS resistance

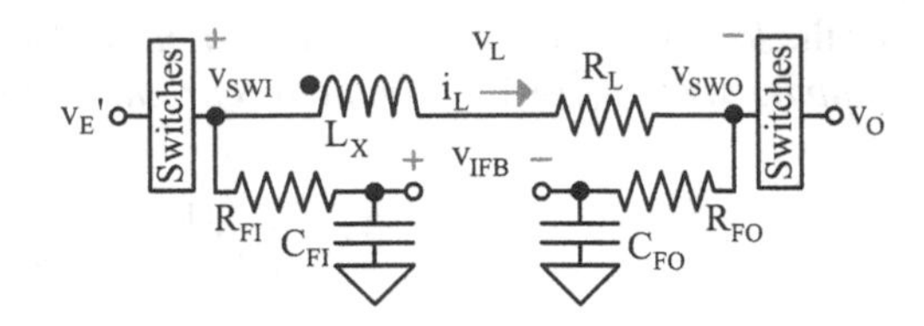

Fig. 8.3 Low-pass-filtered inductor resistance

C. **Inductor Resistance**

L_X's *inductor resistance* R_L is also in the conduction path. Except, R_L is *in* L_X, so R_L is not physically accessible. But since L_X shorts at low frequency, low-pass filtering the *inductor voltage* v_L removes the effects of L_X on current and voltage.

R_F's and C_F's in Fig. 8.3, for example, low-pass filter the *input* and *output switching voltages* v_{SWI} and v_{SWO} that drop v_L. The voltage between C_F's is therefore the difference between v_{SWI}'s and v_{SWO}'s averages. Since L_X shorts and C_F's open at low frequency, this v_{IFB} is R_L's dc voltage $v_{L(DC)}$. So v_{IFB} carries the voltage i_L's average drops across R_L's dc component:

$$\beta_{IFB} = \frac{v_{IFB}}{i_{L(AVG)}} = \frac{v_{SWI(AVG)} - v_{SWO(AVG)}}{i_{L(AVG)}} \approx \frac{v_{L(DC)}}{i_{L(AVG)}} = \frac{i_{L(AVG)}R_{L(DC)}}{i_{L(AVG)}} = R_{L(DC)}. \tag{8.4}$$

R_L is normally low, so A_{IE} or CP_{IE} requires more P_Q when processing the low v_{IFB} that results. But for this, R_F's and C_F's should first suppress the switching noise v_{SWI} and v_{SWO} generate to a fraction of the voltage R_L drops. This is important because v_{SWI} swings across the *input voltage* v_{IN} and v_{SWO} across the *output voltage* v_O which is usually across volts.

To suppress these wide swings, the impedance of C_F should be much lower than R_F. When this happens, Δv_{SWI} slews C_{FI} with v_{IN}/R_{FI} across t_E and Δv_{SWO} slews C_{FO} with v_O/R_{FO} across t_D. The *noise* v_n this generates should be a small fraction of the $v_{L(DC)}$ that $i_{L(AVG)}$ drops across $R_{L(DC)}$:

$$v_n \approx \frac{i_{FI/O}t_{E/D}}{C_{FI/O}} \approx \left(\frac{\Delta v_{SWI/O}}{R_{FI/O}}\right)\left(\frac{d_{E/D}t_{SW}}{C_{FI/O}}\right) \approx \left(\frac{v_{IN/O}t_{SW}}{R_{FI/O}C_{FI/O}}\right)\left(\frac{v_{D/E}}{v_E + v_D}\right) \tag{8.5}$$
$$<< v_{L(DC)} = i_{L(AVG)}R_{L(DC)}.$$

Boosts do not need R_{FI} or C_{FI} because L_X connects directly to v_{IN} and v_{IN} is steady. *Bucks* do not need R_{FO} or C_{FO} for similar reasons, because L_X connects directly to v_O and the *output capacitance* C_O averages v_O. Either way, like their *buck–boost* sibling, β_{IFB} only senses i_L's average $i_{L(AVG)}$, which means this β_{IFB} is a low-bandwidth translation.

Example 1: Determine C_F for a buck–boost so v_n is less than or equal to 10% of $v_{L(DC)}$ when v_{IN} is 4 V, v_O is 1.8 V, t_{SW} is 1 μs, R_F is 500 kΩ, $R_{L(DC)}$ is 250 mΩ, and $i_{L(AVG)}$ is 300 mA.

Solution:

$$v_E = v_{IN} = 4\text{ V and } v_D = v_O = 1.8\text{ V}$$

$$v_{L(DC)} = i_{L(AVG)}R_{L(DC)} = (300\text{m})(250\text{m}) = 75\text{ mV}$$

$$v_n \approx \left(\frac{v_{IN/O}t_{SW}}{R_F C_F}\right)\left(\frac{v_{D/E}}{v_E + v_D}\right) \approx \left[\frac{v_{IN/O}(1\mu)}{(500\text{k})C_F}\right]\left(\frac{v_{D/E}}{4+1.8}\right) \le \frac{v_{L(DC)}}{10} = \frac{75\text{m}}{10} \quad \therefore\ C_F \ge 330\,\text{pF}$$

Note: Two 330-pF occupy substantial silicon area.

Explore with SPICE:

See Appendix A for notes on SPICE simulations.

```
* Buck–Boost: Low-Pass Filtered RL Current Sensor
vin vin 0 dc=4
vde vde 0 dc=0 pulse 0 1 0 1n 1n 323n 1u
sei vin vswi vde 0 sw1v
ddg 0 vswi idiode
lx vswi vl 10u
rl vl vswo 250m
seg vswo 0 vde 0 sw1v
ddo vswo vo idiode
co vo 0 5u
ro vo 0 9
rfi vswi vifbp 500k
cfi vifbp 0 330p
rfo vswo vifbn 500k
cfo vifbn 0 330p
.lib lib.txt
.tran 1m
.end
```

Tip: Plot i(Lx), v(vifbp), and v(vifbn) across 1 ms and i(Lx) and v(vifbp) – v(vifbn) across last 2 μs.

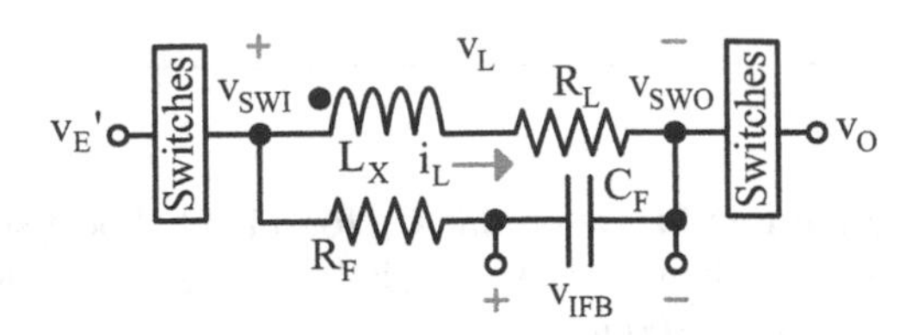

Fig. 8.4 Bypass-filtered inductor resistance

The bypass RC filter in Fig. 8.4 adopts a different approach. In this case, since capacitors complement inductors, C_F and R_F emulate the action of L_X and R_L. In other words, R_F and C_F filter v_L into a voltage v_{IFB} the same way L_X and R_L filter v_L into a current i_L.

i_L is an Ohmic translation of v_L across L_X and R_L. v_{IFB} is the voltage a similar Ohmic translation of v_L across C_F and R_F drops across C_F:

$$i_L = \frac{v_L}{sL_X + R_L} = \frac{v_L/R_L}{sL_X/R_L + 1} \tag{8.6}$$

$$v_{IFB} = \left(\frac{v_L}{1/sC_F + R_F}\right)\left(\frac{1}{sC_F}\right) = \frac{i_L(sL_X + R_L)}{1 + sR_FC_F} = i_LR_L\left(\frac{1 + sL_X/R_L}{1 + sR_FC_F}\right), \tag{8.7}$$

where v_L is also the voltage i_L drops across L_X and R_L. So v_{IFB} drops i_LR_L when R_FC_F matches L_X/R_L. In other words, R_L translates i_L to v_{IFB} when R_F and C_F's *RC frequency* f_{RC} matches L_X and R_L's *RL frequency* f_{RL}:

$$f_{RC} = \frac{1}{2\pi R_FC_F} \equiv f_{RL} = \frac{R_L}{2\pi L_X} \tag{8.8}$$

$$\beta_{IFB} = \frac{v_{IFB}}{i_L} = \frac{i_LR_L}{i_L} = R_L. \tag{8.9}$$

Note this β_{IFB} senses i_L across frequency, which means β_{IFB} is a high-bandwidth translation. But since L_X's *skin effect* crowds i_L toward the edges, R_L in β_{IFB} climbs with frequency. R_L is also low, so A_{IE} or CP_{IE} requires additional P_Q to distinguish the low v_{IFB} that results from the switching noise present.

Another drawback is the switching nature of v_{SWI} and v_{SWO}. Designing A_E or CP_E to tolerate this common-mode variation in v_{IFB} is challenging. But since boosts and bucks only switch their input or output, v_{SWI} or v_{SWO} is steady at v_{IN} or v_O. Connecting C_F to the steady voltage anchors v_{IFB}. In buck–boosts, connecting C_F to the least variable v_{SW} eases A_E's or CP_E's *input common-mode range* (ICMR) requirement.

R_F and C_F can shunt and bypass L_X and R_L in the loop gain that regulates v_O, i_L, or i_O. This happens when R_F and C_F's combined impedance falls below L_X and R_L's. When R_FC_F matches L_X/R_L, R_F and C_F bypass L_X and R_L past f_B when sL_X overcomes R_F:

$$R_F + \frac{1}{sC_F} = R_F + \frac{R_LR_F}{sL_X} = \left(\frac{R_F}{sL_X}\right)(sL_X + R_L)\bigg|_{f_O \geq \frac{R_F}{2\pi L_X} = f_B} \leq sL_X + R_L. \tag{8.10}$$

Since L_X and R_F are usually high in µH's and kΩ's, this usually happens at very high frequency. So this unintended effect on the feedback loop that regulates the output is largely inconsequential across frequencies of interest.

Example 2: Determine C_F, $v_{IFB(AVG)}$, and f_B for a buck–boost when R_F is 500 kΩ, L_X is 10 µH, R_L is 250 mΩ, and $i_{L(AVG)}$ is 300 mA.

Solution:

$$R_F C_F = (500k)C_F \equiv \frac{L_X}{R_L} = \frac{10\mu}{250m} \quad \therefore \quad C_F = 80 \text{ pF}$$

$$v_{IFB(AVG)} = i_{L(AVG)} R_L = (300m)(250m) = 75 \text{ mV}$$

$$f_B = \frac{R_F}{2\pi L_X} = \frac{500k}{2\pi(10\mu)} = 8.0 \text{ GHz}$$

Explore with SPICE:
See Appendix A for notes on SPICE simulations.

```
* Buck–Boost: Bypass-Filtered RL Current Sensor
vin vin 0 dc=4
vde vde 0 dc=0 pulse 0 1 0 1n 1n 323n 1u
sei vin vswi vde 0 sw1v
ddg 0 vswi idiode
lx vswi vl 10u
rl vl vswo 250m
seg vswo 0 vde 0 sw1v
ddo vswo vo idiode
co vo 0 5u
ro vo 0 9
rf vswi vifb 500k
cf vifb vswo 80p
.lib lib.txt
.tran 1m
.end
```

Tip: Plot i(Lx), v(vifb), and v(vswo) across 1 ms and i(Lx) and v(vifb) – v (vswo) across last 2 µs.

Matching time constants is not always necessary. After L_X overcomes R_L (i.e., R_L effectively shorts past f_{RL}) and C_F shorts with respect to R_F (past f_{RC}), v_L/R_F drops v_{IFB} across $1/sC_F$ with the v_L that i_L drops across sL_X. So the *moderate-frequency current-feedback translation* $\beta_{IFB(MF)}$ is L_X/R_FC_F, which can be less or greater than one:

$$\beta_{IFB(MF)} = \frac{v_{IFB}}{i_L}\bigg|_{f_O>>f_{RC}}^{f_O>>f_{RL}} \approx \left(\frac{v_L}{R_F}\right)\left(\frac{Z_C}{i_L}\right) \approx \frac{i_L s L_X}{R_F s C_F i_L} = \frac{L_X}{R_F C_F}. \quad (8.11)$$

Since L_X shorts and C_F opens (which is like R_F effectively shorts) at *low frequency*, $\beta_{IFB(LF)}$ is roughly R_L:

$$\beta_{IFB(LF)} = \frac{v_{IFB}}{i_L}\bigg|_{f_O<<f_{RC}}^{f_O<<f_{RL}} \approx \frac{i_L R_L}{i_L} = R_L. \quad (8.12)$$

$\beta_{IFB(MF)}$ equals R_L when R_FC_F matches L_X/R_L. When $\beta_{IFB(MF)}$ is higher than R_L, β_{IFB} rises over $\beta_{IFB(LF)}$ when sL_X overcomes R_L and reaches $\beta_{IFB(MF)}$ when C_F shorts (i.e., R_F current-limits C_F). So $\beta_{IFB(MF)}$ applies to frequencies over f_{RL} and f_{RC}.

$L_X/\beta_{IFB(MF)}R_F$ sets C_F when designing $\beta_{IFB(MF)}$ for a particular gain. This C_F and R_F bypass L_X and R_L when C_F and R_F's combined impedance falls below L_X and R_L's. This f_B' is higher than f_B when $\beta_{IFB(MF)}$ is greater than R_L. So this new bypass frequency f_B' is also inconsequential across frequencies of interest:

$$R_F + \frac{1}{sC_F} \equiv R_F + \frac{\beta_{IFB(MF)} R_F}{sL_X}\bigg|_{f_O \geq f_B\left(\frac{sL_X+\beta_{IFB(MF)}}{sL_X+R_L}\right)=f_B'} \leq sL_X + R_L. \quad (8.13)$$

Example 3: Determine C_F for Example 2 so $\beta_{IFB(MF)}$ is 10 Ω.

Solution:

$$\beta_{IFB(MF)} \approx \frac{L_X}{R_F C_F} = \frac{10\mu}{(500k)C_F} \equiv 10\ \Omega \quad \therefore \quad C_F = 2\ \text{pF}$$

Explore with SPICE:
Use the SPICE code from Example 2 and adjust C_F to explore.

8.1.2 Sense Transistor

A. Sense FET

Currents through nearby transistors match when their geometries and terminal voltages match. When matched this way, a sense transistor can mirror the current a power switch conducts across temperature and fabrication runs. The challenge is reproducing the voltage dropped across the switch, which is very low. This is important because transistors that drop millivolts are in triode, where their currents are sensitive to voltage.

Sense and *power transistors* M_S and M_P in Fig. 8.5 share gate, source, and body connections and voltages. Q_B carries a steady *bias current* I_B and Q_S carries M_S's *sense current* i_S. Since *base–emitter voltage* v_{BE} is a logarithmic translation of *collector current* i_C, Q_S's and Q_B's v_{BE}'s are roughly the same. So Q_S matches M_S's drain voltage v_S to M_P's v_P.

With all terminal voltages matched, i_S mirrors M_P's *power-transfer current* i_P. To save silicon area and power, M_S is A_I times or 1000× or so smaller than M_P. i_S is therefore lower than the i_L or i_O that i_P carries by a similar factor. This i_S flows through Q_S and R_S to set a v_{IFB} across R_{IFB} that scales with i_L or i_O when M_P conducts:

$$\beta_{IFB} = \left.\frac{v_{IFB}}{i_{L/O}}\right|_{t_{ON}} \approx \left.\frac{(i_{L/O}/A_I)R_{IFB}}{i_{L/O}}\right|_{t_{ON}} = \left.\frac{R_{IFB}}{A_I}\right|_{t_{ON}}. \tag{8.14}$$

Unfortunately, M_S and M_P's matching accuracy suffers when M_S is much smaller than M_P. This is the tradeoff for scaling i_S. R_{IFB} also varies with temperature and

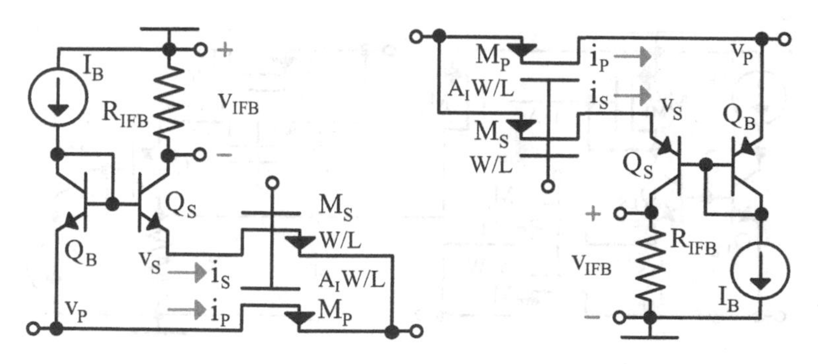

Fig. 8.5 Low- and high-side sense transistors

fabrication runs. But since M_S tracks M_P and R_{IFB} is usually high, these inaccuracies can be manageable, or when considering the alternatives, tolerable.

Q_S's and Q_B's v_{BE}'s match well when i_S and I_B equal. But since i_S scales with i_L or i_O and I_B is constant, i_S does not always equal I_B. This difference causes a mismatch between Q_S's and Q_B's v_{BE}'s that offsets M_S's and M_P's drain voltages v_S and v_P. This nonlinear *error voltage* v_E distorts β_{IFB}:

$$v_E = v_S - v_D = \Delta v_{BE} = V_t \ln\left(\frac{i_S}{I_B}\right). \tag{8.15}$$

Although undesirable, this distortion is not always prohibitive.

Reconstructing i_L across t_{SW} is possible by sensing complementary switches and mirroring one's i_S into the other's R_{IFB}. M_{LS} in Fig. 8.6, for example, senses M_L's low-side current and M_{M1}–M_{M2} mirrors i_{LS} into M_H's high-side R_{IFB}. This way, R_{IFB}'s v_{IFB} scales with i_{LS} and i_{HS}. Since i_{LS} is zero when i_{HS} is not and *vice versa*, v_{IFB} scales with the i_L that i_{LP} and i_{HP} feed. M_{M1}–M_{M2}'s *mirror gain* A_M ensures i_{LS}'s and i_{HS}'s scaling factors A_M/A_{LI} and $1/A_{HI}$ match:

$$\beta_{IFB} = \frac{v_{IFB}}{i_L} \approx \frac{(i_{LS}A_M + i_{HS})R_{IFB}}{i_L} \approx \left(\frac{i_{L(L)}A_M}{A_{LI}} + \frac{i_{H(L)}}{A_{HI}}\right)\left(\frac{R_{IFB}}{i_L}\right) = \frac{R_{IFB}}{A_I}. \tag{8.16}$$

B. **Looped Sense FET**

The feedback loops in Fig. 8.7 eliminate β_{IFB}'s distortion. This is because, with equal I_B's, Q_S's and Q_B's i_C's and v_{BE}'s match. This way, v_S follows v_P.

Q_S, Q_B, and M_{FB} close a feedback loop that mixes and compares i_S with *feedback current* i_{FB}. Resistance at v_S converts the *error current* i_E into a v_S that Q_S level-shifts to *base voltage* v_B and Q_B and M_{FB} amplify and translate into i_{FB}. With enough *loop gain* A_{LG}, the error current i_E is low, i_{FB} nears i_S, and i_{FB} flows through M_{FB} into R_{IFB} to set v_{IFB}. Even when offset by a small mismatch between I_{B2} and I_{B3}, the feedback loop scales i_{FB} with i_S without significant v_S–v_P variations.

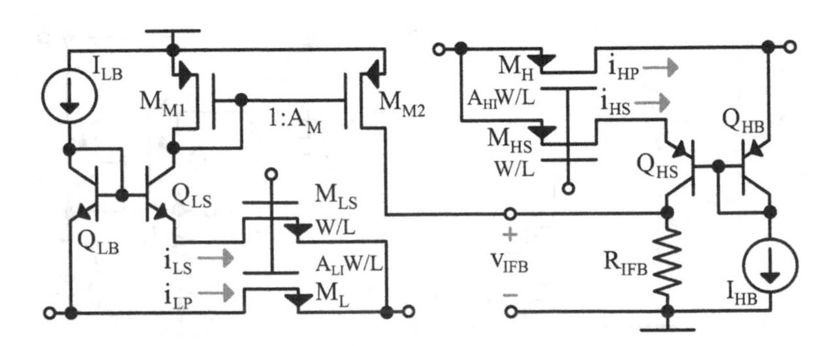

Fig. 8.6 Complementary sense transistors

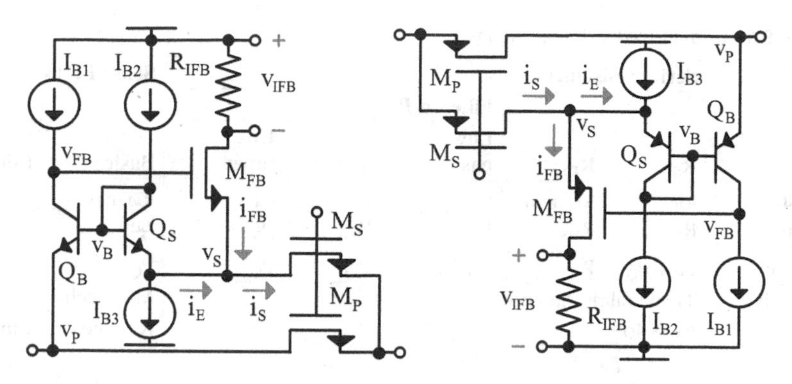

Fig. 8.7 Looped low- and high-side sense transistors

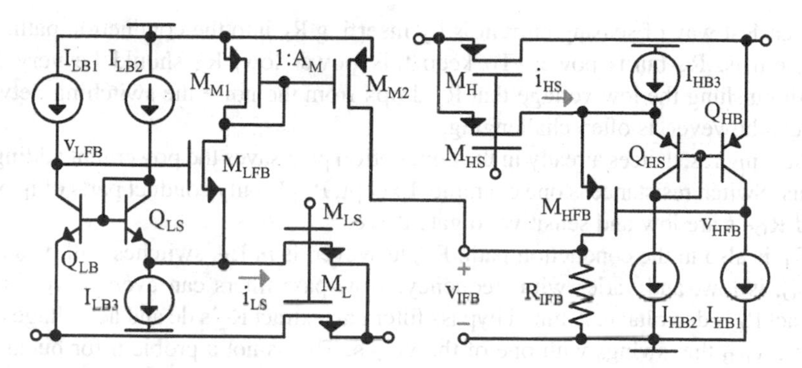

Fig. 8.8 Looped complementary sense transistors

The loop is usually inherently stable because the resistance at v_{FB} is much higher than at v_S (and v_B). So with comparable capacitances, the pole at v_{FB} is much lower than at v_S (and v_B). A_{LG} therefore drops past p_{FB} and reaches the *unity-gain frequency* f_{0dB} near or below p_S and p_B.

The challenge is the bandwidth that f_{0dB} sets. When M_P switches, v_S requires time to match v_P. This *response time* t_R, which f_{0dB} sets, should be a fraction of the t_E or t_D across which M_P conducts. Small Q_S, Q_B, and M_{FB} geometries (for low capacitance) and high I_B's help shorten this t_R (by extending f_{0dB}).

Reconstructing i_L across t_{SW} is possible by sensing complementary switches and mirroring one's i_S into the other's R_{IFB}. M_{LS} in Fig. 8.8, for example, senses M_L's low-side current and M_{M1}–M_{M2} mirrors i_{LS} into M_H's high-side R_{IFB}. This way, R_{IFB}'s v_{IFB} scales with i_{LS} and i_{HS}. Since i_{LS} is zero when i_{HS} is not and *vice versa*, v_{IFB} scales with the i_L that i_{LP} and i_{HP} feed. M_{M1}–M_{M2}'s mirror gain A_M ensures i_{LS}'s and i_{HS}'s scaling factors match.

Table 8.1 Current-feedback translations

	Series resistances					Sense FETs	
			Filtered R_L				
	R_S	R_{DS}	Low pass	Tuned	Un-tuned	Basic	Looped
Input	$i_{O/L}$	$i_{O/L(E/D)}$	$i_{L(DC)}$	i_L	$i_{L(DC/AC)}$	$i_{O/L(E/D)}$	
Gain	R_S	R_{DS}	R_L	R_L	$\frac{L_X}{R_F C_F}$	$\frac{R_{IFB}}{A_I}$	
Power	$P_R + P_Q$	P_Q	P_Q	P_Q	P_Q	P_Q	
Sensitivity	T_J and fabrication runs					Mismatch	
	$R_L \propto f_O$					Nonlinear	Linear

8.1.3 Design Notes

The easiest way of sensing current is by inserting R_S into the conduction path. The problem is, R_S burns power. To keep this power low, R_S should be very low. Distinguishing the low voltage that R_S drops from the noise the switching network injects, however, is often challenging.

Sensing resistances already in the conduction path saves the power that adding R_S burns. Switch resistance is one example. Except, R_{DS}'s only conduct parts of i_L or i_O. And R_{DS}'s are low and sensitive to gate drive.

R_L is also in the conduction path. R_L, however, is *in* L_X, switches with v_{SWI} and v_{SWO}, is low, and varies with frequency. Low-pass filters can average v_{SW}'s and extract R_L's dc voltage. A tuned bypass filter can extract R_L's dc and ac voltages, but with a v_{IFB} that swings with one of the v_{SW}'s. This is not a problem for bucks and boosts because they only have one v_{SW}, which the filter can average. This filter can also amplify R_L's ac voltage, but only after L_X overcomes R_L and C_F shorts with respect to R_F.

R_S, R_{DS}, and R_L all vary with temperature and fabrication runs. Sense FETs are less sensitive because they track the switches they sense. But since they are much smaller, mismatches offset their outputs. Their translations are also nonlinear. A feedback loop can reduce this nonlinearity, but by slowing the response. Although not always, basic sense FETs often offer more favorable tradeoffs than looped FETs and series resistances. Table 8.1 summarizes some of these points.

8.2 Voltage Sensors

8.2.1 Voltage Divider

v_O is typically over the *band-gap voltage* V_{BG} or sub-V_{BG} that sets the *reference voltage* v_R. So in most cases, the *voltage feedback translation* β_{FB} attenuates v_O to the *feedback voltage* v_{FB} that the *error amp* A_E compares to v_R. This is why v_R is normally 1.2 V or lower and β_{FB} is less than one.

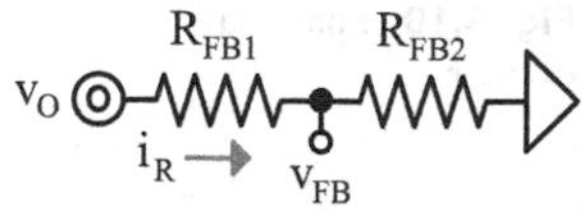

Fig. 8.9 Voltage divider

The *voltage divider* in Fig. 8.9 is the most common way of translating v_O to v_{FB}. R_{FB1} and R_{FB2} translate v_O into a current i_R that drops v_{FB} across R_{FB2}. β_{FB} is the voltage-divided R_{FB2} fraction of R_{FB1} and R_{FB2}:

$$\beta_{FB} = \frac{v_{FB}}{v_O} = \frac{i_R R_{FB2}}{v_O} = \left(\frac{v_O}{R_{FB1} + R_{FB2}}\right)\left(\frac{R_{FB2}}{v_O}\right) = \frac{R_{FB2}}{R_{FB1} + R_{FB2}}. \quad (8.17)$$

This ratio is usually accurate because resistors normally match and track well across temperature and fabrication runs.

i_R should overwhelm the current that noise injects into v_{FB}. This i_R and the targeted v_O dictate the power P_R or $i_R v_O$ that R_{FB1} and R_{FB2} consume. This i_R into R_{FB2} also sets the v_{FB} that reflects v_R and the *voltage-loop offset* v_{VOS}. And with R_{FB1}, R_{FB2} sets v_O's translation to v_{FB}. So when designing the voltage divider, engineers often use v_{FB} and R_{FB2} to set i_R and R_{FB1} to set v_O. This way, R_{FB1} and R_{FB2} account for noise, power, and offset.

Example 4: Determine R_{FB1}, R_{FB2}, and P_R so v_O is 2 V and i_R is 5 μA when v_R is 1.2 V and v_{VOS} is 40 mV.

Solution:

$$P_R = i_R v_O = (5\mu)(2) = 10\ \mu W$$

$$i_R = \frac{v_{FB}}{R_{FB2}} = \frac{v_R - v_{VOS}}{R_{FB2}} = \frac{1.2 - 40m}{R_{FB2}} \equiv 5\ \mu A \quad \therefore \quad R_{FB2} = 232\ k\Omega$$

$$\beta_{FB} = \frac{R_{FB2}}{R_{FB1} + R_{FB2}} = \frac{232k}{R_{FB1} + 232k}$$

$$\equiv \frac{v_R - v_{VOS}}{v_O} = \frac{1.2 - 40m}{2} = 58\% \quad \therefore \quad R_{FB1} = 168\ k\Omega$$

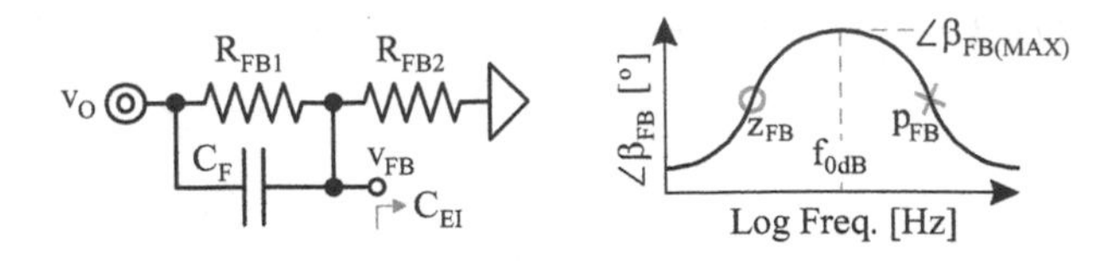

Fig. 8.10 Phase-saving divider

8.2.2 Phase-Saving Voltage Divider

The A_{LG} that controls v_O often includes undesirable poles near or below f_{0dB}. Adding zeros near or below f_{0dB} recovers some of the *phase margin* PM these parasitic poles lose. Unfortunately, the zero that a capacitor induces when it shunts reverses when it shorts. So this zero saves as much PM as the reversal pole that follows it allows.

The *phase-saving divider* in Fig. 8.10 establishes a *zero–pole pair* z_{FB}–p_{FB} in β_{FB}. C_F should, by design, overwhelm the effects of *A_E's input capacitance* C_{EI}, which is parasitic. This way, z_{FB} raises β_{FB} when C_F bypasses R_{FB1}. And since C_O shunts v_O at lower frequency, p_{FB} flattens β_{FB} when C_F and C_{EI} shunt R_{FB1} and R_{FB2}:

$$\beta_{FB} = \frac{v_{FB}}{v_O} = \beta_{FB0}\left(\frac{1 + s/2\pi z_{FB}}{1 + s/2\pi p_{FBX}}\right) = \left(\frac{R_{FB2}}{R_{FB1} + R_{FB2}}\right)\left(\frac{1 + s/2\pi z_{FB}}{1 + s/2\pi p_{FBX}}\right), \quad (8.18)$$

$$z_{FB} = \frac{1}{2\pi R_{FB1} C_F}, \quad (8.19)$$

$$p_{FB} \approx \frac{1}{2\pi(R_{FB1}||R_{FB2})(C_F + C_{EI})} \approx \frac{1}{2\pi(R_{FB1}||R_{FB2})C_F}. \quad (8.20)$$

z_{FB} recovers as much phase shift as p_{FB} allows. The recovery peaks close to 90° when p_{FB} is ten times or more higher than z_{FB}. When within a decade or so, $\angle\beta_{FB}$ maxes halfway between z_{FB} and p_{FB}. So phase margin maxes when z_{FB} and p_{FB} center about f_{0dB}: when z_{FB} is as low as p_{FB} is higher than f_{0dB}:

$$\Delta PM = \angle\beta_{FB}|_{f_{0dB}} = \tan^{-1}\left(\frac{f_{0dB}}{z_{FB}}\right) - \tan^{-1}\left(\frac{f_{0dB}}{p_{FB}}\right) \quad (8.21)$$

$$\frac{f_{0dB}}{z_{FB}} \equiv \frac{p_{FB}}{f_{0dB}}. \quad (8.22)$$

From the perspective of design, the v_{FB} that v_R sets and R_{FB2} establish the quiescent current that R_{FB1} and R_{FB2} conduct. With R_{FB2} known, R_{FB1} sets the β_{FB} that translates v_O to v_{FB}. And with R_{FB1} and R_{FB2} determined, C_F centers z_{FB} and p_{FB} about f_{0dB}.

In practice, v_O is often less than ten times higher than v_R. R_{FB1} is therefore no greater than $9R_{FB2}$. This means, the separation between z_{FB} and p_{FB} is less than a decade of frequency. So ΔPM is usually less than 50° or so.

Example 5: Determine C_F and ΔPM for Example 4 when f_{0dB} is 100 kHz.

Solution:

$$R_{FB1} = 168\ k\Omega \text{ and } R_{FB2} = 232\ k\Omega \text{ from Example 4}$$

$$z_{FB} = \frac{1}{2\pi R_{FB1} C_F} = \frac{1}{2\pi(168k)C_F} = \frac{950n}{C_F}$$

$$p_{FB} \approx \frac{1}{2\pi(R_{FB1}||R_{FB2})C_F} = \frac{1}{2\pi(168k||232k)C_F} = \frac{1.6\mu}{C_F}$$

$$\frac{p_{FB}}{f_{0dB}} = \frac{1.6\mu}{C_F(100k)} \equiv \frac{f_{0dB}}{z_{FB}} = \frac{C_F(100k)}{950n}$$

$$\therefore\ C_F = 12\ pF \ \rightarrow\ z_{FB} = 79\ kHz \text{ and } p_{FB} = 130\ kHz$$

$$\Delta PM = \tan^{-1}\left(\frac{f_{0dB}}{z_{FB}}\right) - \tan^{-1}\left(\frac{f_{0dB}}{p_{FB}}\right) = \tan^{-1}\left(\frac{100k}{79k}\right) - \tan^{-1}\left(\frac{100k}{130k}\right) = 14°$$

Explore with SPICE:
See Appendix A for notes on SPICE simulations.

```
* Phase-Saving Voltage Divider
vo vo 0 dc=2 ac=1
rfb1 vo vfb 168k
cf vo vfb 12p
rfb2 vfb 0 232k
*.op
.ac dec 10 1k 10e6
.end
```

Tip: Plot v(vfb).

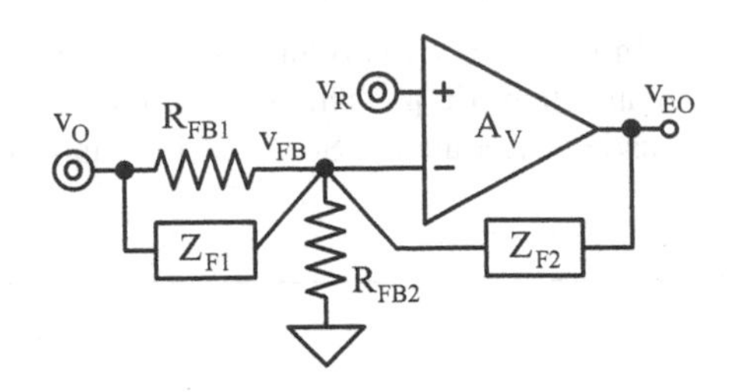

Fig. 8.11 Voltage-dividing error amplifier

8.2.3 Voltage-Dividing Error Amplifier

When the feedback loop includes A_E, integrating β_{FB} into A_E's stabilizer can save power, components, and area. The *voltage-dividing error amplifier* in Fig. 8.11, for example, integrates the voltage divider into the feedback network of an *inverting op amp*. R_{FB1} and R_{FB2} translate v_O to v_{FB} and A_V, Z_{F1}, and Z_{F2} with R_{FB1} set A_E's gain and stabilizing response.

Z_{F1} and Z_{F2} are capacitors C_{F1} and C_{F2} with, on occasion, series resistors R_{F1} and R_{F2}. This way, Z_{F1} and Z_{F2} open at low frequency. So β_{FB} translates v_O to v_{FB} with R_{FB1} and R_{FB2} and A_E's *low-frequency gain* A_{E0} amplifies v_R and v_{FB}'s error with A_{V0}.

Z_{F2} closes a feedback loop that alters v_R's and v_O's translations to A_E's output v_{EO}. As the capacitor in Z_{F2} shunts, these gains change in different ways. The stability of the feedback controller depends on v_O's translation to v_{EO}. This is a *mixed translation* because the effects of feedback surface only at higher frequency, as Z_{F2} closes the loop.

The *forward* and *feedback gains* A_F and A_β of the feedback network set the overall gain. A_F excludes the effects of feedback and A_β excludes the effects of limited forward gain. When combined, the overall gain follows the lowest translation.

In the case of v_O's gain to v_{EO}, A_F voltage-divides with β_{FB0}'s R_{FB1} and R_{FB2}, amplifies with A_V's A_{V0}, and filters with Z_{F1}, Z_{F2}, and the *amplifier's pole* p_A. A_F falls when Z_{F2} shunts R_{FB2}, rises and flattens when Z_{F1} shunts (bypasses) R_{FB1} and shorts with respect to R_{FB1} and R_{FB2}, and falls past p_A. A_β amplifies with the inverting op-amp gain that Z_{F2} and $R_{FB1} \parallel Z_{F1}$ set. This A_β starts infinitely high and falls with Z_{F2} as C_{F2} shorts:

$$A_F = \left[\frac{R_{FB2}||Z_{F2}}{(R_{FB1}||Z_{F1}) + (R_{FB2}||Z_{F2})}\right]\left(\frac{-A_{V0}}{1 + s/2\pi p_A}\right) = -\beta_{FB0}A_{V0}A_X \qquad (8.23)$$

$$A_\beta = \frac{-Z_{F2}}{R_{FB1}||Z_{F1}}, \tag{8.24}$$

where A_X is A_F's filter response with a low-frequency gain of one.

$\beta_{FB}A_E$ in the feedback controller is v_O's overall gain to v_{EO}. Since this gain follows the lowest translation and A_β's low-frequency gain is infinitely high, $\beta_{FB}A_E$ starts with A_{F0}'s $-\beta_{FB0}A_{V0}$ and follows the *stabilizing filter response* A_S that Z_{F1}, Z_{F2}, and p_A set. This A_S starts flat with A_{S0} at 1 or 0 dB:

$$\frac{v_{EO}}{v_O} \equiv \beta_{FB}A_E = A_F||A_\beta = -\beta_{FB}A_{V0}A_S. \tag{8.25}$$

A typical response falls and follows A_β past p_{E1} after A_β falls below A_F. $\beta_{FB}A_E$ can, for example, fall past p_{E1}, flatten when R_{F2} current-limits C_{F2}, climb when C_{F1} bypasses R_{FB1}, and fall when p_A reduces A_F below A_β. For this type of response, p_A should be much higher than p_{E1}.

8.3 Digital Blocks

8.3.1 Push–Pull Logic

A. Inverter

Inverters "invert" their *inputs* v_I's. So v_O in Fig. 8.12 is low or zero when v_I is high or one. And v_O swings high when v_I drops low.

The symbol of an inverter is a triangle. v_I connects halfway across one side of the triangle and v_O to a small circle on the opposing tip. The flat side blocks static current, the sharp joint drives current, and the small circle inverts polarity.

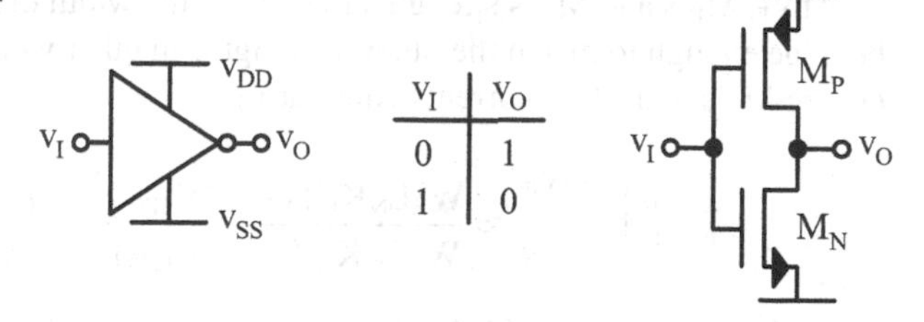

Fig. 8.12 Push–pull inverter

Explore with SPICE:
See Appendix A for notes on SPICE simulations.

```
* 1-V Push–Pull Inverter
vi vi 0 dc=0 pwl 0 0 1n 0 1.001n 1 5n 1 5.001n 0
xi1 vi vo1 inv
xi2 vo1 vo2 inv
.lib lib.txt
.dc vi 0 1 1m
*.tran 10n
.end
```

Tip: Plot v(vi), v(vo1), and v(vo2). Move "*" so it precedes ".dc," re-run simulation, and plot v(vo1) and v(vo2).

Digital highs and lows in CMOS implementations usually reach the *positive* and *negative power supplies* v_{DD} and v_{SS}. Since PFETs are active-low devices that supply current and NFETs are active high that sink current, *push–pull inverters* pull v_O to v_{DD} with a PFET M_P and to v_{SS} with an NFET M_N. Although not always the case, v_{SS} is often ground.

v_O toggles state when v_I crosses the inverter's *trip point* v_T. v_O holds this state as long as v_I, v_{SS}, and v_{DD} noise keeps v_I above or below v_T. Noise immunity is greatest when v_T is halfway across the supplies.

At this v_T, v_I and v_O are halfway between v_{SS} and v_{DD}. This means, M_P's v_{SG} and v_{SD} and M_N's v_{GS} and v_{DS} are all v_T. With this v_T, M_P and M_N invert their channels when $v_{DD} - v_{SS}$ exceeds M_P and M_N's combined *zero-bias threshold voltages* $|V_{TP0}| + V_{TN0}$. They also saturate because, with the same v_T, v_{DS}'s overcome M_P's and M_N's *saturation voltages* $v_{DS(SAT)}$'s, which are a threshold voltage below their v_{GS}'s.

For speed, M_P and M_N's *lengths* L should be the minimum L that can sustain v_{DD} and v_{SS}. The *width* W of the stronger device should also be minimum. This way, with the least capacitance possible, v_I and v_O can transition quickly.

At v_T, M_P's and M_N's strengths match. So the width of the weaker device should be wide enough to match the other's strength. In other words, this weaker W should ensure M_P's and M_N's currents equal at v_T:

$$\left.\frac{i_P}{i_N}\right|_{v_T}^{v_{GS}>V_{T0}} \approx \frac{W_P L_N K_P'(v_T - |V_{TP0}|)^2(1 + v_T\lambda_P)}{W_N L_P K_N'(v_T - V_{TN0})^2(1 + v_T\lambda_N)} \equiv 1, \tag{8.26}$$

where λ's are M_P's and M_N's *channel-length modulation parameters*.

M_P and M_N switch in *sub-threshold* when the voltage across the supplies falls below their combined thresholds $|V_{TP0}|$ V_{TN0}. In these cases, M_P and M_N saturate when v_T is three *thermal voltages* V_t's below v_{DD} and $3V_t$ above v_{SS}. With v_T halfway across the supplies, this corresponds to $v_{DD} - v_{SS}$ exceeding $6V_t$. Saturated this way, strengths match when

$$\left.\frac{i_P}{i_N}\right|_{v_T}^{\substack{v_{GS} < V_{T0} \\ v_{DS} > 3V_t}} \approx \left(\frac{W_P L_N I_{SP}}{W_N L_P I_{SN}}\right) \exp\left(V_{TN0} - |V_{TP0}|\right) \equiv 1, \tag{8.27}$$

where I_{SP} and I_{SN} are their intrinsic *saturation currents* in sub-threshold.

Example 6: Determine W's so v_T is half v_{DD} when v_{DD} is 4 V, W_{MIN} is 3 μm, L_{MIN}'s match, K_N' is 200 μA/V^2, K_P' is 40 μA/V^2, V_{TN0} is 500 mV, V_{TP0} is −700 mV, and λ's match.

Solution:

$$V_{TN0} = 500\text{ mV} < |V_{TP0}| = 700\text{ mV} \quad \therefore \quad W_N \equiv W_{MIN} = 3\ \mu\text{m}$$

$$v_{DD} - v_{SS} = 4 - 0$$
$$> V_{TN0} + |V_{TP0}| = 500\text{m} + 700\text{m} = 1.2\text{ V}$$
$$\therefore M_P \text{ and } M_N \text{ invert}$$

$$v_T = 0.5v_{DD} = 0.5(4) = 2\text{ V} \quad \therefore \quad M_P \text{ and } M_N \text{ saturate at } v_T$$

$$\frac{i_P}{i_N} \approx \frac{W_P L_N K_P'(v_T - |V_{TP0}|)^2}{W_N L_P K_N'(v_T - V_{TN0})^2} = \frac{W_P L_N (40\mu)(2 - 700\text{m})^2}{W_N L_P (200\mu)(2 - 500\text{m})^2} = \frac{15W_P}{100W_N} \equiv 1$$

$$\therefore \quad W_P = 6.7W_N = 20\ \mu\text{m}$$

Explore with SPICE:
See Appendix A for notes on SPICE simulations.

```
* Push–Pull CMOS Inverter
vdd vdd 0 dc=4
vi vi 0 dc=0
mp vo vi vdd vdd pmos2 w=20u l=250n
mn vo vi 0 0 nmos2 w=3u l=250n
.lib lib.txt
.dc vi 0 4 1m
.end
```

Tip: Plot v(vi) and v(vo).

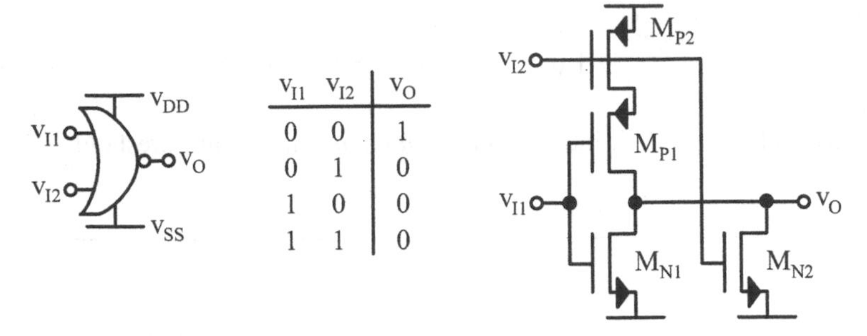

v_{I1}	v_{I2}	v_O
0	0	1
0	1	0
1	0	0
1	1	0

Fig. 8.13 Push–pull NOR gate

B. NOR Gate

OR gates output low only when all inputs are low. *NOR* gates invert the action of OR gates. So the *push–pull NOR gate* in Fig. 8.13 outputs high only when all inputs are low and low when any input is high.

The NOR gate is like an inverter with two inputs. Each input requires two transistors. Since v_O is low when any input is high and NFETs are active high, M_{N1} or M_{N2} pulls v_O low when v_{I1} or v_{I2} is high. And M_{P1} and M_{P2} pull v_O high only when v_{I1} and v_{I2} are both low. Functionally, parallel transistors OR their inputs and series transistors AND their inputs.

Explore with SPICE:
See Appendix A for notes on SPICE simulations.

```
* Push–Pull CMOS NOR Gate
vdd vdd 0 dc=4
vi1 vi1 0 dc=0 pulse 4 0 5p 1p 1p 50p 100p
vi2 vi2 0 dc=0 pulse 4 0 5p 1p 1p 100p 200p
mp1 vx2 vi2 vdd vdd pmos2 w=20u l=250n
mp2 vo vi1 vx2 vx2 pmos2 w=40u l=250n
mn1 vo vi1 0 0 nmos2 w=3u l=250n
mn2 vo vi2 0 0 nmos2 w=3u l=250n
.lib lib.txt
*.dc vi1 0 4 1m
.tran 200p
.end
```

Tip: Plot v(vi1), v(vi2), and v(vo). Move "*" so it precedes ".tran," re-run simulation, and plot v(vi1), v(vi2), and v(vo).

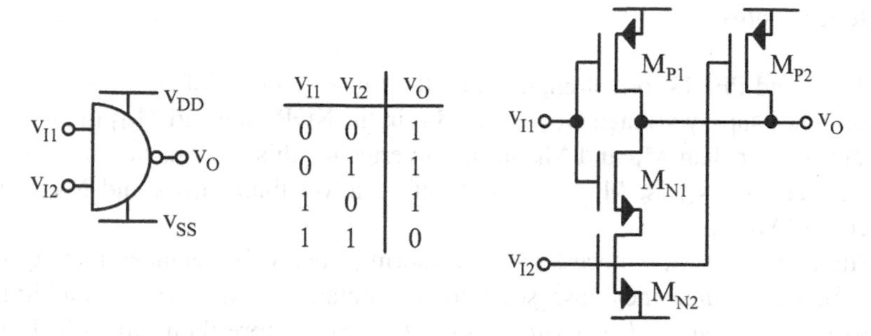

Fig. 8.14 Push–pull NAND gate

C. NAND Gate

AND gates output a high only when all inputs are high. *NAND* gates invert the action of AND gates. So the *push–pull NAND gate* in Fig. 8.14 outputs low only when all inputs are high and high when any input is low.

The NAND gate is basically an inverter with two inputs. Each input requires two transistors. Since v_O is high when any input is low and PFETs are active low, M_{P1} or M_{P2} pull v_O high when v_{I1} or v_{I2} is low. And M_{N1} and M_{N2} pull v_O low only when v_{I1} and v_{I2} are both high. Like in the NOR gate, parallel transistors effectively OR their inputs and series transistors AND their inputs.

Explore with SPICE:
See Appendix A for notes on SPICE simulations.

```
* Push–Pull CMOS NAND Gate
vdd vdd 0 dc=4
vi1 vi1 0 dc=4 pulse 4 0 5p 1p 1p 50p 100p
vi2 vi2 0 dc=4 pulse 4 0 5p 1p 1p 100p 200p
mp1a vo vi1 vdd vdd pmos2 w=20u l=250n
mp1b vo vi2 vdd vdd pmos2 w=20u l=250n
mn1a vo vi1 vx vx nmos2 w=6u l=250n
mn1b vx vi2 0 0 nmos2 w=3u l=250n
.lib lib.txt
*.dc vi1 0 4 1m
.tran 200p
.end
```

Tip: Plot v(vi1), v(vi2), and v(vo). Move "*" so it precedes ".tran," re-run simulation, and plot v(vi1), v(vi2), and v(vo).

D. **Design Notes**

NFETs and PFETs lose strength when their gate–source voltages are lower than the combined supply voltage $v_{DD} + v_{SS}$. M_{P1} in the NOR gate and M_{N1} in the NAND gate are weaker than M_P and M_N in the inverter for this reason, because M_{P2} and M_{N2} reduce their v_{GS}'s. M_{P1}'s v_T is therefore lower than M_{P2}'s and M_{N1}'s v_T is higher than M_{N2}'s.

Although not always necessary, re-centering their v_T's balances their t_P's and noise margins. Since they lose strength to similarly sized devices, doubling the widths of these *source-degenerated transistors* can restore their strength. In other words, v_T's for the inverter, two-input NOR, and two-input NAND circuits are roughly the same when L's and non-degenerated W's match and degenerated W's (M_{P1}'s in Fig. 8.13 and M_{N1}'s in Fig. 8.14) are twice as wide.

When adding inputs, the concepts used to extend the one-input inverter to the two-input gates still apply. Each additional input requires two FETs: one for the parallel combination and another for the series stack. Three- and four-input variations of this sort are not uncommon in systems.

8.3.2 SR Flip Flops

Set–reset (SR) *flip flops* "set" or "reset" their *current state* "Q" high or low when their "S" or "R" input in Fig. 8.15 is high and the other is low. They hold their *previous state* Q^{-1} when both inputs are low. And when both inputs are high, *set-dominant flip flops* output high and *reset-dominant flip flops* output low. Either way, SR implementations normally output Q and Q's *complementary* (*opposite*) *state* $\overline{Q}$.

Flip flops perform four functions: set, reset, hold, and dominate. The *temporal SR flip flops* in Fig. 8.16 use set and reset switches to connect Q to v_{DD} and v_{SS} when S–R is high–low and low–high. The set resistance R_S in the set-dominant case is much

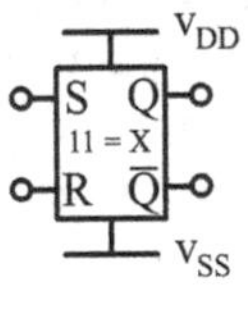

Fig. 8.15 SR flip flop

S	R	Q
0	0	Q^{-1}
0	1	0
1	0	1
1	1	1

S	R	Q
0	0	Q^{-1}
0	1	0
1	0	1
1	1	0

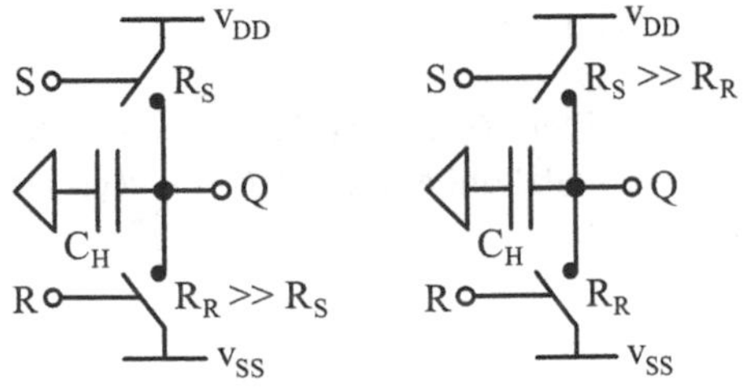

Fig. 8.16 Temporal set/reset-dominant flip flops

Fig. 8.17 Digital set/reset-dominant flip flops

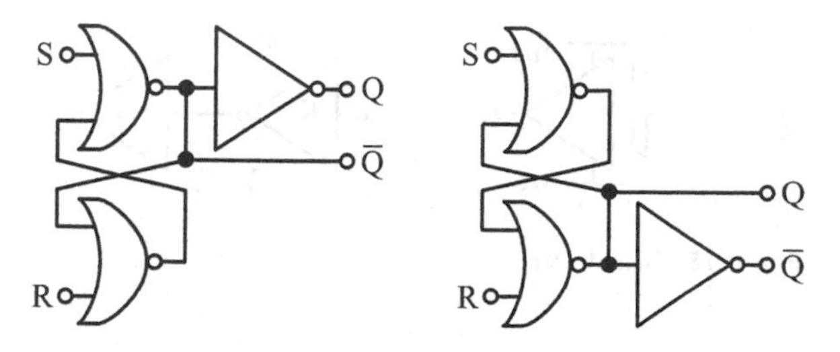

lower than the reset resistance R_R and *vice versa* for the reset-dominant counterpart. This way, Q approaches v_{DD} or v_{SS} when S–R is high–high. And capacitor C_H holds Q^{-1} when S–R is low–low.

Explore with SPICE:
See Appendix A for notes on SPICE simulations.

```
* 1-V Temporal Set-Dominant SR Flip Flop
vs s 0 dc=0 pulse 1 0 5u 1n 1n 12.5u 25u
vr r 0 dc=0 pulse 1 0 5u 1n 1n 25u 50u
x1 s r q qn srs
.lib lib.txt
.tran 50u
.end
```

Tip: Plot v(s), v(r), and v(q).
Note: Use model "srr" to explore the temporal reset-dominant flip flop.

The *digital SR flip flops* in Fig. 8.17 use set and reset NOR gates to set and reset Q when S–R is high–low and low–high. The inverter connects to the set- or reset-dominant NOR gate. This way, Q sets with S or resets with R when S–R is high–high. Positive feedback holds (latches) Q^{-1} when S–R is low–low. Holding Q^{-1} is possible because NOR gates ignore low inputs when one of their inputs is high.

8.3.3 Gate Driver

Power switches are typically large to limit the power they burn when they conduct the i_L that feeds i_O. So the *output gate capacitance* C_{GO} that *gate drivers* feed is usually very high. C_{GO} is so high that a minimum-size inverter requires too much time to charge and discharge C_{GO}.

The chain of increasingly larger inverters in Fig. 8.18 can build the current needed to drive C_{GO}. To unload the circuit that feeds the driver, the first stage K_1 should be a minimum-size inverter. This way, the *input gate capacitance* C_{GI} of the driver is the lowest capacitance possible.

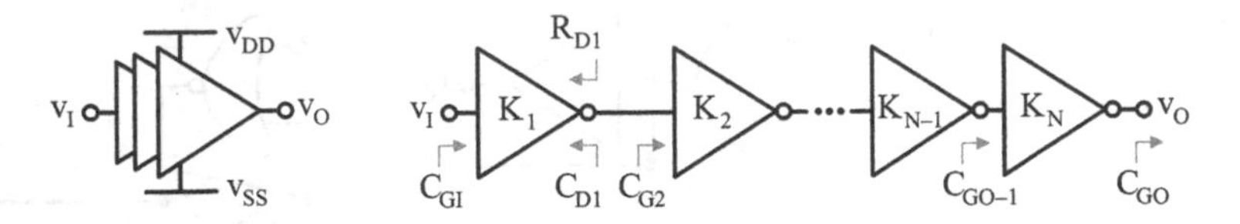

Fig. 8.18 Gate driver

Subsequent stages are f_o times greater than their preceding stage. This f_o is the *interstage fan out* because each *gate* load *capacitance* C_G, which scales with W, is f_o times higher than the preceding gate load C_{G-1}. When f_o is consistent across N stages, C_{GO} is $C_{GI}f_o^N$. The *total fan out* F_O is C_{GO}/C_{GI}, which when L's match, is also the ratio of the W_{GO} and W_{GI} that set C_{GO} and C_{GI}. So in short, F_O is f_o^N, or the other way around, f_o is $F_O^{1/N}$:

$$f_o = \frac{C_G}{C_{G-1}} = \frac{W_G}{W_{G-1}} \quad (8.28)$$

$$F_O = \frac{C_{GO}}{C_{GI}} = \frac{C_{GI}f_o^N}{C_{GI}} = f_o^N = \frac{W_{GO}}{W_{GI}}. \quad (8.29)$$

A. **Minimum Delay**

K_1's *drain resistance* R_{D1} drives K_1's *drain capacitance* C_{D1} and K_2's gate load C_{G2}. K_1's total output capacitance C_{O1} needs 69% of R_{D1} and C_{O1}'s *RC time constant* τ_{RC} to swing v_{O1} halfway across the maximum swing $\Delta v_{O(MAX)}$ that v_{DD} and v_{SS} set. This $69\%\tau_{RC}$ is K_1's *propagation delay* t_{P1}:

$$C_{O1} = C_{D1} + C_{G2} = C_{GI}(k_{SL} + f_o) \approx C_{GI}(1 + f_o), \quad (8.30)$$

$$\Delta v_O = \Delta v_{O(MAX)}\left[1 - \exp\left(\frac{-t_{\%}}{\tau_{RC}}\right)\right] = (v_{DD} - v_{SS})\left[1 - \exp\left(\frac{-t_{\%}}{\tau_{RC}}\right)\right], \quad (8.31)$$

$$t_{P1} \equiv t_{50\%\Delta v_{O(MAX)}} = \tau_{RC}\ln(1 - 50\%)^{-1} = 69\%\tau_{RC} = 69\%R_{D1}C_{O1}. \quad (8.32)$$

k_{SL} relates C_{DI} to C_{GI} because the gate capacitances that set C_{GI} are also present in C_{D1}. Although not necessarily the case, k_{SL} is usually not far from one. And C_{G2} is f_o times the C_{G1} that sets C_{GI}. This k_{SL} is the *self-loading coefficient*.

When f_o is consistent across stages, f_o reduces R_D's by as much as f_o raises C_O's, so t_P's match. t_P across N stages is therefore N times t_{P1}:

$$t_P = Nt_{P1} = N(69\%R_{D1}C_{O1}) \approx 69\%NR_{D1}C_{GI}(1 + f_o). \quad (8.33)$$

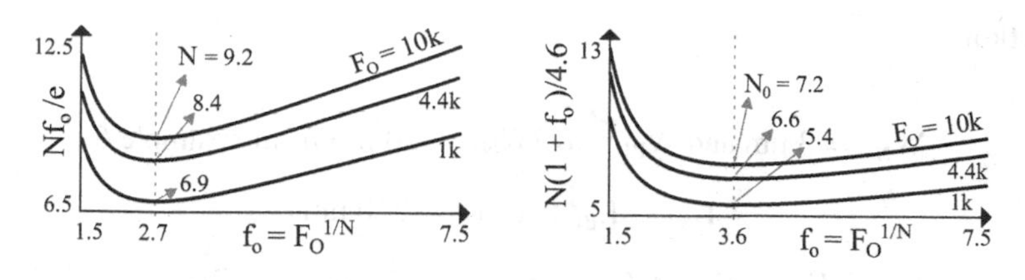

Fig. 8.19 Optimal gate-driver setting for minimum delay

F_O is very high when C_{GO} is much greater than C_{GI}. With this F_O, the f_o that $F_O^{1/N}$ sets is much greater than one with one stage. f_o falls below this level and approaches one as the number of stages increases.

But as N falls for a fixed F_O, f_o climbs and Nf_o and $N(1 + f_o)$ fall, bottom, and rise. Interestingly, Nf_o bottoms when f_o is e and $N(1 + f_o)$ and t_P bottom when f_o is 3.6. When f_o is e, Nf_o/e in Fig. 8.19 is the optimal N when self-loading (k_{SL}) is negligible. $N(1 + f_o)/(1 + 3.6)$ when f_o is 3.6 nears a more optimal N_O because k_{SL} is more realistic near one.

A fractional N is not practicable. Rounding N to the nearest integer without adjusting f_o is better, but not optimal. Adjusting f_o for the F_O targeted and intended integer N selected is better.

Interstage RC M_N in the first inverter stage K_1 pulls v_{O1} low when i_N overcomes i_P, which happens after v_{GS} overcomes v_T. As v_{GS} rises over v_T and v_{O1} falls across $\Delta v_{O(MAX)}$, M_N enters and remains in triode. M_P is similarly in triode as v_{SG} climbs over v_T and v_O rises. When strengths match, R_{D1} is roughly M_N's triode resistance when v_{GS} nears v_T. C_{GI} is roughly the *channel capacitance* C_{CH1} that W's and L's set across the *oxide capacitance* C_{OX}'' (per unit area). So R_{D1} and C_{GI} are

$$R_{D1} \approx R_{N1}\Big|_{v_{DS}<v_{DS(SAT)}}^{v_{GS}>V_{TN0}} \approx \frac{L_{N1} - 2L_{OL}}{W_{N1}K_N'(v_T - V_{TN0})} \tag{8.34}$$

and

$$C_{GI} \approx C_{CH1} = C_{CHN1} + C_{CHP1} \approx (W_{N1}L_{N1} + W_{P1}L_{P1})C_{OX}'', \tag{8.35}$$

where L_{OL} is *overlap length*.

Example 7: Determine N and t_P with the inverter from Example 6 when L_{MIN} is 250 nm, L_{OL} is 30 nm, C_{OX}'' is 6.9 fF/μm^2, and the load is a 100-mm-wide and 250-nm-long NFET.

Solution:

$$W_{N1} = 3\ \mu\text{m and } W_{P1} = 6.7W_{N1} = 20\ \mu\text{m from Example 6}$$

$$L_{N1} \equiv L_{P1} \equiv L_{MIN} = 250\ \text{nm}$$

$$C_{GI} = C_{GN} + C_{GP} = C_{GN} + 6.7C_{GN} = 7.7C_{GN}$$

$$F_O = \frac{C_{GO}}{C_{GI}} = \frac{W_{GO}}{7.7W_{GN}} = \frac{100\text{m}}{7.7(3\mu)} = 4.4\text{k}$$

$$F_O = f_o^{N_O} \equiv 3.6^{N_O} = 4.4\text{k} \quad \therefore\ N_O = 6.6 \ \rightarrow\ N \equiv 7\ \text{Stages} \ \rightarrow\ f_o = 3.3$$

$$R_{D1} \approx \frac{L_{N1} - 2L_{OL}}{W_{N1}K_{N'}(v_T - V_{TN0})} = \frac{250\text{n} - 2(30\text{n})}{(3\mu)(200\mu)(2 - 500\text{m})} = 210\ \Omega$$

$$C_{GI} \approx (W_{N1}L_{N1} + W_{P1}L_{P1})C_{OX}'' = [(3\mu)(250\text{n}) + (20\mu)(250\text{n})](6.9\text{m}) = 40\ \text{fF}$$

$$t_P \approx 69\%NR_{D1}C_{GI}(1 + f_o) = (69\%)(7)(210)(40\text{f})(1 + 3.3) = 170\ \text{ps}$$

Explore with SPICE:
See Appendix A for notes on SPICE simulations.

```
* Seven-Stage Gate Driver
vp vp 0 dc=0 pulse 0 4 0 50p 50p 500p 1n
vdd vdd 0 dc=4
vdd2 vdd2 0 dc=4
mp0 vi vp vdd2 vdd2 pmos2 w=20u l=250n
mn0 vi vp 0 0 nmos2 w=3u l=250n
mp1 vo1 vi vdd vdd pmos2 w=20u l=250n
mn1 vo1 vi 0 0 nmos2 w=3u l=250n
mp2 vo2 vo1 vdd vdd pmos2 w=66u l=250n
mn2 vo2 vo1 0 0 nmos2 w=10u l=250n
mp3 vo3 vo2 vdd vdd pmos2 w=218u l=250n
mn3 vo3 vo2 0 0 nmos2 w=33u l=250n
mp4 vo4 vo3 vdd vdd pmos2 w=720u l=250n
mn4 vo4 vo3 0 0 nmos2 w=108u l=250n
mp5 vo5 vo4 vdd vdd pmos2 w=2370u l=250n
mn5 vo5 vo4 0 0 nmos2 w=356u l=250n
mp6 vo6 vo5 vdd vdd pmos2 w=7800u l=250n
mn6 vo6 vo5 0 0 nmos2 w=1170u l=250n
mp7 vo vo6 vdd vdd pmos2 w=25800u l=250n
```

(continued)

```
mn7 vo vo6 0 0 nmos2 w=3870u l=250n
mpo vo8 vo vdd2 vdd2 pmos2 w=85200u l=250n
mno vo8 vo 0 0 nmos2 w=12780u l=250n
.lib lib.txt
.tran 1n
.end
```

Tip: Plot v(vi), v(vo1), v(vo2), v(vo3), v(vo4), v(vo5), v(vo6), and v(vo).

B. Gate-Charge Power

C_{O1} needs *gate charge* q_{G1} to charge across v_O's swing between v_{SS} and v_{DD}. The power supplies deliver this q_{G1} every t_{SW}. So K_1's *gate-charge power* P_{G1} is $q_{G1}(v_{DD} - v_{SS})f_{SW}$, which is equivalent to $C_{O1}(v_{DD} - v_{SS})^2 f_{SW}$:

$$q_{G1} = C_{O1}\Delta v_{O(MAX)} = C_{O1}(v_{DD} - v_{SS}) \tag{8.36}$$

$$P_{G1} = (v_{DD} - v_{SS})q_{G1}f_{SW} = C_{O1}(v_{DD} - v_{SS})^2 f_{SW}. \tag{8.37}$$

Of the energy P_{G1} supplies, C_{O1} receives $0.5C_{O1}(v_{DD} - v_{SS})^2$, M_P burns the rest, and M_N later burns what C_{O1} holds. So K_1 loses all of P_{G1}. Since C_O's climb with f_o, the P_G lost across the driver increases exponentially with N:

$$P_G = \sum_{1}^{N} P_{G(K)} = P_{G1}\sum_{0}^{N-1} f_o{}^k, \tag{8.38}$$

where P_{G1} is $P_{G1}f_o{}^0$ and P_{GN} is $P_{G1}f_o{}^{N-1}$.

Example 8: Determine P_G for Example 7 when f_{SW} is 1 MHz.

Solution:

$$C_{GI} \approx 40 \text{ fF}, N \equiv 7, \text{ and } f_o = 3.3 \text{ from Example 7}$$

$$C_{O1} \approx C_{GI}(1 + f_o) \approx (40f)(1 + 3.3) = 170 \text{ fF}$$

$$P_{G1} = C_{O1}v_{DD}{}^2 f_{SW} = (170f)(4)^2(1M) = 2.7\ \mu W$$

$$P_G = P_{G1} \sum_{0}^{N-1} f_o^{\,k} \approx (2.7\mu) \sum_{0}^{7-1} 3.3^k = 5.0 \text{ mW}$$

Explore with SPICE:
Use the SPICE code from Example 7.

Tip: Plot I(Vdd)*V(vdd), extract the average, multiply times 1 ns, and divide by 1 μs to determine the power v_{DD} sources across the 1-μs period, which is mostly P_G and partially shoot-through power.

C. Shoot-Through Power

As M_N discharges C_O, M_P conducts *shoot-through current* i_{ST} that M_N sinks. M_N also sinks i_{ST} that M_P supplies when M_P charges C_O. Since this i_{ST} flows from v_{DD} to v_{SS}, v_{DD} and v_{SS} lose *shoot-through power* P_{ST}.

Without C_{O1}, v_I and the *unloaded output* v_{O1}' in Fig. 8.20 crisscross at v_T. M_N and M_P invert their channels and saturate at this point. Since v_{GS}'s and v_{DS}'s match, i_N and i_P conduct the saturated i_T that v_T over V_{T0} sets:

$$i_T = i_{P/N}\big|_{v_T} \approx \left(\frac{W_N}{L_N}\right)\left(\frac{K_N'}{2}\right)(v_T - V_{TN0})^2(1 + v_T\lambda_N). \tag{8.39}$$

i_{ST} is below i_T before v_I reaches v_T because M_N's v_{GS} or M_P's v_{SG} is weaker than the other. i_{ST} is also below i_T after v_I passes v_T because the other's v_{SG} or v_{GS} is weaker than the first's. So as v_I transitions without C_{O1}, i_{ST} rises to i_T (where it maxes) and falls to zero.

But since C_{O1} delays v_{O1}, v_I reaches v_T before v_{O1} does. So at v_T, M_N's v_{DS} or M_P's v_{SD} is lower than the others. As a result, i_P sets i_{ST} with a lower v_{SD} when v_{O1} falls and i_N sets i_{ST} with a lower v_{DS} when v_{O1} rises. In other words, propagation delay keeps i_{ST} from reaching i_T.

When f_o is one (i. e., C_{G2} equals C_{GI} and $1 + f_o$ in C_O is two), i_{ST} can reach 15% of i_T. $i_{ST(MAX)}$ falls with higher C_O's below this point because delay reduces the v_{DS}

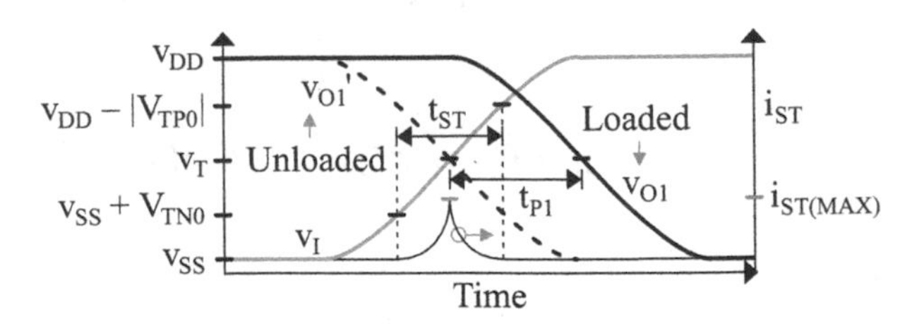

Fig. 8.20 Inverter's shoot-through response

that sets i_{ST}. So across transitions, i_{ST} rises and peaks to about $30\%i_T/(1 + f_o)$ before falling back to zero.

M_N and M_P conduct i_{ST} when v_I is between MOS thresholds: V_{TN0} above v_{SS} and $|V_{TP0}|$ below v_{DD}. The v_O that sets this v_I is from another stage whose R_D and C_O resemble R_{D1} and C_{O1}. So *shoot-through time* t_{ST} is the time v_I needs to traverse between MOS thresholds:

$$t_{ST} = t_{RC}\Big|_{v_{SS}+V_{TN0}}^{v_{DD}-|V_{TP0}|} \approx \tau_{RC} \ln\left[\frac{(v_{DD} - v_{SS}) - (v_{DD} - |V_{TP0}|)}{(v_{DD} - v_{SS}) - (v_{SS} + V_{TN0})}\right]^{-1} = \tau_{RC} \ln\left(\frac{v_{DD} - 2v_{SS} - V_{TN0}}{|V_{TP0}| - v_{SS}}\right). \tag{8.40}$$

i_{ST} averages about one-third of $i_{ST(MAX)}$ because i_{ST} scales quadratically with the v_{GS} that v_I sets and v_I scales linearly with time. P_{ST1} is therefore the power v_{DD} and v_{SS} burn with i_{ST1}'s average $i_{ST1(MAX)}/3$ across two t_{ST} fractions of t_{SW}: once when v_O rises and again when v_O falls:

$$P_{ST1} = i_{ST1}(v_{DD} - v_{SS})\left(\frac{2t_{ST}}{t_{SW}}\right) \approx \left(\frac{i_{ST(MAX)}}{3}\right)(v_{DD} - v_{SS})\left(\frac{2t_{ST}}{t_{SW}}\right) \approx \left[\frac{30\%i_{T1}}{3(1+f_o)}\right](v_{DD} - v_{SS})\left(\frac{2t_{ST}}{t_{SW}}\right). \tag{8.41}$$

P_{ST}'s scale with f_o^k because i_T's are f_o times higher than the preceding stage's i_T. So the P_{ST} lost across the driver climbs exponentially with N:

$$P_{ST} = \sum_{1}^{N} P_{ST(K)} \approx P_{ST1} \sum_{0}^{N-1} f_o^k, \tag{8.42}$$

where P_{ST1} is $P_{ST1}f_o^0$ and P_{STN} is $P_{ST1}f_o^{N-1}$.

Example 9: Determine P_{ST} for Example 7 when λ_N is 5%.

Solution:

$v_T = 2$ V, $W_N = 3$ μm, $L_N = 250$ nm, $R_{D1} = 210$ Ω, $C_{GI} = 40$ fF,
and $f_o = 3.3$ from Example 7

$$i_{T1} \approx \left(\frac{W_N}{L_N - 2L_{OL}}\right)\left(\frac{K_N'}{2}\right)(v_T - V_{TN0})^2(1 + v_T\lambda_N)$$

$$= \left(\frac{3\mu}{190n}\right)\left(\frac{200\mu}{2}\right)(2 - 500m)^2[1 + (2)5\%] = 4.0 \text{ mA}$$

$$\tau_{RC} \approx R_{D1}C_{GI}(1 + f_o) = (210)(40f)(1 + 3.3) = 36 \text{ ps}$$

$$t_{ST} \approx \tau_{RC} \ln\left(\frac{v_{DD} - V_{TN0}}{|V_{TP0}|}\right) = (36p)\ln\left(\frac{4 - 500m}{700m}\right) = 58 \text{ ps}$$

$$P_{ST1} \approx \left[\frac{30\% i_{T1}}{2(1 + f_o)}\right] v_{DD}\left(\frac{2t_{ST}}{t_{SW}}\right) = \left[\frac{30\%(4.0m)}{3(1 + 3.3)}\right](4)\left[\frac{2(58p)}{1\mu}\right] = 43 \text{ nW}$$

$$P_{ST} \approx P_{ST1}\sum_{0}^{N-1} {f_o}^k \approx (43n)\sum_{0}^{7-1} 3.3^k = 80\ \mu\text{W}$$

Note: P_{ST} is 1.6% of P_G.

Explore with SPICE:
Use the SPICE code from Example 7.

Tip: Plot I(Vdd)*V(vdd), extract the average, multiply times 1 ns, and divide by 1 μs to determine the power v_{DD} sources across the 1-μs period, which is about 98% P_G and 1.6% P_{ST}.

D. Design Notes

This analysis shows t_P is minimal when f_o nears 3.6, t_P scales with N, and P_G and P_{ST} climb exponentially with N. With f_o fixed, F_O sets N. And since F_O is usually high in power supplies, this N is also relatively high. So t_P, P_G, and P_{ST} are that much higher than for the lightly loaded minimum-size inverter.

P_{ST} maxes without C_O. P_{ST} is roughly 15% of this maximum when f_o is one and 6.5% when f_o is 3.6. At this level, P_{ST} is a small fraction of P_G, so total *gate-driver power* P_{DRV} is mostly the P_G that C_O's draw from v_{DO}.

Delay scales with $N(1 + f_o)$ and power with $\Sigma {f_o}^k$ up to ${f_o}^{N-1}$. So P_{DRV} is more sensitive to N than t_P is. In this light, reducing N favors P_{DRV} savings over t_P. In power supplies, limiting N to three or five and setting f_o with this N to a level that is higher than 3.6 is fairly routine for this reason.

In practice, R_{D1} and C_{GI} are dynamic parameters. Engineers often skew W's in the last stage to minimize the *i_{DS}–v_{DS} overlap power* P_{IV} lost across the power switch. And power switches are sometimes off chip. Even with these variations, fewer stages often save more power than extend delay. Plus, saving power is often more significant and important than shortening delay.

Example 10: Determine t_P and P_{DRV} for Example 7 when N is 4.

Solution:

$$F_O = 4.4k, R_{D1} = 210\ \Omega, \text{ and } C_{GI} = 40\ fF \text{ from Example 7}$$

$$F_O = f_o^{\ N} \equiv f_o^{\ 4} = 4.4k \quad \therefore \quad f_o = 8.1$$

$$t_P \approx 69\%NR_{D1}C_{GI}(1 + f_o) = (69\%)(4)(210)(40f)(1 + 8.1) = 210\ ps$$

$$C_{O1} \approx C_{GI}(1 + f_o) \approx (40f)(1 + 8.1) = 360\ fF$$

$$P_{G1} = C_{O1}v_{DD}^{\ 2}f_{SW} = (360f)(4)^2(1M) = 5.8\ \mu W$$

$$P_G \approx P_{G1}\sum_0^{N-1} f_o^{\ k} \approx (5.8\mu)\sum_0^{4-1} 8.1^k = 3.5\ mW$$

$$\tau_{RC} \approx R_{D1}C_{GI}(1 + f_o) = (210)(40f)(1 + 8.1) = 76\ ps$$

$$t_{ST} \approx \tau_{RC}\ln\left(\frac{v_{DD} - V_{TN0}}{|V_{TP0}|}\right) = (76p)\ln\left(\frac{4 - 500m}{700m}\right) = 120\ ps$$

$$P_{ST1} \approx \left[\frac{30\% i_{T1}}{3(1 + f_o)}\right]v_{DD}\left(\frac{2t_{ST}}{t_{SW}}\right) = \left[\frac{30\%(4.0m)}{3(1 + 8.1)}\right](4)\left[\frac{2(120p)}{1\mu}\right] = 42\ nW$$

$$P_{ST} \approx P_{ST1}\sum_0^{N-1} f_o^{\ k} \approx (42n)\sum_0^{4-1} 8.1^k = 26\ \mu W$$

$$P_{DRV} = P_G + P_{ST} = 3.5m + 22\mu = 3.5\ mW$$

Note: Reducing N to 4 extends t_P 24% and reduces P_{DRV} 30%.

Explore with SPICE:
See Appendix A for notes on SPICE simulations.

```
* Four-Stage Gate Driver
vp vp 0 dc=0 pulse 0 4 0 50p 50p 500p 1n
vdd vdd 0 dc=4
vdd2 vdd2 0 dc=4
mp0 vi vp vdd2 vdd2 pmos2 w=20u l=250n
mn0 vi vp 0 0 nmos2 w=3u l=250n
mp1 vo1 vi vdd vdd pmos2 w=20u l=250n
mn1 vo1 vi 0 0 nmos2 w=3u l=250n
mp2 vo2 vo1 vdd vdd pmos2 w=162u l=250n
mn2 vo2 vo1 0 0 nmos2 w=24.3u l=250n
mp3 vo3 vo2 vdd vdd pmos2 w=1312u l=250n
mn3 vo3 vo2 0 0 nmos2 w=197u l=250n
mp4 vo vo3 vdd vdd pmos2 w=10627u l=250n
mn4 vo vo3 0 0 nmos2 w=1596u l=250n
mpo vo5 vo vdd2 vdd2 pmos2 w=86079u l=250n
mno vo5 vo 0 0 nmos2 w=12928u l=250n
.lib lib.txt
.tran 1n
.end
```

Tip: Plot v(vi), v(vo1), v(vo2), v(vo3), v(vo), and I(Vdd)*V(vdd), extract the average of the product term, multiply times 1 ns, and divide by 1 μs to determine the power v_{DD} sources across the 1-μs period, which is mostly P_G.

8.3.4 Dead-Time Logic

Power switches that connect to the same switching node should not close at the same time. If they do, they would short v_{IN} or v_O to ground. Even if the switches do not melt, they would burn too much power.

Engineers insert *dead time* t_{DT} between adjacent switches for this reason. This way, even if turn-on gate signals overlap, switches cannot close at the same time. Consider gate command v_G' and *gate voltages* v_G and v_{GX} in Fig. 8.21, for example. v_G rises a t_{DT} after adjacent v_{GX} falls, even when v_G' commands a high before v_{GX} falls.

t_{DT} only applies to closing (turn-on) events, when burning unnecessary power is possible. Opening switches is inherently safe because it stops current flow. This is why v_G falls with v_G' without delay in Fig. 8.21.

t_{DT} should delay turn-on commands only when adjacent commands are high, which is when shorting events are possible. Whenever safe, reducing delays helps

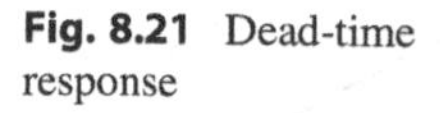

Fig. 8.21 Dead-time response

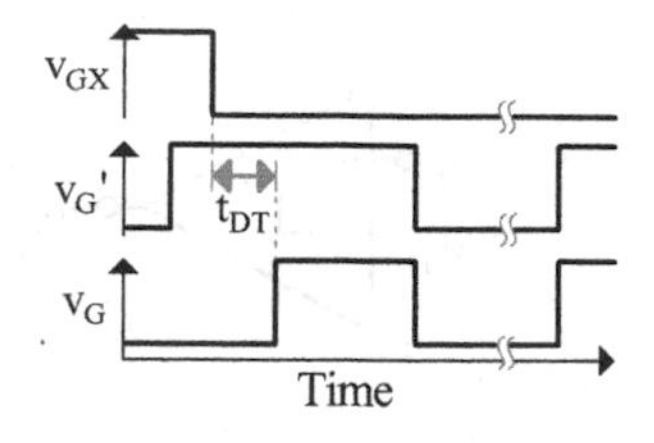

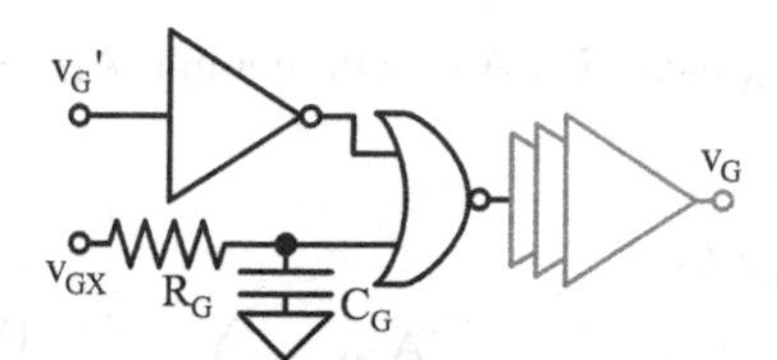

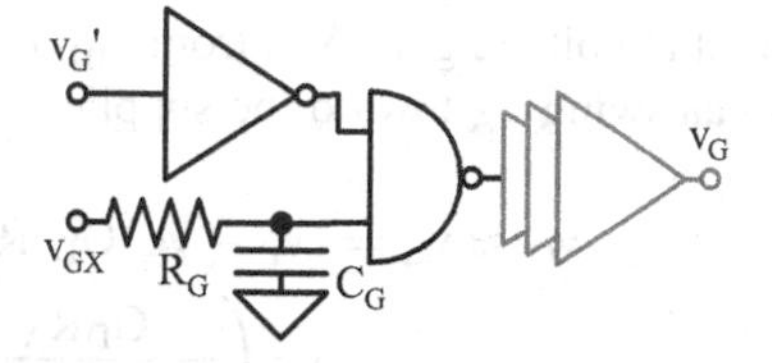

Fig. 8.22 Active-high and active-low dead-time circuits

the system respond and recover more quickly. This is why v_G' in Fig. 8.21 rises with v_G' without delay the second time v_G' transitions high.

To assert these states, *dead-time circuits* should sense, delay, and compare neighboring commands with v_G'. R's and C's in Fig. 8.22, for example, sense and delay neighboring v_G's. And NOR and NAND gates compare these delayed signals with v_G' to determine v_G. R's and C's or inverter chains can sense and delay neighboring v_G's.

The NOR gate outputs an active-high command when v_G' is high after neighboring v_G's fall. The NAND gate outputs an active-low command when v_G' is low after neighboring v_G's rise. The NOR gate deactivates v_G as soon as v_G' falls and the NAND gate deactivates v_G as soon as v_G' rises.

8.4 Comparator Blocks

8.4.1 Comparators

A *comparator* is an *analog–digital converter* (ADC) that compares two analog voltages and outputs a digital voltage to indicate which is higher. v_O is high when the *positive input* v_P overcomes the *negative input* v_N and low when v_N overcomes v_P. In effect, v_O is high when the *differential input voltage* v_{ID} between v_P and v_N is positive and low when v_{ID} is negative. So comparators are also *polarity detectors*.

For this functionality, comparators usually sense and amplify v_{ID} until v_O saturates near v_{DD} or v_{SS}. In this respect, comparators are like op amps. The only difference is comparators do not need the stabilizing components op amps use to close stable feedback loops around them.

So like op amps, most comparators sense v_{ID} in Fig. 8.23 with a differential transconductor G_D and amplify with resistance R_A. This way, the low-frequency

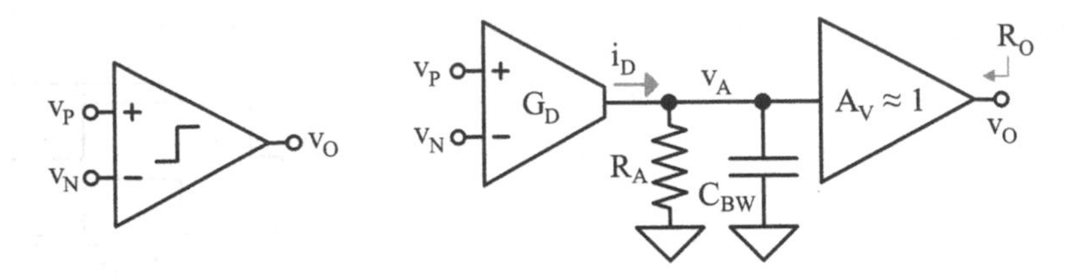

Fig. 8.23 Typical comparator

differential voltage gain A_{D0} from v_{ID} to v_A is G_DR_A. And with enough A_{D0}, small v_{ID}'s can swing v_A toward the supplies:

$$\begin{aligned} v_O \approx v_A &= (v_P - v_N)G_D(R_A \| Z_C) \\ &= v_{ID}\left(\frac{G_DR_A}{1 + R_AC_{BW}s}\right) = v_{ID}\left(\frac{A_{D0}}{1 + s/2\pi p_{BW}}\right). \end{aligned} \tag{8.43}$$

C_{BW} embeds the unintended capacitances that shunt R_A past the *bandwidth-setting pole* p_{BW} that sets t_P. The voltage buffer A_V sets the *output resistance* R_O with which v_O establishes a state. Actual implementations of G_D and A_V cannot swing v_A or v_O past their supplies v_{SS} and v_{DD}.

Explore with SPICE:
See Appendix A for notes on SPICE simulations.

```
* Comparator
vdd vdd 0 dc=2
vss vss 0 dc=-2
vp vp 0 dc=495m pwl 0 490m 10n 490m 10.01n 510m 500n 510m
+ 500.01n 490m
vn vn 0 dc=500m
x1 vp vn vo vdd vss cp
.lib lib.txt
.dc vp 450m 550m 1u
*.tran 1u
.end
```

Tip: Plot v(vp) and v(vn) on one graph and v(vo) on another. Move "*" so it precedes ".dc," re-run simulation, and plot v(vp) and v(vn) on one graph and v(vo) on another.

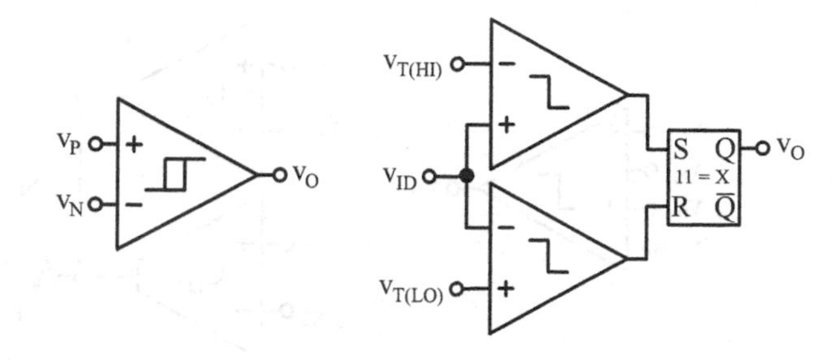

Fig. 8.24 Flip-flopped hysteretic comparator

8.4.2 Hysteretic Comparators

Hysteretic comparators trip high when v_{ID} rises over the *rising threshold* $v_{T(HI)}$ and low when v_{ID} falls under the *falling threshold* $v_{T(LO)}$. Between these thresholds, v_O holds the previous state. The difference between these thresholds is the *hysteresis* Δv_T of the comparator.

The two comparators in Fig. 8.24 realize a hysteretic comparator by driving the S and R terminals of an SR flip flop. The top and bottom comparators set and reset the flip flop when v_{ID} rises over $v_{T(HI)}$ and falls under $v_{T(LO)}$. Between thresholds, when the comparator outputs are low, the flip flop holds the previous state. Since v_{ID} cannot both rise over $v_{T(HI)}$ and fall under $v_{T(LO)}$ at the same time, set/reset dominance is irrelevant.

Some implementations use positive feedback to latch the state of the comparator past its trip point. Holding the state this way effectively splits the trip point apart into separate thresholds. This way, the output flips state when the input crosses separate upper and lower thresholds. And like in digital flip flops, positive feedback holds the previous state.

Explore with SPICE:
See Appendix A for notes on SPICE simulations.

```
* Hysteretic Comparator: Temporal Model
vdd vdd 0 dc=2
vss vss 0 dc=-2
vp vp 0 dc=450m pwl 0 450m 1m 550m 2m 450m
vn vn 0 dc=500m
x1 vp vn vo vdd vss cphys50m
.lib lib.txt
.tran 2m
.end
```

Tip: Plot v(vp) and v(vn) on one graph and v(vo) on another.

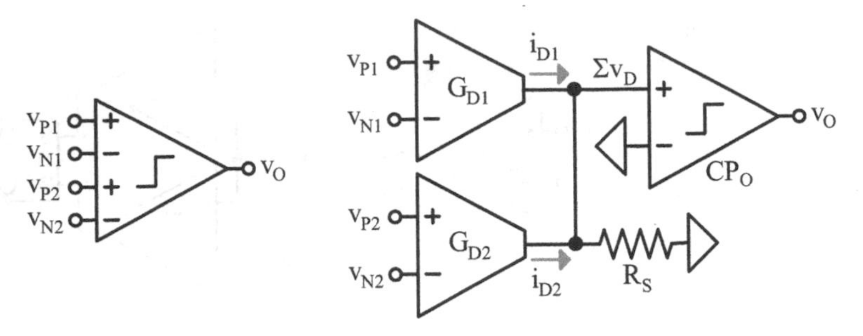

Fig. 8.25 Summing comparator

8.4.3 Summing Comparators

Summing comparators add their inputs. So v_O trips high when v_P's overcome v_N's and low when v_N's surpass v_P's. And v_O approaches v_{DD} when the *differential sum* Σv_{ID} is positive and v_{SS} when Σv_{ID} is negative.

Paralleling differential transconductors into a comparator is one way of adding inputs. G_{D1} and G_{D2} in Fig. 8.25, for example, add i_{D1} and i_{D2}, R_S drops Σv_D, and CP_O compares Σv_D to zero. When G_D's match, i_{D1} and i_{D2} are $\Sigma v_{ID}G_D$ and Σv_D is $\Sigma v_{ID}G_DR_S$. This way, Σv_{ID}'s polarity matches Σv_D's. So CP_O trips v_O high and low when Σv_{ID} and Σv_D are positive and negative:

$$\begin{aligned}\Sigma v_D &= (i_{D1} + i_{D2})R_S = [(v_{P1} - v_{N1}) + (v_{P2} - v_{N2})]G_DR_S \\ &= (v_{ID1} + v_{ID2})G_DR_S = \Sigma v_{ID}G_DR_S. \end{aligned} \tag{8.44}$$

Explore with SPICE:
See Appendix A for notes on SPICE simulations.

```
* Three-Input Comparator
vdd vdd 0 dc=2
vss vss 0 dc=-2
vp1 vp1 0 dc=495m pwl 0 490m 1m 510m 2m 490m
vn1 vn1 0 dc=500m
vp2 vp2 0 dc=710m
vn2 vn2 0 dc=715m
vp3 vp3 0 dc=405m
vn3 vn3 0 dc=400m
x1 vp1 vn1 vp2 vn2 vp3 vn3 vo vdd vss cp3vid
.lib lib.txt
*.tran 2m
```

(continued)

```
.dc vp1 450m 550m 0.1m
.end
```

Tip: Plot v(vp1) – v(vn1), v(vp2) – v(vn2), and v(vp3) – v(vn3) on one graph and v(vo) on another. Move "*" so it precedes ".dc," re-run simulation, and plot v(vp1) – v(vn1), v(vp2) – v(vn2), and v(vp3) – v(vn3) on one graph and v(vo) on another.

Summing hysteretic comparators add inputs the same way. In Fig. 8.26, for example, G_{D1} and G_{D2} add i_{D1} and i_{D2} into R_S and a hysteretic comparator compares the resulting Σv_D to zero. When G_D's match $1/R_S$, the $\Sigma v_{ID} G_D R_S$ that sets Σv_D reproduces Σv_{ID}. This way, v_O trips when Σv_{ID} climbs over or below CP_Σ's $v_{T(HI)}$ or $v_{T(LO)}$. Other implementations integrate CP_Σ's v_T and feedback mechanics together with G_{D1} and G_{D2}.

Explore with SPICE:
See Appendix A for notes on SPICE simulations.

```
* Three-Input Hysteretic Comparator: Temporal Model
vdd vdd 0 dc=2
vss vss 0 dc=-2
vp1 vp1 0 dc=450m pwl 0 450m 1m 550m 2m 450m
vn1 vn1 0 dc=500m
vp2 vp2 0 dc=750m
vn2 vn2 0 dc=740m
vp3 vp3 0 dc=600m
vn3 vn3 0 dc=605m
x1 vp1 vn1 vp2 vn2 vp3 vn3 vo vdd vss cphys3vid50m
.lib lib.txt
.tran 2m
.end
```

Tip: Plot v(vp1) – v(vn1), v(vp2) – v(vn2), and v(vp3) – v(vn3) on one graph and v(vo) on another.

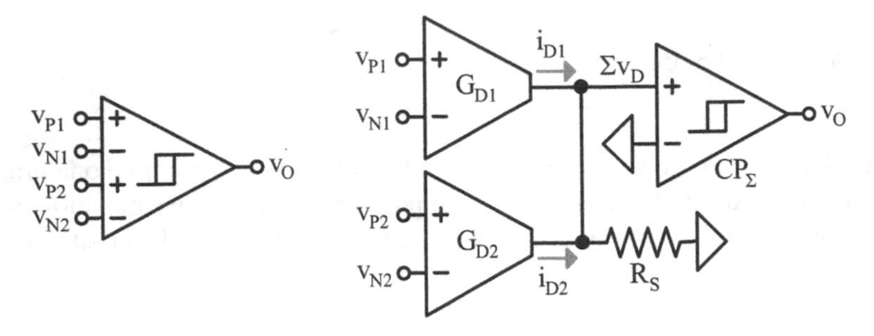

Fig. 8.26 Summing hysteretic comparator

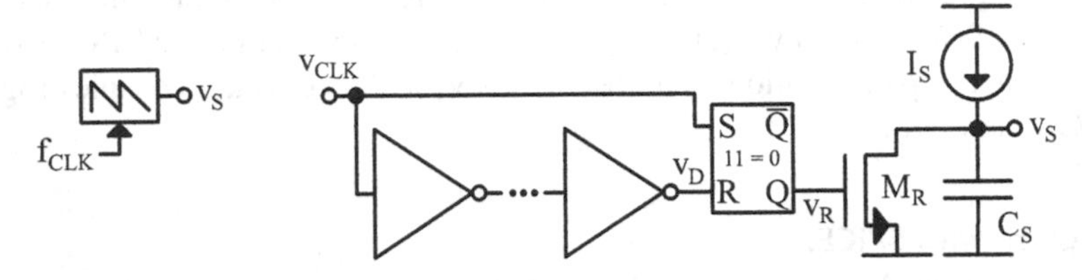

Fig. 8.27 Clocked sawtooth generator

8.5 Timing Blocks

8.5.1 Clocked Sawtooth Generator

A *clocked sawtooth generator* outputs a *sawtooth voltage* v_S that ramps across every *clock period* t_{CLK}. This type of circuit needs a pulsed clock input v_{CLK}, a ramp generator, and a reset circuit. In Fig. 8.27, I_S and C_S ramp v_S and the inverter chain, SR flip flop, and M_R reset v_S.

The inverters output with v_D a delayed version of v_{CLK}. v_{CLK} sets the flip flop that starts a reset event and the inverters end the "reset" when v_D resets the flip flop. This way, the flip flop pulses v_R across the *inverters' combined delay* t_I.

M_R should discharge C_S before t_I elapses. Since v_R reaches v_{DD} when M_R resets v_S and v_{DD} is typically much higher than $v_{S(HI)}$, the v_{GS} that v_R establishes usually overwhelms the v_{DS} that $v_{S(HI)}$ sets. So M_R's triode resistance R_R collapses most of v_S. The resulting *reset time* t_R should be shorter than t_I:

$$t_R \approx t_{90\%\Delta v_{O(MAX)}} = \tau_{RC} \ln(1 - 90\%)^{-1} \approx 2.3\tau_{RC} = 2.3R_R C_S. \tag{8.45}$$

v_R falls quickly across v_{DD} and v_{SS} after t_I elapses. This transition is so fast that M_R's C_{GD} injects noise into C_S. The voltage-divided noise that C_{GD} drops across C_S shifts $v_{S(LO)}$ below v_{SS}. Since M_R is in triode when this happens, C_{GD} is roughly half of C_{CH}, which is usually much lower than C_S:

$$v_{S(LO)} \approx v_{SS} - \Delta v_R\left(\frac{C_{GD}}{C_{GD} + C_S}\right) \approx v_{SS} - (v_{DD} - v_{SS})\left(\frac{0.5C_{CH}}{C_S}\right). \quad (8.46)$$

After v_R falls, I_S ramps v_S from this $v_{S(LO)}$ to $v_{S(HI)}$ across the part of t_{CLK} that excludes t_I:

$$v_{S(HI)} \approx v_{S(LO)} + \left(\frac{I_S}{C_S}\right)(t_{CLK} - t_I). \quad (8.47)$$

Example 11: Determine I_S and W_R so $v_{S(HI)}$ is 300 mV when v_{DD} is 1 V, C_S is 5 pF, t_{CLK} is 1 μs, t_I is 2 ns, V_{TN0} is 500 mV, K_N' is 200 μA/V^2, C_{OX}'' is 6.9 fF/μm^2, L_R is 250 nm, and L_{OL} is 30 nm.

Solution:

$$t_R \approx 2.3R_RC_S = 2.3R_R(5p) \equiv 25\%t_I = 25\%(2n) = 500\text{ ps} \quad \therefore \quad R_R \leq 44\ \Omega$$

$$R_R \approx \frac{L_R - 2L_{OL}}{W_RK_N'(v_{DD} - V_{TN0})} = \frac{250n - 2(30n)}{W_R(200\mu)(1 - 500m)} \leq 44\ \Omega$$

$$\therefore \quad W_R \geq 43\text{ nm} \quad \rightarrow \quad W_R \equiv 43\ \mu\text{m}$$

$$v_{S(LO)} \approx -v_{DD}\left(\frac{0.5W_RL_RC_{OX}''}{C_S}\right) = -(1)\left[\frac{(0.5)(43\mu)(250n)(6.9m)}{5p}\right] = -7\text{ mV}$$

$$v_{S(HI)} \approx v_{S(LO)} + \left(\frac{I_S}{C_S}\right)(t_{CLK} - t_I) \approx -7m + \left(\frac{I_S}{5p}\right)(1\mu - 2n) \equiv 300\text{ mV}$$

$$\therefore \quad I_S = 1.5\ \mu\text{A}$$

Explore with SPICE:
See Appendix A for notes on SPICE simulations.

```
* Clocked Sawtooth Generator
vclk vclk 0 dc=0 pulse 0 1 0 0.1n 0.1n 500n 1u
xi1 vclk v1 inv
```

(continued)

```
xi2 v1 vd inv
xsrr vclk vd vr qn srr
is 0 vs dc=1.5u
cs vs 0 5p
mr vs vr 0 0 nmos2 w=43u l=250n
.ic v(vs)=0
.lib lib.txt
.tran 4u
.end
```

Tip: Plot v(vclk), v(vr), and v(vs) across 4 μs and between 0.995 and 1.010 μs.

8.5.2 Sawtooth Oscillator

A *sawtooth oscillator* does not need a clock to generate v_S. In fact, it is a clock. And in the case of Fig. 8.28, it is also a *relaxation oscillator*: a delayed (by C_S), unity-gain (by v_T) inverting (across M_R) feedback loop.

Here, I_S and C_S ramp v_S and comparator CP_R resets v_S after v_S overcomes v_T. v_R determines $v_{S(LO)}$ when v_R falls. v_R trips after v_S climbs over v_T and across CP_R's rising t_P^+ to $v_{S(HI)}$. This way, v_S rises to

$$v_{S(HI)} = v_T + \left(\frac{I_S}{C_S}\right) t_P^+. \tag{8.48}$$

Since v_S falls below v_T when M_R resets v_S, M_R should collapse v_S before CP_R's falling t_P^- elapses.

t_{CLK} is the time C_S and CP_R need to ramp v_S from $v_{S(LO)}$ to $v_{S(HI)}$ and the time CP_R needs to start another cycle:

$$t_{CLK} = \left(\frac{C_S}{I_S}\right)\left(v_{S(HI)} - v_{S(LO)}\right) + t_P^- = \left(\frac{C_S}{I_S}\right)\left(v_T - v_{S(LO)}\right) + t_P^+ + t_P^-. \tag{8.49}$$

Since I_S, C_S, and t_P vary with temperature and fabrication runs, t_{CLK}'s tolerance is usually fairly high. Trimming and balancing variations across temperature reduce this variability, but only with more silicon area and test time.

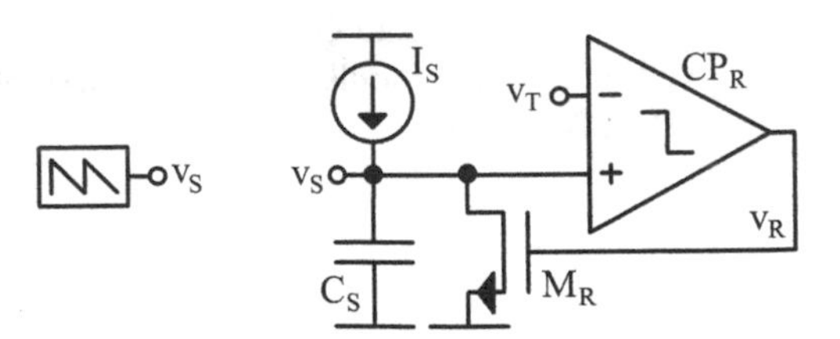

Fig. 8.28 Sawtooth oscillator

Example 12: Determine v_T, I_S, and W_R so $v_{S(HI)}$ is 300 mV and t_{CLK} is 1 μs when v_{DD} is 1 V, C_S is 5 pF, t_P is 100 ns, V_{TN0} is 500 mV, K_N' is 200 μA/V^2, C_{OX}'' is 6.9 fF/μm^2, L_R is 250 nm, L_{OL} is 30 nm, and W_{MIN} is 3 μm.

Solution:

$$t_R \approx 2.3R_RC_S = 2.3R_R(5p) \equiv 25\%t_P = 25\%(100n) = 25\,\text{ns} \quad \therefore \quad R_R \le 2.2\,\text{k}\Omega$$

$$R_R \approx \frac{L_R - 2L_{OL}}{W_RK_N'(v_{DD} - V_{TN0})} = \frac{250n - 2(30n)}{W_R(200\mu)(1 - 500m)} \le 2.2\,\text{k}\Omega$$

$$\therefore \quad W_R \ge 860\,\text{nm} \quad \to \quad W_R \equiv 3\,\mu\text{m}$$

$$v_{S(LO)} \approx -v_{DD}\left(\frac{0.5W_RL_RC_{OX}''}{C_S}\right) = -(1)\left[\frac{(0.5)(3\mu)(250n)(6.9m)}{5p}\right] = -520\,\mu\text{V}$$

$$t_{CLK} = \left(\frac{C_S}{I_S}\right)\left(v_{S(HI)} - v_{S(LO)}\right) + t_P^-$$

$$= \left(\frac{5p}{I_S}\right)[300m - (-520\mu)] + 100n \equiv 1\mu\text{s} \quad \therefore \quad I_S = 1.7\,\mu\text{A}$$

$$v_{S(HI)} = v_T + \left(\frac{I_S}{C_S}\right)t_P^+ = v_T + \left(\frac{1.7\mu}{5p}\right)(100n) \equiv 300\,\text{mV} \quad \therefore \quad v_T = 270\,\text{mV}$$

Explore with SPICE:

See Appendix A for notes on SPICE simulations.

```
* Sawtooth Oscillator
is 0 vs dc=1.7u
cs vs 0 5p
mr vs vr 0 0 nmos2 w=3u l=250n
v1v v1v 0 dc=1
xcpd vs vt vr v1v 0 cp
vt vt 0 dc=270m
.ic v(vs)=0
.lib lib.txt
.tran 4u
.end
```

Tip: Plot v(vs), v(vt), and v(vr) across 4 μs and between 0.5 and 1.5 μs.

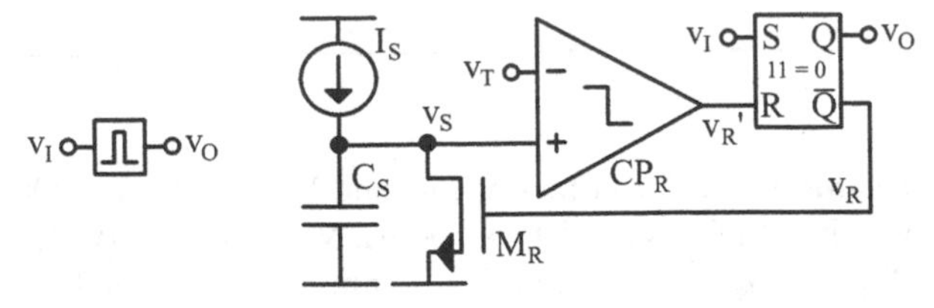

Fig. 8.29 One-shot oscillator

8.5.3 One-Shot Oscillator

One-shot oscillators are *pulse generators*. But more specifically, they are interruptible relaxation oscillators. They pulse once when pulsed and pulse continuously when kept enabled.

Like *ring oscillators*, they usually close unity-gain inverting feedback loops that internal components delay. In the case of Fig. 8.29, CP_R, the flip flop, and M_R close the loop, v_T and CP_R ensure v_S peaks match (so cycle–cycle gain is one), and I_S and C_S delay the loop. This way, v_O pulses each time v_I pulse-sets the flip flop and oscillates between states when v_I stays high.

Raising v_I sets v_O and opens M_R to start a pulse. I_S charges C_S across the pulse until M_R resets v_S. The *pulse width* t_{PW} is the time C_S, CP_R, and the reset-dominant flip flop need to ramp v_S to v_T, trip, and reset. So across this t_{PW}, I_S charges C_S over v_T across t_P^+ and t_{SR}^+:

$$v_{S(HI)} = v_T + \left(\frac{I_S}{C_S}\right)\left(t_P^+ + t_{SR}^+\right) \tag{8.50}$$

$$t_{PW} = \left(\frac{C_S}{I_S}\right)\left(v_{S(HI)} - v_{S(LO)}\right) = \left(\frac{C_S}{I_S}\right)\left(v_T - v_{S(LO)}\right) + t_P^+ + t_{SR}^+. \tag{8.51}$$

Keeping v_I high sets another pulse when the reset command v_R' falls. This happens after v_S falls below v_T. The time between pulses t_{OFF} is the time CP_R and the flip flop need to trip v_R' and set v_O, which corresponds to their falling delays t_P^- and t_{SR}^-. So the *oscillation period* t_{OSC} is

$$t_{OSC} = t_{PW} + t_{OFF} \approx t_{PW} + t_P^- + t_{SR}^-. \tag{8.52}$$

To collapse v_S before t_{OFF} elapses, M_R's t_R should be shorter than t_{OFF}.

Example 13: Determine v_T, I_S, W_R, and t_{OSC} so $v_{S(HI)}$ is 300 mV and t_{PW} is 750 ns when v_{DD} is 1 V, C_S is 5 pF, t_P is 100 ns, t_{SR} is 1 ns, V_{TN0} is 500 mV, K_N' is 200 μA/V^2, C_{OX}'' is 6.9 fF/μm^2, L_R is 250 nm, L_{OL} is 30 nm, and W_{MIN} is 3 μm.

Solution:

$$t_{OFF} = t_P{}^- + t_{SR}{}^- = 100n + 1n = 100 \text{ ns}$$

$$t_{OSC} = t_{PW} + t_{OFF} = 750n + 101n = 850 \text{ ns}$$

$$t_R \approx 2.3R_RC_S = 2.3R_R(5p) \equiv 25\%t_{OFF} = 25\%(100n) = 25 \text{ ns}$$

$$\rightarrow \text{Same as Example 12} \quad \therefore \quad W_R \equiv 3\ \mu\text{m}, v_{S(LO)} = -520\ \mu\text{V}$$

$$t_{PW} = \left(\frac{C_S}{I_S}\right)\left(v_{S(HI)} - v_{S(LO)}\right) = \left(\frac{5p}{I_S}\right)(300m + 520\mu) \equiv 750 \text{ ns}$$

$$\therefore \quad I_S = 2.0\ \mu\text{A}$$

$$v_{S(HI)} = v_T + \left(\frac{I_S}{C_S}\right)(t_P{}^+ + t_{SR}{}^+) = v_T + \left(\frac{2.0\mu}{5p}\right)(100n + 1n)$$

$$\equiv 300 \text{ mV}\pi \quad \therefore \quad v_T = 260 \text{ mV}$$

Explore with SPICE:
See Appendix A for notes on SPICE simulations.

```
* One-Shot Oscillator
vi vi 0 dc=0 pwl 0 0 100n 0 100.1n 1
*vi vi 0 dc=0 pwl 0 0 100n 0 100.1n 1 200n 1 200.1n 0
is 0 vs dc=2.0u
cs vs 0 5p
mr vs vr 0 0 nmos2 w=3u l=250n
vt vt 0 dc=260m
v1v v1v 0 dc=1
xcpd vs vt vx v1v 0 cp
xsrr vi vx vo vr srr
.lib lib.txt
.tran 4u
.end
```

Tip: Plot v(vi), v(vo), v(vs), v(vt), and v(vr) across 4 μs and between 0.5 and 1.5 μs. Move "*" so it precedes the first "vi" statement, re-run simulation, and plot v(vi), v(vo), v(vs), v(vt), and v(vr).

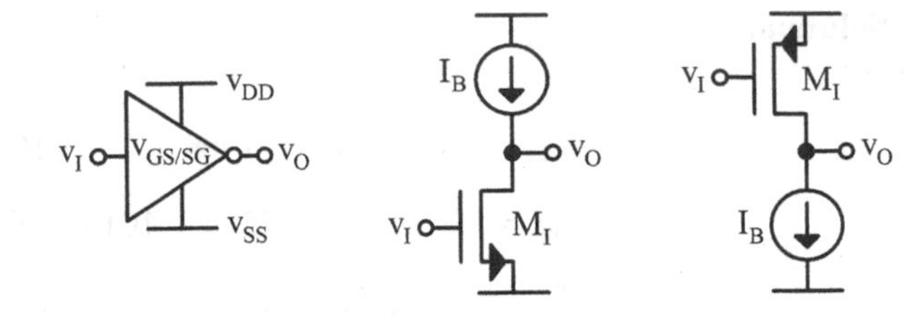

Fig. 8.30 Class-A pull-down/pull-up inverters

8.6 Switch Blocks

8.6.1 Class-A Inverters

Inverters are essentially comparators that compare v_I to a v_P that v_T sets. In the case of *class-A inverters*, v_O trips when v_I climbs over or below the v_T that M_I's v_{GS} or v_{SG} in Fig. 8.30 sets with v_{SS} or v_{DD}. These inverters are useful when transitioning analog v_{GS}-level swings to digital supply-level swings.

These circuits balance when v_I is the v_G needed to sustain I_B with v_O between the supplies. Excess M_I current or I_B pulls v_O low when v_I is higher and excess I_B or M_I current pulls v_O high when v_I is lower. This balancing v_G, which sets v_T, is V_{GS} over v_{SS} or V_{SG} below v_{DD} when M_I carries I_B in saturation:

$$v_T = V_{GS}\Big|_{I_B}^{V_{DS}>V_{DS(SAT)}} + v_{DD/SS}. \tag{8.53}$$

Example 14: Determine W and L so v_I is 650 mV when v_{DD} is 4 V, I_B is 10 μA, W_{MIN} is 3 μm, L_{OL} is 30 nm, K_N' is 200 μA/V^2, V_{TN0} is 500 mV, and λ_N is 5%.

Solution:

$$v_I \equiv 650\text{ mV} > V_{TN0} = 500\text{ mV} \quad \therefore \quad \text{Inversion}$$

$$V_{DS} = 0.5v_{DD} = 0.5(4) = 2\text{ V}$$

$$I_B = \left(\frac{W_{CH}}{L_{CH}}\right)\left(\frac{K_N'}{2}\right)(v_T - V_{TN0})^2(1 + \lambda_N V_{DS})$$

$$= \left(\frac{W_{CH}}{L_{CH}}\right)\left(\frac{200\mu}{2}\right)(650\text{m} - 500\text{m})^2[1 + 5\%(2)] = 10\ \mu\text{A}$$

$$\therefore \quad \frac{W_{CH}}{L_{CH}} = 4.0$$

$$\rightarrow \quad W = W_{CH} \equiv W_{MIN} = 3\ \mu m$$

$$L_{CH} = L - 2L_{OL} = L - 2(30n) \equiv \frac{W_{CH}}{4.0} = \frac{3\mu}{4.0} = 750\ nm$$

$$\therefore \quad L = 810\ nm$$

Explore with SPICE:
See Appendix A for notes on SPICE simulations.

```
* Class-A NMOS Inverter
vdd vdd 0 dc=4
vi vi 0 dc=0 pwl 0 0 1p 0 1.1p 4 100p 4 100.1p 0
ib vm 0 dc=10u
mm1 vm vm vdd vdd pmos2 w=20u l=250n
mm2 vo vm vdd vdd pmos2 w=20u l=250n
mi vo vi 0 0 nmos2 w=3u l=810n
.lib lib.txt
.dc vi 0 1.3 1m
*.tran 10n
.end
```

Tip: Plot v(vi) and v(vo). Move "*" so it precedes ".dc," re-run simulation, and plot v(vi) and v(vo).

8.6.2 Supply-Sensing Comparators

A. Low Side

Low-side comparators sense and compare v_{SS}-level voltages. In the case of Fig. 8.31, v_{GS}'s and v_{DS}'s for M_L, M_N, and M_B match when the *low input* v_L equals v_N and v_O equals v_M. So M_L, M_N, and M_B sink I_B and M_{M2} mirrors the I_B that M_N pulls from M_{M1}. With all currents matched, the circuit balances.

When v_L climbs over v_N, M_L's v_{GS} shrinks. This reduces M_L's current, allowing excess I_B from M_{M2} to charge capacitances at v_O towards v_{DD}. The operation reverses when v_L falls under v_N: M_L's v_{GS} grows, so excess M_L current pulls v_O low.

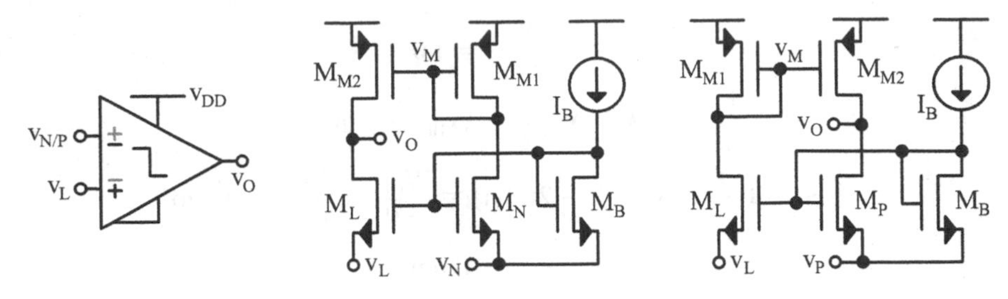

Fig. 8.31 Positive and negative low-side comparators

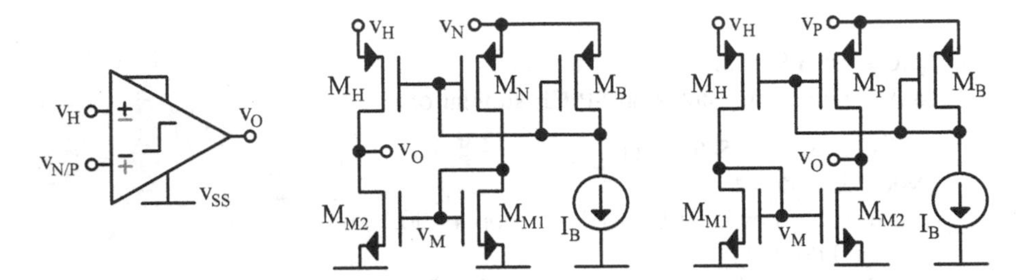

Fig. 8.32 Positive and negative high-side comparators

M_B connects to v_N to "crush" I_B when v_N nears v_{DD}. A high enough v_N shuts off M_B, M_N, and M_{M1} and raises M_L's v_{GS}, so M_L pulls v_O low towards v_L. I_B reactivates M_B, M_N, and M_{M1} automatically when v_N falls.

Feeding M_L's current to M_{M1} reverses the polarity of the comparator. In this case, M_L's v_{GS} shrinks when v_L climbs over v_P. This decreases the current M_L pulls from M_{M1}, which in turn reduces the current M_{M2} sources. So excess I_B from M_P pulls v_O low towards v_P. The operation reverses when v_L falls under v_P: M_L's v_{GS} grows, M_{M1}'s and M_{M2}'s currents rise, and excess M_{M2} current pulls v_O high.

B. **High Side**

High-side comparators sense and compare v_{DD}-level voltages. In the case of Fig. 8.32, v_{SG}'s and v_{SD}'s for M_H, M_N, and M_B match when the *high input* v_H equals v_N and v_O equals v_M. So M_H, M_N, and M_B conduct I_B and M_{M2} mirrors the I_B that M_N feeds M_{M1}. With all currents matched, the circuit balances.

When v_H climbs over v_N, M_H's v_{SG} grows. This raises M_H's current i_H, charging capacitances at v_O towards v_H. The operation reverses when v_H falls under v_N: M_H's v_{SG} shrinks and i_H falls, so excess I_B from M_{M2} pulls v_O low.

M_B connects to v_N to "crush" I_B when v_N nears v_{SS}. A low enough v_N shuts off M_B, M_N, and M_{M1} and raises M_I's v_{SG}, so M_H pulls v_O high towards v_H. I_B reactivates M_B, M_N, and M_{M1} automatically when v_N rises.

Feeding M_H's current to M_{M1} reverses the polarity of the comparator. In this case, M_H's v_{SG} climbs when v_H climbs over v_P. This raises the current M_H feeds M_{M1}, which in turn increases the current M_{M2} pulls. So excess M_{M2} current pulls v_O low towards v_{SS}. The operation reverses when v_H falls under v_P: M_H's v_{SG} shrinks, M_{M1}'s and M_{M2}'s currents fall, and excess M_P current pulls v_O high towards v_P.

C. **Offset**

Integrating a *systemic offset* $V_{OS(S)}$ into the comparator is often useful. This way, v_O trips when $v_{L/H}$ climbs $V_{OS(S)}$ over or falls $V_{OS(S)}$ under v_N. Favoring one input over the other this way amounts to widening the transistor or reducing the current it conducts, which is to say, reducing current density. The resulting $V_{OS(S)}$ is the difference of v_{GS}'s when the circuit balances.

$V_{OS(S)}$ is the difference of logarithms in sub-threshold and saturation voltages in inversion. So $V_{OS(S)}$ is the ratio or square-root difference of current densities. When currents match I_B, widening $M_{L/H}$'s $W_{L/H}$ sets a $V_{OS(S)}$ that favors $v_{L/H}$ and trips v_O high when $v_{L/H}$ climbs $V_{OS(S)}$ over v_N:

$$V_{OS(S)}\Big|_{V_{DS}>3V_t}^{V_{GS}<V_T} = V_{GSL/H} - V_{GSN} \approx n_I V_t \ln\left[\frac{I_{L/H}(W/L)_N}{(W/L)_{L/H} I_N}\right] \tag{8.54}$$

$$V_{OS(S)}\Big|_{V_{DS}>V_{GST}}^{V_{GS}>V_T} = V_{GSL/H} - V_{GSN} \approx \sqrt{\frac{2I_{L/H}}{K'(W/L)_{L/H}}} - \sqrt{\frac{2I_N}{K'(W/L)_N}}. \tag{8.55}$$

Since these comparators balance when v_O matches v_M, $M_{L/H}$ and M_N should saturate (by design) when v_O equals v_M.

Example 15: Determine W_L so $V_{OS(S)}$ favors v_L with 10 mV when I_B is 10 μA, W_N is 3 μm, L's are 1 μm, L_{OL} is 30 nm, and K_N' is 200 μA/V^2.

Solution:

$$L_{CH} = L_L - 2L_{OL} = 1\mu - 2(30n) = 940\text{ nm}$$

$$V_{OS(S)} \approx \sqrt{\frac{2I_B L_{CH}}{K_N'}}\left(\sqrt{\frac{1}{W_L}} - \sqrt{\frac{1}{W_N}}\right)$$

$$= \sqrt{\frac{2(10\mu)(940n)}{200\mu}}\left(\sqrt{\frac{1}{W_L}} - \sqrt{\frac{1}{3\mu}}\right) \equiv 10\text{mV}$$

$$\therefore \quad W_L = 2.7\ \mu\text{m}$$

Explore with SPICE:
See Appendix A for notes on SPICE simulations.

```
* Low-Side Comparator
vdd vdd 0 dc=4
vi vi 0 dc=0
vn vn 0 dc=-10m pwl 0 -10m 5n -10m 5.1n 30m 100n 30m 100.1n -10m
mi vo vb vi vi nmos3 w=2.7u l=1u
mn vm vb vn vn nmos3 w=3u l=1u
mm1 vm vm vdd vdd pmos3 w=10u l=1u
mm2 vo vm vdd vdd pmos3 w=10u l=1u
ib vdd vb dc=10u
mb vb vb vn vn nmos3 w=3u l=1u
.lib lib.txt
.dc vn -25m 25m 0.1m
*.tran 200n
.end
```

Tip: Plot v(vi) and v(vn) on one graph and v(vo) on another. Move "*" so it precedes ".dc," re-run simulation, and plot v(vi) and v(vn) on one graph and v(vo) on another.

D. **Design Notes**

These comparators are compact, fast, and low power. With only two nodes and four transistors in its core, silicon area, stray capacitances, and power consumption are low. The circuit also shuts down and reactivates automatically. This on-demand feature saves quiescent power.

In practice, v_{ID} can require 5–15 mV to trip the comparator. Cascading an inverter can reduce this five to ten times. But since the circuit trips when it balances, v_O's tripping point equals the v_M that M_{M1}'s v_{GS} sets with I_B and v_{SS} or v_{DD}. So v_O should feed a class-A inverter with a similarly v_{GS}-set threshold.

$v_{L/H}$ need not always be the positive input. Cascading an inverter is one way of inverting the polarity. Flipping the mirroring transistors M_{M1} and M_{M2} is another. With this last method, M_L or M_H feeds M_{M1}, M_P and M_{M2} set v_O, and v_O trips low or high when v_L or v_H rises over or falls under v_P.

8.6.3 Zero-Current Detectors

Systems enter *discontinuous-conduction mode* (DCM) when *zero-current detectors* (ZCD) determine the *inductor drain current* $i_{L(D)}$ reaches zero. ZCDs normally monitor the voltage across a drain switch for this purpose. Since its v_{DS} scales with $i_{L(D)}$, $i_{L(D)}$ crosses zero when v_{DS} crosses zero.

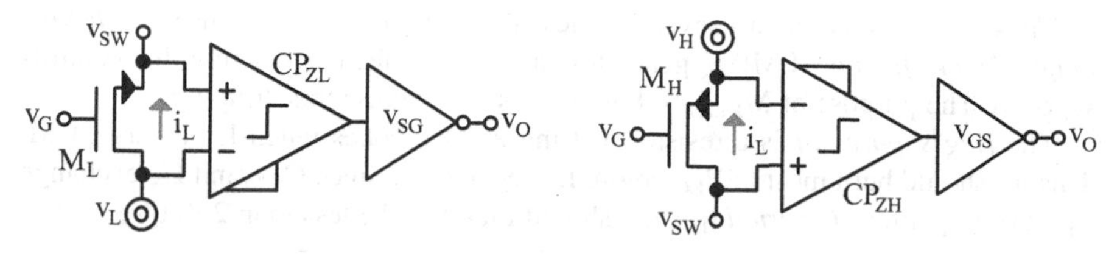

Fig. 8.33 Low- and high-side zero-current detectors

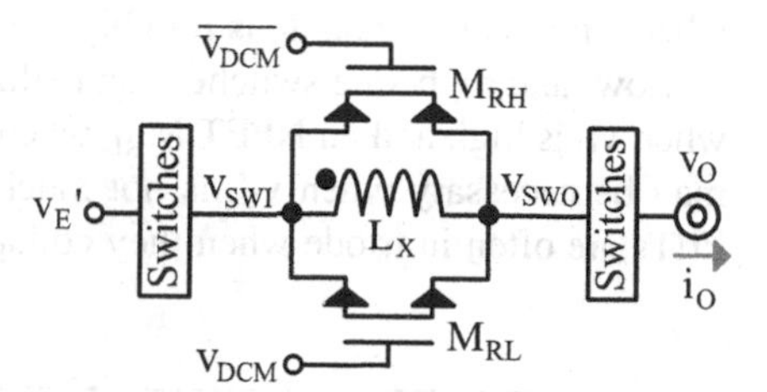

Fig. 8.34 Ring suppressor

Buck-based supplies use a *low-side switch* M_L or M_{DG} to drain L_X. $i_{L(D)}$ flows from v_L towards the v_{SW} that connects to L_X. So v_L in Fig. 8.33 is usually higher than v_{SW}. This prompts the low-side comparator CP_{ZL} to trip the class-A inverter low. v_O trips high when $i_{L(D)}$ reverses v_{ID}'s polarity.

Boost-based supplies use a *high-side switch* M_H or M_{DO} to drain L_X. $i_{L(D)}$ flows from the v_{SW} that connects to L_X towards v_H. So v_{SW} is usually higher than v_H in Fig. 8.33. This prompts the high-side comparator CP_{ZH} to trip the class-A inverter low. v_O trips high when $i_{L(D)}$ reverses v_{ID}'s polarity.

In practice, ZCDs trip t_P after $i_{L(D)}$ reverses. Some engineers add an offset to CP_Z that favors $v_{L/H}$ to compensate for this delay. This way, the ZCD detects when $v_{L/H}$ is within $V_{OS(S)}$ of v_{SW} (before $v_{L/H}$ reaches v_{SW}) so v_O can trip later when $v_{L/H}$ is closer to v_{SW}. Anticipating the transition this way can keep $i_{L(D)}$ from reversing and consuming unnecessary power.

8.6.4 Ring Suppressor

When the ZCD opens the switches in DCM, v_{SWI} in Fig. 8.34 nears ground and v_{SWO} is close to v_O. This voltage across L_X induces an i_L that charges and discharges the capacitances C_{SWI} and C_{SWO} at v_{SWI} and v_{SWO}. L_X and C_{SW}'s exchange this energy E_{LC} at their transitional *LC resonant frequency* f_{LC} until R_L in L_X burns it.

As R_L dampens oscillations, v_{SWI} and v_{SWO} approach one another until v_L is zero. Since bucks and boosts exclude input or output switches, v_{SW} approaches v_O in bucks and v_{IN} in boosts. The C_{SW} that sets the *LC time constant* τ_{LC} in buck–boosts is the series combination of C_{SWI} and C_{SWO}.

These oscillations can last several cycles. This is unfortunate because the *electromagnetic interference* (EMI) i_L generates can alter the feedback action that controls v_O or i_O. The purpose of $M_{RL/H}$ in Fig. 8.34 is to suppress this "ringing."

The *ring suppressor* is a resistor that the ZCD invokes when L_X enters DCM. This R_R should burn most of E_{LC} before L_X receives it. Since C_{SW} and L_X exchange E_{LC} every quarter *LC period* t_{LC}, t_{RC} should therefore be less than $25\% t_{LC}$:

$$t_S = t_{RC} \approx 2.3\tau_{RC} = 2.3R_RC_{SW} < \frac{t_{LC}}{4} = \frac{2\pi\tau_{LC}}{4} = \frac{2\pi/4}{\sqrt{L_XC_{SW}}}, \tag{8.56}$$

where *suppression time* t_S is roughly the t_{RC} needed to collapse 90% of v_L.

Low- and high-side switches can realize this R_R. A PFET M_{RH} may be sufficient when v_O is high and an NFET M_{RL} when v_{IN} is greater than v_O and v_O is low. Both may be necessary when v_{IN} is not much greater than a low v_O. Regardless, these FETs are often in triode when they collapse v_L.

Example 16: Determine W_R when v_O for a buck is 1.8 V, L_X is 10 μH, C_{SW} is 5 pF, W_{MIN} is 3 μm, L_R is 250 nm, L_{OL} is 30 nm, K_P' is 40 μA/V², and V_{TP0} is −700 mV.

Solution:

$$L_{CH} = L_R - 2L_{OL} = 250\text{n} - 2(30\text{n}) = 190 \text{ nm}$$

$$t_S \approx 2.3R_RC_{SW} = 2.3R_R(5\text{p}) < \frac{2\pi\sqrt{L_XC_{SW}}}{4} = \frac{2\pi\sqrt{(10\mu)(5\text{p})}}{4} \quad \therefore \quad R_R < 1.9 \text{ k}\Omega$$

$$R_R \approx \frac{L_{CH}}{W_RK_P'(v_O - |V_{TP0}|)} = \frac{190\text{n}}{W_R(40\mu)(1.8 - |-700\text{m}|)} < 1.9 \text{ k}\Omega$$

$$\therefore \quad W_R > 2.3 \ \mu\text{m} \quad \rightarrow \quad W_R \equiv 5 \ \mu\text{m}$$

Explore with SPICE:

See Appendix A for notes on SPICE simulations.

```
* Buck: Ring Suppressor
vin vin 0 dc=4
vde vde 0 dc=0 pwl 0 1 300n 1 300.1n 0
```

(continued)

```
si vin vsw vde 0 sw1v
dg 0 vsw idiode
csw vsw 0 5p
lx vsw vo 10u
co vo 0 5u
vdcmb vl 0 dc=1.8 pwl 0 1.8 550n 1.8 550.1n 0
*mr vsw vdcmb vo vin pmos2 w=3u l=250n
.ic v(vo)=1.8 i(lx)=0
.lib lib.txt
.tran 3u
.end
```

Tip: Plot i(Lx) and v(vsw) with and without "*" before "mr."

8.6.5 Switched Diodes

Asynchronous power supplies use diodes to drain L_X. Since diodes block reverse current, L_X transitions into DCM automatically when i_O falls (without a ZCD). The drawback is the power they burn with the 500–800 mV they drop when they conduct $i_{L(D)}$.

Switched diodes are transistors that behave like ideal diodes. They close when their voltage is positive and open when their voltage reverses. This way, with millivolts across them when they conduct $i_{L(D)}$, they dissipate little power. And with current flowing in one direction only, $i_{L(D)}$ cannot reverse and burn unnecessary power.

A. Low Side

Buck-based supplies use a *low-side diode* D_L or D_{DG} to drain L_X. In Fig. 8.35, the low-side comparator CP_{DL} switches M_L like an ideal D_L. CP_{DL} trips v_G high or low to close or open M_L when v_{SW} falls under or climbs over v_L.

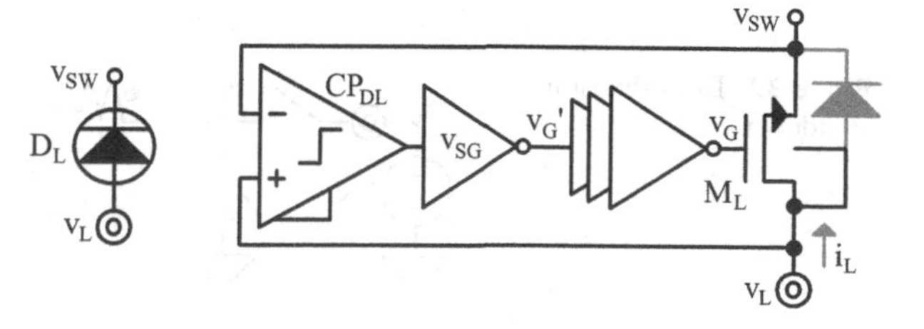

Fig. 8.35 Switched low-side diode

In practice, CP_{DL}, the inverter, and gate driver require time t_P to react. M_L's body diode should conduct i_L across this t_P. This is why M_L's body connects to v_L, so i_L can flow through the body diode into v_{SW} across t_P.

B. **High Side**

Boost-based supplies use a *high-side diode* D_H or D_{DO} to drain L_X. In Fig. 8.36, the high-side comparator CP_{DH} switches M_H like an ideal D_H. CP_{DH} trips v_G low or high to close or open M_H when v_{SW} climbs over or under v_H.

In practice, CP_{DH}, the inverter, and gate driver require time to react. M_H's body diode should conduct i_L across this t_P. This is why M_H's body connects to v_H, so i_L can flow through the body diode out of v_H across t_P.

8.6.6 Starter

A. **Shutdown**

A *shutdown* command should open all power switches. This way, the drain diodes deplete L_X into C_O and the *load* R_{LD} and R_{LD} discharges C_O. So i_L and v_O end at zero.

In buck–boosts, *ground* and *output diodes* D_{DG} and D_{DO} in Fig. 8.37 *drain* L_X into C_O and R_{LD}. And as R_{LD} discharges C_O, R_{LD} and R_L drain leftover energy that L_X and C_{SW}'s exchange. So v_{SW}'s oscillations shrink with v_O as long as D_{DO} conducts, and cease altogether when R_L burns E_{LC}.

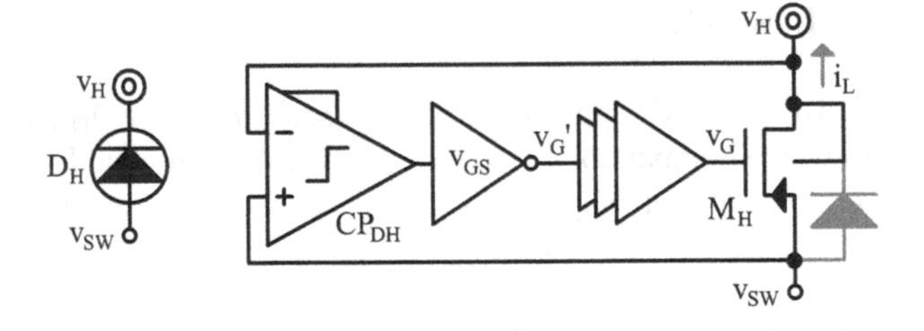

Fig. 8.36 Switched high-side diode

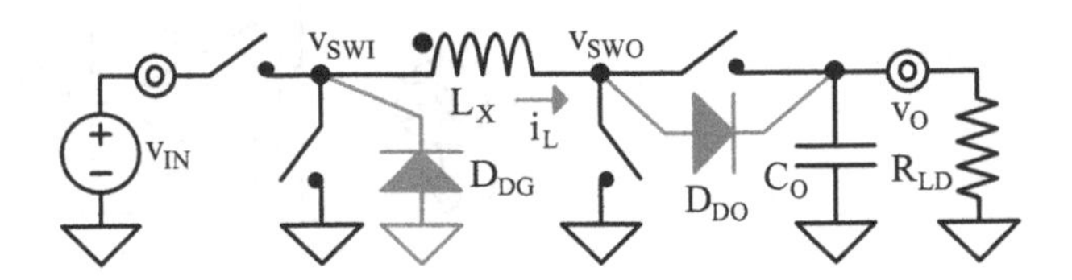

Fig. 8.37 Buck–boost in shutdown

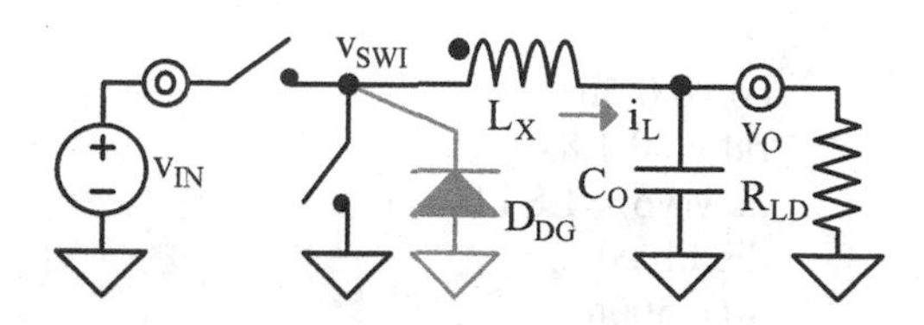

Fig. 8.38 Buck in shutdown

Explore with SPICE:
See Appendix A for notes on SPICE simulations.

```
* Buck–Boost in Shutdown
ddg 0 vswi diode1
cswi vswi 0 10p
lx vswi vl 10u
rl vl vswo 250m
cswo vswo 0 10p
ddo vswo vo diode1
co vo 0 5u
rld vo 0 1.8
.ic v(vo)=1.8 i(lx)=1
.lib lib.txt
.tran 300u
.end
```

Tip: Plot i(Lx), v(vswi), v(vswo), and v(vo).

The buck excludes the output switches that the buck–boost uses to connect L_X to v_O. So the shutdown process is the same, but without D_{DO}. D_{DG} in Fig. 8.38 drains L_X into C_O and R_{LD}, R_{LD} discharges C_O, R_{LD} and R_L drain leftover energy that L_X and C_{SWI} exchange, and v_{SWI} oscillates until R_L burns E_{LC}.

Explore with SPICE:
See Appendix A for notes on SPICE simulations.

```
* Buck in Shutdown
ddg 0 vswi diode1
cswi vswi 0 10p
lx vswi vl 10u
rl vl vo 250m
```

(continued)

```
co vo 0 5u
rld vo 0 1.8
.ic v(vo)=1.8 i(lx)=1
.lib lib.txt
.tran 500u
.end
```

Tip: Plot i(Lx), v(vswi), and v(vo).

The boost excludes the input switches that the buck–boost uses to connect v_{IN} to L_X. So the shutdown process is similar, but without D_{DG}. D_{DO} in Fig. 8.39 drains L_X into C_O and R_{LD}, R_{LD} discharges C_O, and R_{LD} and R_L drain leftover energy that L_X and C_{SWO} exchange.

But since L_X eventually shorts, v_{SWO} oscillates until D_{DO} clamps v_O to a diode below the v_{SWO} that v_{IN} and R_L set. So $R_{L(DC)}$ and R_{LD} load v_{IN} and i_L climbs to an Ohmic translation of v_{IN} and v_{DO}. The shutdown current that results is substantial when R_{LD} is low:

$$i_{L(SHUT)} = \frac{v_{IN} - v_{DO}}{R_{L(DC)} + R_{LD}} = \frac{v_{IN} - V_t \ln\left(i_{L(SHUT)}/I_S\right)}{R_{L(DC)} + R_{LD}} \tag{8.57}$$

$$\begin{aligned} v_{O(SHUT)} &= v_{IN} - i_{L(SHUT)}R_{L(DC)} - v_{DO} \\ &= v_{IN} - i_{L(SHUT)}R_{L(DC)} - V_t \ln\left(\frac{i_{L(SHUT)}}{I_S}\right). \end{aligned} \tag{8.58}$$

Although not always possible, disabling the load during shutdown is more efficient and reliable. This way, *load current* i_{LD} reduces to leakage, i_L and v_{DO} fade, and v_O approaches v_{IN}. Reducing this standby current extends the operational life of the system.

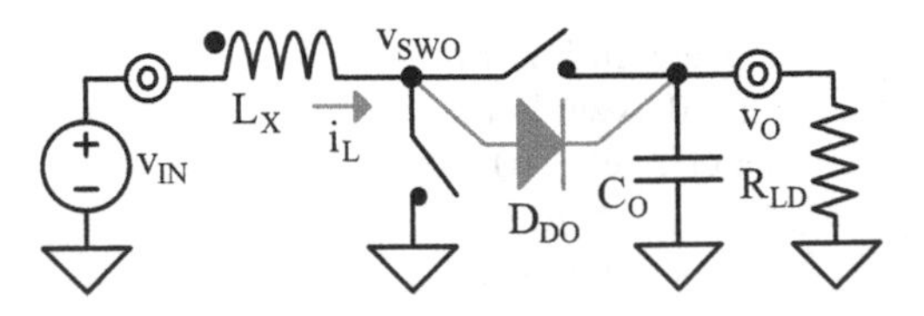

Fig. 8.39 Boost in shutdown

Example 17: Determine $v_{O(SHUT)}$ and $i_{L(SHUT)}$ for a boost when v_{IN} is 2 V, R_L is 250 mΩ, R_{LD} is 4 Ω, and I_S is 50 fA.

Solution:

$$i_{L(SHUT)} = \frac{v_{IN} - V_t \ln\left(i_{L(SHUT)}/I_S\right)}{R_L + R_{LD}} \approx \frac{2 - (26m)\ln\left(i_{L(SHUT)}/50f\right)}{250m + 4}$$

$$\therefore \quad i_{L(SHUT)} = 290 \text{ mA} \quad v_{O(SHUT)} = v_{IN} - i_{L(SHUT)}R_L - V_t \ln\left(\frac{i_{L(SHUT)}}{I_S}\right)$$

$$= 2 - (270m)(250m) - (26m)\ln\left(\frac{270m}{50f}\right) = 1.2 \text{ V}$$

Explore with SPICE:
See Appendix A for notes on SPICE simulations.

```
* Boost in Shutdown
vin vin 0 dc=2
lx vin vl 10u
rl vl vswo 250m
cswo vswo 0 10p
ddo vswo vo diode1
co vo 0 5u
rld vo 0 4
.ic v(vo)=4 i(lx)=1
.lib lib.txt
.tran 200u
.end
```

Tip: Plot i(Lx), v(vswo), and v(vo).

B. Startup

Starters wake power supplies from their shutdown state. They first wake the comparators and amplifiers that monitor and manage the system. And once armed, these blocks control and command i_L to supply the load.

Since C_O is usually very high and $v_{O(SHUT)}$ is well below the targeted v_O, the feedback action of the system is to supply a very high i_L. Unfortunately, such a high inrush of inductor current can burn components. So once blocks are ready, the starter should impede their action until i_O and v_O near their targets.

Over-current protection (OCP) is one way to keep i_L from damaging the system. OCP disables the power stage when i_L reaches $i_{L(MAX)}$ and re-enables it when i_L falls below $i_{L(OK)}$. This on-and-off process repeats until i_O and v_O reach their targets. Note OCP duals as *short-circuit protection. Thermal shutdown* is another form of OCP because oversupplying i_O heats the system over the thermal threshold that invokes a shutdown event.

Limiting the *duty-cycle command* d_E' that energizes L_X during *startup* is another way. But since d_E' also limits v_{IN}'s translation to v_O, $d_{E(START)}$ should be greater than the d_E needed to set v_O, but still lower than 90% or 95%. Or $d_{E(START)}$ can ramp slowly to d_E. This way, i_L never reaches $i_{L(MAX)}$.

The reference voltage can also *slow-* or *soft-start* the system. v_R can start at zero and ramp slowly to the v_R needed to set i_O or v_O. This way, the feedback controller ramps i_L with v_R slowly.

8.7 Summary

Switched-inductor power supplies are mixed-signal systems. They embed analog, analog–digital, and digital functions that set and regulate i_O or v_O. Some of the building blocks needed for this functionality are sensors, feedback translations, amplifiers, comparators, timers, digital logic, and switch controllers.

Current sensors normally hinge on resistance. Using switch and inductor resistance already in the network for this purpose is more power efficient than adding resistance. This resistance, however, varies widely with temperature and fabrication runs. Sense transistors often offer more favorable tradeoffs. Their weakness is mismatch.

Voltage dividers sense and translate voltages well because resistors usually match well. Paralleling a capacitor across the top resistor is an easy way of inserting a phase-saving zero–pole pair. Combining the voltage divider with the stabilizing error amplifier is also possible when the error amplifier implements a mixed feedback translation.

Amplifiers and comparators compare analog inputs to produce an output that is as high as the power supplies and the difference between the inputs allow. Comparators are basically amplifiers without the stabilizing features amplifiers need to close stable feedback loops. Adding flip flops or positive feedback establishes hysteresis. And paralleling transconductors adds inputs so the output trips with the polarity of their sum.

Constant-time loops rely on timing blocks to operate. Capacitors are essential here. In sawtooth generators, current into a capacitor ramps a voltage that a flip flop resets. Sawtooth and one-shot oscillators ramp a voltage that a comparator resets. And a flip flop in the one shot can interrupt the oscillations to pulse once or any number of times.

Digital blocks control switching events. SR flip flops use digital gates and positive feedback to decouple on–off commands. Inverter chains with increasingly larger stages drive large power switches. And dead-time logic keeps adjacent power switches from shorting their inputs.

Switch blocks help manage switching events. Supply-sensing comparators are useful in this respect. They help zero-current detectors invoke DCM operation and switched diodes behave like ideal diodes. They can also trigger the ring suppressor, which subdues DCM oscillations.

The starter is also important. It keeps initial shutdown conditions from spiking i_L. Current and duty-cycle limiters are useful guardrails during startup. Ramping the reference that sets the output is another safety measure that can soften the impact of power-up on i_L and the components that conduct i_L.

Appendix: SPICE Simulations

Note on Simulations Some simulations require more time than others to complete. If the simulation requires too much time to compute or cannot solve an operating point, adjusting SPICE options, the on–off thresholds of switches, initial conditions, or the sequence and timing of events can help the simulator converge on a solution sooner. Closing positive feedback loops (e.g., in comparators and flip flops) can delay the processor. Adding small capacitors to nodes without capacitance can keep voltages from spiking uncontrollably, which could otherwise delay, freeze, or halt simulations.

Note on MOSFET Model Distinguishing MOS channel current from gate–drain and gate–source capacitance currents in time-domain (transient) simulations is usually involved. This is because drain current in SPICE simulators often includes gate–drain capacitance C_{GD} current, source current includes gate–source capacitance C_{GS} current, and gate current includes both C_{GD} and C_{GS} currents. One way of separating these currents is by eliminating the effects of C_{GD} and C_{GS} from the model (with very thick oxide parameters, like when oxide thickness t_{OX} is 1 m) and adding C_{GD} and C_{GS} separately to the circuit (outside the model). The challenge with this approach is that SPICE will not adjust C_{GD} and C_{GS} when the MOSFET transitions between operating regions.

Online Access to SPICE Code Most of the SPICE code listed in this textbook is available online at rincon-mora.gatech.edu under the link titled "SPICE."

Basic Structure

ASCII Text File:	[name].cir
First (Title) Line:	[text]
Comment Lines:	* [text]
Net List:	[circuit: list of connected components]
Model Lines:	.model [model definition]

G. A. Rincón-Mora, *Switched Inductor Power IC Design*,
https://doi.org/10.1007/978-3-030-95899-2

Command Lines: .[command]
Last Line (End of File): .end

Useful Components

Resistor:

R[name] node1 node2 [value]

Capacitor:

C[name] node1 node2 [value]

Inductor:

L[name] node1 node2 [value]

Inductor Coupling (for transformers):

K[name] L[name] L[name] [coupling value]

Switch:

S[name] node1 node2 vnode+ vnode– [model]

Diode:

D[name] anode cathode [model] [area multiplier]

Bipolar Junction Transistor (BJT):

Q[name] collector base emitter [model] M=[multiplier]

Junction Field-Effect Transistor (FET):

J[name] drain gate source [model] M=[multiplier]

Metal–Oxide–Semiconductor (MOS) FET:

M[name] drain gate source body [model] W=[width] L=[length] M=[multiplier]

Voltage Source:

V[name] node+ node– dc=[value] ac=[value] [stimulus]

Current Source:

I[name] source output dc=[value] ac=[value] [stimulus]

Voltage Amplifier:

E[name] node+ node– vnode+ vnode– [gain]

Current Amplifier:

F[name] source output [controlling voltage source] [gain]

Transconductance Amplifier:

G[name] source output vnode+ vnode– [gain]

Transimpedance Amplifier:

H[name] node+ node– [controlling voltage source] [gain]

Sub-circuit:

X[name] node1 node2 [other nodes] [name]

Useful Stimuli

Piecewise Linear:

PWL time1 value1 time2 value2 time3 value3 ...

Pulse:

Pulse value1 value2 tdelay trise tfall value2width tperiod

Sine:

Sin offset amplitude frequency tdelay damping_factor phase

Useful Device Models (for Library File)

Switch:

.Model [name] VSwitch Roff=1e12 Ron=1m Voff=490m Von=510m

Ideal Diode:

.Model [name] D Is=1p n=0.001

Fast Diode:

.Model [name] D Is=1p n=1

Nominal Diode:

.Model [name] D Is=1p n=1 Tt=1n Cjo=100f Vj=600m M=500m Bv=7

Fast BJT:

.Model [name] [NPN or PNP] Bf=100 Va=50 Is=1f

Nominal BJT:

.Model [name] [NPN or PNP] Bf=100 Va=50 Is=1f
+ Tf=100p Cjc=100f Vjc=600m Mjc=0.5

Nominal JFET:

.Model [name] [NJF or PJF] Vto=-2 Beta=50u + Lambda=50m

Fast MOS:

.Model [name] [NMOS or PMOS] Vto=0.5 Kp=200u Lambda=10m

Nominal MOS:

.Model [name] [NMOS or PMOS] Vto=0.5 Kp=200u

+ Lambda=100m Gamma=600m Phi=600m Tox=5n Cgso=200p Cgdo=200p

Useful Commands

Temperature	.temp [value]
Initial Conditions:	.ic i([inductor name])=[value] v([node name])=[value]
Library File:	.lib [text file name]
Operating Point:	.op
Static (DC) Sweep:	.dc [source1] [start] [end] [step] [source2] [start] [end] [step]
Small-Signal (AC) Response:	.ac dec [data points per decade] [start freq.] [end freq.]
Time-Domain (Transient) Response:	.tran [end]

Useful Behavioral (Sub-circuit) Models (for Library File)

A. 1-V Push–Pull Inverter

```
.subckt inv vi vo
v1v v1v 0 dc=1
ei va 0 vi 0 1
* tP = R(Cdly + Cfb) time to 50% = 69.3%RC = 1 ns
rdly va vb 1
cdly vb 0 722p
cfb vfb vb 722p
sh1 vo v1v v1v vb dig_sw
sl1 vo 0 vb 0 dig_sw
sh2 vfb v1v v1v vo dig_sw
sl2 vfb 0 vo 0 dig_sw
.model dig_sw vswitch roff=1e12 ron=1m voff=499m von=501m
.ends
```

B. 1-V Delay Block

```
.subckt dly vi vo
v1v v1v 0 dc=1
sih va v1v vi 0 dig_sw
sil va 0 v1v vi dig_sw
* Rise tX+
dr va vr idiode
rr vr vb 1
* Fall tX-
df vf va idiode
* tP(F) = tP(R) * (rf/rr) = 90 ns
rf vf vb 200m
* tP(R) = R(Cdly + Cfb) time to 50% = 69.3%tRC = 450 ns
cdly vb 0 325n
cfb vfb vb 325n
sh1 vo v1v vb 0 dig_sw
sl1 vo 0 v1v vb dig_sw
sh2 vfb v1v vo 0 dig_sw
sl2 vfb 0 v1v vo dig_sw
.model dig_sw vswitch roff=1e12 ron=1m voff=490m von=510m
.model idiode d is=1f n=0.001
.ends
```

C. 1-V Set-Dominant Set–Reset Flip Flop

```
.subckt srs s r q qn
v1v v1v 0 dc=1
sh va v1v s 0 dig_sw
sl va 0 r 0 wk_sw
rdly va vb 1
* tP = R(Cdly + Cfb) time to 50% = 69.3%tRC = 1 ns
cdly vb 0 722p
cfb q vb 722p
sqh q v1v vb 0 dig_sw
sql q 0 v1v vb dig_sw
snh qn v1v v1v vb dig_sw
snl qn 0 vb 0 dig_sw
.model wk_sw vswitch roff=1e15 ron=100m voff=499m von=501m
.model dig_sw vswitch roff=1e12 ron=1m voff=490m von=510m
.model idiode d is=1f n=0.001
.ends
```

D. 1-V Reset-Dominant Set–Reset Flip Flop

```
.subckt srr s r q qn
v1v v1v 0 dc=1
sh va v1v s 0 wk_sw
```

```
sl va 0 r 0 dig_sw
rdly va vb 1
* tP = R(Cdly + Cfb) time to 50% = 69.3%tRC = 1 ns
cdly vb 0 722p
cfb q vb 722p
sqh q v1v vb 0 dig_sw
sql q 0 v1v vb dig_sw
snh qn v1v v1v vb dig_sw
snl qn 0 vb 0 dig_sw
.model wk_sw vswitch roff=1e15 ron=1000m voff=499m von=501m
.model dig_sw vswitch roff=1e12 ron=1m voff=490m von=510m
.model idiode d is=1f n=0.001
.ends
```

E. Comparator

```
.subckt cp vp vn vo vdd vss
v1v v1v 0 dc=1
g1 0 va vp vn 1
ra va 0 100k
dp va v1v idiode
dn 0 va idiode
ebw vbwx 0 va 0 1
rbw vbwx vbw 1
* tP = R(Cbw + Cfb) time to 50% = 69.3%tRC = 100 ns
cbw vbw 0 72.2n
cfb vfb vbw 72.2n
sh vo vdd vbw 0 dig_sw
sl vo vss v1v vbw dig_sw
sh2 vfb v1v vbw 0 dig_sw
sl2 vfb 0 v1v vbw dig_sw
.model idiode d is=1f n=0.001
.model dig_sw vswitch roff=1e12 ron=1m voff=499m von=501m
.ends
```

F. Summing Comparator

```
.subckt cp3vid vp1 vn1 vp2 vn2 vp3 vn3 vo vdd vss
v1v v1v 0 dc=1
g1 0 va vp1 vn1 1
g2 0 va vp2 vn2 1
g3 0 va vp3 vn3 1
ra va 0 100k
dp va v1v idiode
dn 0 va idiode
ebw vbwx 0 va 0 1
rbw vbwx vbw 1
```

```
* tP = R(Cbw + Cfb) time to 50% = 69.3%tRC = 100 ns
cbw vbw 0 72.2n
cfb vfb vbw 72.2n
sh vo vdd vbw 0 dig_sw
sl vo vss v1v vbw dig_sw
sh2 vfb v1v vbw 0 dig_sw
sl2 vfb 0 v1v vbw dig_sw
.model idiode d is=1f n=0.001
.model dig_sw vswitch roff=1e12 ron=1m voff=499m von=501m
.ends
```

G. Hysteretic Comparator: Temporal Model (Transient Simulations Only)

```
.subckt cphys vp vn vo vdd vss
v1v v1v 0 dc=1
v1vn v1vn 0 dc=-1
g1 0 vid vp vn 1
ra vid 0 1
dp vid v1v idiode
dn v1vn vid idiode
sh1 va v1v vid 0 hys_sw
sl1 va 0 0 vid hys_sw
ca va 0 1n
ebw vbwx 0 va 0 1
rbw vbwx vbw 1
* tP = R(Cbw + Cfb) time to 50% = 69.3%tRC = 100 ns
cbw vbw 0 72.2n
cfb vfb vbw 72.2n
sh vo vdd vbw 0 dig_sw
sl vo vss v1v vbw dig_sw
sh2 vfb v1v vbw 0 dig_sw
sl2 vfb 0 v1v vbw dig_sw
* Hysteresis = +-25 mV
.model hys_sw vswitch roff=1e12 ron=1m
+ voff=24.9m von=25.1m
.model dig_sw vswitch roff=1e12 ron=1m voff=499m von=501m
.model idiode d is=1f n=0.001
.ends
```

H. Summing Hysteretic Comparator: Temporal Model (Transient Simulations Only)

```
.subckt cphys3vid vp1 vn1 vp2 vn2 vp3 vn3 vo vdd vss
v1v v1v 0 dc=1
v1vn v1vn 0 dc=-1
g1 0 vid vp1 vn1 1
g2 0 vid vp2 vn2 1
```

```
g3 0 vid vp3 vn3 1
rid vid 0 1
dp vid v1v idiode
dn v1vn vid idiode
sh1 va v1v vid 0 hys_sw
sl1 va 0 0 vid hys_sw
ca va 0 1n
ebw vbwx 0 va 0 1
rbw vbwx vbw 1
* tP = R(Cbw + Cfb) time to 50% = 69.3%tRC = 100 ns
cbw vbw 0 72.2n
cfb vfb vbw 72.2n
sh vo vdd vbw 0 dig_sw
sl vo vss v1v vbw dig_sw
sh2 vfb v1v vbw 0 dig_sw
sl2 vfb 0 v1v vbw dig_sw
* Hysteresis = +-25 mV
.model hys_sw vswitch roff=1e12 ron=1m
+ voff=24.9m von=25.1m
.model dig_sw vswitch roff=1e12 ron=1m voff=450m von=550m
.model idiode d is=1f n=0.001
.ends
```

Type-I and -II Stabilizers (Not Included in Chap. 6)

A. Type-I OTA

```
* Type I OTA
vi vi 0 dc=0 ac=1
g1 0 vo vi 0 80u
rf vo 0 500k
cf vo 0 320p
.ac dec 1000 1 10e6
.end
```

B. Type-II OTA

```
* Type II OTA
vi vi 0 dc=0 ac=1
g1 0 vo vi 0 80u
rf vo 0 500k
cf vo vx 0 290p
rc vx 0 56k
co vo 0 32p
```

```
.ac dec 1000 1 10e6
.end
```

C. Type-I Inverting Mixed Translation

```
* Type I Inverting Fwd Translation
vi vi 0 dc=0 ac=1
rf vi vn 500k
cf vn vo 8p
eavi va1 0 0 vn 40
* Internal pA
ra va1 va2 1
ca va2 0 15.92u
eavo vo 0 va2 0 1
.ac dec 1000 1 10e6
.end
```

C. Type-II Inverting Mixed Translation

```
* Type II Inverting Fwd Translation
vi vi 0 dc=0 ac=1
rf vi vn 500k
rc vn vx 2e6
cf vx vo 8p
eavi va1 0 0 vn 40
* Internal pA
ra va1 va2 1
ca va2 0 13.27u
eavo vo 0 va2 0 1
.ac dec 1000 1 10e6
.end
```

Index

G. A. Rincón-Mora, *Switched Inductor Power IC Design*, https://doi.org/10.1007/978-3-030-95899-2

P

Printed in the United States
by Baker & Taylor Publisher Services

Printed in the United States
by Baker & Taylor Publisher Services